CHEMISTRY IN
MICROTIME

Selected Writings on

Flash Photolysis, Free Radicals,

and the Excited State

CHEMISTRY IN MICROTIME

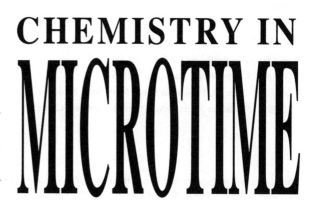

Selected Writings on

Flash Photolysis, Free Radicals,

and the Excited State

Lord George Porter

Department of Chemistry and Biochemistry
Centre for Photomolecular Sciences
Imperial College
London

Imperial College Press

ICP

Published by

Imperial College Press
516 Sherfield Building
Imperial College
London SW7 2AZ

Distributed by

World Scientific Publishing Co. Pte. Ltd.
P O Box 128, Farrer Road, Singapore 912805
USA office: Suite 1B, 1060 Main Street, River Edge, NJ 07661
UK office: 57 Shelton Street, Covent Garden, London WC2H 9HE

The editors and publisher would like to thank the following organisations and publishers of the various journals and books for their assistance and permission to reproduce the selected reprints found in this volume:

Academic Press
Almqvist & Wiksell
American Institute of Physics
Elsevier
Macmillan
National Academy of Sciences (USA)
The Nobel Foundation
The Royal Society of Chemistry
The Royal Society of London
VCH Publishers

While every effort has been made to contact the publishers of reprinted papers prior to publication, we have not been successful in some cases. Where we could not contact the publishers, we have acknowledged the source of the material. Proper credit will be accorded to these publishers in future editions of this work after permission is granted.

British Library Cataloguing-in-Publication Data
A catalogue record for this book is available from the British Library.

CHEMISTRY IN MICROTIME
Selected Writings on Flash Photolysis, Free Radicals and the Excited State
Copyright © 1997 by Imperial College Press

ISBN 1-86094-015-3
ISBN 1-86094-021-8 (pbk)

Printed in Singapore by Uto-Print

To my wife Stella, who followed and encouraged our research from its beginning, nearly half a century ago, and also to the hundred and more colleagues with whom it has been my privilege to work as a collaborator and fellow student.

PREFACE

This volume is a collection of papers, beginning with the first description of flash photolysis and followed by applications of the method to photochemistry, spectroscopy, the excited state and photosynthesis, on time scales of milliseconds down to femtoseconds. Only papers of which I am the author, or a coauthor, are included and I have added a few notes to each section indicating what advances have been made and why they were of interest. In making the selection I have tried to include examples of each of the main areas of the research and many of the publications appear as references only. Some of these may be more important than some of those that are published in full but the titles given in the references will indicate how they fit into the whole story and relate to the other papers. A few papers have been abridged or are short abstracts; where this has taken place it is clearly indicated.

The prefix "micro" has two definitions in the dictionary. When it is a prefix to a unit, like second as in microsecond, or gram or litre, it means "one millionth of" or 10^{-6} times that unit. In this it resembles other prefixes like "nano" meaning 10^{-9} or "femto" meaning 10^{-15}. When micro, (but not the other prefixes), is a prefix to a word with unspecified units, like time, or chemistry, or economics or technology or an instrument, as in microscope, it signifies "very small scale". Unless the unit, such as second or year or gram, is specified, other prefixes than micro, like milli or femto, do not define a quantity. Microtime, which appears as "Chemistry in microtime" in the title of this book, therefore means chemistry in very short times, be they milliseconds or femtoseconds.

George Porter

CONTENTS

CHEMISTRY IN
MICROTIME

Selected Writings on

Flash Photolysis, Free Radicals,

and the Excited State

Chapter 1

FLASH PHOTOLYSIS

INTRODUCTION

Extract 1 from Nobel Symposium 5, © 1967 Almqvist & Wiksell, pp. 141–142

Flash photolysis and primary processes in the excited state
By G. Porter, F.R.S.

The Royal Institution, London

In 1851, Fox Talbot, one of the pioneers of photography, gave a discourse at the Royal Institution during which he performed a demonstration which deserves to be better known because it was a landmark in the recording of very rapid events. He pasted the front page of a newspaper, *The Times*, on to a board which was rotated as rapidly as possible. He then discharged several Leyden jars across a spark gap whilst the shutter of a camera remained open. He developed the photographic plate on the spot and showed a quite readable record of the rapidly rotating print. The duration of the spark which Fox Talbot used was probably not much more than a few microseconds and, prior to this, photographic records, whether by sunlight or by artificial light such as limelight, required at least several seconds. Electronic methods of recording and even fast pen recorders were not available at the time and therefore Fox Talbot had, in one step, extended the time resolution of recorded events by a factor of a million. Since the light was made to produce a photochemical effect, this experiment may also be regarded as a forerunner of the flash photolysis technique. Some credit should be given to Wheatstone who had earlier, by visual examination, shown that the duration of an electric spark was in the microsecond range. Fox Talbot realised the significance of his demonstration and reported it in the *Athenaeum* as follows: 'From this experiment the conclusion is inevitable that it is in our power to obtain pictures of all moving objects, no matter in how rapid motion they may be, provided we have the means of sufficiently illuminating them with a sudden electric flash.'

He was right and, although development of these potentialities was very slow, subsequent studies of rapid change, in the fields of ballistics and zoology in particular, have been carried out almost entirely by spark photography and, since about 1940, by the more actinically valuable pulsed gaseous discharge.

In all these applications the flash is used merely to record an event and, indeed, one assumes that the flash has no effect on the change itself. In 1949, the flash photolysis technique was introduced (Norrish & Porter, 1949; Porter, 1950) in which the light flash was used to produce the change as well as to record it. Since all substances absorb light and almost any physical and chemical process can be initiated in this way in gaseous, liquid and solid systems, it was clear that this development was one with very wide applications in the study of rapid change at the atomic and molecular level. The development of the sister technique of pulse radiolysis has recently extended the initiation processes to include those produced by ionising radiations.

The increasing realisation that most fundamental processes in molecular systems are very fast, and many of the important intermediates have very short lifetimes, caused a great deal of attention to be given to all methods of producing and recording rapid change and many elegant techniques were developed. Particularly valuable were improved (stopped flow) mixing techniques, shock waves for microsecond temperature jumps in gases, and the perturbation methods, involving rapid change in an external parameter such as temperature, pressure or electric field and the consequent displacement of an equilibrium or steady state condition (see Eigen, 1963).

The particular power of flash photolysis techniques is their ability to produce large perturbations in very short times and, although they can of course be used to effect small perturbations as well, they have found their most important applications in fields where a high absolute rate of perturbation is essential. The most valuable of these are the following.

1. The production of free radicals and other transients in concentrations high enough to allow their structure and physical properties to be determined by absorption spectroscopic and other direct observation methods.
2. The study of kinetics by time resolved recording of the concentrations of the short lived intermediates themselves.
3. The study of primary processes in excited states by direct observation of these states and of the primary products to which they give rise.

Extract 2 from Les Prix Nobel en 1967, pp. 1–4, Nobel Foundation 1968.

FLASH PHOTOLYSIS AND SOME OF ITS APPLICATIONS

by

GEORGE PORTER

The Royal Institution, 21 Albemarle Street, London, W.1.

Nobel Lecture, December 11, 1967

One of the principal activities of man as scientist and technologist has been the extension of the very limited senses with which he is endowed so as to enable him to observe phenomena with dimensions very different from those he can normally experience. In the realm of the very small, microscopes and microbalances have permitted him to observe things which have smaller extension or mass than he can see or feel. In the dimension of time, without the aid of special techniques, he is limited in his perception to times between about one twentieth of a second (the response time of the eye) and about 2×10^9 seconds (his lifetime). Yet most of the fundamental processes and events, particularly those in the molecular world which we call chemistry, occur in milliseconds or less and it is therefore natural that the chemist should seek methods for the study of events in microtime.

My own work on "the study of extremely fast chemical reactions effected by disturbing the equilibrium by means of very short pulses of energy" was begun in Cambridge twenty years ago. In 1947 I attended a discussion of the Faraday Society on "The Labile Molecule". Although this meeting was entirely concerned with studies of short lived chemical substances, the four hundred pages of printed discussion contain little or no indication of the impending change in experimental approach which was to result from the introduction, during the next few years, of pulsed techniques and the direct spectroscopic observation of these substances. In his introduction to the meeting H. W. Melville referred to the low concentrations of radicals which were normally encountered and said "The direct physical methods of measurement simply cannot reach these magnitudes, far less make accurate measurements in a limited period of time, for example 10^{-3} sec."

Work on the flash photolysis technique had just begun at this time and details of the method were published two years later (1, 2). Subsequent developments were very rapid, not only in Cambridge but in many other laboratories. By 1954 it was possible for the Faraday Society to hold a discussion on "The study of fast reactions" which was almost entirely devoted to the new techniques introduced during the previous few years. They included, as well as flash photolysis, other new pulse techniques such as the shock wave, the stopped flow method, and the elegant pressure, electric field density and temperature pulse methods described by Manfred Eigen. Together with pulse radiolysis, a sister technique to flash photolysis which was developed around 1960, these methods have made possible the direct study of nearly all fast reactions and transient substances which are of interest in chemistry and, to a large extent, in biology as well.

The various pulse methods which have been developed are complementary to each other, each has its advantages and limitations, and the particular power of the flash photolysis method is the extreme perturbation which is produced, making possible the preparation of large amounts of the transient intermediates and their direct observation by relatively insensitive physical methods. Furthermore, the method is applicable to gases, liquids and solids and to systems of almost any geometry and size, from path lengths of many metres to those of microscopically small specimens.

My original conception of the flash photolysis technique was as follows: the transient intermediates, which were, in the first place, to be gaseous free radicals, would be produced by a flash of visible and ultra-violet light resulting from the discharge of a large condenser bank through an inert gas. The flash would be of energy sufficient to produce measurable overall change and of short duration compared with the lifetime of the intermediates. Calculation showed that an energy of 10,000 J dissipated in a millisecond or less, in lamps of the type which were being developed commercially at that time, would be adequate for most systems. The bank of condensers was given by my friends in the Navy, and, although I am grateful to them for saving us much expense, it consisted of a motley collection of capacitors which, owing to their high inductance, gave a flash of rather longer duration than was desirable. The detection system was to consist of a rapid-scanning spectrometer and much time was wasted in the development of this before I realised that to demand high spectral resolution, time resolution and sensitivity in a period of a few milliseconds was to disregard the principles of information theory. Subsequent applications of flash photolysis, with a few exceptions, have been content to record, from a single flash, either a single spectrum at one time or a small wavelength range at all times. The use of a second flash, operating after a time delay, to record the absorption spectrum of the transients must now seem a very obvious procedure but it was many months before it became obvious to me. The double-flash procedure was an important step forward and is still, in principle, the most soundly based method for the rapid recording of information.

In the first apparatus the delay between the two flashes was introduced by a rotating sector, with two trigger contacts on its circumference, a photograph of which is shown in Figure 1. The reason for using this, in preference to an electronic delay, was in anticipation of difficulties with scattered light from the high energy photolysis flash, which could be eliminated by the shutter incorporated in the rotating wheel. As flash durations were reduced it became necessary to resort to purely electronic methods but the apparatus worked well for several years and provided, for the first time, the absorption spectra of many transient substances and a means for their kinetic study.

There have been many reviews of the flash photolysis method (3, 4, 5) and in this lecture I should like to illustrate our work by describing four rather

Fig. 1.
Part of the original flash photolysis apparatus showing the rotating sector with shutter and trigger contacts.

different problems: the first two are simple gas phase reactions which were the earliest to be investigated in detail and which illustrate rather clearly the two principal variations of the flash photolysis technique; the second two examples are studies of the principal types of transient which appear in photochemical reactions, i.e. radicals and triplet states, with special reference to aromatic molecules.

The first free radical to be studied in detail by flash photolysis methods,

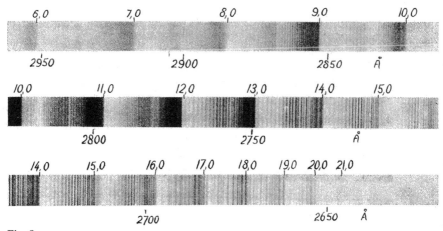

Fig. 2.
Absorption spectrum of the ClO radical.

both spectroscopically and kinetically, was the diatomic radical ClO, and this study provided a proving exercise for the technique. The spectrum was discovered, somewhat accidentally, in the course of a study of the chlorine—carbon monoxide—oxygen system and provided one of the first of many lessons on the limitation of predictions based on conventional studies. This new spectrum, which is shown in Figure 2, was produced by flash photolysis of mixtures of oxygen and chlorine, in which no photochemical reaction had previously been suspected.

The next two papers (Nos. 3 and 4), submitted in August 1949, describe the first work on flash photolysis. The Nature paper, with R. G. W. Norrish, describes the effects of very high light intensities, some of them quite dramatic, on the course of photochemical change. The Royal Society paper describes the method of flash photolysis and kinetic spectroscopy, in which two flashes are used, one to intitiate the photochemical change and one to record it a short time later. A sequence of such "pulse and probe" records provides a "movie" of molecular change. Although the techniques described in this paper are primitive by modern standards they are reproduced here as an illustration of the early developments of the technique, over 45 years ago.

Paper 3

(Reprinted from *Nature*, Vol. 164, p. 658, October 15, 1949)

Chemical Reactions Produced by Very High Light Intensities

IT has been a matter of general experience that photochemical reactions are not much altered in their courses by change of light intensity. The range of light intensity hitherto available, however, has been limited by that obtainable from the sun and from such sources as high-pressure mercury vapour lamps, the total usable output of which in the region between 2,000 and 5,000 A. does not exceed 10^{20} quanta/second. There are many cases, particularly those photochemical reactions where free radicals are involved, where it would be desirable to extend investigations to much higher intensities into the region where the concentration of intermediates is comparable with that of the reactants.

The gas-discharge flash-lamp which has been developed recently for photographic purposes seemed to offer great possibilities for experiments of this kind, and we have found that this type of lamp is a very efficient source of light in the photochemically useful region. The flash is produced by the discharge of a large condenser through a tube filled with rare gas, and up to energies of about 100 Joules almost any of the usual types of discharge lamp may be used. By using quartz tubes and large tungsten electrodes, we have worked with energies up to 10,000 Joules/flash in a tube 1 metre long. The duration of this flash was 4 m.sec., and the useful light output in the region of absorption of uranyl oxalate was 10^{24} quanta/second. The output is continuous down to at least 2,000 A. and appears to be fairly evenly distributed, and the output at wave-lengths less than 3,500 A. may be further increased by the addition of a little liquid mercury to the gas filling.

The following figures give the percentage decomposition obtained in a few typical substances with a single flash of 4,000 Joules lasting less than 2 m.sec., the substance being contained in a quartz tube, one metre long, lying parallel to the lamp :

Gas	Pressure (cm.)	% Decomposition
NO$_2$	4	Nearly 100
CH$_3$COCH$_3$	3	15
CH$_3$COBr	3	15
CH$_3$COCOCH$_3$	2·5	50
CH$_2$CO	10	40

The gas volume was 500 c.c. in each case, and mercury had been added to the lamp in the cases of acetone and ketene. In addition, with chlorine at a pressure of 1 cm., the enormous Budde effect of 2 cm. was observed.

Three examples of the different behaviour of well-known photochemical systems under these high light intensities will be given.

Acetone and ketene. Instead of the hydrocarbons and carbon monoxide which are normally produced with these compounds, a complex mixture of substances including hydrogen and a high proportion of carbon was obtained. The carbon appeared in an unusual fine cobweb-like formation which hung across the tube from wall to wall.

Chlorine and carbon tetrachloride. Carbon tetrachloride is usually considered to be the final product of the photochemical reaction of chlorine with methane. At these high light intensities, however, when the partial pressure of atoms and radicals present is relatively high, a mixture of chlorine at 1 cm. pressure and carbon tetrachloride at 2 cm. pressure showed an increase of pressure of about 0·3 cm., and a considerable amount of solid product was formed.

Chlorine and carbon monoxide. This well-known photochemical system exhibits one of the most striking examples of the effect of intensity on the course of a reaction. Illumination in the ordinary way of a mixture of chlorine and carbon monoxide produces phosgene quite rapidly, and the reaction can be taken almost to completion. We find that irradiation by a 4,000-Joule flash, however, produces no detectable pressure change in this mixture, and that if phosgene is added it is rapidly decomposed to chlorine and carbon monoxide. The reaction is quite reversible and can be made to proceed either way at will by changing the intensity of the illumination. That the important primary act is the photochemical decomposition of chlorine in both cases was shown by the very small pressure changes produced when pure phosgene was used.

Two effects seem to play a part in changing the mechanism of photochemical reactions at high intensities. First, since the concentration of the intermediates is comparable with that of the reactants, inter-atomic and inter-radical collisions are as frequent as collisions of these intermediates with more stable molecules. Thus, for example, the reversibility of the phosgene reaction may be explained by the reversibility of the reaction COCl + Cl$_2$ = COCl$_2$ + Cl followed by COCl + Cl = Cl$_2$ + CO, in full agreement with the schemes of Bodenstein, and Lehner and Rollefson. Secondly, the process is nearly adiabatic, and very high instantaneous temperatures may be reached, due in part to the exothermic reactions of the labile intermediates with each other and in part to the thermal degradation of the absorbed light energy.

The mechanism of some reactions at high light intensity will be discussed in more detail elsewhere ; the above examples serve to show that this type of source, which extends the available upper limit of intensity some 10^4 times, produces photochemical effects of unusual magnitude, and in some cases may alter profoundly the normal course of the reaction.

R. G. W. NORRISH
G. PORTER

Department of Physical Chemistry,
Cambridge.   Aug. 6.

Extract 4 from the *Proceedings of the Royal Society, A*, volume 200, 1950, pp. 284–286, 292–299.

Flash photolysis and spectroscopy
A new method for the study of free radical reactions

By G. Porter, *Department of Physical Chemistry, University of Cambridge*

(*Communicated by* R. G. W. Norrish, *F.R.S.*—Received 9 August 1949)

(Plates 6 to 8)

Photochemistry provides us with one of the most generally useful methods of studying the reactions of free radicals and atoms, but the concentration of these intermediates in the usual photochemical systems is too low to allow the use of direct physical methods of investigation such as absorption spectroscopy.

To overcome this difficulty a new technique of flash photolysis and spectroscopy has been developed, using gas-filled flash discharge tubes of very high power. The properties of these lamps as spectroscopic and photochemical sources have been studied and details are given of their construction, spectra, duration of flash, and luminous efficiency in the photochemically useful region. An apparatus is described which produces a very great photochemical change, in some cases over 80 %, in one-thousandth of a second and in a gas at several cm. pressure contained in an absorption tube 1 m. long, and which photographs the absorption spectrum at high resolution in one twenty-thousandth of a second at short intervals afterwards.

Examples of the rapidly changing spectra of substances undergoing reaction, including the spectra of some of the intermediate radicals involved, are shown. These include the recombination of chlorine atoms, the absorption spectra of S_2 and CS obtained during the photochemical decomposition of carbon disulphide and new spectra attributed to the ClO and CH_2CO radicals.

Introduction

Investigations into the mechanism of chemical reactions have revealed that in very many cases intermediate substances are involved which, although they exist only for a very short time, determine the course and rate of the changes which take place. Much information has been obtained about these intermediates, which are usually free radicals or atoms, by induction from the kinetics of the overall change, but the indirectness of this method has so far rendered it incapable of giving the kinetic details of some of even the simplest radical reactions. We have much to learn, for example, about the combination of two methyl radicals. The mirror technique, developed by Paneth and his colleagues, is another powerful means of study, but suffers from similar limitations in so far as it depends on inference from the final products with the metal mirror, and it is also accompanied by some rather severe experimental difficulties (Norrish & Porter 1947).

The only direct methods of investigation which have been applied to the problem are spectroscopy and mass spectroscopy, and, of the two, the former is potentially the more powerful because it enables the reaction to be studied in a static system as well as giving information about the structure of the radical, and more certain identification. Furthermore, spectroscopy is ideally suited to free radical studies, as the majority of free radicals have a transition involving the ground state in the easily accessible visible or near ultra-violet regions and, as pointed out by Wieland (1947), this often forms an extremely sensitive method of detection.

[284]

Although the spectra of a large number of diatomic radicals are known from their emission bands in flames and discharge tubes, very little information about their chemical reactions can be obtained in this way because of the difficulty of estimating concentrations from emission spectra. Absorption spectroscopy, on the other hand, makes it possible to estimate concentrations and follow the reactions of the absorbing molecule without interfering with the reacting system, but unfortunately to observe a radical in absorption the concentration must be higher than is usually obtained by any method other than the electrical discharge. High equilibrium concentrations can be obtained by thermal decomposition, but it is not possible to change the temperature of the gas rapidly enough to use this method for kinetic studies. The electrical discharge method has been used very successfully by Oldenberg (1935) in a detailed study of the OH radical and also by White (1940). For the study of the chemistry of intermediate compounds, however, the method is limited to radicals which survive the violent conditions in a discharge tube, that is, virtually, to diatomic radicals, and to the pressures under which the discharge will take place. A further disadvantage is that the complexity of the reactions occurring in the discharge tube makes interpretation of the results impossible in all but the very simplest systems. Very few polyatomic radical spectra are known, and even in these cases the identity of the radical is doubtful.

Photochemical decomposition provides the best general method of preparation of free radicals as the overall reaction is relatively quite simple, the initial act and the radicals produced are better known than in any other type of reaction and nearly all radicals, both simple and complex, can be produced by photochemical means. No free radical has ever been detected in absorption in a photochemical reaction however, because the method has one great disadvantage; the concentration of radicals produced by even the most intense light sources is very low indeed. If this difficulty could be overcome, and free radicals could be followed spectroscopically in photochemical systems, an ideal combination would result. This communication describes how such a method has been developed, making use of the fact that illumination of the system for a longer time than the half-life of the radicals is not necessary so that a flash technique can be used. It is shown that a modification of a type of discharge lamp now in use as a photographic source is capable of producing a partial pressure of free radicals higher than has ever been obtained by photochemical or any other methods.

There is one other difficulty associated with kinetic absorption spectroscopy of all kinds, that of producing an image on the photographic plate in a short time, from a source of continuum. Other workers have resorted to the integration of a large number of short exposures, but this is not feasible in the apparatus described here, first, because with such a high percentage decomposition the absorption tube has to be removed and cleaned after each flash with many of the substances used, and secondly, because the high energy makes it necessary to cool the lamp between flashes, so that if several thousand flashes had to be used each spectrum would require an experiment of several days.

An attempt was made to overcome this difficulty by making use of the great sensitivity and rapid response time of photomultiplier tubes, and a rapid scanning system was built into a 10·5 ft. grating spectrograph for this purpose. If this method is to be suitable, a high resolution coupled with good signal/noise ratio is necessary. The noise, caused by statistical fluctuations in the photocurrent, can only be decreased by decreasing the band width and hence the resolution, and even with the brightest light sources available no satisfactory compromise was reached. Fortunately, investigations into the properties of flash discharge lamps as photochemical sources showed that they could be made to produce a very intense continuum over the whole spectral region, and a lamp was eventually designed which would record photographically down to 2000 Å in a large Littrow spectrograph in less than 10^{-4} sec.

The procedure adopted was to produce a high percentage decomposition of the reactant by a high intensity flash in a lamp alongside the 1 m. absorption tube and to record the absorption spectrum of the products as the reaction progressed by means of flashes from another lamp at the opposite end of the tube from the spectrograph. The requirements for these two lamps are slightly different. Both must be of very high intensity and must be capable of accurate synchronization, but whereas a maximum energy output in the region producing chemical change is the main consideration with the photolysis lamp, a continuous spectrum covering the whole ultra-violet region is required from the spectrographic source. The duration of the photolysis flash should be not greater than the half-life of the radicals which are to be studied, and the spectroflash must be shorter still if several snapshots of the changing radical spectra are to be obtained.

Of the several ways of obtaining a brief flash of light, high pressure gas-filled discharge tubes showed the best promise for this type of experiment. Very high energies can be dissipated in one flash, and by arranging the pressure to be high enough to prevent breakdown at the operating voltage until a trigger pulse is applied, accurate synchronization can be obtained without any power loss at a switch or spark gap. No previous investigation of the possibilities of this type of lamp for photochemical and spectroscopic purposes seems to have been carried out, and there follows a brief account of the properties of high-energy flash discharge lamps as emitters in the photographic ultra-violet.

DESCRIPTION OF THE EXPERIMENTAL ARRANGEMENT

The complete apparatus is shown schematically in figure 7. The 1 m. photolysis flash lamp lies alongside the quartz reaction vessel in the reflector already described, which is constructed in two semi-cylinders to facilitate inspection and removal. The spectroflash lamp is at one end of the reaction vessel and the spectrograph slit at the other, and a small detachable mirror placed near the slit enables the iron arc or other comparison spectrum to be taken without disturbing the alinement. A current of air is blown into the centre of the reflecting cylinder and escapes at either end, and the high-voltage trigger pulse is led on to the centre electrode of the lamp via a glass insulator through this case.

The vacuum apparatus consists of the usual pumping arrangements and pressure gauges, storage bulb, purification train and gas-analysis apparatus. Provision is made for filling the lamps *in situ*, although frequent refilling is necessary unless they are heated vigorously during evacuation.

Between the reaction vessel and spectrograph is the wheel responsible for synchronization of the shutter, photolysis flash, spectroflash and oscilloscope time base. The scattered light from the photolysis flash which recorded on the plate was about one-quarter as intense as that from the spectroflash, so that for exploratory work it is possible to work without a shutter, but for intensity measurements it is necessary

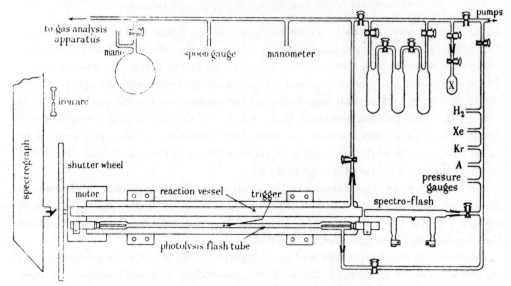

FIGURE 7. Diagram of the experimental arrangement.

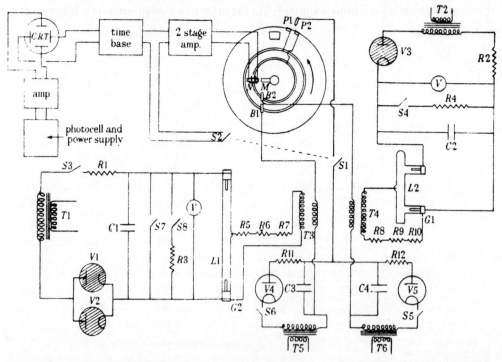

FIGURE 8. Electrical circuit diagram. Description of components: $C\,1$, $1200\,\mu$F, 4000V, in units of 1 to $4\,\mu$F; $C\,2$, $125\,\mu$F, 4500V, in units of 1 to $4\,\mu$F; $C\,3$ and $C\,4$, $1\,\mu$F, 1000V; $V\,1$, $V\,2$ and $V\,3$, mercury rectifiers, 8000V, R.M.S. type CV 128; $V\,4$ and $V\,5$, rectifiers, 2000V R.M.S.; V, electrostatic voltmeter, 5000V; $T\,1$, $200/3000$, 4000, 5000V, 10A; $T\,2$, $200/5000$V, 4A; $T\,3$ and $T\,4$, high-ratio induction coils; $T\,5$ and $T\,6$, $200/1000$V, 2A; $R\,1$, $2000\,\Omega$, 400W; $R\,2$, $40,000\,\Omega$; 200W; $R\,3$, $30,000\,\Omega$, 400W; $R\,4$, $4000\,\Omega$, 400W; $R\,5$, to $R\,10$, $60,000\,\Omega$, 5W; $R\,11$ and $R\,12$, $30,000\,\Omega$, 3W.

G. Porter 294

to eliminate it completely. A synchro-motor rotates the wheel via gears giving 8, 60, 200 and 600 r.p.m., and fine adjustment of the time interval between flashes is obtained by altering the distance apart of the two contacts on the wheel which are responsible for the trigger pulses.

The electrical arrangement is shown in figure 8. Apart from brushes, direct contacts to the wheel have been avoided so that the speed of rotation is uniform. The contacts at $P1$ and $P2$, which supply trigger pulses to the photolysis flash and spectro-flash respectively, are 0·1 mm. apart and platinum tipped, and the 1000 V primary pulse is sufficient to cross the air gap between them. $P1$ is fixed but $P2$

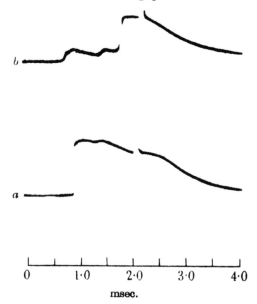

FIGURE 9. Typical oscillograph traces of the pulse and shutter timing.

moves radially over the circumference of the wheel and has a scale attached enabling time intervals between 2×10^{-4} and 0·5 sec. to be selected, longer intervals being obtained by manual switching. The synchronization pulse for the cathode-ray tube time base is induced by a small permanent magnet M on the wheel shaft in a small telephone coil, N, amplified, differentiated and rectified and passed to the hard-valve time base.

A long slot, the height of the spectrograph slit, is cut in the wheel and a thin aluminium slide moves on a scale over this enabling the whole or part of the photolysis flash to be eliminated. Calibration of the time scales and regular checking was necessary as there is a considerable interval introduced between contact and the initiation of the flash. This was carried out by means of a photocell inside the spectrograph connected via an amplifier and switch $S2$ to the cathode-ray tube. The oscilloscope trace is photographed, and, so that only one scan shall appear, the switch $S2$ is ganged to the main trigger switch $S1$. The timing of two typical experiments is illustrated in figure 9. In $9a$ the timing between the beginning of the two flashes is 1·1 msec. and in $9b$ it is 1·5 msec. The latter also shows how the photolysis flash can be eliminated quite sharply by introducing the shutter.

The trigger pulses for the two flash tubes are obtained by discharging a $1\,\mu$F condenser charged to $1000\,$V through the primary of an induction coil. The condenser $C\,3$ produces the trigger pulse for the photolysis flash tube when it is discharged via the primary of $T\,3$, the brush $B\,1$, the gap $P\,1$ and the switch $S\,1$. The pulse from the secondary of transformer $T\,3$ discharges through the lamp $L\,1$ via the gap $G\,2$ and resistances $R\,6$, 5 and 7. The gap avoids the necessity of a direct connexion of the lamp electrodes to the rest of the circuit, and the resistance chain is necessary to prevent the discharge short-circuiting to earth via the trigger electrode and $T\,3$. The spectroflash lamp has identical trigger arrangements employing $C\,4$, $T\,4$, $B\,2$, $P\,2$, $S\,1$, $G\,1$ and $R\,8$, 9 and 10.

$C\,1$ is the main condenser for the photoflash which is a variable capacity having a maximum value of $1200\,\mu$F. It is made up of units of 1 to $4\,\mu$F and is housed in a separate room with remote controls. The connecting leads are of $1\cdot27\,$cm. ($\frac{1}{2}$ in.) copper strip, the main leads from the whole bank are of $2\cdot54\,$cm. (1 in.) diameter copper pipe in order to keep all resistances down to a minimum, and precautions were taken during construction to reduce the inductance as far as possible. $C\,2$ is a similar condenser of $125\,\mu$F maximum capacity which supplies the power for the spectroflash. The master switch which fires both lamps and triggers the time base is the ganged switch $S\,1$ and $S\,2$.

The procedure used to photograph the spectrum a fraction of a second after photolysis is as follows. The reaction vessel is filled with the substance to be studied and the spectrograph shutter is opened. If a time check is required, the oscilloscope camera shutter is opened and the photocell, oscilloscope and trigger circuits are switched on. The shutter slide and spectroflash trigger contacts are set at the positions required to produce the desired time interval, the filament circuits of all valves are made and switches 4, 7 and 8 are opened. $S\,7$ is a safety-door switch, and $S\,4$ and $S\,8$ are remote discharge switches. The synchromotor is started, and switches 5 and 6 made for a few seconds to charge condensers 3 and 4. Switches 3 and 4 are now made and the condenser banks charged to exactly the operating voltage. All is now ready, and switch $S\,1/2$ is pressed when the following train of events occurs. The oscilloscope scan is tripped by the pulse induced in N and the wheel turns until $P\,1$ is opposite the standing contact when $C\,3$ discharges through $T\,3$ and fires $L\,1$. As the wheel turns farther the shutter clears the slit of the spectrograph, and when $P\,2$ comes opposite the standing contact $C\,4$ discharges through $T\,4$ and fires $L\,2$. A series of spectra at increasing time intervals is obtained by repeating this procedure for different positions of $P\,2$.

Several different spectrographs have been used, but the most generally suitable is the large Littrow type as it combines high dispersion with short exposures. One flash of the type described is sufficient to record over the whole photographic region with a slit width of $0\cdot02\,$mm.

RESULTS

The amount of photochemical decomposition obtained with a few typical compounds is given below. The mercury/krypton filling was used with acetone and ketene and pure krypton with the other substances; in each case the volume of gas was 500 ml. and the illumination a single flash of 4000 J:

gas	pressure (cm.)	% decomposition
NO_2	4	nearly 100
CH_3COCH_3	3	15
CH_3COBr	3	15
$CH_3COCOCH_3$	2·5	50
CH_2CO	10	40

If the lifetime of the intermediates involved in these reactions is comparable with the duration of the flash, the amount of decomposition obtained in this way makes possible the direct study of radical reactions by pressure change and the other physical methods which, hitherto, it has only been possible to apply to the overall change. The effect of high intensities on the nature of photochemical change has been mentioned in a previous note (Norrish & Porter 1949) and will be the subject of another communication.

The spectrographic technique has been applied to several photochemical reactions, and in most cases the amount of change was sufficient to be easily measurable by absorption changes. Three such systems will be mentioned here as illustrations of the potentialities of the method.

Atomic chlorine and its reactions

With a flash of 4000 J, 1 cm. of chlorine showed a Budde effect of 2·2 cm. Control experiments with inert gas showed that there was no detectable direct heating effect from the lamp. In 0·5 cm. of chlorine the absorption spectrum of the chlorine molecules almost completely disappeared, a direct demonstration of dissociation into atoms. The pressure of atoms must have been nearly 1 cm. and over 80 % of the total, a much higher concentration of atoms than has ever been obtained by photochemical or any other methods. The recombination of chlorine atoms may be studied in this way and figure 10, plate 7, shows the spectrum at increasing times after a flash of 2000 J, with a pressure of 1 cm. chlorine. The original decomposition is about 50 % in this case, and the half-life of the chlorine atom immediately after the flash, judged by visual comparison with the spectrum of the chlorine molecule at different pressures, is about 30×10^{-3} sec. Accurate estimations will have to take into account the temperature dependence of the absorption coefficient (Gibson & Baylis 1933).

When carbon monoxide was added to the system there was no permanent pressure change even after several flashes, and no phosgene or intermediate radicals were detected spectroscopically. If the mixture was illuminated in the ordinary way by a small mercury lamp, phosgene was rapidly formed, and the reaction could be taken almost to completion. On flash illumination of the mixture containing phosgene the pressure increased again and the phosgene was once more decomposed. The pressure change/flash was small at first, increased as more chlorine was formed and finally decreased when most of the phosgene had disappeared. The reaction was quite reversible and could be taken almost to completion either way by simply changing the intensity of the illumination. Two factors probably play a part in changing the mechanism of the reaction at high intensities. First, owing to the much higher concentration of atoms and radicals, interradical reactions become very frequent, so that the relative probability of the reaction

$$COCl + Cl = CO + Cl_2 \tag{1}$$

to the reaction

$$COCl + Cl_2 = COCl_2 + Cl \tag{2}$$

will be greatly increased. Secondly, owing to the higher temperature, the reverse of reaction (2) will occur more readily.

No evidence of $COCl$ or Cl_3 radicals was obtained in the chlorine or chlorine-carbon monoxide systems, but when phosgene was present an intense continuous absorption over the whole region studied appeared. It has not yet been decided whether this is due to an intermediate or to absorption by 'hot' phosgene molecules.

When oxygen was present a new banded spectrum appeared in the region of 2800Å. This spectrum was also obtained with chlorine and oxygen alone, but not when nitrogen or inert gases were substituted for the oxygen. It is shown in figure 11a, plate 7, and consists of a regular system of bands degraded to the red with a few weaker bands probably belonging to the $v'' = 1$ progression. The bands appear to be single headed and have a very simple rotational structure, and it seems most probable that a diatomic molecule is responsible. The most reasonable choice under these conditions is the ClO radical which has been frequently discussed as an intermediate in chlorine-sensitized oxidations. The half-life of the radical was found to be about 4×10^{-3} sec. in the presence of 1 cm. chlorine and 10 cm. oxygen. A Birge-Sponer extrapolation gives a dissociation energy to the products in the upper state of 108 kcal./g.mole. If the products of dissociation are normal chlorine atoms and ^1D oxygen atoms this would lead to a dissociation energy of the radical to atoms in the ground state of 63 kcal./g.mole.

Diacetyl

Figure 12a, plate 8, shows the absorption spectrum of diacetyl at 2·5 cm. pressure before photolysis, there being two distinct regions of absorption, one between 3800 and 4500 Å and the other below 3000 Å. The amount of decomposition can be judged from figure 12e, which is taken several minutes after a flash of 4000 J. Figure 12b is taken during the flash when photochemical decomposition is not complete. The reappearance of the absorption in the long wave-length region may be due to recombination of radicals, but is more probably due to a temperature effect. The increased absorption at lower wave-lengths is of too short a duration to be a temperature effect, however, the half-life being less than 1 msec., and it is difficult to find any other interpretation than that it is due to some intermediate substance of short life formed in the decomposition. The acetyl radical, which is probably formed in good yield in the reaction, might be expected to give an absorption in this region without any obvious fine structure. An analysis of the products not condensed in solid CO_2/ether gave: ethane 27 %, carbon monoxide 62 %, and methane 11 %.

Acetyl bromide on photolysis showed a similar increase of absorption in the same region, but in this case, owing to the smaller percentage decomposition, it was not possible to discriminate between this and the long-duration temperature effect on the acetyl bromide spectrum sufficiently readily to obtain a lifetime measurement.

Carbon disulphide

This substance is an example of those whose decomposition mechanism cannot easily be elucidated by analysis of the products alone. The final products of photolysis are solid sulphur and a polymer of composition $(CS)_n$, little being known with certainty about the existence of gaseous, unpolymerized CS.

The spectrum of carbon disulphide at 2 cm. pressure after illumination with a flash of 4000 J is shown in figure 13, plate 8. The decrease in the intensity of the continuum in *c* is due to clouding of the window by the solid products. The bands of the S_2 molecule are seen in *b* and continued at lower wave-lengths in *i*. The latter spectrum also shows three clear new bands with prominent heads at 2575·5, 2509·2 and 2444·5 Å, which agree closely with the wave-lengths of the 0, 0, 1, 0 and 2, 0 emission bands of carbon monosulphide measured by Jevons (1928). The 0, 0 and 0, 1 bands in absorption are shown enlarged in figure 14 *a* and *b*, plate 8, respectively. Weak bands around 2588·6 and 2523·2 Å correspond to the wave-lengths of the first two bands of the $v'' = 1$ progression, and the band head at 2504·8 Å is probably the 0, 0 band of the $^1\Sigma-^1\Sigma$ system observed by Crawford & Shurcliffe (1934), though its occurrence in absorption is not in agreement with their contention that a different lower state is involved in this system; as both are observed here the transition must be from the ground state in each case.

The spectra so far discussed were all taken a few msec. after the flash, and when the time interval was extended in order to measure the lifetime of the CS molecule no decrease in absorption was observable until several seconds had elapsed. The spectra in figure 13*d* to 13*h* show the decreasing absorption with time in 1 cm. of carbon disulphide, and the 0, 0 band is still faintly visible after 5 min. So persistent was this spectrum that it was observed as an impurity due to a little carbon disulphide which had dissolved in the tap grease, and it is surprising that it has not previously been observed in absorption.

The spectrum of S_2 appeared at maximum intensity immediately after the flash and disappeared much more rapidly than CS. These observations suggest the following as the most likely reaction scheme, the times given being those measured when the pressure of carbon disulphide originally present was 0·5 cm.:

		approximate time of half-reaction
$CS_2 + h\nu$	$= CS + S$	—
$S \ + CS_2$	$= S_2 \ + CS$	less than $1·5 \times 10^{-3}$ sec.
nS_2	$= S_{2n}$	10^{-1} sec.
nCS	$= (CS)_n$	60 sec.

As the ground states of S and CS are triplet and singlet respectively one must be produced in the excited state from singlet carbon disulphide, and the above lifetimes indicate that it is the sulphur atom which is so liberated.

REMARKS

Only one important difficulty in the method has appeared in the course of a wide range of investigations, the fact that the adiabatic nature of the reaction produces a change in the spectra of the parent molecules themselves. It is usually possible to discriminate between the two effects by lifetime measurements as was shown in the case of diacetyl, which is a particularly difficult example, but for quantitative absorption measurements it will often be necessary to measure the temperature and make allowance for the changing absorption coefficient. On the other hand, the phenomenon suggests the possibility of thermal and kinetic measurements on radical reactions by direct pressure observations.

The results which have been described are preliminary only, and the conclusions must be verified by more detailed investigations. The main purpose here has been to illustrate the power of this method for the investigation of fast reactions, and for this reason a wide range of substances has been studied. It is believed that the results are sufficient to show that the methods of flash photochemistry and spectroscopy provide a valuable weapon for the study of the more elusive of chemical compounds.

I am extremely grateful to Professor R. G. W. Norrish, F.R.S., for his support in this work from the beginning and for much encouragement and valued advice throughout. Thanks are also due to the Anglo Iranian Oil Company for a grant for research, part of which was applied to this work.

REFERENCES

Crawford, F. H. & Shurcliffe, W. A. 1934 *Phys. Rev.* **45**, 860.
Gibson, G. E. & Baylis, N. S. 1933 *Phys. Rev.* **44**, 188.
Jevons, W. 1928 *Proc. Roy. Soc.* A, **117**, 351.
Leighton, W. G. & Forbes, G. S. 1930 *J. Amer. Chem. Soc.* **52**, 3139, 5309.
Murphy, P. M. & Edgerton, H. E. 1941 *J. Appl. Phys.* **12**, 848.
Norrish, R. G. W. & Porter, G. 1947 *Faraday Soc. Discussion. The Labile Molecule*, **2**, 142.
Norrish, R. G. W. & Porter, G. 1949 *Nature*, **164**, 658.
Oldenberg, O. 1935 *J. Chem. Phys.* **3**, 266.
White, J. U. 1940 *J. Chem. Phys.* **8**, 79.
Wieland, K. 1947 *Faraday Soc. Discussion. The Labile Molecule*, **2**, 172.

Chemistry in Microtime

George Porter

The figure on p. 19 should be

Porter   *Proc. Roy. Soc. A, volume* **200**, *plate* 8

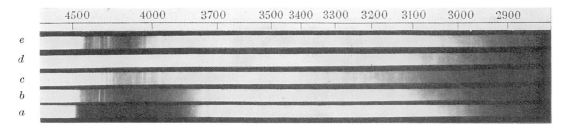

FIGURE 12

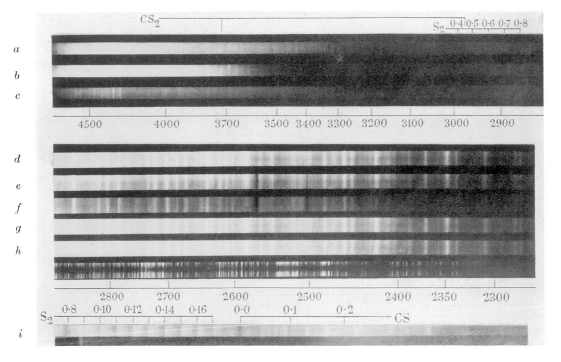

FIGURE 13

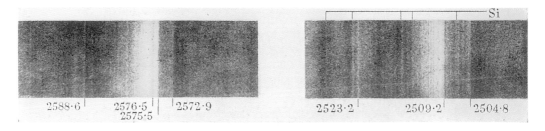

a   *b*

FIGURE 14

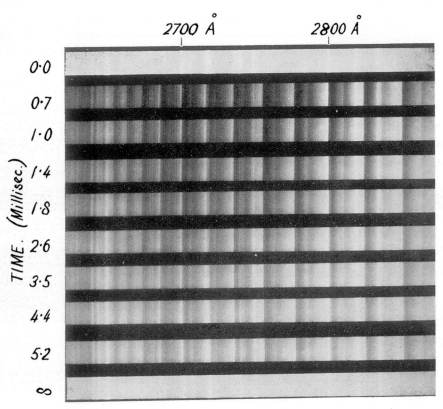

Bimolecular Disappearance of Chloric Oxide (ClO)

PLATE 1.

Porter   *Proc. Roy. Soc. A, volume* 200, *plate* 8

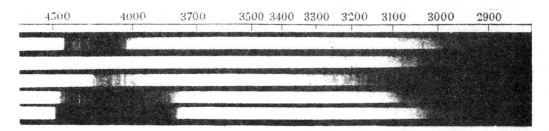

FIGURE 12

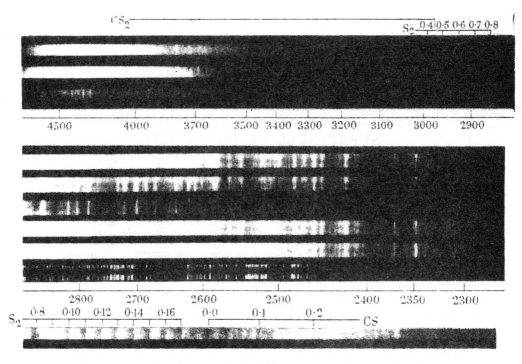

FIGURE 13

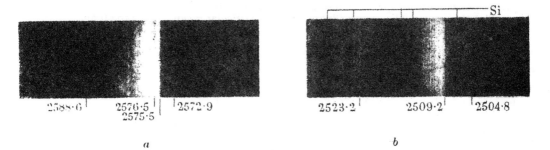

a   *b*

FIGURE 14

Paper 5 describes the absorption spectra of the first simple free radicals recorded by flash photolysis. All were diatomic except that attributed to the triatomic radical HS_2. Of particular interest was the free radical ClO, observed after flash photolysis of a mixture of Cl_2 and O_2. Its analysis led to a determination of dissociation energy and other spectroscopic constants and, more interestingly, the spectrum provided an excellent means for study of the reactions by which ClO was formed and by which it disappeared (paper 6). These investigations became of great importance some 25 years later when ClO was found in the stratosphere and was shown to be a key intermediate in the reactions leading to the depletion of ozone in the Antarctic spring.

Paper 5

Reprinted from the *Faraday Society Discussion*, 1950, No. 9

THE ABSORPTION SPECTROSCOPY OF
SUBSTANCES OF SHORT LIFE

By George Porter

Received 17th July, 1950

The experimental methods available for the absorption spectroscopy of free radicals are critically discussed with special reference to flash photolysis and spectroscopy. Some free radical spectra which have been obtained in this way are described and values are given for the dissociation energies and vibration frequencies in the upper and lower states of the ClO, SH and SD radicals.

Most of our knowledge of the molecular structure of free radicals and similar short-lived substances is a result of the interpretation of emission spectra obtained from flames and electrical discharges. About 100 di-atomic radicals have now been recognized in this way and many of their molecular constants determined, but there are very many other radicals whose emission spectra have not been observed and about whose structure virtually nothing is known. Chemical methods do not lend themselves readily to high-speed manipulation and other physical methods such as electron diffraction are not at present applicable. The method of absorption spectroscopy is almost alone in its suitability for investigations of this kind and has many advantages ; in particular it can be used to obtain spectra which cannot be observed in emission and it has the great advantage that it makes possible the determination of concentration. Furthermore, as more complex radicals are studied, it will be necessary to use other spectral regions such as the infra-red and far infra-red where absorption techniques become essential.

The difficulties associated with free radical absorption spectroscopy are almost entirely experimental ones and in the first part of this paper a brief account is given of the limitations of the existing methods with particular reference to the flash technique developed by the author. The second part deals with a few of the radical spectra which have been obtained by this method.

Experimental Methods

There are two problems associated with the kinetic absorption spectroscopy of short-lived substances in addition to those encountered with stable molecules. Firstly, the labile molecules must be produced rapidly in high concentration, and secondly, the spectrum must be recorded in a time which is short compared with their average life.

The Preparation of Free Radicals.—The partial pressure of radicals necessary for their observation in a path of 10 cm. varies from 10^{-5} mm. for a radical such as CN with high resolving power to 10 mm. or more under less favourable conditions. In many cases, and especially in the infra-red, a high relative concentration is also necessary if the spectrum is not to be masked by that of the parent molecule. There are four general methods available for free radical preparation and their uses for our purpose are summarized below.

THERMAL DECOMPOSITION.—This is only available when the dissociation of the radical does not occur much more readily than that of the parent molecule. It has not so far been possible to heat a gas rapidly enough to produce radicals instantaneously at high concentrations for kinetic studies and the method is limited to systems in thermodynamic equilibrium which have, however, one great advantage ; if the thermochemical constants are known the radical concentration and hence its absorption coefficients can be determined.[1, 2, 3]

CHEMICAL REACTION.—Owing to the fact that diffusion is slow compared with the rate of most radical reactions it is difficult to produce mixing rapid enough to give a long absorption path and except in the special case of flames this method has been unsuccessful.[4] In flames also the reaction zone is narrow and sensitive methods, such as emission line reversal, are usually necessary.[5]

ELECTRICAL DISCHARGE.—By this means accurate synchronization and high percentage decomposition are readily attained and it was, until recently, the only successful method of instantaneous preparation.[6] Its main disadvantages are that it is limited to the low pressures, less than about 1 mm., at which the discharge can be passed, and the fact that the violence of the method results in the production of almost every possible molecular species making the interpretation of the kinetics a matter of great difficulty, except in the very simplest systems.

PHOTOCHEMICAL DISSOCIATION.—This method has many advantages ; the system is relatively simple and well understood, a wide range of conditions may be used and almost · all radicals can be produced photochemically. With ordinary light sources, however, the concentration of radicals which can be obtained is so low that it has been impossible to detect them by means of their absorption spectra. The use of high intensity flash sources has completely overcome this difficulty and partial pressures of atoms and radicals of the order of cm. Hg have been produced, making this the most powerful of all methods of preparation.[7]

The Rapid Recording of Absorption Spectra.—If the kinetics of the radical disappearance are to be studied the exposure time must be only a fraction of the radical lifetime which may mean 10^{-4} sec. or less. With continuous sources such as the hydrogen lamp the exposure time required is several seconds at least but three methods of high speed recording are available.

[1] Bonhoeffer and Reichardt, *Z. physic. Chem.*, 1928, **139**, 75.
[2] Oldenberg, *J. Chem. Physics*, 1938, **6**, 439.
[3] White, *ibid.*, 1940, **8**, 439.
[4] Geib and Harteck, *Trans. Faraday Soc.*, 1934, **30**, 139.
[5] Kondratjew and Ziskin, *Acta Physicochim.*, 1937, **6**, 307.
[6] Oldenberg, *J. Chem. Physics*, 1934, **2**, 713.
[7] Porter, *Proc. Roy. Soc. A*, 1950, **200**, 284.

ELECTRONIC METHODS.—The response time of radiation detectors such as the photocell may be as low as 10^{-8} sec. but for high speed scanning a limit is set by the statistical fluctuations in the photocurrent. The relative fluctuations F_r are given by

$$F_r = 1/(nt)^{\frac{1}{2}},$$

where n = no. of electrons/sec. from the cathode $\simeq$ no. of photons/sec. and t = time interval resolved. For example, if the rate of scanning were 1000 Å/m. sec. and the s/n ratio were 50, the high output current of 100 μA from the 1P. 28 photomultiplier would only give a resolution of 1 Å. Only if resolution or scanning speed can be sacrificed is this method suitable with present sources, and similar considerations apply to infra-red detectors.

PHOTOGRAPHIC INTEGRATION.—If ordinary light sources are used the intensity may be increased by repeating the process many times, 60,000 such exposures being used by Oldenberg for each spectrum.[8] Maeder and Miescher have used a spinning mirror arrangement so that the decay of the radical is recorded at the same time.[9] Such methods are satisfactory if the interval between exposures is short but in flash photolysis the time required for lamp cooling, and for refilling and cleaning the reaction vessel owing to the high decomposition, makes a large number of exposures impracticable.

FLASH SPECTROGRAPHY.—The very high energy which can be dissipated in a single flash, the short duration and easy synchronization coupled with the fact that it gives a very good continuous spectrum makes the high-pressure rare gas filled flash-tube ideal for this purpose. A single flash from a 70 μF condenser at 4000 V across a 15 cm. tube, lasting 5×10^{-5} sec. gives a sufficient exposure on the plate of a Littrow (Hilger E.1) spectrograph down as far as the quartz absorption limit.[7]

Experimental Limitations of the Method of Flash Photolysis.— The construction and properties of flash tubes for photochemical and spectroscopic purposes have been described previously [7] and the present limitations of the method will now be given.

ENERGY DISSIPATION.—Energies of 10,000 J per flash have been dissipated in a 1 m. long tube, corresponding to a useful output of 2×10^{21} quanta in the quartz ultra-violet region. At the higher energies occasional refilling is necessary, for example, about every 100 flashes at 5000 J, and it has been found useful to have a separate pumping system for this purpose. It has also been found that xenon filling is the least troublesome in this respect. In most photochemical systems the pressure of intermediates produced by these energies is of the order of several mm. Hg.

FLASH DURATION.—It has become clear that the flash time of 1 msec. is longer than the lifetime of some of the radicals studied and it will be necessary to use the following methods of reducing its duration.

(a) DECREASE OF CAPACITY.—This reduces the duration of the flash without decreasing the intensity but the total output is thereby reduced.

(b) INCREASE OF VOLTAGE.—Up to the highest voltage used (8000 V) the energy dissipation of the tube does not appear to vary if CV^2 is kept constant by reducing the capacity, and as there is no consequent reduction in output this is the most useful method.

(c) DECREASE OF TUBE RESISTANCE.—This may be accomplished by increasing the diameter or decreasing the length of the tube, the latter being approximately proportional to flash duration. It has been found that the maximum energy dissipation of the lamp is also proportional to its length and again the total output must be reduced, but the output/unit length remains unchanged and, if the length of the reaction vessel can be reduced proportionately, the *percentage* decomposition is also unchanged.

The Determination of Radical Concentrations.— The approximate estimation of the lifetime of a radical by noting the disappearance of its absorption spectrum presents no difficulty and yields much interesting qualitative information about its chemical reactions, but there are many important data, such as rate constants and absorption coefficients, which cannot be obtained without a knowledge of absolute concentrations. At present such information is limited to the few radicals which can be obtained in thermodynamic equilibrium and it is important therefore that the relative simplicity of the photochemical system makes possible, in many cases, the estimation of radical concentration indirectly from the decrease in the absorption of the parent molecule

[8] Oldenberg, *J. Chem. Physics*, 1935, **3**, 266.
[9] Maeder and Miescher, *Helv. physic. Acta*, 1942, **15**, 511 ; 1943, **16**, 503.

and subsequently the appearance of the absorption by the stable products. A complication is introduced by the changing temperature, which is unavoidable in systems containing intermediate products at high concentrations, but the effect of temperature on the absorption spectrum of stable molecules can be determined and suitable corrections applied.

The flash technique was designed primarily for the study of the kinetics of radical reactions but much information is also obtained from the interpretation of the spectra themselves, and two such spectra, of the radicals ClO and SH, will now be considered.

The ClO Radical

A preliminary report and photograph of the spectrum attributed to this radical have been given elsewhere.[7] It appears whenever chlorine is photolyzed in the presence of oxygen and has a half-life of a few milliseconds. The simple vibrational structure strongly suggests a diatomic molecule and, under the circumstances, the only possibility is the ClO radical whose occurrence in chemical reactions has frequently been postulated. Very little is known about the diatomic compounds of Group 6 of the periodic table with Group 7 and the interpretation of this spectrum is therefore of some interest.

VIBRATIONAL STRUCTURE.—The main feature of the spectrum is a series of bands with fairly sharp heads degraded to the red which is clearly a $v'' = 0$ progression and the measurements of these heads with intensities on a scale of 10 are given in Table I. Most of the measurements are accurate to better than 10 cm.$^{-1}$ except for a few bands around $v' = 18$ where the overlapping rotational structure makes the heads difficult to locate.

TABLE I.—$v'' = 0$ PROGRESSION OF ClO

v'	Int.	λ Å	ν cm.$^{-1}$	v'	Int.	λ Å	ν cm.$^{-1}$
4	1	3034·5	32945	14	7	2729·4	36627
5	2	2993·0	33402	15	7	2711·1	36874
6	3	2954·3	33839	16	6	2695·0	37095
7	5	2918·0	34261	17	5	2682·5	37267
8	5	2884·0	34664	18	5	2671·2	37425
9	6	2851·8	35056	19	5	2661·0	37569
10	8	2822·4	35421	20	5	2652·5	37689
11	8	2796·0	35755	21	3	2645·8	37784
12	10	2771·6	36070	22	2	2640·6	37859
13	8	2749·5	36360	23	2	2636·3	37920

An assignment of vibrational quantum numbers in the upper state is not possible on this information alone but a band system attributed to ClO has been obtained in emission from flames containing chlorine or methyl chloride by Pannetier and Gaydon [10] and assuming their interpretation to be correct we may proceed as follows. The average value of the second difference of the wave numbers of the $v' = $ constant progressions in the emission bands is 15 cm.$^{-1}$ and the corresponding value for the upper state, calculated from the absorption spectrum, is 22 cm.$^{-1}$. Starting with the last observed bands of the $v'' = 0$ and $v' = 0$ progressions the positions of further bands heads can be estimated, assuming linear convergence, to give the following values :

Absorption, v = 0 progression
cm.$^{-1}$

Obs.   32945

Calc. $\begin{cases} 32466 \\ 31965 \\ 31442 \\ 30897 \end{cases}$

Emission, v' = 0 progression
cm.$^{-1}$

Obs.   27598

Calc. $\begin{cases} 28406 \\ 29227 \\ 30067 \\ 30920 \end{cases}$

The value for the last observed emission band has been adjusted slightly and the progression starts in effect from the previous band head which is

[10] Pannetier and Gaydon, *Nature*, 1948, **161**, 242.

probably the more accurate measurement. It will be seen that there is a coincidence within the accuracy of the extrapolations between the last two figures given and a continuation in this way as far as the observed bands of the other system shows no other coincidence within 100 cm.$^{-1}$. The vibrational quantum numbers obtained in this way are given in Table I and the interpretation is supported by intensity considerations and the close agreement between the values for ω_e' obtained from the two spectra. The value of x in the Table of Pannetier and Gaydon is seen to be 4.

These data lead to the following values for the constants of the two states with a probable error in ω_e of 3 %. Account has been taken of a small cubic term in v' in the estimation of ω_e'.

Ground state. $\omega_e'' = 868$ cm.$^{-1}$ $x_e''\omega_e'' = 7\cdot5$ cm.$^{-1}$.
Upper state $\omega_e' = 557$ cm.$^{-1}$ $x_e'\omega_e' = 11$ cm.$^{-1}$.
        $\nu_e = 31,077$ cm.$^{-1}$.

FURTHER DETAILS OF BAND STRUCTURE.—Owing to the extended rotational structure the chlorine isotope effect is difficult to observe, the weaker isotope system being obscured by the stronger, and it has not been possible to confirm the above assignment of quantum numbers in this way.

In addition to the bands already discussed a second progression appears at about one-quarter of the intensity, the heads showing a fairly constant separation to higher frequencies from those of the main system. The bands appear to belong to a second multiplet component, the whole forming a doublet system and over the nine measured heads in Table II the doublet splitting decreases from 200 cm.$^{-1}$ to about 185 cm.$^{-1}$.

TABLE II.—SECOND SYSTEM OF ClO

v'	λ Å	ν cm.$^{-1}$	v'	λ Å	ν cm.$^{-1}$
5	2975·0	33603	10	2807·0	35615
6	2937·6	34032	11	2781·4	35943
7	2902·0	34449	12	2758·0	36248
8	2866·7	34874	13	2735·3	36548
9	2835·7	35255	—		

A full analysis of the rotational structure has not yet been carried out, the resolution being rather too low. The P and R branches appear to be the most prominent feature of the bands and the structure suggests a $^2\Pi - ^2\Pi$ or $^2\Delta - ^2\Delta$ transition with near case (a) coupling. The former is supported by theoretical considerations which predict a $^2\Pi$ ground state for ClO.[11]

Determination of the Dissociation Energy.—The bands of the main system are observed almost to the convergence limit and a very good extrapolation can be made, the $\Delta\nu - v$ plot showing slight positive curvature. The value of the dissociation energy to products in the upper state obtained in this way is 37,930 cm.$^{-1}$ or 108·4 kcal./mole, ± about 0·1 %.

To obtain the dissociation energy to unexcited products the only low-lying levels of the atoms to be considered, ignoring multiplet splitting, are the ^2P ground state of chlorine and the ^3P ground and the ^1D and ^1S excited states of oxygen, the latter lying 15,868 and 33,793 cm.$^{-1}$ above the ground state. These lead to three possible values for D_0' and the lower one is immediately eliminated by the fact that vibrational levels of the ground state are observed which lie above it. The upper one is confirmed by a linear extrapolation of the constants for the ground state which gives a value for D_0'' of 24,680 cm.$^{-1}$. It is clear therefore that the oxygen atom is liberated from the upper level in the ^1D state and the following values for the dissociation energies can now be given unambiguously.

Ground state. Dissociates to Cl ^2P and O ^3P.
        $D_0'' = 22,062$ cm.$^{-1}$ = 63·04 kcal./mole ± 0·2 %.

Upper state. Dissociates to Cl ^2P and O ^1D.
        $D_0' = 7,010$ cm.$^{-1}$ = 20·0 kcal./mole ± 1 %.

[11] Perring (private discussion).

These values assume dissociation to the lowest atomic multiplet levels in each case as the actual levels are unknown. If the atoms are liberated in higher multiplet levels the value of D_0'' would need slight adjustment but the value of D_0' is unchanged. The normal dissociation energy of ClO to chlorine $^2P_{1\frac{1}{2}}$ and oxygen $^3P_2$ depends only on the multiplet level of the Cl atom liberated from the upper state, being the ground state value given above if Cl $^2P_{1\frac{1}{2}}$ is liberated and 881 cm.$^{-1}$ less if the product is Cl $^3P_{\frac{1}{2}}$.

Discussion.—The only existing information about the dissociation energy of ClO is obtained from the spectrum of the ClO_2 molecule.[12] The predissociation at 3750 Å, corresponding to an upper limit of 76 kcal./mole, is not inconsistent with the above value but much lower energies have been suggested from interpretations of the extrapolated convergence limit.

The only existing information about the dissociation energy of ClO is obtained from the spectrum of the ClO_2 molecule.[12] The predissociation at 3750 Å, corresponding to an upper limit of 76 kcal./mole, is not inconsistent with the above value but much lower energies have been suggested from interpretations of the extrapolated convergence limit. The postulated dissociation to normal ClO and O(^1D) now seems quite probable, and although there is some doubt about the heat of formation of ClO_2 it can hardly be high enough to account for this discrepancy. The explanation probably lies in the potential energy surface involved in the dissociation of polyatomic molecules which may lead to products with excess vibrational energy.

It is interesting to compare the bond energy of ClO with the average bond energy in the other chlorine oxides calculated from the heats of dissociation as follows : [13]

Cl_2O	Cl_2O_7	ClO_3	ClO_2	ClO
47·0	50·3	55·6	59·5	63·0 kcal./mole.

The influence of the odd electron in strengthening the bond is clear and the last four molecules form a group of inorganic free radicals showing increasing chemical reactivity as the odd electron becomes localized.

The formation of ClO by the reaction of Cl atoms with O_2 suggests that a radical Cl—O—O· takes part as an intermediate, no trace of other chlorine oxides such as ClO_2 being found in the spectra. Subsequent reactions are then probably

$$Cl—O—O + Cl = (Cl_2O_2 =) \ 2ClO,$$

and at higher temperatures,

$$Cl—O—O + Cl_2 = (Cl_2O_2 + Cl =) \ 2ClO + Cl.$$

Similar experiments with bromine indicate that the bromine atom does not react with O_2.

The SH and SD Radicals

The SH radical is the only diatomic hydride of the first two periods whose vibrational constants are completely unknown and one of the few common hydrides for which not even an approximate value of the dissociation energy is available. Its spectrum does not appear readily and only one band, the o—o band of the $^2\Sigma - ^2\Pi$ transition, has ever been observed.[14] It is well known that hydrogen sulphide can be decomposed photochemically into its elements, and although other mechanisms have been proposed [15] it is probable that the primary decomposition is to H and SH.[16] For this reason the flash photolysis of H_2S was studied as a probable source of the SH radical.

[12] Finkelnberg and Schumacher, *Z. physic. Chem.* (Bod. Fest.), 1931, 704.
[13] Goodeve and Marsh, *J. Chem. Soc.*, 1939, 1332.
[14] Lewis and White, *Physic. Rev.*, 1939, **55**, 894. Gaydon and Whittingham, *Proc. Roy. Soc. A*, 1947, **189**, 313.
[15] Goodeve and Stein, *Trans. Faraday Soc.*, 1931, **27**, 393.
[16] Herzberg, *ibid.*, 1931, **27**, 402.

Hydrogen sulphide was prepared from ferric sulphide and sulphuric acid and also by the action of water on an intimate mixture of calcium sulphide and phosphorous pentoxide. It was dried over $CaCl_2$ and P_2O_5 and fractionally distilled *in vacuo*. At pressures of 40 mm. of H_2S one flash produced a partial pressure of 9 mm. H_2, sulphur being deposited on the wall, and there was no overall pressure change. Spectra were taken of the products on a Littrow (Hilger E.I.) spectrograph at increasing time intervals, after the flash. Immediately after the flash the absorption spectrum showed, in addition to the continuum of H_2S and the $^3\Sigma - {}^3\Sigma$ system of S_2, the 0—0 band of SH at 3236·6 Å identical with that described by Lewis and White,[14] another similar band at 3060 Å, and a diffuse band system between 3168 and 3797 Å.

TABLE III.—SH 1—0 BAND

ν (cm.$^{-1}$)	Int.	Branch (J)
32664·8	8	R_1 head
32634·4	6	$Q_1(1\frac{1}{2})$
32618·4	8	$Q_1(2\frac{1}{2})$. $P_1(1\frac{1}{2})$
32601·0	5	$Q_1(3\frac{1}{2})$
32587·9	3	$P_1(2\frac{1}{2})$
32580·2	4	$Q_1(4\frac{1}{2})$
32555·7	6	$Q_1(5\frac{1}{2})$. $P_1(3\frac{1}{2})$
32528·1	2	$Q_1(6\frac{1}{2})$
32518·4	2	$P_1(4\frac{1}{2})$
32498·1	2	$Q_1(7\frac{1}{2})$
32480·0	2	$P_1(5\frac{1}{2})$
32498·1	2	$Q_1(8\frac{1}{2})$
32469·6	1	$Q_1(9\frac{1}{2})$
32267·5	3	Q_2 head

SD 0—0 BAND

ν (cm.$^{-1}$)	Int.	Branch (J)
30977·4	10	R_1 head
30959·6	3	$Q_1(1\frac{1}{2})$
30951·9	3	$Q_1(2\frac{1}{2})$
30943·6	3	$Q_1(3\frac{1}{2})$
30934·4	5	$Q_1(4\frac{1}{2})$
30923·6	2	$Q_1(5\frac{1}{2})$
30912·1	2	$Q_1(6\frac{1}{2})$
30899·6	3	$Q_1(7\frac{1}{2})$
30594·3	5	Q_2 head

SD 1—0 BAND

ν (cm.$^{-1}$)	Int.	Branch (J)
32294·0	10	R_1 head
32277·0	4	$Q_1(1\frac{1}{2})$
32269·2	5	$Q_1(2\frac{1}{2})$
32260·7	3	$Q_1(3\frac{1}{2})$
32251·1	4	$Q_1(4\frac{1}{2})$
32238·8	8	$Q_1(5\frac{1}{2})$
32224·3	5	$Q_1(6\frac{1}{2})$
32211·4	5	$Q_1(7\frac{1}{2})$
32195·4	5	$Q_1(8\frac{1}{2})$
32174·8	5	$Q_1(9\frac{1}{2})$
32154·4	4	$Q_1(10\frac{1}{2})$
31907·1	3	Q_2 head

THE 3060 Å BAND.—The rotational lines of this band appeared with greater intensity than those the 0—0 band, the relative intensity of the two bands always being the same, and both bands had a half-life of about 1 msec. The resolving power of the spectrograph in this region was 1·5 cm.$^{-1}$ which was not sufficient to separate the satelite $^QP_{21}$ and $^RQ_{21}$ branches from the main Q_1 and R_1 branches nor in some cases the main branch lines from each other. For this reason some of the lines were very broad and the line measurements which are given in Table III are the observed maxima with the probable quantum number assignment for the main branch lines. It is found that, apart from the different spacing owing to the higher value of $B'' - B'$, this band is identical in structure with the

3236·6 Å band and it is fairly certain that it is the 1—0 band of the same system. This was confirmed by comparing the SH spectrum with the spectrum of the SD radical obtained from D_2S when it was found that there was a small isotope shift to shorter wavelengths with SD for the 3236·6 Å band and a greater one to longer wavelengths for the 3060 Å band, the order of magnitude leaving no doubt as to the correctness of the above interpretation. The measurements of the strongest lines of the 0—0 and 1—0 bands of SD are given in Table III.

There are two unusual features about the appearance of the 1—0 band of SH ; firstly, it has a greater intensity than the 0—0 band though a careful search for further bands of the progression shows that they are absent and secondly, Lewis and White observed the weaker 0—0 band only. Both these anomalies, as well as the difficulty experienced in obtaining the bands in emission, would be explained if there were a potential energy curve leading to a lower dissociation limit which crossed the $^2\Sigma$ curve at about the second vibrational level. In this case the 1—0 band might appear stronger at low resolution owing to the broadening of the lines even if the transition were of lower probability, whilst higher transitions might be completely diffuse.

VIBRATIONAL CONSTANTS AND DISSOCIATION ENERGY.—Apart from the upper limit of 93 kcal./mole set by the above-mentioned predissociation these constants cannot be obtained from the spectrum of SH alone as at least three vibrational bands are necessary for their derivation. By using the different zero point energy of the isotopic molecule, however, another relationship is introduced which makes the calculation possible. If we assume the same force constant for the two molecules it can be shown that

$$\Delta^2\nu = (\nu_{1-0} - \nu^i_{1-0}) - (\nu_{0-0} - \nu^i_{0-0}) = \omega_e'(1 - \rho) - 2x_e'\omega_e'(1 - \rho^2)$$

where $\rho = \sqrt{\mu/\mu^i}$, ν_{1-0} is the wave number of the 1—0 band, etc. and the superscript i refers to the SD molecule. We also have

$$\Delta\nu'_{1-0} = \nu_{1-0} - \nu_{0-0} = \omega_e' - 2x_e'\omega_e'$$

and therefore if the separations of the bands are known the values of ω_e' and $x_e'\omega_e'$ can be obtained. The values $\Delta^2\nu = 471$ cm.$^{-1}$ and $\Delta\nu'_{1-0} = 1787$ cm.$^{-1}$ are obtained from the origins of the $^2\Sigma - ^2\Pi_{3/2}$ subbands estimated from the Q_1 branches, and substitution in the above equations gives for the $^2\Sigma$ state,

$$\omega_e' = 1950 \text{ cm.}^{-1} \quad \text{and} \quad x_e'\omega_e' = 81 \text{ cm.}^{-1}.$$

A linear extrapolation of these values gives the upper state dissociation energy $D_0' = 10,800$ cm.$^{-1}$ and in deriving a value for the normal dissociation energy two analogies with the $^2\Sigma$ state of OH, which might be expected to be very similar, will be made. The first is that the products in the upper state are H(^2S) and S(^1D) and this is fairly safe as the only alternative of S(^1S) would give a very low value for the normal dissociation energy.

The second analogy is that the linear extrapolation comes about 25 % too high and the final value obtained in this way will be assumed to have a possible error of 20 %. Applying this correction, subtracting the energy of promotion of S from the ^3P to the ^1D state and adding ν_{0-0} we get, for the normal dissociation energy, $D_0'' = 29,700$ cm.$^{-1}$ = 84·9 kcal./mole with an error probably less than 5 %.

From the relation

$$\nu_{v-0} - \nu^i_{v-0} = \omega_e'(v' + \tfrac{1}{2})(1 - \rho) - x_e'\omega_e'(v' + \tfrac{1}{2})^2(1 - \rho^2)$$
$$- [\tfrac{1}{2}\omega_e''(1 - \rho) - \tfrac{1}{4}x_e''\omega_e''(1 - \rho^2)]$$

the quantity in square brackets is found to be 369·5 cm.$^{-1}$ and to find the value of ω_e'' a rough estimate of $x_e''\omega_e''$ may be made from the value of $D_0'' + 10$ % (as in OH) the term in $x_e''\omega_e''$ being small. Using the value $x_e''\omega_e'' = 52$ cm.$^{-1}$ obtained in this way we get $\omega_e'' = 2670$ cm.$^{-1}$. The corresponding values of the vibrational constants of the SD radical are $\omega_e' = 1400$ cm.$^{-1}$ and $\omega_e'' = 1910$ çm.$^{-1}$.

THE DIFFUSE BAND SYSTEM.—These bands form a regularly spaced system showing no fine structure, most of them being degraded to the red with fairly sharp heads, the measurements of which are given in Table IV. They appear and disappear with the SH bands and in all the spectra taken, using pressures of H_2S between 1 cm. and 10 cm. Hg their intensity was proportional to that of the SH bands. That they are not bands of sulphur is shown by the fact that they do not appear along with the S_2 bands in the photolysis of CS_2 or S_2Cl_2 and also that they show a shift when D_2S is used in place of H_2S. At first sight they might be another system of SH, their simple vibrational structure suggesting a diatomic molecule, but this is not supported by the isotope shift which occurs to shorter wavelengths with deuterium and is roughly the same, about 50 cm.$^{-1}$ throughout the system.

TABLE IV.—DIFFUSE BAND SYSTEM

λ Å	Int.	ν cm.$^{-1}$	λ Å	Int.	ν cm.$^{-1}$
3168·0	3	31557	3443·9	10	29028
3195·6	4	31284	2479·5	9	28732
3222·1	4	31027	3519·5	8	28405
3249·7	6	30764	3562·5	6	28062
3278·5	8	30493	3604·4	6	27736
3308·0	8	30222	3647·8	6	27406
3340·0	9	29932	3696·5	4	27045
3373·5	10	29634	3745·6	3	26691
3407·0	10	29343	3796·5	1	26332

These facts suggest that the system is that of a molecule, probably polyatomic, containing S and H atoms only. Of these the most likely are the HS_2 radical and the H_2S_2 molecule, the latter compound being quite stable but without a recorded spectrum. One might expect the spectrum of hydrogen persulphide to be entirely continuous by analogy with hydrogen peroxide and it also seems likely that some of it would survive the reaction and be detected in the products but it cannot be entirely ruled out on these grounds. The vibration frequency of something over 350 cm.$^{-1}$ is a reasonable value for the frequency of the —S—S— bond in either molecule, and the only other information which can be obtained from the spectrum is the dissociation energy which a fairly good linear extrapolation gives as 35,600 cm.$^{-1}$ or 102 kcal./mole. The energy of the —S—S— bond in S_8 is 52 kcal.[17] and if it is this bond which is broken dissociation of HS_2 to SH ($^2\Pi$) and S(^1S) gives fair agreement whereas dissociation of H_2S_2 to two SH radicals would not give this value unless there were another state of SH much lower than the $^2\Sigma$ state. Pending an investigation of the spectrum of H_2S_2 it seems more likely that the bands are those of the HS_2 radical which could be formed by the union of H atoms with the S_2 molecule.

Other Diatomic Hydrides of Group 6.—The absorption bands of the $^2\Sigma - ^2\Pi$ systems of OH and OD have been obtained very strongly by the reaction of H or D atoms, prepared by photolysing a small amount of Cl_2 or Br_2, in the presence of H_2 or D_2, with oxygen. The first three bands of the $v'' = 0$ progression appeared at high intensity and this seems to be the first report of the OD bands in absorption. A very complex system of lines, which is the same in both cases and therefore attributable to oxygen

[17] Siskin and Dyatkina, *Structure of Molecules* (Butterworth, 1950), p. 255.

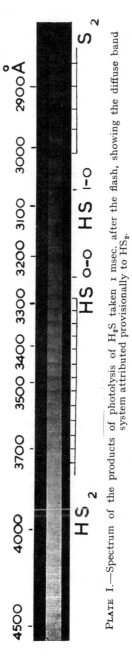

PLATE I.—Spectrum of the products of photolysis of H_2S taken 1 msec. after the flash, showing the diffuse band system attributed provisionally to HS_2.

alone, appears from 3000 Å to shorter wavelengths but no diffuse band system similar to that obtained with H_2S is present.

Attempts have been made to obtain the spectrum of the SeH radical in the photochemical decomposition of H_2Se but the only spectrum recorded on the plate was that of Se_2 despite the fact that the amount of decomposition was considerably greater than with H_2S. The absence of the spectrum of SeH may be explained if the predissociation occurs below the first vibrational level in this case, or if the SeH radical is chemically less stable, the latter explanation being probable in view of the decrease in stability from OH to SH, the former being observed for as long as 1/10th sec. after the flash.

The author wishes to thank Prof. R. G. W. Norrish for many helpful discussions and suggestions in connection with this work.

The University,
  Cambridge.

Dr. G. Porter (*Cambridge*) (*communicated*) : Like Dr. Herzberg and Dr. Ramsay we have made several attempts to obtain the spectrum of methylene in absorption, some of these experiments being mentioned at a previous Discussion of the Society,[35] and in fact the flash technique was originally developed for this purpose. All attempts have been unsuccessful despite the fact that a partial pressure of methylene of several mm. Hg was probably present in some cases ; it is possible that the higher resolution of the apparatus mentioned by Herzberg and Ramsay will be of assistance in this problem. The long life of CF_2 reported by Dr. Barrow is interesting in this connection, we have been unsuccessful in attempts to find the spectrum of CCl_2 in the photochemical decomposition of CCl_4, both alone and sensitized by Cl_2, and it seems improbable that it has a comparable stability. Incidentally CF_2 and CS should not, by the usual definition, be described as free radicals as both have singlet ground states.

The diffuse band spectrum obtained in the photolysis of H_2S appeared quite strongly as is shown in Plate I which is taken through 4-cm. H_2S, 1 msec. after the beginning of the flash, the total decomposition being about 25 %. It may have been missed by the other workers owing to the relative weakness of diffuse spectra at high resolution. Owing to this diffuseness the isotope shift is difficult to detect and it will be useful to have measurements of the band system mentioned by Leach, which occurs in the same region, to see whether there are any coincidences.

In reply to the question of Prof. Goldfinger, the only direct evidence about the energy of removal of the first H atom from H_2S is the low frequency limit of continuous absorption at 37,000 cm.$^{-1}$ which is of no assistance in deciding between the three suggested values for $D(S_2)$. All the known data show a close resemblance between the H_2O and H_2S molecules and the OH and SH radicals and although analogies of this kind are somewhat uncertain one would not expect the energy of removal of the first H atom from H_2S to be less than that of the second in view of the great difference in H_2O. If this argument is accepted we are led to $D(S_2) = 4\cdot4$ eV rather than the lower values.

[35] Norrish and Porter, *Faraday Soc. Discussions,* 1947, **2,** 97.

Paper 6

Reprinted from the *Faraday Society Discussion*, 1953, No. 14

STUDIES OF FREE RADICAL REACTIVITY BY THE METHODS OF FLASH PHOTOLYSIS

THE PHOTOCHEMICAL REACTION BETWEEN CHLORINE AND OXYGEN

By George Porter and Franklin J. Wright

Department of Physical Chemistry, University of Cambridge

Received 15th April, 1952

The reaction of chlorine atoms with oxygen has been studied by the flash photolysis and flash spectroscopy method of Porter.[1] The chloric oxide radical ClO is readily formed at 293° K and its concentration has been followed throughout the reaction by quantitative measurements of its absorption.

The initial reactions are $Cl + O_2 = ClOO$ and $Cl + ClOO = 2ClO$. The rate constant of removal of chlorine atoms by oxygen, to form both Cl_2 and ClO, is 46 times the rate of removal in nitrogen to form Cl_2.

The decomposition of ClO to Cl_2 and O_2 occurs relatively slowly, is bimolecular with respect to ClO, and the rate is independent of Cl_2, O_2 and total gas pressure. The rate constant of this reaction is

$$7 \cdot 2 \times 10^4 \ \epsilon_s \exp (0 \pm 650/RT) \text{ l. mole}^{-1} \text{ sec}^{-1},$$

where ϵ_s is the molar extinction coefficient of ClO at 2577 Å. The value of ϵ_s is greater than 310 and probably less than 3000. The mechanism of this reaction is discussed in terms of the intermediate Cl_2O_2.

The extensive literature on the reactivity of chlorine atoms, produced by photochemical dissociation of the molecule, gives little evidence of a direct reaction with oxygen. The well-known inhibiting effect of oxygen on photochemical chlorinations is usually attributed to the removal by oxygen of hydrogen atoms, COCl radicals, hydrocarbon radicals, etc., rather than of chlorine atoms, but the rate expressions do not allow an unequivocal choice of mechanism. In a mixture of chlorine and oxygen alone there is no apparent photochemical change and no transient reactions have been detected.

It was therefore somewhat unexpected when a new absorption spectrum, which was clearly that of a diatomic molecule, was discovered at high intensity during the investigation of chlorine + oxygen mixtures by the flash technique.[1] The spectrum was attributed to the ClO radical, and a vibrational analysis, as well as a determination of the dissociation energy of the molecule have already been given.[2] The methods of flash photolysis and flash spectroscopy were originally developed to make possible the production of labile molecules in a concentration high enough for absorption spectroscopy to be applied to the study of their kinetics, and we here describe such an investigation of the formation and reactions of the ClO radical. Direct information is also obtained in this way about the role of the oxygen in the photochemical reactions of chlorine.

EXPERIMENTAL

The absorption spectra of 20 free radicals, about half of them new, or previously unknown in absorption, have now been obtained in this laboratory. A detailed kinetic investigation by the flash technique involves more difficulties, however, than simply recording the spectrum of a free radical. Many hundred spectra must be taken under different conditions, flash intensities and time intervals must be accurately reproduced and the temperature rise, which is a result of the adiabatic nature of the reaction, must be eliminated.

The number of intensity measurements necessary for a complete kinetic investigation by flash photolysis is best reduced by the use of photocell recording. The relative merits of this method, which is being used for other problems, have already been discussed,[2] but in the present case, owing to the fact that the mechanism of the reaction was quite unknown, and also that a rather complex band system was being investigated, it was thought advisable to examine the whole spectral region throughout the investigation so as to be able to detect the presence of other chlorine oxides, changes in the intensity distribution of the ClO spectrum and intensity changes in the spectrum of the chlorine molecule. The use of photographic recording introduces more scatter into intensity measurements but the relationships eventually obtained are particularly significant as they are the result of many experiments carried out in a quite arbitrary order.

A full description of the method, and of the apparatus used for this investigation has been given elsewhere.[1] The only modification necessary was to enclose the reaction vessel and the photolysis flash tube in a furnace, so that the temperature dependence of the reaction rates could be determined. The furnace, which was of the same dimensions as the original reflector, completely enclosed the reaction vessel, and was coated internally with magnesium oxide. To facilitate removal it consisted of two semi-cylindrical portions wound separately, and the temperature was measured by three thermocouples at the centre and at either end. An investigation of the properties of the photolysis flash lamp at temperatures up to 350° C showed that the firing characteristics are a function of the concentration of the inert gas filling rather than of the pressure and that the output is not noticeably affected by the rise in temperature. This, fortunately, makes it possible to use the same gas filling throughout and therefore to avoid intensity variations which occur from one filling to another.[3]

PREPARATION OF GASES.—Chlorine was taken from a cylinder and redistilled several times *in vacuo*. Oxygen was prepared electrolytically, dried over $CaCl_2$, freed from traces of hydrogen by passing over 30 cm of platinized asbestos at 350° C and finally dried over P_2O_5. Nitrogen in which not more than 0·05 % oxygen could be tolerated, was prepared by heating sodium azide which gives a pure product.[4]

INTENSITY AND CONCENTRATION MEASUREMENTS.—The spectrograph was a Littrow (Hilger E.1) instrument ; 25 spectra were recorded on each plate, and the intensities were measured on a non-recording microphotometer. The output of the spectroflash was constant within the accuracy of the microphotometric measurements. Each plate was calibrated separately and methyl ethyl ketone, which has a continuous spectrum in the same region as the ClO radical, was used for this purpose. The ketone was made up to about 1 atm. pressure with carbon dioxide and a range of ketone partial pressures chosen to give the same densities as the particular ClO concentrations being studied. The intensity of the ClO spectrum could then be expressed in terms of the pressure of ketone having the same extinction at a given wavelength, the plate sensitivity, path length and incident intensities being constant. The extinction curve of methyl ethyl ketone vapour was measured on a Unicam spectrophotometer.

Chlorine itself has a significant absorption in the region of the ClO spectrum, and this must be allowed for in the intensity determination. It will be shown later that the amount of ClO formed is so small that the chlorine concentration does not change significantly during the experiment, and therefore this correction is a constant one. If Beer's law is obeyed we have, for a given wavelength,

$$\ln(I_0/I_1) = \epsilon_1 c_1 l + \epsilon_2 c_2 l,$$

where I_0 is the incident intensity, I_1 is the intensity transmitted by the mixture of Cl_2 and ClO, l is the path length, ϵ_1 and c_1 are the extinction coefficient and concentration of ClO, and ϵ_2 and c_2 the same quantities for chlorine. Now if ketone, at concentration c_3 also transmits the intensity I_1, the incident intensity being unchanged, and the extinction coefficient being ϵ_3,

$$\epsilon_1 c_1 + \epsilon_2 c_2 = \epsilon_3 c_3.$$

If I_2 is the intensity through chlorine alone, and also through ketone at concentration c_4

$$\ln (I_0/I_2) = \epsilon_2 c_2 l = \epsilon_3 c_4 l$$

and

$$\epsilon_3(c_3 - c_4) = \epsilon_1 c_1. \tag{1}$$

The values of $c_3 - c_4$, which are determined experimentally from the calibration spectra, are therefore proportional to the ClO concentration if Beer's law applies.

ϵ_3 being known, $\epsilon_1 c_1$ can be determined but the values of ϵ_1 and c_1 cannot at present be obtained separately. It is therefore necessary to express the concentrations of ClO as some function of its extinction coefficient and all concentration measurements are expressed as $\epsilon_s c$, where c is the true pressure of ClO in mm/Hg and ϵ_s is the molar decadic extinction coefficient at 2577 Å. This wavelength is in the continuous region and should therefore be unchanged if a different resolving power is used.

VALIDITY OF BEER'S LAW.—Beers' law has been assumed to apply both to the ketone and the ClO in the derivation of eqn. (1). No deviations were expected for methyl ethyl ketone, which was chosen for the lack of fine structure in its spectrum,[5] and the law was verified by measuring the extinction spectrophotometrically over the range of total pressures used in the calibrations. The following precautions were taken to assure the applicability of the law to the ClO spectrum. Firstly, possible deviations due to pressure effects were eliminated by keeping the total pressure many hundred times greater than the pressure of ClO. Except in one case, which is discussed separately, no comparisons are made at different temperatures. Finally, errors which might be caused by incomplete resolution of fine structure were first reduced by taking all measurements in the continuum or near to the heads of the predissociated bands where most of the absorption is due to lines of a width greater than the resolving power of the spectrograph. Each position was then checked by plotting $c_3 - c_4$ for a number of ClO spectra taken at different concentrations against the same quantity for a different wavelength. In one case only, the 13, 0 band at 2751 Å, there was a small deviation at high concentrations. In all other cases a linear plot was obtained and as this included measurements in the continuum, where the spectrum is completely continuous, the validity of Beer's law is confirmed for the measurements to follow, in which the 13, 0 band is not used.

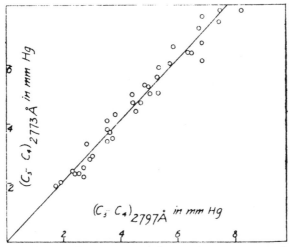

FIG. 1.—Plot of $c_3 - c_4$ at 2773 Å against $c_3 - c_4$ at 2797 Å.

From the gradient of these plots, one of which is shown in fig. 1, the relative extinction coefficients at the different wavelengths were obtained so that all measurements could be expressed in terms of ϵ_s. For the three bands which were used in addition to the continuum they are as follows :

band	12,0	11,0	10,0	continuum
wavelength λ (Å)	2773	2797	2824	2577
$\epsilon_\lambda/\epsilon_s$	1·50	1·44	1·44	1·00

REDUCTION OF THE ADIABATIC TEMPERATURE EFFECT.—Unless precautions are taken, the heat liberated by the reactions of the atoms and radicals formed may produce a temperature rise of over 1000°, the rate of thermal diffusion to the walls being less than that of the chemical reactions. Preliminary experiments showed that the ClO radical was still present in pressures of oxygen or nitrogen as high as 1 atm, and in most of the experiments to follow the total pressure used was 600 mm the pressure of chlorine being about 5 mm. Under these conditions the temperature rise, estimated by calculation and by use of the concentration effect, is 1 or 2 C and it will be shown later that a temperature rise of 100° C has no effect on the measured constants.

The photolysis flash was operated from 168 μF at 4000 V unless otherwise stated and the absorption tube was 1 m in length and 2 cm diam.

26   FLASH PHOTOLYSIS

RESULTS

A typical series of spectra, taken at increasing times after the flash, is shown in plate 1. It was first nesessary to ascertain that the whole of the spectrum being measured was that of ClO and that there was no overlying spectrum of another molecule. It was found that the rates determined in each part of the spectrum were the same, as would be expected in view of the relationships between the extinction coefficients mentioned previously. A search was then made for spectra in other regions, in particular those of the ClO_2 and Cl_2O molecules which have high extinction coefficients in the region of 2800 Å. None was found. Finally there was no difference between the results of experiments on newly mixed gases and those which had been flashed up to 50 times.

Under all conditions, the rate of reaction of ClO was very much less than the rate of its formation during the flash. Further, at the shortest times, immediately after the flash, the observed rate of reaction of ClO at a given concentration was the same as at longer times at the same concentration, showing that the reactions by which ClO is formed occur in a time which is short compared with the duration of the flash. It is therefore possible to divide the investigation into two parts : the dark reactions of ClO occurring after the flash and the photochemical reactions by which ClO is formed during illumination. The times involved in a typical experiment are illustrated by fig. 2 which gives the

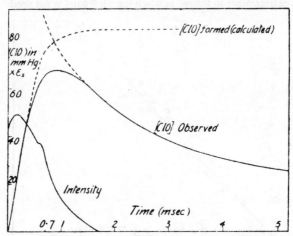

FIG. 2.—Light intensity and ClO concentration against time.

intensity against time and the ClO concentration against time curves. The latter, at times greater than 1·4 msec gives the rate law for the dark reaction and will be studied first, the reactions during the first msec being discussed later.

THE KINETICS OF ClO DISAPPEARANCE.—The concentration of ClO as a function of time was determined for a given mixture by recording a number of spectra, in an arbitrary order, at different times after the flash. Measurements were made at several wavelengths and concentrations determined, via the ketone calibrations, in the manner described.

DEPENDENCE ON ClO CONCENTRATION.—Fig. 3 shows the relationship between the reciprocal of the ClO concentration and the time, using a mixture of 10 mm Cl_2 and 600 mm O_2. A linear plot is obtained and, in the course of this work, over 60 such graphs were drawn from measurements on a wide variety of mixtures, all of which showed a direct proportionality within the experimental error. We conclude that, under all conditions described below,

$$- \, d(ClO)/dt = k(ClO)^2.$$

If k' is defined by the equation

$$-- \, d((ClO)\epsilon_s)dt = k'((ClO)\epsilon_s)^2,$$

the value of k' obtained from the gradient of fig. 3 is 4·9 mm⁻¹ sec⁻¹ ϵ_s units⁻¹, and the absolute rate constant is given by

$$k = k'\epsilon_s \text{ mm}^{-1} \text{ sec}^{-1} = 1·70 \times 10^4 \, \epsilon_s \text{ l. mole}^{-1} \text{ sec}^{-1}.$$

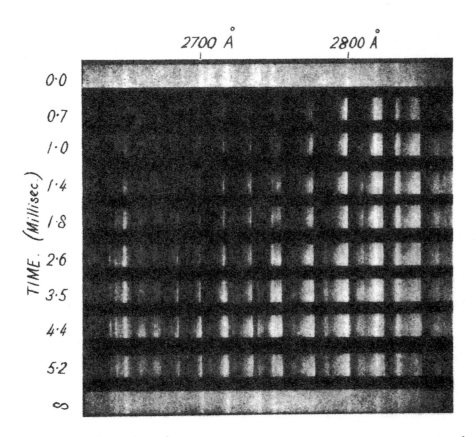

Bimolecular Disappearance of Chloric Oxide (ClO)

PLATE 1.

DEPENDENCE ON OXYGEN PRESSURE.--The total pressure of oxygen plus nitrogen was kept constant and the relative pressure of oxygen varied. Owing to the similar specific heats and collision diameters of the two gases, oxygen pressure is the only significant variable. The rate constants for different mixtures (pressures in mm Hg), determined as before, are given in table 1. About 10 spectra were used in each determination of these and all subsequent rate constants. The values of k' are constant to within 10 % whilst the oxygen pressure is changed by a factor of 60.

TABLE 1.—RATE CONSTANT k' AT DIFFERENT OXYGEN PRESSURES (P IN MM Hg)

$P(Cl_2)$	$P(O_2)$	$P(N_2)$	$P(O_2 + N_2)$	k' (mm^{-1} sec^{-1} ϵ_s units^{-1})				mean k'
10	600	0	600	4·0	3·7	3·8	4·9	4·1
10	100	500	600	4·3	4·1	4·3	—	4·2
10	10	590	600	3·8	4·7	3·5	4·1	4·0

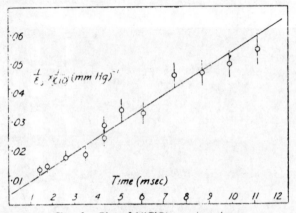

FIG. 3.—Plot of $1/(ClO)\epsilon_s$ against time.

DEPENDENCE ON CHLORINE PRESSURE.—The results given in table 2 show that the rate constants are invariant over a tenfold range of chlorine pressure.

TABLE 2.—RATE CONSTANT k' AT DIFFERENT CHLORINE PRESSURES (P IN MM Hg)

$P(Cl_2)$	$P(O_2)$	k' (mm^{-1} sec^{-1} ϵ_s units^{-1})				mean k'
20	600	4·3	3·7	3·8	—	3·9
10	600	4·0	3·7	3·8	4·9	4·1
5	600	4·5	4·4	4·0	4·0	4·2
2	600	3·9	3·7	4·3	—	4·0

DEPENDENCE ON TOTAL PRESSURE.—At low total pressures the reaction is no longer even approximately isothermal but reference to the next section shows that the effect of temperature rise on the rate can be ignored over a wide range and it was therefore possible to vary the total pressure by a factor of 10 without introducing a temperature change large enough to affect the rate. The results given in table 3 show that over this range the value of k' is independent of total pressure.

TABLE 3.—RATE CONSTANT k' AT DIFFERENT TOTAL PRESSURES (P IN MM Hg)

$P(Cl_2)$	$P(O_2)$	$P(N_2)$	$P(total)$	k' (mm^{-1} sec^{-1} ϵ_s units^{-1})			mean k'
10	100	500	610	3·8	4·3	5·2	4·4
10	100	300	410	4·9	4·9	6·1	5·3
5	50	100	155	4·0	4·8	4·4	4·4
10	100	0	100	3·1	3·2	3·4	3·2
5	50	0	55	4·8	4·5	5·2	4·8

DEPENDENCE ON TEMPERATURE.—The rate constants measured at higher temperatures are given in table 4, the pressures being referred to 293° K.

TABLE 4.—RATE CONSTANT k' AT DIFFERENT TEMPERATURES (P IN MM Hg)

$T°$ K	$P(Cl_2)$	$P(O_2)$	k' (mm^{-1} sec^{-1} ϵ_s units^{-1})			mean k'
293		mean of values in tables 1, 2 and 3				4·2
433	5	300	4·0	4·2	4·0	4·1
473	10	400	6·2	6·3	7·3	6·6

The values of k' are constant in the temperature range 293° K to 433° K but increase slightly at 473° K. Measurements at higher temperatures showed a further increase but they are of doubtful significance owing to the changed intensity distribution (see results on temperature dependence of ClO formation). If there were a true rate increase at higher temperatures it might imply a change in mechanism at about 450° K but it is also possible that the ratio of ϵ_s to $\int_0^\infty \epsilon_\lambda d\lambda$ is no longer constant. In view of this uncertainty we cannot say anything about the higher temperature mechanism at present.

In the range between 293° K and 433° K there is no significant change in the intensity distribution and the measured rates are constant. Assuming a possible variation in k' of $\pm$ 40 %, in order to allow for any slight changes in the intensity distribution, we conclude that if the temperature coefficient of ClO removal is expressed as exp ($- E/RT$) then the activation energy in the range of temperature between 293° K and 433° K is 0 $\pm$ 650 cal. The introduction of a $T^{\frac{1}{2}}$ dependence of the non-exponential term would reduce the value of E by 350 cal.

The results of this section are summarized as follows :
the rate law of ClO removal in the dark reaction is

$$- d(ClO)/dt = k(ClO)^2(O_2)^0(Cl_2)^0(N_2)^0 ;$$

the mean value of $k' = k/\epsilon_s$ at 293° K is 4·2 mm^{-1} sec^{-1} ϵ_s units^{-1} ;
the rate constant k, in the temperature range 293° K to 433° K, is

$$7·2 \times 10^4 \epsilon_s \exp (0 \pm 650/RT) \text{l. mole}^{-1} \text{sec}^{-1}.$$

ABSOLUTE VALUE OF ϵ_s AND k.—In view of the very intense absorption by the ClO radical it seemed probable that a proportional decrease in the intensity of the chlorine molecule spectrum would be observed and that this would give the absolute concentrations of ClO and values for ϵ_s and k. Experiments designed for this purpose have shown no such decrease and therefore, at present, it is only possible to give a lower limiting value for ϵ_s.

In order to increase the sensitivity of the method 10 similar spectra were taken of chlorine in the presence of a high concentration of ClO, each accompanied by a blank of the chlorine alone, the voltage being 6000 V in this case. In mixtures of 5 mm Cl_2 with 700 mm O_2 a slight decrease in Cl_2 absorption was observed which control experiments, using nitrogen in place of oxygen, showed to be due to the temperature-concentration effect alone. A mixture of 1 mm Cl_2 with 700 mm O_2 showed no decrease whilst calibration spectra with known chlorine pressures showed that 0·05 mm pressure decrease would have been detected. The average value of $(ClO)\epsilon_s$ was 31 mm which gives the minimum value for ϵ_s of 310.

It is helpful to have some idea of the maximum possible value of ϵ_s for the purpose of discussion and we have two reasons for supposing that it is not greatly different from the above minimum. Firstly an examination of the known extinction coefficients of similar molecules and radicals having partly continuous spectra make it improbable that the value of ϵ_s, which is by no means the maximum extinction coefficient, would exceed 3000. Secondly, if chlorine atoms recombine at a rate equal to or less than the three-body collision rate in the presence of nitrogen the later investigations on the relative rate constants of this reaction and ClO formation, coupled with the fact that the formation of ClO is not observable after 1·5 msec lead to a lower limit for the ClO concentration which is again within a factor of 10 of the above minimum. We hope to be able to determine ϵ_s experimentally by other methods ; it seems very probable, however, that the true values of the constants are not more than a factor of 10 greater than the following experimental minima :

$$\epsilon_s > 310, \quad k_{293} > 2·2 \times 10^7 \text{ l. mole}^{-1} \text{ sec}^{-1}.$$

THE INITIAL PHOTOCHEMICAL REACTION

The times involved are too short for the investigation of the ClO concentration changes during the flash, though a simple modification of the apparatus would make this possible. A different approach was used here, however, the reactions during the flash being studied by measurements of the concentration $((ClO)\epsilon_s)$ formed during a given time, the time chosen being 0·7 msec. During this time a fraction of the ClO formed will have reacted by the mechanism already studied and, using the known rate constant, it is possible to correct for this and to calculate, for any measured concentration at time 0·7 msec, the total ClO formed during this period. The calculation was performed as follows :

(i) The intensity against time curve was taken from an oscillograph of the flash and a graph of total light output against time was constructed, from the area beneath, in arbitrary units.

(ii) As the concentration changes are small the light absorbed, and therefore, owing to the high rate of ClO formation, the ClO present at a given time will follow the same curve if the reactions by which the ClO is removed are ignored. Taking these reactions into account it is possible to calculate the true ClO concentration at any time by using the known rate constant k'. A step method was found most convenient and the calculation was simplified by the almost linear nature of the intensity curve up to 0·7 msec. A corrected curve is shown in fig. 2.

(iii) For the series of observed concentrations at time 0·7 msec the total ClO formed is determined in this way.

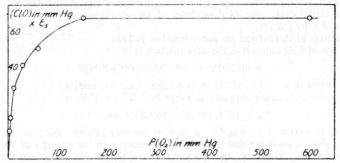

FIG. 4.—(ClO) formed as a function of oxygen pressure.

It was found that even at the highest concentrations the correction was only about 15 % and no significant errors are likely to be involved in these corrections.

RELATION BETWEEN ClO FORMED AND OXYGEN PRESSURE.—The concentration of ClO formed during the first 0·7 msec was determined for a range of oxygen pressures, the total pressure being kept constant at 600 mm by the addition of nitrogen. The chlorine pressure was 5 mm and the flash intensity and other conditions were kept as constant as possible. The concentrations, after correction, are plotted as a function of oxygen pressure in fig. 4. It will be seen that only when the oxygen concentration is reduced to about 1/50 of the nitrogen concentration is the ClO formed reduced by ½. This suggests that the rate constants of the reactions in oxygen and nitrogen respectively are also in this ratio and a more detailed derivation of this relation will be given later.

EFFECT OF TEMPERATURE ON ClO FORMATION.—The intensity of ClO at 0·7 msec was investigated as a function of temperature between 293° K and 593° K using a mixture of 5 mm of chlorine and 300 mm of oxygen measured at 293° K. At the highest temperatures the rotational structure became very extended and the bands were less distinct, but intensity measurements in the continuum and even in the diffuse bands at our standard wavelengths were independent of temperature to within ± 20 %. The only true measure of relative concentrations under these conditions is $\int_0^\infty \epsilon_\lambda d\lambda$ which could not be evaluated exactly, but intensity comparisons at a number of wavelengths showed that the value of the integral was almost constant and could hardly have varied by a factor of more than 2 over the whole temperature range. If the temperature coefficient of ClO formation is expressed as $\exp(-E/RT)$ the value of E is therefore 0 ± 0.8 kcal.

DISCUSSION

The reactions studied occur exclusively in the homogeneous gas phase. This is shown by the nature of the observations, which are made on the gas near to the centre of the vessel, the independence of the rates on the total pressure and by simple calculation, which gives times of diffusion to the wall greatly in excess of the time intervals observed. The problem is further simplified by our accurate knowledge of the bond energies of ClO (63 kcal) [2] Cl_2 (57 kcal) and O_2 (117 kcal) [6] and the heats of formation of the stable chlorine oxides. [7] Consider, for example, the reactions

$$Cl + O_2 = ClO + O \qquad - 54 \text{ kcal} \qquad (1)$$

$$Cl + ClO = Cl_2 + O \qquad - 6 \text{ kcal} \qquad (2)$$

$$2ClO = ClO_2 + Cl \qquad - 7 \text{ kcal} \qquad (3)$$

$$2ClO = Cl_2O + O \qquad - 32 \text{ kcal} \qquad (4)$$

Reaction (1) is very endothermic and cannot be a significant mechanism of ClO formation which has nearly zero temperature dependence. Reaction (3) and (4) are similarly eliminated as mechanisms of ClO removal for which the activation energy is zero within a few 100 cal. Reaction (2) must be considered a little more carefully for, if the oxygen atom formed reacts with ClO, a chain mechanism is possible and with suitable activation energies for the termination reaction, this might lead to a temperature independent rate constant for ClO removal. Apart from the fact that no such termination step can be found which gives the observed rate law, and also that the rate constant is too high to involve reaction (2) as a propagation step, the use of chlorine atoms in this mechanism is entirely incompatible with the fact that ClO is formed very rapidly by their reaction with oxygen but the rate of removal of ClO is completely independent of oxygen pressure. The probable fate of any oxygen atoms formed by this or any other reaction would be a reaction with chlorine by the reverse of (2).

We are now in a position to consider the few remaining possibilities which, in conjunction with the observed rate expressions, give the mechanism in some detail.

THE MECHANISM OF ClO FORMATION.—Reaction (1) having been excluded, the only possible reaction by which ClO can be formed is

$$2Cl + O_2 = 2ClO, \qquad (5)$$

although this may proceed in two stages as follows :

$$Cl + O_2 = ClOO \qquad (6)$$

$$Cl + ClOO = 2ClO. \qquad (7)$$

The radical ClOO is not to be confused with the stable radical O—Cl—O.

Using this mechanism it should now be possible to interpret the results of the experiments on $O_2 + N_2$ mixtures given in fig. 4. As nitrogen can play no chemical role its effect must be ascribed to the reaction

$$2Cl + N_2 = Cl_2 + N_2^* \qquad (8)$$

and we must also consider the reaction

$$2Cl + O_2 = Cl_2 + O_2^* \qquad (9)$$

which again may proceed via the intermediate ClOO.

The reactions by which ClO is removed are automatically eliminated by the method of calculating the total ClO formed. Ignoring the intermediate ClOO for the moment the only reactions by which Cl atoms are removed will now be (5), (8) and (9). Then

$$d(2ClO)/dt = k_5(Cl)^2(O_2) \text{ and } d(Cl_2)/dt = k_9(Cl)^2(O_2) + k_8(Cl)^2(N_2).$$

When the removal of Cl atoms is complete we have

$$(2ClO) = \int_{t=0}^{t=\infty} k_5(Cl)^2(O_2)dt$$

and

$$(Cl_2)_f = \int_{t=0}^{t=\infty} k_9(Cl)^2(O_2)dt + \int_{t=0}^{t=\infty} k_8(Cl)^2(N_2)dt,$$

where $(Cl_2)_f$ is the total (Cl_2) formed from Cl atoms. For any given mixture (O_2) and (N_2) are constants and $\int_{t=0}^{t=\infty} (Cl)^2 dt$ has the same value in both expressions. Therefore

$$\frac{(2ClO)}{(Cl_2)_f} = \frac{k_5(O_2)}{k_9(O_2) + k_8(N_2)}. \qquad (i)$$

If $(2ClO)$ and $(Cl_2)_f$ refer to the concentrations at a given time, say 0·7 msec, then the same expression is obtained by eliminating the integral

$$\int_{t=0}^{t=0·7} (Cl)^2 dt.$$

For all $(O_2)/(N_2)$ ratios, twice the number of Cl atoms formed during the flash is a constant and is equal to

$$(2ClO) + (Cl_2)_f = K \text{ (say)}. \qquad (ii)$$

This again applies to any time interval if the reactions removing Cl atoms are fast.

When $(N_2) = 0$, and putting $(2ClO) = (2ClO)_{max}$,

then

$$\frac{(2ClO)_{max}}{K - (2ClO)_{max}} = \frac{k_5}{k_9}. \qquad (iii)$$

Eliminating K and $(Cl_2)_f$ from (i), (ii) and (iii) we obtain

$$\frac{(ClO)_{max}}{(ClO)} = 1 + \frac{k_8(N_2)}{k_5 + k_9(O_2)}.$$

A plot of $(ClO)_{max}/(ClO)$ against $(N_2)/(O_2)$, where $(ClO)_{max}$ is the ClO formed in the absence of nitrogen, should therefore give a straight line of slope $k_8/(k_5 + k_9)$. The results already shown in fig. 4 are plotted in this way in fig. 5 and a linear plot is obtained confirming the mechanism suggested. Further, the gradient of this line gives the above ratio of rate constants and is found to be 1/46. The sum of k_5 and k_9 is the rate constant of removal of Cl atoms by oxygen and k_8 by nitrogen, i.e.

$$- d(Cl)/dt = 2(k_5 + k_9)(Cl)^2(O_2) \text{ in oxygen,}$$

and

$$- d(Cl)/dt = 2(k_8)(Cl)^2(N_2) \text{ in nitrogen.}$$

Therefore the reaction of chlorine atoms with oxygen to form Cl_2 and ClO occurs at a rate 46 times that of the reaction in nitrogen to form Cl_2.

This result has several interesting consequences. Firstly, there is every reason to suppose that the recombination of chlorine atoms in a gas such as nitrogen occurs at every termolecular collision. The collision diameters of oxygen and nitrogen being very similar, for example in the recombination of other halogen atoms,[8] it follows that reactions (5) and (9) as written are insufficient to account for the rate and these reactions must involve the formation of a relatively stable complex, which can only be ClOO. The inclusion of this intermediate leads to the same rate expressions if reaction (6) is reversible and the equilibrium maintained. It is not possible to say whether this implies an activated complex ClOO*

32   FLASH PHOTOLYSIS

with a lifetime 46 times greater than the ClN_2^* complex or equilibrium involving stabilization by a third body and bimolecular dissociation. In either case, if the collision efficiency of Cl atom recombination in nitrogen at 1 atm. pressure is taken as 1/900, the equilibrium constant $(Cl)(O_2)/(ClOO)$ is less than 20 atm even if the reaction of Cl with ClOO occurs at every collision.

The very small temperature coefficient of ClO formation is fully in accordance with our mechanism and confirms that reactions (1), (2), (3) and (4) are unimportant in the photochemical part of the reaction. If reaction (9) occurs to a significant extent it follows that $E_9 = E_5 \pm 0.8$ kcal.

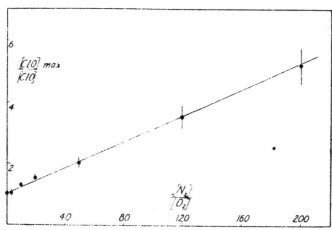

FIG. 5.—$(ClO)_{max}/(ClO)$ against the $(N_2)/(O_2)$ ratio.

THE MECHANISM OF ClO REMOVAL.—The rate law shows conclusively that the subsequent reactions by which ClO is removed are unaffected by the pressure of any gas other than ClO. Reasons have already been given for excluding chlorine atoms, and reactions (3) and (4), and the only remaining possibility is

$$2ClO = Cl_2 + O_2. \tag{10}$$

There is a strong objection to this reaction in its simple form in that all known double decompositions of this type are associated with high energies of activation, a useful empirical rule being that the activation energy is about 1/4 of the sum of the energies of the bonds broken.[9] There is one important difference in the radical reaction, however, and that is the possibility of dimerization. We must therefore consider the reactions

$$2ClO = Cl_2O_2 \tag{11}$$

$$Cl_2O_2 = 2ClO \tag{12}$$

$$Cl_2O_2 = Cl_2 + O_2 \tag{13}$$

which lead to the rate expression

$$- d(ClO)/dt = \frac{k_{11}k_{13}}{k_{12} + k_{13}} (ClO)^2,$$

if we assume a stationary concentration of Cl_2O_2. Two limiting cases may be considered.

(i) $k_{13} \gg k_{12}$.

Reaction (11) now becomes rate determining and a zero activation energy is quite probable. We then have the difficulty of explaining the very low rate constant, for it has been shown that the rate is independent of total pressure and a third-body collision cannot therefore be necessary. We should have to explain the low rate entirely by means of a steric factor which could hardly be greater than 10^{-3}, whereas such rate constants as are known for simple radical recombinations have steric factors very near to unity.[10] This explanation therefore seems improbable.

In this connection it should be mentioned that the possibility of formation of a " stable " Cl_2O_2 molecule is not entirely eliminated by the experimental evidence, in which case the observed rate would be that of formation of Cl_2O_2 which might then decompose more slowly to Cl_2 and O_2. This would only be true if Cl_2O_2 had a very low extinction over the whole region investigated and had a lifetime of the order of seconds. In addition exactly the same objections apply as were given in the last paragraph.

(ii) $k_{12} \gg k_{13}$.

The effective rate constant is now $k_{11}k_{13}/k_{12}$ and the observed activation energy will be $E_{11} - E_{12} + E_{13}$. The difference $E_{12} - E_{11}$ is equal to the heat of formation of Cl_2O_2 from two ClO radicals and must be very nearly equal to E_{13}, the activation energy of dissociation to Cl_2 and O_2 in order to explain the observed temperature independence. This is not unlikely if both energies are small and the low rate is then explicable.

We therefore think that the reaction is best explained in terms of an equilibrium between ClO and Cl_2O_2, the latter decomposing to Cl_2 and O_2. It is then possible to see why ClO might behave differently from both NO and OH which are in many ways similar radicals. The former has a very unstable dimer which dissociates to 2NO much more readily than to N_2 and O_2 whilst the latter forms a stable dimer which dissociates only very slowly. ClO is an intermediate, semi-stable radical because its dimer is also of intermediate stability. We are investigating this reaction further by the transition state method but it seems doubtful whether, in the absence of data about the Cl_2O_2 molecule, anything much more quantitative can be said.

THE CHLORINE + OXYGEN REACTION IN PHOTOSENSITIZED OXIDATIONS.—Our results show unequivocally that Cl atoms react rapidly with oxygen, and these reactions must occur to some extent in the presence of other gases. This is in accordance with the complete rate expressions of Thon[11] and of Bodenstein and Schenk[12] for the $H_2 + O_2 + Cl_2$ system. We have carried out some preliminary experiments on ClO formation in the presence of hydrogen and have found that in excess oxygen (10 mm Cl_2, 10 mm H_2 and 300 mm O_2) the ClO formed and the rate of its reactions are not greatly changed and that little HCl is formed. In excess hydrogen (10 mm Cl_2, 25 mm O_2 and 350 mm H_2) no ClO was observed and most of the Cl_2 reacted permanently.

Excess carbon monoxide, on the other hand, resulted in a very high ClO concentration and ClO is almost certainly the long-lived intermediate which resulted in the slow approach to the steady state observed by Bodenstein, Brenschede and Schumacher.[13] In this, and similar oxidations photosensitized by chlorine, the ClOO radical may play an important part, in addition to ClO. For example, the following mechanism for the sensitized oxidation of CO_2 would now appear probable:

$$Cl + O_2 \rightleftharpoons ClOO$$

$$ClOO + CO \rightarrow CO_2 + ClO$$

$$ClO + CO \rightarrow CO_2 + Cl.$$

A more detailed discussion of these reactions will be given elsewhere along with an investigation of the reactions of ClO with other gases.

One of us (F. J. W.) is indebted to the Anglo-Iranian Co. for financial support during the tenure of which this work was carried out.

[1] Porter, *Proc. Roy. Soc. A*, 1950, **200**, 284.
[2] Porter, *Faraday Soc. Discussions*, 1950, **9**, 60.
[3] Christie and Porter, *Proc. Roy. Soc. A*, 1952, **212**, 398.
[4] Justi, *Ann. Physik*, 1931, **10**, 983.
[5] Duncan, Ells and Noyes, *J. Amer. Chem. Soc.*, 1936, **58**, 1454.
[6] Gaydon, *Dissociation Energies* (Chapman and Hall, 1947).
[7] Goodeve and Marsh, *J. Chem. Soc.*, 1939, 1332.
[8] Rabinowitch and Wood, *Trans. Faraday Soc.*, 1936, **32**, 907.
[9] Glasstone, Laidler and Eyring, *The Theory of Rate Processes* (1941).
[10] see, for example, Dodd, *Trans. Faraday Soc.*, 1951, **47**, 56.
[11] Thon, *Z. physik. Chem.*, 1926, **124**, 327.
[12] Bodenstein and Schenk, *Z. physik. Chem. B*, 1933, **20**, 420.
[13] Bodenstein, Brenschede and Schumacher, *Z. physik. Chem. B*, 1937, **35**, 382.

In order to obtain the absolute rates of these reactions it was necessary to determine the extinction coefficients of ClO. This was done by observation of the photolysis of ClO_2 (paper 7). In the course of this work another spectral band system was observed which was not known and not immediately identified. It turned out to be the well known Schumann-Runge system of the oxygen molecule in its electronic ground state but excited to very high vibrational levels. The abstract number 8 describes how this type of excitation arises and how it has been observed in other molecules.

Paper 7

Reprinted from *Nature*, Vol. 174, p. 785, October 23, 1954.

PHOTOLYSIS OF CHLORINE DIOXIDE AND ABSOLUTE RATES OF CHLORINE MONOXIDE REACTIONS

By F. J. LIPSCOMB, Prof. R. G. W. NORRISH, F.R.S., and Dr. G. PORTER

Department of Physical Chemistry, University of Cambridge

THE absorption spectrum of chlorine monoxide was first detected in the photochemical reaction between chlorine and oxygen[1,2] and all direct studies of its reactions have so far been carried out on this system. The mechanisms of these reactions have been thoroughly elucidated, but[3] it has been necessary to express the absolute rates and concentrations in terms of an unknown constant which is numerically equal to the extinction coefficient of chlorine monoxide at a given wave-length. The evaluation of this constant would lead to absolute extinction coefficients of chlorine monoxide at all wave-lengths, absolute concentrations under any required conditions and absolute rate constants of the reactions of chlorine monoxide.

The extinction coefficient appears in these quantities because the observed parameter is an optical density, D_λ, of the reaction mixture at a given wave-length λ. We then have

$$D_\lambda = \varepsilon_\lambda c l,$$

where ε_λ is the extinction coefficient of chlorine monoxide at this wave-length, c is its concentration and l is the length of the absorption path. It is therefore possible only to measure the product $\varepsilon_\lambda c$ and not the separate quantities. Since the relative extinctions at two wave-lengths are easily obtained, all can be referred to a standard extinction, ε_s, at a wave-length of 2577 A., which is in the continuous region of the spectrum and is therefore insensitive to the exact wave-length and independent of the resolving power of the spectrograph.

The ClO radical is removed by the bimolecular process :

$$2\text{ClO} = \text{Cl}_2 + \text{O}_2 ; \qquad (1)$$

and the rate constant of this reaction is given by

$$k_1 = 7\cdot2 \times 10^4 \times \varepsilon_s \exp\frac{(0 \pm 650)}{RT} \text{ litre mole}^{-1} \text{ sec.}^{-1}.$$

To evaluate ε_s, the concentration of chlorine monoxide must be known and, in principle, this can be found by difference if the concentrations of the other species present are known. Thus, in the reaction between chlorine and oxygen, the only species which contain chlorine and which are present in significant concentrations are chlorine monoxide and chlorine, and the decrease in concentration of chlorine should lead directly to the concentration of chlorine monoxide. Unfortunately, owing to the low extinction of chlorine, this decrease of concentration was below the limit of detectability and it was therefore only possible to give a lower limit for ε_s which was 310. Reasons were given for supposing that ε_s was less than 3,000, but no further limits could be fixed by these experiments.

The extinction coefficient of chlorine dioxide is much greater than that of chlorine and is of the same order of magnitude as the above predicted values for chlorine monoxide. It follows that, if chlorine monoxide were detected during the photolysis of chlorine dioxide, the decrease in concentration of chlorine dioxide should be readily measurable. We have therefore investigated this reaction by the same experimental methods as were used to study the chlorine–oxygen reaction. The photolysis flash was operated with an energy of 250 J and had an effective duration of 100 μsec. Pressures of chlorine dioxide were $\frac{1}{2}$ mm. mercury or less, and the experiments were conducted in the presence of up to half an atmosphere of inert gas (argon, nitrogen or carbon dioxide) in order to reduce the rise in temperature to a few degrees. Plate densities were measured by reference to calibrations made with a neutral step wedge.

It was found that a single flash decomposed about 90 per cent of the chlorine dioxide and that this spectrum was replaced by that of chlorine monoxide, which then decayed slowly in the same manner as had been observed in previous work. Spectra of other chlorine oxides were absent at all times, and microphotometric measurements at 2780 A. established that chlorine trioxide (ClO_3), which has an extinction coefficient of 1,210 at this wave-length[4], could not have been formed in an amount greater than 1 per cent of the total chlorine dioxide decomposed. This is different from the photolysis of chlorine dioxide at normal intensities, where chlorine trioxide is a major product[5], but the difference can be understood in terms of the following competing reactions :

$$O + ClO_2 = ClO_3 \qquad (2)$$

$$O + ClO = ClO_2 \qquad (3)$$

$$O + O + M = O_2 + M \qquad (4)$$

Since the ratio of the concentration of atomic oxygen plus chlorine monoxide to the dioxide is many thousand times greater in the flash experiments, the probability of reaction 2 relative to 3 and 4 will be correspondingly reduced.

The primary photolysis undoubtedly occurs by the process :

$$ClO_2 + h\nu = ClO + O ; \qquad (5)$$

and therefore, if measurements are made before the chlorine monoxide has reacted to form chlorine, the concentration of chlorine monoxide will be given simply by the difference between the corresponding concentration of chlorine dioxide and the concentration originally present. Allowance was made for the chlorine monoxide which had reacted before measurement by carrying out a number of determinations at pressures of chlorine dioxide of 0·1, 0·2 and 0·5 mm. mercury, and extrapolating to zero concentration. The correction was very small, the observed value at the lowest chlorine dioxide pressure being only 4 per cent less than the extrapolated value, while the mean deviation from the mean at a given pressure was 7 per cent. The final value of the extinction coefficient at 2577 A. obtained in this way was 1,190 $\pm$ 100 litre mole^{-1} cm.$^{-1}$. Substituting this numerical value into the expression of Porter and Wright, we obtain :

$$k_1 \ (20° \ C.) = 8·6 \times 10^7 \ \text{litre mole}^{-1} \ \text{sec.}^{-1}.$$

We have also measured the rate constant of removal of chlorine monoxide independently, using the same methods but preparing it by photolysis of chlorine dioxide. The mean value of the bimolecular constant obtained at 20° C. was $5·7 \times 10^7$ litre mole^{-1} sec.$^{-1}$. The difference between these two values is probably within the combined errors of the two completely independent determinations, which used different density standards for plate calibration. It follows that chlorine dioxide, like chlorine and oxygen, has no effect on the rate of disappearance of the ClO radical.

It has been pointed out that reaction 1 has a low frequency factor and the reasons for this have already been discussed[3]. It is one of the simplest bimolecular radical reactions for which quantitative rate data and spectroscopic constants are now available for more detailed theoretical treatment.

[June 15.

[1] Porter, G., Proc. Roy. Soc., A. 200, 284 (1950).
[2] Porter, G. Discuss. Farad. Soc., 9, 60 (1950).
[3] Porter, G., and Wright, F. J., Discuss. Farad. Soc., 14, 23 (1953).
[4] Goodeve, C. F., and Richardson, F. D., Trans. Farad. Soc., 33, 453 (1937).
[5] Bodenstein, Harteck and Padelt. Z. anorg. allgem. Chem., 147, 233 (1925).

Printed in Great Britain by Fisher, Knight & Co., Ltd., St. Albans.

46

Extract 8 from Radiation Research, Supplement 1, p. 489

Application of Flash Photolysis in Irradiation Studies

GEORGE PORTER

Department of Chemistry, University of Sheffield, Sheffield, England

Vibrationally excited molecules may also be observed out of equilibrium in certain cases. A particularly interesting example is vibrationally excited oxygen which was first observed during the flash photolysis of ClO_2 (*11, 12*). Although the average temperature of the system does not change by more than one or two degrees, absorption is observed from vibrational levels of the ground state as high as $v'' = 8$. The excited oxygen is formed by the reaction

$$O + ClO_2 \rightarrow O_2^* + ClO$$

and similar behavior has been found by Norrish, R. G. W., *Proc. Chem. Soc.* (1958), 247; (*18*) in the nitrogen dioxide and ozone systems. The specific distribution of energy in the products of a reaction and the rates of collisional deactivation of individual vibrational levels can be studied directly.

Extensions of the method to other regions of the spectrum are being developed. Perhaps the most interesting possibility from the point of view of radiation chemistry is the application of similar pulsed techniques to other forms of radiation, e.g., electron or X-ray pulses. The recording methods already in use are directly applicable if a suitable pulse of ionizing radiation is applied. To be of comparable utility to flash photolysis methods such a pulse should be of energy at least 100 joules and have a duration of 50 μsec or less, if we assume a comparable yield of the radiation chemical reaction with typical photochemical ones, i.e., a *G*-value of the order of 10. These requirements are by no means unattainable, and pulsed radiation techniques of this kind are under active investigation in a number of laboratories.

A reference is also made in extract 8 to the forthcoming extension of flash photolysis principles to pulse radiolysis. Extensions of the flash energies to shorter wavelengths were also attempted since the radiation from the gas discharge tubes extended well into the ultra violet but these shorter wavelengths were absorbed by the quartz reaction vessel. It was possible to eliminate the lamp envelope so as to transmit radiation in the far ultra-violet region capable of splitting water. This enabled the study of the kinetics of hydroxyl radicals (see G. Black and G. Porter reference).

Millisecond and microsecond flash photolysis made possible spectroscopic and kinetic studies of free radicals and atoms and also of long-lived electronically excited levels of the metastable triplet states, which will be described in later chapters.

The time resolution of the flash photolysis technique remained in the region of a few microseconds throughout the 1950s in spite of several attempts (see references) to reduce the flash duration. When the laser was discovered in 1960 it was clear that this was the new source that we had been waiting for, but it was not until the Q-switched laser was introduced that times in the nanosecond region became available.

LASER FLASH PHOTOLYSIS

Use of a laser for photochemical studies is described in paper 9 which gives one of the first examples of a biphotonic photochemical process. Before these short pulses could be used for nanosecond flash photolysis, other problems had to be solved. A probe flash, preferably of white light, and a synchronised but variable delay were needed. These problems were solved by using only one lamp, for both pulse and probe, split into two parts with an optical delay between them. (Papers 10 and 11.) In the nanosecond delay region this entailed several metres of path length but the split-beam pulse-probe principle came into its own as mode-locked lasers were developed providing pulses in the picosecond and femtosecond time regions.

The first important achievement of flash photolysis in the sub-microsecond region was the detection of the absorption spectra of the excited singlet states of a number of aromatic molecules.

Paper 9

Reprinted from THE JOURNAL OF CHEMICAL PHYSICS, Vol. 45, No. 9, 3456–3457, 1 November 1966

Giant-Pulse-Laser Flash Photolysis of Phthalocyanine Vapor

G. PORTER* AND J. I. STEINFELD†

Department of Chemistry, The University, Sheffield 10, England

(Received 8 August 1966)

THERE has been considerable recent interest in the primary photochemical processes of porphyrinlike molecules, in part because of the biological significance of these species. Since many of these molecules are stable at high temperatures, it has been possible to carry out some of this work in the vapor phase, thus eliminating the need to consider solvent interactions. In particular, recent observations on the fluorescence of phthalocyanine gas[1] has raised questions about the radiationless processes involved, which we wished to investigate by carrying out flash photolysis of these materials.

The metal-free phthalocyanine samples, contained in 1-m quartz tubes, and pressurized with inert buffer gas, were heated in an oven[2] to a temperature of 420°–430°C, which was sufficient to produce an optical density of about 0.7 at the vapor absorption peak at 6860 Å. The samples could be heated and cooled many times without appreciable decomposition. Conventional flash photolysis, using a 2200-J discharge through flash tubes filled with 3 torr of oxygen, produced a depletion of singlet absorption followed by extremely rapid recovery, faster than could be resolved by the apparatus. No additional transient absorption could be detected, and there was no indication of photodecomposition.

Because of the apparent rapidity of the processes under investigation, we decided to attempt the flash photolysis of the samples with a Q-switched ruby laser. We used a Type 350 laser (G. and E. Bradley Ltd., London) incorporating a saturable cryptocyanine solution Q switch, which delivered 2–3 J at 6943 Å in a 20-nsec pulse, giving a peak power of approximately 100 MW. This output wavelength is strongly absorbed by the phthalocyanine vapor. The absorption of the sample during the laser flash was monitored at 6860 and 6200 Å by a continuous "quartz–iodine" incandescent source, dispersed by a small Bausch & Lomb monochromator, detected by an RCA 931A photomultiplier, and displayed on a Tektronix 545A oscilloscope.

When the laser flash passed through the vapor, we observed fast, large depletion (approximately 50%) of absorbing molecules, followed by a slow recovery in a time of the order of 0.1–1.0 sec. The slow recovery is due to the vaporization of excess solid in the sample tube and subsequent diffusion of the vapor into the absorbing region; thus, the laser flash irreversibly decomposes a large fraction of the phthalocyanine molecules in its path. If the Q switch is removed from the laser and the full 40-J pulse (at approximately 50-kW

peak power) is passed through the sample, no decomposition is observed. This strongly suggests that a process involving two photons is responsible for the photodissociation. After a number of flashes, a small amount of crystalline white product could be isolated from the cell, having ultraviolet absorption maxima at 2915 and 2320 Å. This is similar to an indole spectrum, but a full identification of this material has not yet been made.

When a similar experiment is carried out in chloronaphthalene solution at 25°C,[3] the singlet absorption recovers in about 1 μsec, and there is apparently no photodissociation. A possible explanation for this, suggested by the observation[1] that the fluorescence yield decreases by a factor of 77 in going from solution at 25°C to vapor at 450°C, is that the second photon is absorbed by a metastable state M_1, which could be the lowest triplet or a singlet of different electronic symmetry,[4] producing a state M_2, which dissociates. Since the crossing rate has an apparent activation energy of 4.5–7.5 kcal/mole, this process will be favored at high temperatures. An additional effect which might be operative is the rapid relaxation of M_1 by solvent.

This mechanism is supported by our results for conventional flash photolysis of tetraphenylporphyrin at 310°–350°C, which absorbs at 4010 A.[5] We find an upper limit to the half-time for disappearance of triplet and regeneration of singlet of 30–40 μsec, while in toluene solution at 25°C this time is of the order of 1 msec, again implying a temperature-dependent intersystem crossing rate.[6]

Work is being continued on this problem, with particular emphasis on determining the nature and decay kinetics of the intermediate state.

We thank Professor M. Gouterman for furnishing samples of metal-free phthalocyanine and tetraphenylporphyrin, and one of us (J.I.S.) thanks the National Science Foundation for the award of a Fellowship.

* Present address: The Royal Institution, 21 Albemarle Street, London W.1., England.
† Present address: Department of Chemistry, Massachusetts Institute of Technology, Cambridge, Mass. 02139.

[1] D. Eastwood, L. Edwards, M. Gouterman, and J. I. Steinfeld, J. Mol. Spectry. (to be published).
[2] S. K. Hussain, Ph.D. thesis, University of Sheffield, 1965.
[3] W. F. Kosonocky, S. E. Harrison, and R. Stander, J. Chem. Phys. **43**, 831 (1965).
[4] C. Weiss, H. Kobayashi, and M. Gouterman, J. Mol. Spectry. **16**, 415 (1965), and references therein.
[5] J. A. Mullins, A. D. Adler, and R. M. Hochstrasser, J. Chem. Phys. **43**, 2548 (1965).
[6] H. Linschitz and L. Pekkarinen, J. Am. Chem. Soc. **82**, 2407 (1960).

Paper 10

(*Reprinted from Nature*, Vol. 220, No. 5173, pp. 1228–1229,
December 21, 1968)

Nanosecond Flash Photolysis and the Absorption Spectra of Excited Singlet States

THE conventional flash photolysis technique is limited, by the flash duration, to times which are usually greater than 1 μs (ref. 1). There is a need to reach times shorter than this, particularly to observe and study the absorption of excited singlet states which typically have lifetimes in the nanosecond range. The appearance of the Q switched laser has provided a means to this end, and photoelectronic detection methods have already been used in several laboratories to observe transients in the nanosecond region. Because of the high bandwidth required, such methods have serious limitations and we have therefore sought to extend the original double flash technique[2], which has been so useful in microsecond flash photolysis, into the nanosecond region. A brief description of our method was given a year ago[3] and here we describe an improved form of the apparatus and its application to the observation of the absorption spectra of excited singlet states of a number of molecules.

The experimental arrangement which has been found most useful for the present application is shown in Fig. 1. A ruby laser with vanadyl phthalocyanine Q switch delivers a pulse which is frequency doubled by an ADPH crystal, filtered to reduce red light and light scattered from the flash lamp, and eventually delivers to the beam splitter a pulse of 3471 Å light, lasting 18 ns and of 70 mJ energy. The beam is divided into two parts, the first of which passes directly to the reaction vessel and acts as the photolysis flash. The second part passes first to an optical delay system which consists of a light path terminated by a movable plane mirror giving a variable delay of up to 100 ns. After reflexion this pulse returns to the beam splitter and thence to a cell containing a fluorescent solution (1,1',4,4' tetraphenyl buta 1,3-diene in cyclohexane) which emits a pulse lasting 18 ns, having a continuous spectrum in the region from 400 to 600 nm. This pulse passes through the irradiated region of the reaction vessel to the slit of the spectrograph. A single laser pulse is adequate both for photolysis and the recording of an image on the spectrograph (HPS) plate. Accurately reproducible delays between excitation and monitoring flash are readily obtained in this way, the time resolution being limited only by the duration of the laser pulse.

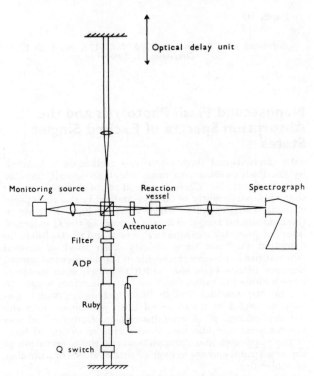

Fig. 1. Nanosecond flash photolysis apparatus.

Transient absorption spectra recorded in this way are shown in Figs. 2 and 3. The solutions were triphenylene and 3,4 benzpyrene in cyclohexane and were outgassed. Both series of spectra show the rise of triplet absorption and the simultaneous decay of new bands. These new transients are assigned to absorption by the lowest excited singlet states, because (a) the lifetimes are equal to the fluorescence lifetimes in the same conditions, and (b) the decay rate is equal to the rate of growth of the triplet states[4] which are simultaneously recorded on the plates. The wavelengths and lifetimes of these spectra and those of five other aromatic hydrocarbons which we have studied in the same way are recorded in Table 1. The solvent was cyclohexane in all cases except coronene for which the solvent was dioxane.

The shortest lifetime transients so far recorded are the singlet decay and triplet growth of phenanthrene, which has a half-life of 19 ns. Although this is comparable with the lifetime of the two flashes, the almost perfect reproducibility of the time delays obtained by using a fluor-

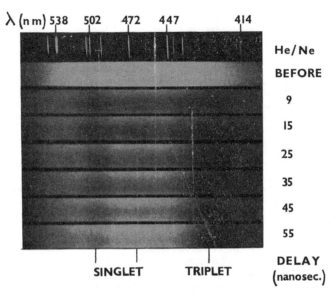

Fig. 2. Sequence of spectra after flash photolysis of triphenylene.

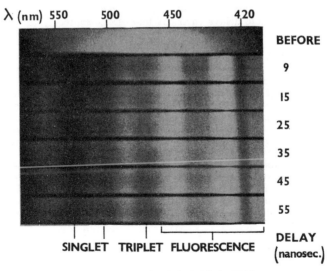

Fig. 3. Sequence of spectra after flash photolysis of 3,4-benzpyrene.

Table 1. WAVELENGTHS AND LIFETIMES OF SPECTRA OF SOME AROMATIC HYDROCARBONS

Molecule	Singlet S_1 absorption (nm)	Triplet T_1 absorption (nm)	Singlet absorption decay time (nsec)	Fluorescence life-time (nsec)
Phenanthrene	505	485, 460, 425	20 ± 5	19 (ref. 7)
Triphenylene	500, 465	430	35 ± 5	37 (ref. 8)
1,2 benzanthracene	550	480	45 ± 5	44 (ref. 7)
3,4 benzpyrene	535, 510	500, 475	45 ± 5	49 (ref. 7)
1,2,3,4 dibenz-anthracene	535, 495	440	50 ± 5	50 (ref. 5)
Coronene	525	490	> 100	300 (ref. 5)
Pyrene	470	510, 490, 415	> 100	380 (ref. 6)

escent source makes it possible to study the concentration changes during the decay of the excitation flash, and the quantitative kinetic study of even shorter transients should therefore be possible by this method even without any reduction in the duration of the flash.

The triplet state absorption spectra are readily observed in aerated solutions with the shorter lifetimes expected as a result of oxygen quenching. The upper states of the singlet–singlet absorptions cannot be immediately assigned on the basis of the normal absorption spectra because parity and other selection rules allow strong transitions from the first excited state to levels which are forbidden from the ground state. Both spectroscopic and kinetic investigation of these transients will therefore provide a rich field for future investigation.

In place of the fluorescent source, we have also used a second spark lasting 20 ns, triggered by ionizing the spark gap with the focused laser beam or, alternatively, a Marx bank circuit in which the first spark gap is ionized by the focused laser beam. Both these systems gave a spectrum of sufficient intensity for single-shot recording on the spectrographic plate but a poorer continuum and less reproducible delays than the fluorescent source.

During the course of this work Novak and Windsor (ref. 5 and personal communication) have described another approach to the problem in which the time resolved spectra are recorded by means of an image converter tube using a laser spark as source. They recorded singlet absorption spectra of coronene, 1,2 benzanthracene and 1,2,3,4 dibenzanthracene which agree with those reported here. Nakato et al.[6] have recently reported a short lived transient after flash photolysis of pyrene which they assigned to an excited singlet state. Our results confirm this assignment.

As in conventional flash photolysis, each of these various modifications of technique will have its advantages for specific problems. The method used here has the general advantage of simplicity and the complete elimination of electronic devices other than the laser itself. The use of optical in place of electronic methods for delay and time measurement will become increasingly advantageous as the work is extended to even shorter times.

We thank the Science Research Council and the European Office of Aerospace Research (USAF) for support of this work.

GEORGE PORTER
MICHAEL R. TOPP

Davy Faraday Research Laboratory,
The Royal Institution,
London.

Received November 15, 1968.

[1] Porter, G., *Z. Electrochem.*, **64**, 59 (1960).

[2] Porter, G., *Proc. Roy. Soc.*, A, **200**, 284 (1950).

[3] Porter, G., and Topp, M. R., in *Nobel Symposium 5—Fast Reactions and Primary Processes in Reaction Kinetics*, 158 (Interscience, London and New York, 1967).

[4] Porter, G., and Windsor, M. W., *Proc. Roy. Soc.*, A, **245**, 238 (1958).

[5] Novak, J. R., and Windsor, M. W., *J. Chem. Phys.*, **47**, 3075 (1967); *Science*, **161**, 1342 (1968).

[6] Nakato, Y., Yamamoto, N., and Tsubomura, H., *Chem. Phys. Lett.*, **2**, 57 (1968).

[7] Birks, J. B., and Munro, H., *Progress in Reaction Kinetics*, **4**, 239 (Pergamon Press, 1967).

[8] Berlman, I. B., *Handbook of Fluorescence Spectra of Aromatic Molecules* (Academic Press, 1965).

Paper 11

Proc. Roy. Soc. Lond. A. **315**, 163–184 (1970)

Nanosecond flash photolysis

By G. Porter, F.R.S. and M. R. Topp

*Davy Faraday Research Laboratory of The Royal Institution,
21 Albemarle Street, London W 1X 4BS*

(*Received 23 July* 1969)

[Plates 1–3]

A flash photolysis system, using a pulsed laser as source, has been designed and used to study events having a duration of a few nanoseconds; an improvement over conventional flash techniques by a factor of a thousand.

The apparatus incorporates both spectrographic and photoelectric monitoring techniques which are easily interchangeable and, apart from the laser itself, it is readily constructed from standard components.

Its applications to the observation of the absorption spectra of excited singlet states, short-lived excited triplet states and chemical events in the nanosecond time region are described.

Introduction

As flash photolysis techniques have been increasingly applied to a variety of problems, it has become clear than at extension of the method to shorter times would have very wide applications.

The time resolution of a flash photolysis system is limited first by the duration of the initiation flash and secondly by the response time of the diagnostics. In the double flash spectrographic recording system (Porter 1950) both limitations are primarily determined by the flash duration, although another factor—which becomes increasingly important at shorter times—is the reproducibility of the delay between the initiating and monitoring flashes. Many attempts have been made in recent years to reduce the duration without a consequent decrease in the energy of the conventional electronic flash discharge tube; they have met with only marginal success (Boag 1968). The advent of the pulsed laser has provided a new source with great potentialities for flash photolysis studies.

The principal components of a flash photolysis apparatus, apart from a spectrograph or monochromator, are an initiating photolysis flash source, a monitoring flash source, a unit which introduces a delay between these two flashes and, for subsequent kinetic work at selected wavelengths, a monitoring source with photoelectric detection system. The substantial changes which have been made in each of these components in the present work will now be described.

EXPERIMENTAL

The laser flash

In order that useful information may be derived from a flash photolysis experiment, it is usually necessary that the photolytic and monitoring pulses be of shorter duration than the processes to be studied. For nanosecond events, the giant-pulsed laser provides a means of generation of a burst of ultraviolet radiation sufficiently energetic for the purposes of flash photolysis, of duration less than or equal to 20 ns.

The ruby laser employed in this work was equipped with a $6\frac{1}{2}$ in $\times \frac{1}{2}$ in (165 mm $\times$ 13 mm) parallel-ended ruby rod, and was originally obtained from G. and E. Bradley Limited. With vanadyl phthalocyanine solution in nitrobenzene used as a Q-switch (Sorokin, Luzzi, Lankard & Pettit 1964) and with a totally internally reflecting quartz prism as the 100 % reflector, the laser generated a pulse containing about 1.5 J of red light of wavelength 694.3 nm (10^{18} photons), whose half-peak duration was less than 20 ns, and whose pulse shape was roughly Gaussian (Müller & Pflüger 1968). The exponential 'tail' of the gas discharge flash was absent so that the real improvement in the time resolution is considerably greater than is indicated by a comparison of half-peak durations.

The red laser pulse was passed through a crystal of ammonium dihydrogen phosphate, from which emerged, in addition to residual red light, a pulse containing about 80 mJ of ultraviolet radiation at 347.1 nm, the second harmonic of the ruby frequency. It was of approximately the same duration as the red laser pulse.

We have found this pulse to be quite adequate for the creation of transient concentrations of up to 5×10^{-4} mol l^{-1} in a 1 cm path which is sufficient for spectroscopic purposes.

The monitoring source

Initially, we modified a commercially available nanosecond spark source for our spectroscopic flash (The Fischer Nanolite (Fischer 1961)). This source gave a pulse of 13 ns duration at half-peak intensity, in an atmosphere of pressurized oxygen, sufficiently bright to expose completely an Ilford HPS or Kodak Royal-X Pan film with a single shot, using a small Hilger spectrograph (Porter & Topp 1967).

This source was triggered by focusing the red laser beam onto the spark gap, but the irreproducibility of the firing was quite large unless the electrodes were freshly cleaned (Pendleton & Guenther 1965). Subsequently it was found that if the Nanolite were replaced by a spark-gap powered by a more conventional low-inductance capacitor, with pin-heads as the electrodes, the discharge could be triggered photoelectrically using the ultraviolet laser pulse, probably due to the presence of zinc in the electrode material. (Work function of zinc = 3.32 eV or wavelengths shorter than 372 nm (Dillon 1931).)

With this arrangement, the pulse duration in pressurized oxygen was 25 to 30 ns with a jitter of only a few tens of nanoseconds. This was reproducible up to about 100 flashes, when burning of the electrodes became appreciable.

The larger delays in this spark apparatus were furnished by a passive delay unit—a small spark gap (1 kV) was triggered by the laser pulse, from which the resulting electrical pulse was fed into a system of B.I.C.C. delay cables (55 ns m^{-1}). These were capable of producing delays in steps of 50 ns from 50 to 1500 ns. The output end of the delay system was connected to a third electrode in the primary spark gap of a Marx–Bank cascade capacitor unit (Marx 1924; Lewis, Jung, Chapman, Van Loon & Romanovski 1966) whereupon the pulse was amplified to about 12 kV, and could be used to trigger the main monitoring spark. The over-all minimum delay was 20 ns and the jitter of the order of 20 to 30 ns.

Using this system, we were able to observe the excited singlet state absorption spectrum of coronene (half-life 280 ns). Although a useful apparatus in some ways, the spark apparatus lacked the nanosecond sensitivity and reproducibility that we were seeking.

Apart from the problem of a rather large jitter, on a nanosecond scale at least, the spark apparatus suffered from another major drawback. There were present, in the emission spectrum of the monitoring spark, many gaseous emission lines together with several from the electrode material. Owing to the necessity for a short duration and intense light output, we were unable to use the inert gases, and had to use oxygen, which is notable for its line emission.

It was not possible, unless fairly substantial absorption was present in our sample, to distinguish between real absorption and regions of low-intensity emission between the lines. The narrower the absorption, the worse the problem, and microdensitometry of the photographic plates obtained on this apparatus was difficult to interpret. It should be borne in mind that this is, to a greater or lesser degree a disadvantage of most gaseous spectroscopic flashes, although oxygen is a particularly extreme example.

The necessity to eliminate this line structure in the monitoring background spectrum led us to consider the possibility of using a fluorescent substance as a monitoring source. Fluorescence emission can be extremely broad in its spectral distribution and is, of course, free from sharp emission lines. The use of the fluorescent monitoring technique has solved both this problem, and the problem of jitter, without the addition of any electronic component to the laser unit, and has enabled us to dispense with the rather troublesome spark system. In this technique, a part of the laser beam is split off from the photolysis beam, and is used to excite fluorescence in the monitoring cell which, in its turn, is used as the background continuum for the spectrographic technique.

The monitoring source is self-synchronizing with the exciting laser pulse, the inherent delay being easily calculable, and usually about equal to the fluorescence lifetime of the monitor substance (Berlman 1965). The duration of this monitoring flash depends upon the fluorescence lifetime of the monitor and the laser pulse lifetime. Since there is no need for external synchronization, the relatively trouble-free and inexpensive passive Q-switch may be used. A table of some common scintillators and their relevant properties is given in table 1. It can be seen that,

166   G. Porter and M. R. Topp

TABLE 1. PROPERTIES OF SCINTILLATORS

scintillator	$\Delta\lambda$	ϕ_F	τ_F	ϵ_{347}	ϵ_{265}
PPF (Cx)	420–340	1.0	1.2	15 000	3 500
terphenyl (Cx)	400–315	0.93	0.95	0	28 000
PPO (Cx)	440–333	1.0	1.4	0	8 000
PBD (Cx)	430–310	0.89	1.35	0	20 000
POPOP (MeOH)	510–380	0.9	1.65	52 000	6 000
TPB (Cx)	590–390	0.5	1.76	36 000	(3 000)
BBOT (Cx)	530–395	0.74	1.1	40 000	(2 000)
DPS (benzene)	480–370	0.8	1.2	50 000	(3 000)

$\Delta\lambda$, fluorescence bandwidth at 10 % of maximum intensity (nm); ϕ_F, fluorescence quantum yield; τ_F, fluorescence decay time (ns); ϵ_{347}, ϵ_{265}, molar decadic extinction coefficients at 347 and 265 nm. Cx, cyclohexane; MeOH, methanol.

with a suitable choice of scintillators, alone or mixed, it should be possible to cover the whole of the spectral range of interest in the photographic work. The figures given in the extreme right-hand column show that the technique is also suitable for use with a far-u.v. laser such as the fourth harmonic from a Nd (III) glass laser (265 nm). An allowance has to be made for the re-absorption of the emitted fluoresence by the ground-state scintillator, a reservation which places some limit on the scintillators which may and may not usefully be used together. Tetraphenyl-butadiene (TPB) in cyclohexane solution is very stable to large amounts of laser radiation at 347.1 nm and, in addition, has a short fluorescence decay time, a broad emission spectrum, and a fairly high quantum yield of fluorescence.

The delay unit

In the normal flash spectrographic technique, two flashes are used, usually of several microseconds duration, the delay between which is usually accomplished by an active electronic delay unit. For delays of less than about 1 μs, however, the reproducibility in the delays deteriorates owing to limitations in circuit response time, and, for delays of such an order, a passive delay unit is to be preferred. For nanosecond processes, one requires a delay system which can impose reproducible delays separated by only a few nanoseconds.

Since light travels at the finite rate of 30 cm ns^{-1}, it is possible to delay one half of the laser beam behind the other, by interposing a suitable path difference behind them, using the first half for photolysis and the second half to trigger a monitoring flash. In this way, we have been able to achieve our nanosecond delays, from zero to 150 ns; for times longer than this the flash photoelectric technique was used. The accuracy of the optical delay itself is determined by the positioning of the mirrors, which can be performed simply by using a metre rule for all but picosecond work. Figure 1 shows an oscillograph record of the two flashes separated by an optical delay of 60 ns.

The optical delay would be less valuable if used with a spectroscopic source which had an appreciable uncertainty in triggering. Because of the extreme reproducibility of the delays, and of the laser pulse shape and energy, it has been possible to observe transient absorption changes in solution over times considerably shorter than the pulse duration.

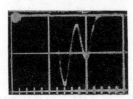

FIGURE 1. Oscillograph of an optical delay of 60 ns. Horizontal scale 100 ns/division.

The flash spectrographic apparatus

The complete apparatus is shown in figure 2. The mixed laser beams are passed through a filter of copper sulphate solution (F_2) which transmits only the ultra-violet pulse. This is then passed through the quartz beam splitter (S) which is set for 50/50 reflexion transmission, the optimum for the operation of the apparatus.

The beam splitter consists of two right-angle quartz prisms pressed together in dry contact so that they form a cube. A beam passing normally through a side of this cube will be totally internally reflected by the hypotenuse face, as it impinges at an angle greater than the critical angle. However, at just below the critical angle of incidence, the beam emerges from the hypotenuse face at a glancing angle, after considerable internal reflexion in the prism, and impinges on the corresponding face of the second prism. At such angles of incidence, the faces are very highly reflective, and several reflexions between them contribute to both the reflected and transmitted beams through the beam splitter, in addition to the initial partial internal reflexion in the first prism.

The reflected beam is passed through a weak (25 cm) lens into the reaction vessel (V), a quartz irradiation cell (usually 1.25 cm) situated on the 'waist' of the light beam focused by the lens. The focusing is by no means sharp, the angle of focusing being 0.03 radians. For photolysis with the red beam, the copper sulphate filter can be diluted slightly to allow 10 % transmission.

The transmitted laser beam is passed from the beam splitter through a telescope system of quartz lenses and is fed into the optical delay line. For the very shortest delays, a mirror was placed before the telescope. The returning laser beam from the optical delay line is collected by the same telescope and, after being turned by the beam splitter, is focused strongly into the cell containing the scintillator solution (C). The fluorescence, reinforced by reflexion off a plane mirror fused to the rear wall of the 1 mm cell, is collected by the same strong lens and, by symmetry, it is channelled through the irradiation cell in the same manner as the original

168   G. Porter and M. R. Topp

photolysis beam. The spectrograph (G) placed behind the cell receives light transmitted through the irradiated volume of the cell. Surplus laser radiation is eliminated by the use of a filter of concentrated biphenylene solution (F_4) before the spectrograph. This compound neither fluoresces nor has any detectable transient absorption to interfere with the spectroscopic measurements; furthermore it is transparent to wavelengths greater than 360 nm.

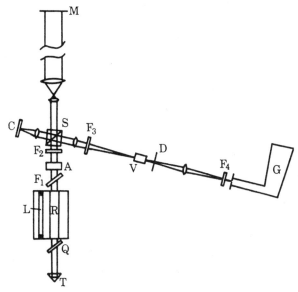

FIGURE 2. Nanosecond spectrographic apparatus. A, ADPh crystal; C, scintillator cell; D, aperture stop; F, filters; F_1, Wratten 29, transmits $\lambda \geqslant 630$ nm; F_2, $CuSO_4$, attenuates 694 nm; F_3, biphenylene, control of intensity of 347 nm, F_4, biphenylene, u.v. laser cut-off; G, spectrograph; L, xenon-filled flash lamp; M, mirror; Q, passive Q-switch; S, beam splitter; T, quartz t.i.r. prism; V, reaction vessel; R, ruby.

It is possible to record, with a single shot, a complete absorption spectrum within the limits set by the scintillators. By this means, we have been able, by simple variation of the optical path difference, to time-resolve the absorption spectra of the excited singlet states of several aromatic molecules (Porter & Topp 1968).

One great advantage of the scintillator-monitor is that one can create an effective point source of light without having to comply with any of the stringent mechanical requirements which would be necessary for an active spectral source such as a spark or a flash lamp. The point source allows optimum recollimation by the collector lens, thus allowing more light to be focused onto the spectrograph slit.

The flash photoelectric apparatus

Owing to the limitation of the length of the optical delay system, and its unsuitability for long delays, as well as the need for kinetic measurements at a single wavelength once the nanosecond transients have been assigned, the apparatus was developed further to allow photoelectric monitoring. For ease of conversion it was desirable to use, as far as possible, the existing components of the photographic apparatus.

Since the bandwidth of response of the photomultiplier system must be about 1000 times greater than that used for conventional microsecond flash photolysis, it follows that, in order to maintain the same signal/noise ratio in the detection system, the light intensity falling onto the photomultiplier cathode must be of the order of 1000 times greater than that of a conventional monitoring source. This intensity is conveniently provided by the laser flash-pumping lamp which has a duration of $1200\,\mu s$. The laser pulse occurs at the peak intensity of the lamp, or a little after it.

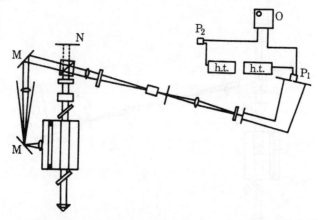

FIGURE 3. Nanosecond photoelectric apparatus. h.t., power supplies; M, mirrors; N, safety screen; O, oscilloscope; P_1 and P_2, photomultipliers.

Figure 3 shows the arrangement of our apparatus for photoelectric recording. Light from the laser flashlamp is leaked out of a hole drilled in the reflector, through the beam splitter and into the spectrograph. The scintillator cell and the lens of the spectrographic system are removed, and the photographic plate holder on the rear of the spectrograph is replaced by a travelling photomultiplier (P_1) on a calibrated slide. The photomultiplier has its own slit arrangement, which gives a resolution of 1.5 nm on the small Hilger spectrograph. The oscilloscope (O) is triggered by a second photomultiplier (P_2) which is activated by the laser pulse.

The application of these techniques to the two principal types of excited molecule —the lowest excited singlet and triplet states—will now be described.

ABSORPTION SPECTRA AND KINETICS OF EXCITED SINGLET STATES

The lowest excited singlet (S_1) states of polyatomic molecules have lifetimes too short for detection by conventional flash methods though they can, of course, often be observed in emission as fluorescence. The positions of higher singlet states to which strong transitions could occur in absorption are also generally unknown since, because of the parity selection rule, these would generally be those states to

G. Porter and M. R. Topp

which transitions from the ground state are forbidden. We have therefore sought, throughout the spectra, new absorptions having decay rates identical with those of the corresponding fluorescence and with the growth rate of the triplet state.

Using the spectrographic apparatus, we have investigated solutions of polynuclear aromatic hydrocarbons. In most cases, a new absorption was observed which decayed within the 70 ns time-spread of the delay sequences; the simultaneous growth in several cases of the well-known triplet absorption was observed. Examples of the records obtained are presented in figures 4 and 5, plates 1 and 2.

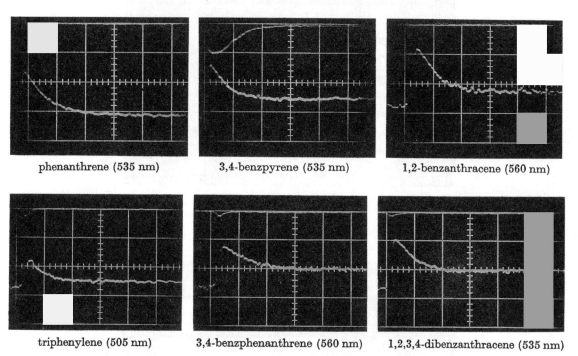

| phenanthrene (535 nm) | 3,4-benzpyrene (535 nm) | 1,2-benzanthracene (560 nm) |

| triphenylene (505 nm) | 3,4-benzphenanthrene (560 nm) | 1,2,3,4-dibenzanthracene (535 nm) |

FIGURE 6. Photoelectric records of decay of excited singlet absorptions to residual triplet. Horizontal scale, 100 ns/division; vertical scale, transmitted light intensity.

Using the kinetic apparatus, the decays of these same absorptions were monitored at wavelengths corresponding to the singlet absorption maxima; examples of these decays are presented in figure 6. In every case, the decays to the new base corresponding to residual triplet absorption were found to be accurately first order, with lifetimes corresponding, within experimental error, to the fluorescence lifetimes observed from the same samples. In each case, the lower trace is the absorption and the upper trace shows the total amount of fluorescence and scattered laser light received by the spectrograph with the monitoring light shuttered off. The low level of this extraneous light demonstrates the usefulness of the monochromatic exitation beam. Only in the case of 3,4-benzpyrene is the fluorescence of comparable intensity to the monitoring light. Subtraction of the scattered light from the total light signal enables accurate calculation of the decay kinetics to be made even in this case.

Porter & Topp   *Proc. Roy. Soc. Lond. A, volume* 315, *plate* 1

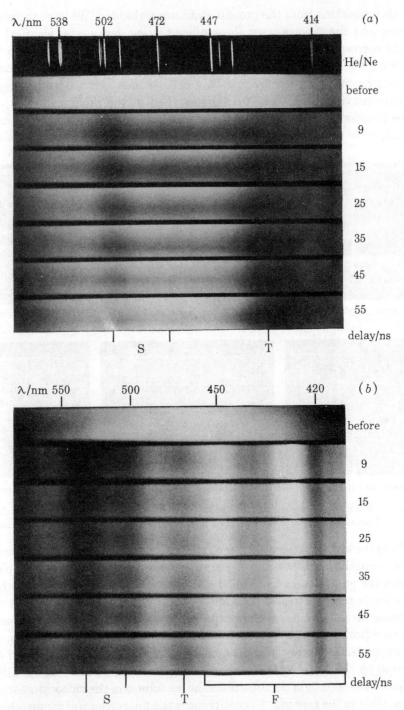

FIGURE 4. Time-resolved excited singlet and triplet absorption spectra:
(a) triphenylene in benzene; (b) 3,4-benzpyrene in cyclohexane.

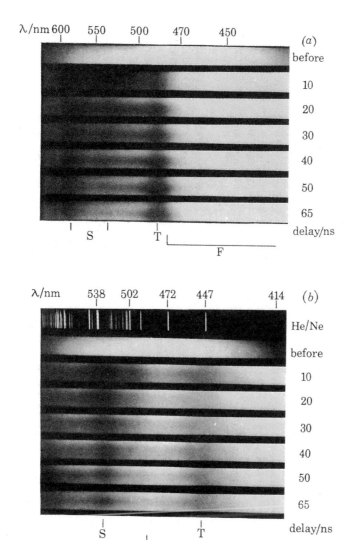

FIGURE 5. Time-resolved excited singlet and triplet absorption spectra:
(a) 1,2-benzanthracene; (b) 1,2,3,4-dibenzanthracene.

TABLE 2. HIGHER SINGLET LEVELS AND CORRELATION OF LIFETIMES WITH FLUORESCENCE DATA

molecule	solvent	singlet S_1 absorption/nm	level of higher S state/cm^{-1}	triplet T_1† absorption/nm	singlet absorption decay time/ns	fluorescence decay time/ns
phenanthrene (d_{10})	Cx	(545), 515	48200	520, 510 / 481, 454	61.1	63.5
phenanthrene (h_{10})	PMMA	(545), 515	48200	520, 510, 481 / 454	65	67.2
triphenylene	Benzene	500, 465, 433	48800	428	44.2	43.0
	PMMA	500, 465, 433	48800	428	45.0	44.0
1,2-benzanthracene	Cx	550, 495	44200	540, 490, 461	52.7	49.4
	PMMA	550, 495	44200	540, 490, 461	51.7	52.5
1,2,3,4-dibenzanthracene	Cx	540, 500	45700	445‖	51.2	53.5
	PMMA	540, 500	45700	445‖	50.3	52.5
pyrene	Cx	515, (480), 470	48000	520, 483, 416	296‡	261‡
	PMMA	515, (480), 470	48000	520, 483, 416	326	319
coronene	Dx	(600, 570), 530, / 495, (465)	42500	525, 480	319	307
	PMMA	(600, 570), 530, / 495, (465)	42500	525, 480	390	380
3,4-benzpyrene	Cx	(590), 535, 510	43700	480§	49.1	57.5
	PMMA	(590), 535, 510	43700	480§	39.4	40.5
3,4-benzphenanthrene	Cx	595, 525	43200	517	68.5	70
	PMMA	595, 525	43200	517	76	81
3,4,9,10-di-benzpyrene	Cx	575, 552	40400	495‖	140	143
1,2,5,6-di-benzanthracene	PMMA	(570)	(42900)	532, 480	(40.0)¶	37.5

† Except where stated, data correlate with those from Porter & Windsor (1958).
‡ 3×10^{-4} mol l^{-1}.
§ (Craig & Ross 1954).
‖ Experimentally determined.
¶ Growth of triplet absorption.
Dx, dioxan; PMMA, polymethylmethacrylate.

172   G. Porter and M. R. Topp

Table 2 summarizes the results of these investigations in both solid and fluid solution. The value for the singlet lifetime of phenanthrene is larger than that reported previously (Birks & Munro 1967; Porter & Topp 1968). In general, the spectra in polymethylmethacrylate solution were indistinguishable from those in liquid solution although, as will be seen, it has not been possible to assign accurately the absolute spectra. The levels of the higher states have been calculated by direct addition of the S_1–S_n transition energies to the known values of the energies of S_1.

Fluorescence lifetime determinations

Owing to the presence of strong transient absorptions in solution, the wavelength of monitoring of the fluorescence emission profile had to be carefully selected.

time/ns
0   200   400   600   800   1000

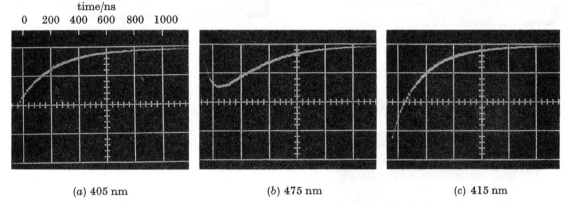

(a) 405 nm   (b) 475 nm   (c) 415 nm

FIGURE 7. Effects of reabsorption on the fluorescence profile of pyrene.

Figure 7 shows the fluorescence profile of pyrene observed at three different wavelengths in cyclohexane solution. The pyrene concentration was sufficiently low $(3 \times 10^{-4}\,\mathrm{mol\,l^{-1}})$ to exclude appreciable excimer formation. At 405 nm, which is close to the peak of the fluorescence emission spectrum, the observed fluorescence lifetime is close to the value obtained with low irradiation intensities. However, at 475 nm the presence of excited singlet state absorption distorts the profile, and the observed lifetime is longer. At the triplet absorption peak of 415 nm the decay time is shorter, owing to the ingrowth of reabsorption with time.

Effect of oxygen

When our hydrocarbon solutions were air-saturated, lifetimes of both singlet and triplet were reduced. Fluorescence decay times were shortened to several tens of nanoseconds, except in viscous solvents, but the triplet decay times were greater than this by a factor varying between three and ten. An example is shown in figure 8. The upper trace (a) depicts the decay of absorption at 535 nm due to the first excited singlet state of perdeutero-phenanthrene in aerated cyclohexane. The singlet absorption is seen to follow approximately the laser pulse profile, while the triplet (b) decays with a pseudo-first order lifetime of 262 ns. Details of oxygen

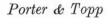

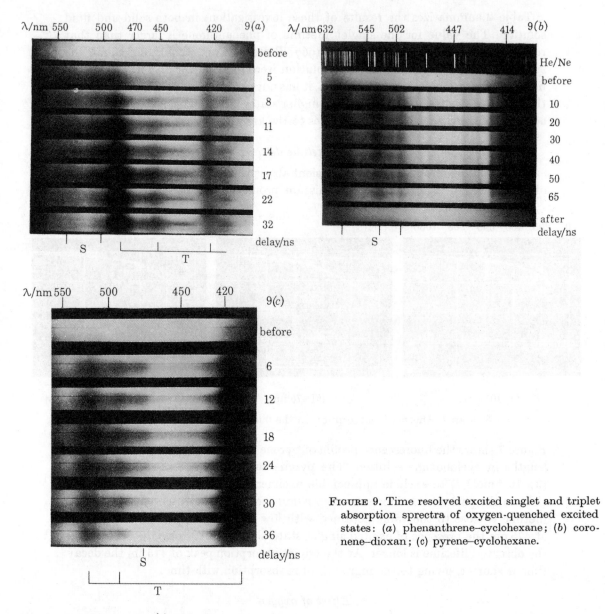

FIGURE 9. Time resolved excited singlet and triplet absorption sprectra of oxygen-quenched excited states: (a) phenanthrene–cyclohexane; (b) coronene–dioxan; (c) pyrene–cyclohexane.

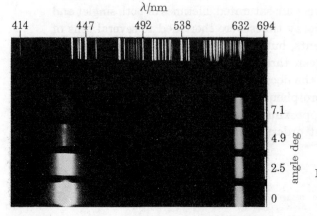

FIGURE 14. Laser emission from scintillator BBOT (ethanol). Variation of laser output with angle of cell windows to direction of excitation beam.

quenching rates of singlet and triplet states and their dependence on energy levels will be given elsewhere.

Figure 9(*a*), plate 3, shows the spectrographic sequence of these changes in phenanthrene. The triplet absorption increases in intensity whilst the singlet absorption at longer wavelengths decays within the laser pulse duration of about 20 ns. In this way, the singlet–singlet absorption decays have been time-resolved for those compounds whose natural fluorescence decay times are too long for the spectrographic apparatus. Figures 9(*b*) and (*c*), plate 3, show the decays of coronene and pyrene singlet absorptions under these conditions.

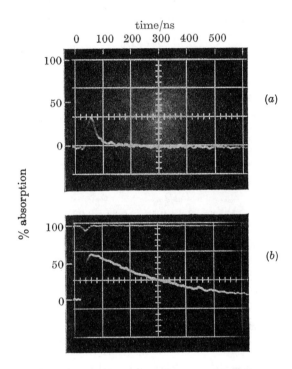

FIGURE 8. Demonstration of oxygen-quenching of (*a*) excited singlet (535 nm), (*b*) triplet (480 nm) absorptions in phenanthrene-d_{10} in cyclohexane. Horizontal scale 100 ns/ division.

Extinction coefficients

The residual absorption seen in figure 6 is due to the triplet state which has been formed by decay of the excited singlet state. In several cases, the extinction co-efficient of the triplet–triplet absorption is known at these wavelengths and, using literature values of triplet yields formed from the singlet, values of the extinction coefficients of the singlet–singlet absorptions may be calculated.

We have assumed that, to a first approximation, the total amount of excitation has been applied instantaneously at a point in time ('zero time') corresponding to the peak of the actual laser pulse. The logarithm of the optical density when plotted

174   G. Porter and M. R. Topp

against time yields a straight line from whose slope has been calculated the absorption decay time. The intercept of this line with 'zero time' chosen as above yields a value of the initial singlet state optical density, corresponding to an initial singlet concentration (S_0), with an error not greater than 20 %. At very long times (about $1\,\mu s$) after the photolysis pulse, all the absorption is assigned to the triplet, concentration (T_∞). From these values of optical density, the following relationship may be calculated:

initial optical density $= D_0 = (S_0)\,\epsilon_S$ ($\epsilon_S =$ singlet extinction coefficient),

residual optical density $= D_\infty = (T_\infty)\,\epsilon_T$ ($\epsilon_T =$ triplet extinction coefficient),

and $(T_\infty) = \phi_T(S_0)$ where ϕ_T is the quantum yield of triplet formation. Thus

$$\epsilon_S = \frac{\phi_T\,\epsilon_T\,D_0}{D_\infty}.$$

The experimental results are averaged over several determinations. The ratio $(D_0)/(D_\infty)$ determined for several solutions is listed in table 3. Where possible, calculation of the singlet extinction coefficient has been carried out.

TABLE 3. RELATIVE EXTINCTION COEFFICIENTS OF TRANSIENT
SINGLET AND TRIPLET ABSORPTIONS

molecule	λ_{max}	D_0/D_∞	ϕ_T[†]	ϵ_T[‡] (λ_{max})	ϵ_S
phenanthrene	535	8.9 ± 0.4	0.85 ± 0.02	2000	$1.5 \pm 0.2 \times 10^4$
1,2-benzanthracene	560	9.6 ± 0.3	0.77 ± 0.02	3450	$2.5 \pm 0.2 \times 10^4$
1,2,3,4-dibenz-anthracene	535	4.4 ± 0.4	—	—	—
pyrene	475	7.0 ± 0.7	0.38 ± 0.02	—	—
coronene	525	3.3 ± 0.4	0.56 ± 0.01	320	590
3,4-benzphenanthrene	560	7.2 ± 0.7	—	300	—
triphenylene	505	6.9 ± 0.6	(0.93)	~ 0	—

† Horrocks & Wilkinson (1968)
‡ Land (1968); Porter & Windsor (1958).

It can be seen from the ratio of initial and final optical densities that there is considerable overlap between the singlet and triplet spectra and this has prevented the observation of triplet growth versus singlet decay in most of these solutions. In 1,2,5,6-dibenzanthracene, no absorption decay was discernible on either the spectrographic or the kinetic apparatus but the triplet absorption was seen to increase with time at precisely the same rate as the fluorescence decay. Recent work by Thomas (1969) has established, in the case of 1,2-benzanthracene, that the singlet and triplet absorption spectra are very similar.

Microdensitometry

The spectrographic records obtained from the laser apparatus were analysed on a microdensitometer. From these records were obtained the total absorption spectra of the transient species in solution.

Where the singlet lifetime is far greater than the laser pulse duration, then the absorptions recorded immediately after it had passed were taken as true absorption spectra for the excited singlet states. In such cases, the triplet contribution to the total amount of absorption is considered negligible. The absorption spectra of pyrene and coronene excited singlet states obtained in this way are shown in figure 10.

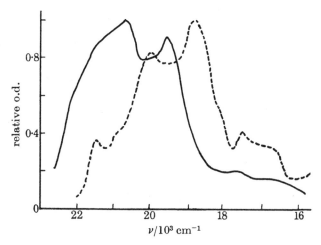

FIGURE 10. Absorption spectra from microdensitometer traces:
———, pyrene; - - - -, coronene.

In the majority of the molecules studied, this assumption was not valid. Because of shorter singlet lifetime, the absorption spectra could not be corrected for triplet absorption. Figure 11 shows uncorrected spectra of the singlet absorption of 1,2-benzanthracene, 3,4-benzpyrene, 3,4-benzphenanthrene, triphenylene and 1,2,3,4-dibenzanthracene.

Microdensitometry of sequences of absorption spectra taken at different times after photolysis show clearly the redistribution of absorption which takes place as the singlet absorption is replaced by the triplet. Figure 12 shows three absorption spectra taken at different times during the lifetime of the oxygen-quenched singlet state of perdeuterophenanthrene. The data were taken directly from figure 9. The presence of an isosbestic point at about $20\,300\,\mathrm{cm^{-1}}$ shows that the decaying transient gives rise directly to the growing one. Moreover, the presence of the isosbestic point makes possible an estimation of the singlet absorption extinction coefficient at this point. The value obtained is $\epsilon_{560} = 9000\,\mathrm{l\,mol^{-1}\,cm^{-1}}$.

Difference spectra

In order to obtain an estimate of the shape of the singlet–singlet absorption peaks in the vicinity of large triplet absorptions, difference spectra were employed. By subtracting the absorption spectrum recorded immediately after the laser pulse

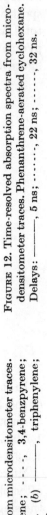

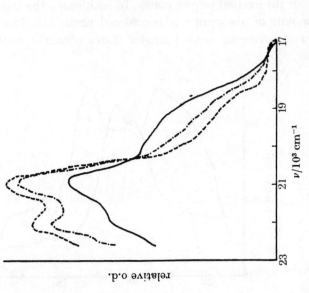

FIGURE 12. Time-resolved absorption spectra from micro-densitometer traces. Phenanthrene-aerated cyclohexane. Delays: ——, 5 ns; ·····, 22 ns; —·—·—, 32 ns.

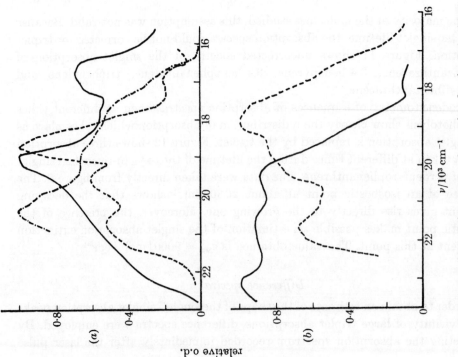

FIGURE 11. Absorption spectra from microdensitometer traces. (a) ——, 3,4-benzphenanthrene; -·-·-, 3,4-benzpyrene; ·······, 1,2-benzanthracene. (b) ——, triphenylene; ----, 1,2,3,4-dibenzanthracene.

Nanosecond flash photolysis

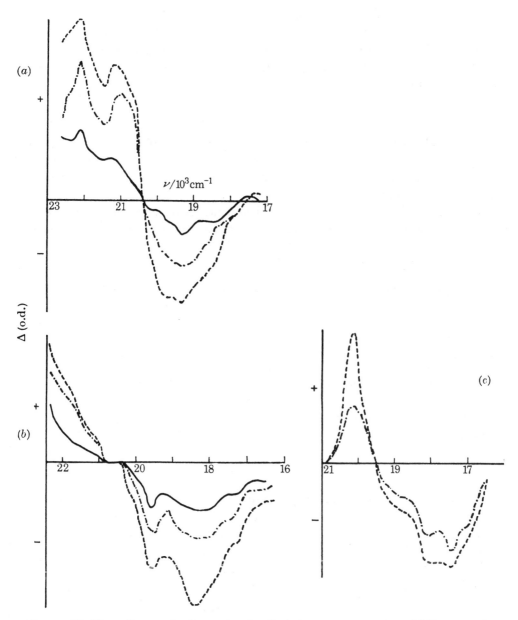

FIGURE 13. Absorption spectra from microdensitometer traces; sequences of difference spectra. (a) Phenanthrene-aerated cyclohexane: ——, o.d. (11 ns)–o.d. (5 ns); ·····, o.d. (22 ns)–o.d. (5 ns); -----, o.d. (32 ns)–o.d. (5 ns); (b) 1,2,3,4-dibenzanthracene–cyclohexane: ——, o.d. (30 ns)–o.d. (10 ns); ·····, o.d. (40 ns)–o.d. (10 ns); -----, o.d. (65 ns)–o.d. (10 ns). (c) 3,4,9,10-dibenzpyrene-aerated cyclohexane: ·······, o.d. (40 ns)–o.d. (10 ns); -----, o.d. (65 ns)–o.d. (10 ns).

G. Porter and M. R. Topp

from a spectrum taken at a later time, one may plot a difference spectrum, whose ordinate is a measure of the difference between the triplet and singlet extinctions. A sequence of such difference spectra for (a) the same phenanthrene example, (b) 1,2,3,4-dibenzanthracene, and (c) 3,4,9,10-dibenzpyrene (oxygen quenched), is shown in figure 13.

Intersystem crossing rate constants

These may be measured with some accuracy from traces obtained on a kinetic apparatus, although the present laser pulse duration does not permit accurate extrapolation. However, this limitation is only temporary, as will be discussed at the end of the paper.

The ratios of the initial and residual optical densities of the singlet states have already been discussed, and several results presented in table 3. Consider the relations

$$D_0/D_\infty = \epsilon_S(S_0)/\epsilon_T(T_\infty) \quad \text{and} \quad D_0/D_\infty = \epsilon_S/\epsilon_T\,\phi_T = R.$$

By altering ϕ_T, for example by adding oxygen, it is possible to obtain two values of R.

Hence
$$R/R' = \phi'_T/\phi_T.$$

But
$$\phi'_T = (k_{\text{i.s.c.}} + k')/(k_D + k') \quad \text{and} \quad \phi_T = k_{\text{i.s.c.}}/k_D;$$

where k' is the resultant increase in intersystem crossing rate constant as a result of the perturbation applied, $k_{\text{i.s.c.}}$ is the normal intersystem crossing rate constant and k_D is the normal decay rate constant of singlet.

Whence

$$k_{\text{i.s.c.}} = \frac{k'R'}{R(k_D + k')/(k_D) - R'}. \tag{1}$$

Such a perturbation occurs when air is admitted to the solution. Coronene, under anaerobic conditions, has a fluorescence decay rate constant of $2.5 \times 10^6\,\text{s}^{-1}$. The ratio R is found to be 3.3 (see table 3). Owing to the finite triplet decay on the time scale used, in aerated solution, it is necessary to take the triplet optical density, D_∞, as that at which the decay ceases to have any contribution due to singlet. In aerated solution, the ratio R' is found to be 2.1, whereas the fluorescence decay rate constant is $2.5 \times 10^7\,\text{s}^{-1}$. From equation (1), we find that $k_{\text{i.s.c.}} = 1.55 \times 10^6\,\text{s}^{-1}$. Division of this rate constant by the observed natural decay rate constant of the singlet state gives the triplet quantum yield as $1.55/2.5 = 0.6 \pm 0.04$. This agrees with published data (Horrocks & Wilkinson 1968). Another example, even though less accurate, as a result of the short fluorescence lifetime in aerated solution, is that of 1,2-benzanthracene. In this case, $k_D = 2.0 \times 10^7\,\text{s}^{-1}$; $k' = 3.9 \times 10^7\,\text{s}^{-1}$; $R = 9.6$ and $R' = 8.7$. Equation (1) gives, for this set of results, $k_{\text{i.s.c.}} = 1.7 \times 10^7\,\text{s}^{-1}$. The quantum yield of triplet formation is found to be 0.85 with an error of $\pm 10\,\%$ (literature value 0.77 ± 0.22).

Materials

All aromatic compounds and solvents could be obtained sufficiently pure for use in our experiment with one exception. The phenanthrene to be used in the fluid solvents resisted attempts at removal of anthracene as impurity. We are grateful to Dr D. Lavalette for the pure sample of perdeuterophenanthrene which we used for our fluid solution work. All solutions in polymethylmethacrylate were provided by Dr M. West.

Triplet state of ketones and quinones

Although the lifetimes of the triplet state of many organic molecules are long enough, in both gases and liquids, for easy detection by conventional flash methods, there are other triplet states, particularly the reactive triplets of some carbonyl compounds, which have submicrosecond lifetimes and have not, therefore, hitherto been observed. The wavelength of the frequency doubled laser flash at 347 nm is very suitable for photolysis of most aromatic ketones and quinones and we have investigated two important cases, benzophenone and derivatives of benzoquinone.

Benzophenone

Benzophenone photochemistry has been studied extensively by various workers, and is fairly well understood. The lowest triplet state, which is of $n - \pi^*$ character, readily abstracts hydrogen from solvent molecules, forming the long-lived ketyl radical with unit quantum yield:

$$\phi_2 CO \xrightarrow{h\nu} {}^1\phi_2 CO^* \rightarrow {}^3\phi_2 CO^*,$$
$$^3\phi_2 CO^* + RH \rightarrow \phi_2 \dot{C}OH + \dot{R}.$$

It has been calculated that the lifetime of the triplet state in an alcoholic solvent is of the order of several tens of nanoseconds (Beckett & Porter 1963), and the direct observation of the triplet state under these conditions should, therefore, be possible. There is a difficulty inherent in benzophenone transient spectra in that both the triplet and the transient reaction products have similar spectra (Bell & Linschitz 1963). Fortunately, however, the extinction coefficient of the triplet state is the larger.

Using the flash-photoelectric apparatus for studies of benzophenone in various solvents we observed an absorption at 535 nm, the peak of the triplet–triplet absorption. We have determined the lifetimes of the benzophenone triplet state in various solvents which are listed in table 4. Beckett & Porter (1963) assigned a lifetime for the benzophenone triplet state in isopropanol of 60 ns using results based on naphthalene quenching and assuming diffusion controlled rates. Our direct measurement agrees moderately well with their estimate.

From earlier argument, the ratio of the initial and final optical densities is equal to

$$D_0/D_\infty = \epsilon_A/\epsilon_B \phi_B = R,$$

180   G. Porter and M. R. Topp

where A is the initial transient species and B is the residue. It can be seen from the table that there are two apparent modes of reaction of the benzophenone triplet. The top four solvents have comparable values of R, and decay rates which closely correlate with predicted relative rates of the hydrogen abstraction reactions. The benzenoid solvents, however, exhibit quite different behaviour. In the case of benzene-h_6 and benzene-d_6, there is the possibility that the hydrogen abstraction may still occur, although it is energetically unfavourable, but then the absorption should appear, as in the other cases, as ketyl radical. It seems that chemical reaction other than hydrogen abstraction is taking place. This reaction has been interpreted as electrophilic attack upon the aromatic ring by the benzophenone triplet to form an addition product (Schuster & Topp 1969).

TABLE 4. BENZOPHENONE TRIPLET IN VARIOUS SOLVENTS

	triplet state		
solvent	lifetime/ns	decay constant, k/s^{-1}	D_0/D_∞
isopropanol	46 ± 6	2.2×10^7	1.7 ± 0.2
ethanol	104 ± 15	9.7×10^6	1.5 ± 0.2
dioxan	200 ± 20	5.0×10^6	1.5 ± 0.2
cyclohexane	300 ± 20	3.3×10^6	2.0 ± 0.3
benzene-h_6	2500 ± 300	4.0×10^5	9.0 ± 0.5†
benzene-d_6	3450 ± 300	2.9×10^5	12.5 ± 1.0†
benzene-f_6	435 ± 40	2.3×10^6	15.0 ± 2.0†

† Results obtained in collaboration with Professor D. I. Schuster.

The similarity of the R values for the first four solvents, in spite of widely different triplet lifetimes, can only be rationalized if the quantum yield of reaction is approximately unity in all four cases. The quantum yields of final product vary by as much as a factor of 4 (Beckett & Porter 1963) but this must be due to subsequent reactions. R is, therefore, a measure of the ratio of the extinction coefficients of the triplet and ketyl radical and the value of R in cyclohexane is 2.0 ± 0.3. Estimates of the extinction coefficient of the radical at 535 nm of 5000 (Beckett & Porter 1963) and 3220 (Land 1968) lead to triplet extinction coefficients at the same wavelength of 10 000 and 6440.

Energy transfer

The relatively high energy of the benzophenone triplet ($24\,400\,\mathrm{cm}^{-1}$) compared with triplets of many aromatic hydrocarbons has made possible its use as a donor in the study of intermolecular energy transfer reactions. Porter & Wilkinson (1961) investigated the transfer of energy from benzophenone to naphthalene (triplet energy $21300\,\mathrm{cm}^{-1}$) in benzene. The transfer reaction competes with the chemical reaction in benzene for the triplet energy. A study of the dependence of the relative rates of these processes on naphthalene concentration enabled Porter & Wilkinson to estimate the second-order rate constant for the energy transfer process.

This type of system is ideal for study with the ruby laser, as the donor molecule (benzophenone) can be selectively excited by monochromatic light at 347 nm. Moreover, the process, competing with the chemical reaction of the benzophenone triplet in benzene, has a submicrosecond duration.

We studied the energy transfer process with both the spectrographic and the photoelectric apparatus, and found a strong absorption due to the benzophenone triplet (535 nm) present immediately after the photolysis pulse. This decayed rapidly and was replaced by an absorption due to the naphthalene triplet (425 nm). We used a naphthalene concentration substantially higher than that of the benzophenone triplet produced on photolysis and this resulted in the observation of pseudo-first-order kinetics. For example, on photolysis an initial benzophenone triplet optical density of 0.28 (3×10^{-5} mol l^{-1}), in the presence of 8.15×10^{-4} mol l^{-1} naphthalene, decayed with a first-order rate constant of 4.3×10^{6} s^{-1}. The total decay constant of the benzophenone triplet may be expressed as $k = k_d + k_q(N)$ where k_d is the decay constant observed in the absence of naphthalene (4.0×10^{5} s^{-6}), k_q is the second-order constant for energy transfer and (N) is the naphthalene concentration, assumed constant. We found $k_q = 4.7 \pm 0.4 \times 10^{9}$ l mol^{-1} s^{-1}, a value higher than that of Porter & Wilkinson who indicated that their indirect estimation might be too low.

Derivatives of benzoquinone

Benzoquinone and some of its derivatives were studied by Bridge & Porter (1958) who found the semiquinone radical at 410 nm in all cases. With duroquinone they also observed a short lived transient with band maxima at 490 nm which they assigned to the triplet state. Some subsequent workers have questioned this assignment and given good reasons for attributing the 490 nm band to a reversible tautomeric rearrangement (Herman & Schenck 1968; Wilkinson, Seddon & Tickle 1967).

TABLE 5. DUROQUINONE AND CHLORANIL TRIPLET DECAY CONSTANTS

quinone	solvent	λ_{max}/nm triplet	k/s^{-1}
duroquinone	benzene	490	3.2×10^{5}
chloranil	ethanol	500	8.4×10^{5}
chloranil	cyclohexane	500	5.0×10^{5}

Apart from this, there is little information about the triplet state of benzoquinones and since phosphorescence is very weak or absent, but quantum yields of reaction high, it appears that triplet state lifetimes are likely to be small and probably in the nanosecond region.

In collaboration with D. R. Kemp we have investigated the transient absorption spectra of benzoquinone derivatives in various solvents. While the possibility remains that with duroquinone in some solvents the 490 nm transient is that of a tautomer, we have established with reasonable certainty that the triplet state is the species responsible for this band from duroquinone in benzene. Furthermore, we have observed an almost identical absorption from chloranil in ethanol and cyclohexane where tautomerisation is impossible and have shown in the ethanol solution, by energy transfer experiments, that this is a triplet state and that it reacts to from the radical. The triplet state maxima and first-order decay constants are given in table 5 and full details of this work will appear elsewhere (Kemp & Porter 1969).

DISCUSSION

In the spectrographic and in the photoelectric apparatus, the limiting factor of time resolution is the duration of the laser pulse, which is, at present, about 20 ns. Should the duration of the laser pulse be reduced, the limiting factor in the spectrographic technique would be the fluorescence lifetime of the scintillator solution. Similarly, the photoelectric technique would be time-limited by the response time of the photomultiplier (1.5 ns) and the rise time of the oscilloscope circuits (2 ns).

It is now possible, through the techniques of regeneration switching, or pulse transmission mode operation (p.t.m.) of a laser to generate pulses sufficiently short, and of sufficient energy (1 J) for the above parameters to become the limiting factor. Such techniques are also applicable for use with neodymium lasers, and deep u.v. (265 nm) irradiation with ultra-short pulses obtained by quadrupling the fundamental frequency is available.

In order to produce the same amount of transient absorption with shorter flashes and without changing the sensitivity of the monitoring techniques, a corresponding increase in the pulse intensity is necessary.

While many of the consequences of increased laser intensity are quite straightforward, there is one which merits further discussion.

Stimulated emission in the scintillator cell

It is apparent that stimulated emission plays some part in the output of light from the scintillator cell, even with the 20 ns laser pulse. For example, when POPOP is irradiated in the scintillator cell position shown in figure 2, the emitted pulse duration is about 25 % shorter than the irradiating pulse duration. In the cases of the scintillators POPOP, BBOT and DPS, it has been found that the emission spectrum is strongly dependent on the laser intensity.

When high laser intensities are used, there is a marked dependence of the fluorescent output on the angle of the scintillator cell windows to the direction of the exciting laser beam. In orientations where the cell faces are far from normal to the laser beam direction, then the fluorescent output received at the spectrograph is of normal efficiency as determined by the collection ratio of the lens system. Moreover, the emission spectrum under such conditions agrees closely with that published by Berlman (1965), obtained at low light intensities. However, as the cell face is brought into a position normal to the laser beam direction, then the emission spectrum develops a very intense, and very sharp line at the wavelength of the 1–0 vibronic fluorescence transition band, in the case of BBOT in ethanol situated at about 435 nm as is shown in figure 14, plate 3. This phenomenon is attributed to 'lasing' in the cell, such as has already been obtained in other cases by Sorokin et al. (1967). These scintillators do not emit the 'laser' beam in the 0–0 band, presumably owing to reduction of gain by reabsorption of emitted light by the ground state. The effect is illustrated by our observations for these scintillators given in table 6.

These emissions were obtained without the use of a special laser cavity around the scintillator cell. Even when the scintillator cell reflector was removed, the windows possessed sufficient reflectivity to obtain laser threshold level, owing to the high fluorescence efficiency of these molecules.

It might have been hoped that the limitation imposed on time resolution by the fluorescence lifetime of the scintillator would be overcome by use of stimulated emission, especially with a picosecond laser pulse. However, if the spectrum always shows the extreme narrowing which has been observed in these three cases, then sacrifice of the polychromaticity may prove not worthwhile. Stimulated emission under such circumstances can easily be prevented by mixing two scintillators with similar characteristics. In this way the total amount of light emitted from the scintillator remains unchanged.

TABLE 6. LASER WAVELENGTHS OF SCINTILLATORS

scintillator	solvent	1–0 band/nm	laser wavelength/nm	ϵ_0 at 0–0
POPOP	EtOH	418	419	500–1000
BBOT	EtOH	433	435	500–1000
DPS	Cx	408	409	1000

ϵ_0 is the ground state extinction coefficient.

During the last two years, several other laboratories have developed laser techniques for nanosecond flash photolysis. Novak & Windsor (1967, 1968) used a laser-induced spark in a gaseous medium as a spectroscopic background continuum. Time-resolution was achieved by an image-converter scanning technique. In this way the apparatus combined both spectroscopic and photoelectric techniques. Apparatus using only the point by point photoelectric method but capable of more quantitative measurements was used by Bonneau, Faure & Joussot-Dubien (1968) and Thomas (1969). The former obtained kinetics and a point by point plot of the naphthalene transient singlet–singlet absorption spectrum. Thomas used a laser flash apparatus to verify observations made using a nanosecond pulse radiolysis technique.

Each of these modifications of the applications of lasers to flash photolysis has its particular advantages. The method described here is remarkably simple and reproducible. Apart from the laser itself, the equipment is available in most laboratories and is even less complex than that used in a conventional flash photolysis apparatus. In the spectrographic double-flash method, electronic components, with their intrinsic limitations on time resolution, are completely eliminated. There are of course, other limitations at the present time, particularly of wavelengths available for pulsed lasers, but in view of the rapid development of laser technology, these limitations may reasonably be regarded as only temporary.

184   G. Porter and M. R. Topp

We thank the Science Research Council and the U.S. Air Force Office of Scientific Research, for support of this work.

REFERENCES

Beckett, A. & Porter, G. 1963 *Trans. Faraday Soc.* **59**, 2038.

Bell, J. A. & Linschitz, H. 1963 *J. Am. Chem. Soc.* **85**, 528.

Berlman, I. B. 1965 *Handbook of fluorescence spectra of aromatic molecules.* New York: Academic Press.

Birks, J. B. & Munro, H. 1967 *Progr. Reaction Kinetics*, **4**, 239.

Boag, J. W. 1968 *Photochem. Photobiol.* **8**, 565.

Bonneau, R., Faure, J. & Joussot-Dubien, J. 1968 *Chem. Phys. Lett.* **2**, 65.

Bridge, N. K. & Porter, G. 1958 *Proc. Roy. Soc. Lond.* A **244**, 259, 276.

Craig, D. P. & Ross, I. 1954 *J. chem. Soc.* p. 1589.

Dillon, J. H. 1931 *Phys. Rev.* **38**, 408.

Fischer, H. 1961 *J. opt. Soc. Am.* **51**, 543.

Herman H. & Schenck, G. O. 1968 *Photochem. Photobiol.* **7**, 255.

Horrocks, A. R. & Wilkinson, F. 1968 *Proc. Roy. Soc. Lond.* A **306**, 257.

Kemp, D. R. & Porter, G. 1969 To be published.

Land, E. J. 1968 *Proc. Roy. Soc. Lond.* A **305**, 457.

Lewis, R. N., Jung, E. A., Chapman, G. L., Van Loon, C. S. & Romanovski, T. A. 1966 *I.E.E.E. Trans. Nuc. Sci.* April, p. 84.

Marx, E. 1924 *Elektrotech. Z.* **45**, 652.

Müller, A. & Pflüger, E. 1968 *Chem. Phys. Lett.* **2**, 155.

Novak, J. R. & Windsor, M. W. 1967 *J. chem. Phys.* **47**, 3075.

Novak, J. R. & Windsor, M. W. 1968 *Proc. Roy. Soc. Lond.* A **308**, 95.

Pendleton, W. K. & Guenther, A. H. 1965 *Rev. scient. Instrum.* **36**, 1546.

Porter, G. 1950 *Proc. Roy. Soc. Lond.* A **200**, 284.

Porter, G. & Topp, M. R. 1967 *Nobel Symposium 5—Fast Reactions and Primary Processes in Reaction Kinetics*, p. 158. London and New York: Interscience.

Porter, G. & Topp, M. R. 1968 *Nature, Lond.* **220**, 1228.

Porter, G. & Wilkinson, F. 1961 *Proc. Roy. Soc. Lond.* A **264**, 1.

Porter, G. & Windsor, M. W. 1958 *Proc. Roy. Soc. Lond.* A **245**, 238.

Schuster, D. I. & Topp, M. R. To be published.

Sorokin, P. P., Luzzi, J. J., Lankard, J. R. & Pettit, G. D. 1964 *IBM Jl Res. Dev.* **11**, 130.

Thomas, J. K. 1969 *J. chem. Phys.* **51**, 770; also private communication.

Wilkinson, F., Seddon, G. M. & Tickle, K. 1968 *Ber. Bunsen-Ges. phys. Chem.* **72**, 315.

Examples of picosecond and femtosecond flash photolysis will be found throughout this book and I conclude the first chapter with a recent summary of the development of flash photolysis over the last 45 years, through the time regions from milliseconds to femtoseconds.

Extract 12 from "Femtosecond Chemistry", Vol. I, pp. 3–13
Eds. Jörn Manz and Ludger Wöste, VCH Publishers Inc.

1 Flash Photolysis into the Femtosecond – a Race against Time

George Porter

1.1 Introduction

The study of chemical events that occur in the femtosecond timescale is the ultimate achievement in half a century of development of techniques for the study of fast reactions and, although many future events will be run over the same course, chemists are near the end of the race against time.

Since 1949, when millisecond, pulse-probe, flash photolysis techniques were introduced, the time intervals accessible to measurement have been reduced by a factor of 10^{11}, which may be compared with an improvement of about 10^4 in spatial resolution over the same period. The key advances that made this possible were the arrival of the laser in 1960 and the replacement of electronic devices with a limited response time by split-beam pulse-probes with optical delays.

The principal steps in the advance towards femtosecond time-resolved spectroscopy are described in Sections 1.2–1.5.

1.2 Millisecond and Microsecond Flash Photolysis

Two separate (gas discharge) flashes were employed, one for excitation (the 'pulse' or photolysis flash) and one for monitoring (the probe or spectroflash), synchronized and delayed electronically [1]. The setup is shown schematically in Figure 1.1. The reaction vessel and excitation lamp were 1 m in length and the probe

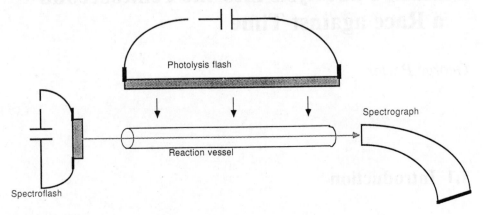

Figure 1.1 The principles of microsecond flash photolysis.

lamp, which was 20 cm long, was viewed end-on. The spectrum of both lamps was close to white light; the excitation pulse was sometimes filtered by using a solution in an outer jacket but the full spectrum of the probe-pulse was usually examined by photographic spectrosopy. The very high energies of the excitation pulse that were first used (energy/pulse = $\frac{1}{2} CV^2$ = 8 kJ with C = 1000 μF and V = 4 kV) limited the pulse duration to a few milliseconds. The pulse duration was reduced to a few microseconds within a year by the use of (a) higher voltages, (b) low-impedance capacitors, (c) shorter lamps and cells, (d) lower energies, and (e) electronic synchronization.

1.2.1 Free Radicals

The technique was originally designed for the spectroscopic detection and study of free radicals in the gas phase. The first experiment, on the flash photolysis of acetone, resulted in complete decomposition of the acetone and the deposition of fine filaments of carbon throughout the reaction vessel! This 'adiabatic flash photolysis' was used to initiate explosion and to investigate the mechanism of carbon formation, but for most purposes a nearly isothermal system was required and this was achieved by operating with a large excess of inert gas.

The first free radical to be studied in detail was ClO [2] which later turned out to be of considerable importance in the solar driven stratospheric ozone depletion process [3].

Other radicals of great interest followed; among them one may particularly note the methylene (CH_2) and methyl (CH_3) radicals [4], the aromatic radicals benzyl ($C_6H_5CH_2$), anilino (C_6H_5NH), and phenoxyl (C_6H_5O) [5], phenyl (C_6H_5) [6], and many of their derivatives such as the ketyls and semiquinones (7). Any free radical of interest was now open to spectroscopic observation and kinetic study.

An early observation was that chlorine and other halogens were temporarily bleached by a flash and restored over a few milliseconds. The three-body, and therefore slow, recombination of iodine atoms formed by photolysis of iodine molecules was studied in detail [8]. In this case it was convenient to initiate dissociation by a single flash and follow the recovery by continuous monitoring of absorption using photomultiplier detectors.

1.2.2 The Triplet State

Along with free-radical absorption spectra, the most notable achievement of microsecond flash photolysis was observation of the absorption spectra of molecules in the excited electronic state. The flash duration limited such studies to the triplet state and the absorption spectra of several dozen triplet state spectra, mostly of aromatic molecules, were recorded for the first time in fluid solvents at normal temperatures [9], as well as in the vapor phase [10]. This opened up the direct study of the chemistry of the triplet state which became the largest branch of photochemistry. The photochemistry of ketones, quinones and other carbonyl compounds, one of the most intensively studied areas of organic photochemistry, gave the first direct observation of an organic photochemical reaction − reaction of the excited state of anthraquinones and simultaneous formation of semiquinone radicals [11].

1.3 Nanosecond Flash Photolysis

During the 1950s many attempts were made to produce shorter high-energy flashes by electrical discharge methods. Low-impedance capacitors with short high-voltage sparks provided flash energies too low for transient absorption work but adequate for fluorescence lifetime studies with a sensitive photomultiplier detector, but then the time resolution was limited by the transit-time spread of the cathodes. The discovery of the laser by Maiman in 1960 was immediately recognized as a revolutionary tool for fast reaction techniques, though its application to these problems came rather slowly owing to the need to develop parallel fast detection and recording techniques.

The original laser was the electronic flash-excited solid-state ruby with a duration equal to that of the flash lamp, which was originally about 1ms. If the lasing action is held back until a substantial population inversion has been attained (Q-switched), either by means of a Pockels cell or by a saturable absorber (e.g., a cryptocyanine solution), a giant pulse of duration about 10 ns was obtained. Its first photochemical application was by Jeff Steinfeld, who demonstrated nonlinear photochemistry in phthalocyanine [12]. An unswitched 40 J pulse had no effect on phthalocyanine vapor (though it was strongly absorbed at 694.3 nm), whilst after switching to give a more intense pulse, but of lower (2–3 J) energy, 50% of the phthalocyanine was decomposed with the formation of indole.

Before these nearly monochromatic pulses could be used for nanosecond flash photolysis with transient absorption spectroscopy, several other developments were necessary.

1.3.1 Introduction of the Split-Beam Pulse and Probe

The single pulse split into pulse and probe, with optical synchronization and optical delay, eliminated the problem of synchronization of two separate flash sources and the need for fast-response instrumentation since the delay time was preset for each experiment by the optical path difference. Although first used for nanosecond delays, which often meant several transits across the room (30 cm/ns^{-1}), the technique became easier as well as essential as the work progressed into the picosecond and femtosecond regions.

1.3.2 White-Light Monitoring Source

Although a monochromatic source could be used for monitoring, using nonlinear conversion techniques to shift the wavelength of the laser source, white light was preferable for the photographic recording that was mainly used at that time. One method used by our laboratory and by M. W. Windsor employed the focused laser probe beam to produce a spark in air, but timing of a catastrophic breakdown is variable and a spark in air has a complex multiline spectrum The method eventually adopted was to excite the short-lived fluorescence of a dye in solution by means of a 347 nm pulse obtained from a frequency-doubled ruby laser. Diphenylbutadiene, for example, gave an excellent, almost white, continuum through the near ultraviolet and much of the visible region, with a lifetime of a few nanoseconds.

The nanosecond arrangement of Porter and Topp [13], incorporating these principles, is shown in Fig 1.2.

It was immediately possible with this approach to detect and study the time-resolved absorption spectra of singlet excited states, the first recorded being those of aromatic hydrocarbons in solution shown in Fig 1.3.

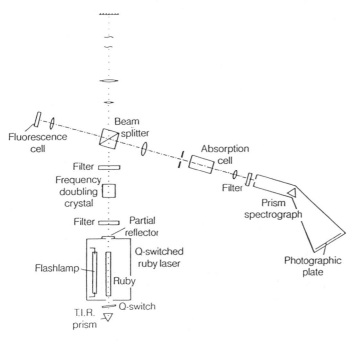

Figure 1.2 Split-beam arrangement for nanosecond flash photolysis.

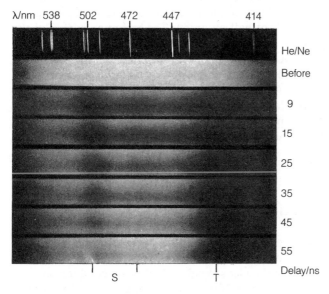

Figure 1.3 Nanosecond flash photolysis, using a fluorescence continuum as probe source, showing decay of the singlet state and growth of the triplet state of triphenylene.

1.4 Picosecond Flash Photolysis

In the late 1960s it became known that, under suitable conditions, the cavity modes of the laser could be locked together so that the Q-switched pulse became a train of much shorter pulses, typically a few picoseconds long, separated from each other by a time interval of $2L/c$, where L is the cavity length and c the velocity of light. This process of 'mode locking' was the key to the generation of shorter pulses of energy than are available by any other means.

The pulse-probe technique was of course even more suitable for the picosecond region (1 ps $\equiv$ $^1/_3$ mm optical path) than for nanoseconds. However, picosecond white-light pulses were not available, so various combinations of monochromatic pulses derived from the same laser by nonlinear techniques were used. Although photoelectric devices in general did not have picosecond time resolution, the streak camera, which deflected the electron beam liberated from a photocathode by means of a sawtooth voltage and projected it onto a cathode-ray tube or image intensifier, was capable of measuring decay times as short as 3 ps in a single shot. This was particularly useful for fluorescence lifetime measurements and it still is, because the only alternative for picosecond and subpicosecond lifetime measurements is to use the pulse-probe method and to upconvert the fluorescence by mixing with the probe-pulse.

One early example of the use of time-resolved picsecond fluorescence was the study of energy transfer in the red alga. *Porphyridium cruentum* [14]. This alga has a light harvesting particle (phycobilisome) attached to the membrane of the photosynthetic organism. The particle is composed of three different pigments arranged in layers, rather like the shells of half an onion. The excitation energy of the pigment decreases from that in the outer layer (phycoerythrin, BPE, 578 nm max.) through the intermediate layer (phycocyanin, RPC, 645 nm) to that in the inner layer (allophycocyanin, APC, 660nm) and thence to chlorophyll which is in the membrane. By excitation of the first pigment (BPE) with a frequency-doubled single pulse at 532 nm from a mode-locked Nd laser, these transfers were shown to occur successively with transfer times of 70 ps, 90 ps, and 120 ps. These lifetimes, compared with the natural lifetimes of the isolated pigments, show that the overall transfer is more than 98% efficient.

1.4.1 White-Light Generation

Flash photolysis using transient absorption spectrosopy in the picosecond region was too fast for fluorescence light sources and there seemed to be little chance of obtaining synchronous white-light pulses in the picosecond and subpicosecond region until Alfano and Shapiro [15] showed that such pulses are readily obtained by the simplest of methods. It is merely necessary to focus a very high-intensity laser pulse, which will be nearly monochromatic, in a dispersive medium such as water, and the beam is converted into a polychromatic pulse covering wavlengths throughout much of the

UV, visible and IR regions. The mechanism by which this occurs is complex but it involves the nonlinear processes of self-focusing and self-phase modulation. The white-light generation occurs only above a threshold power and amplification of the pulse from the mode-locked train is necessary to reach this threshold. The amplification occurs by passage through a pumped amplifier dye cell; typically, four stages of amplification would be used to give a gain of about 10^6.

Physicochemical events that often occur in the picosecond timescale and have been much studied include molecular rotation in solution, isomerization, proton transfer, vibrational relaxation, internal conversion, electron transfer, and energy transfer.

1.5 Femtosecond Flash Photolysis

The next step, from picosecond to femtosecond flash photolysis was a gradual process made possible by progressively shortening the pulse along with a parallel improvement in sensitivity to compensate for the lower energy in the shortened pulse. The time resolution of the pump-probe detection system was, as always, limited only by the pulse duration. There is of course no qualitative difference between a pulse of 1 ps and one of 990 fs duration, just as there is little significance in running one mile in a slightly shorter time than four minutes. But scientists are as competitive as sportsmen in the race against time and their description of the time regime of their experiments has always been on the generous side. For example, no femtosecond flash has yet been produced.

The development of subpicosecond pulses from mode-locked lasers owed much to the work of Shank, Ippen, and their colleagues at Bell Laboratories. Their colliding-pulse mode-locked (CPM) laser immediately reduced pulses well into the subpicosecond range. Pulses tend to be chirped because red light travels faster than blue through a dispersive medium; this was corrected by means of an anomalously dispersive grating or prism pair, leading to pulses of 50 fs or less. Further pulse compression was achieved by increasing the band width using self phase modulation in an optical fiber, thereby eventually reducing pulse durations to 6 fs.

A detailed investigation of a complex reaction requires more than the appropriate time resolution. Reproducibility from pulse to pulse is vital when data are collected over many thousands of pulses. If second-order effects such as exciton annihilation are to be avoided, the relative changes must be small and correspondingly small optical density changes must be employed. Even after pulse amplification by a factor of 10^6, pulse energies are often in the microjoule range. For all these reasons little can be done with single pulses and integration over many thousands of pulses is the norm. Unless data collection times are to be inordinately long, the repetition rate must be high and the limiting factor here is usually the repetition rate of the laser used for pumping the amplifiers. We use a copper vapor laser giving 1 mJ pulses with a repetition rate of 6.5 kHz. An outline of our femtosecond flash photolysis apparatus is shown in Fig 1.4.

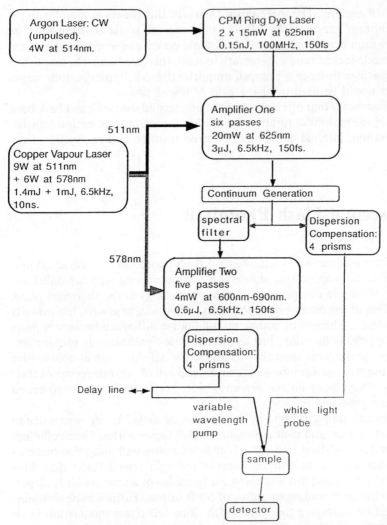

Figure 1.4 Femtosecond flash photolysis arrangement with tunable excitation and probes pulses and 6.5 kHz repetition rate-used for studies of photosynthesis in PS2.

The femtosecond era of times is rich in new phenomena. Over periods of some tens of femtoseconds coherence is retained in a number of molecules and the coherent wave packet produces phase-related excitation, e.g., of vibrations. One of the applications most satisfying to a physical chemist has been the direct observation of the ultimate in fast reactions, the transit of molecules through the transition state described by Zewail and others in this volume.

Since molecular diffusion is excluded in subpicosecond times the phenomena observed will usually occur within a single molecule or in molecular complexes, including interaction of molecules with thè solvent. The transfer of energy and electrons between neighboring molecules in supramolecular complexes and biological systems embraces what is perhaps the most extensive area of subpicosecond events in physics, chemistry, and biology.

But perhaps the most intensively studied of all fast reactions are the primary processes of photosynthesis . Because of the need to attain high efficiency in competition with fast (nanosecond) radiative and (picosecond) radiationless energy dissipation, the primary events of photosynthesis are among the fastest known to chemistry or biology. They are able to occur rapidly because the molecules are adjacent to each other (often near van der Waals distances) and diffusion is not involved.

The femtosecond events that occur in the photosynthetic process serve as an excellent illustration of the present state of the art of femtosecond flash photolysis and are described separately in Chapter 2.1.

1.6 Conclusions

The results just described illustrate what is now possible in flash photolysis experiments. Measurements made at 13 fs time intervals, repetition rates above 5 k s^{-1}, measurement of optical density changes of $< 10^{-4}$, and excitation pulses of 2nm bandwidth at 100fs and continuously variable over the spectrum are adequate to satisfy most requirements. In our present apparatus [16, 17] a multichannel detector is used to record 200 wavelength intervals simultaneously with no loss of precision, increasing further the rate of of accumulation of data. Analysis of the data rather than the rate of data recording is now the bottleneck.

The uncertainty limit of time measurement has been reached for chemical, biological, and most physical processes other than those involving high-energy particles. A 10 fs pulse contains only five wavelengths of red light and the uncertainty in the corresponding energy is greater than the total energy of the chemical bond itself.

The progress that has been made in the time resolution available is illustrated in the Fig. 1.5.

Times encountered in nature extend over some 38 decades in a timescale of seconds, from about 10^{18} to 10^{-20} s, from the age of the universe to the time interval corresponding to the passage of light across an elementary particle. The extension of the time intervals studied by flash photolysis by 11 decades over the last half-century is a significant advance even viewed against this majestic time expanse of nature.

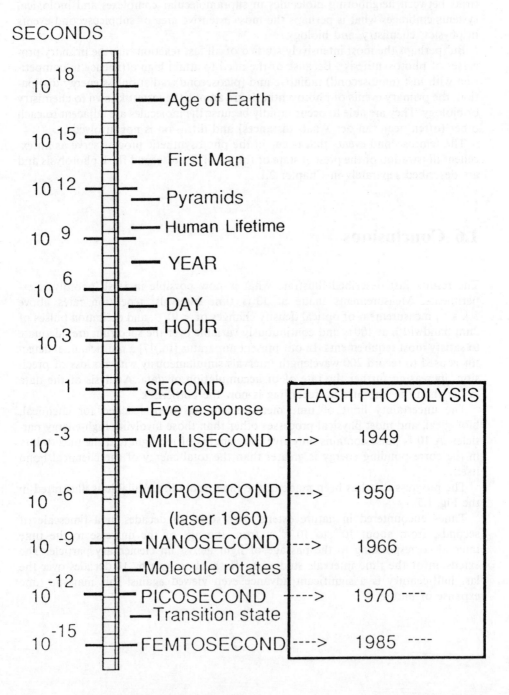

Figure 1.5 Timescales of flash photolysis compared with the cosmic scale.

1.7 Acknowledgements

A full acknowledgement to those involved in the developments described here would include over a hundred names of my colleagues. Many of these appear in the references cited. I should particularly like to thank my present collaborators who have carried out the more recent work that I have described. I am especially grateful to David Klug, Linda Giorgi and James Durant. who have worked with me on femtosecond flash photolysis over the last ten years, both at the Royal Institution and Imperial College, London.

Final version received: 30th January 1994.

1.8 References

[1] G. Porter, *Proc. R. Soc.*, **1950**, *London, Ser. A 200*, 284.
[2] G. Porter, *Disc. Faraday Soc.*, **1950**, *9*, 60.
[3] G. Porter, *Science and Public Affairs* **1988**, *5* (1), 101
[4] G. Herzberg, J. Shoosmith. *Nature (London)* **1959**, *183*, 1801.
[5] G. Porter, F. J. Wright, *Trans. Faraday Soc.* **1955**, *51*, 1469.
[6] G. Porter, B. Ward, *Proc. R. Soc London*, **1965**, Ser. A *287*, 457.
[7] A. Beckett, A. D. Osborne, G. Porter, *Trans. Faraday Soc.*, **1964**, *60*, 873.
[8] M. I. Christie, R. G. W. Norrish, G. Porter, *Proc. R. Soc. London*, **1952**, Ser. A, *216*, 152; G. Porter, J. A. Smith, *Nature (London)*, **1959**, *184*, 446
[9] G. Porter, M. W. Windsor, *Disc. Faraday Soc.* **1954**, *17*, 178.
[10] G. Porter, F. J. Wright, *Trans. Faraday Soc.* **1955**, *51*, 1205
[11] N. K. Bridge, G. Porter, *Proc. R. Soc. London*, **1958**, Ser. A *244*, 259, 276.
[12] G. Porter, J. I. Steinfeld, *J. Chem. Phys.*, **1966**, *45*, 3456.
[13] G. Porter, M. R. Topp, *Nature (London)*, **1968**, *220*, 1228; *idem, Proc. R. Soc. London,* **1970**, Ser. A *315*, 163.
[14] G. Porter, C. J. Tredwell, G. F. W. Searle, J. Barber, *Biochim. Biophys. Acta*, **1978**, *501*, 232. 246.
[15] R. R. Alfano, S. L. Shapiro. *Phys. Rev. Lett.* **1970**, *24*, 592
[16] J. R. Durrant, G. Hastings, D. M. Joseph, J. Barber, G. Porter, D. R. Klug, *Proc Natl. Acad. Sci. U.S.A. (Biophys.)* **1992**, *89* 11632-11636
[17] G. Hastings, J. R. Durrant, J. Barber, G.Porter, D. R. Klug, *Biochemistry* **(1992)** *31*, 7638-7647.
[18] R. V. Bensasson, E. J. Land, T. G. Truscott, *Flash Photolysis and Pulse Radiolysis.* Pergamon, Oxford, **1983**.
[19] L. M. P. Beckman, R. W. Visshers, K. J. Vissher, P. Althuis, B. Barz, D. Oesterhelt, V. Sundström, R. van Grondelle, *Ultrafast Phenomena,* Vol. VIII. Springer, Berlin, **1992**.
[20] See also the other chapters in this book, and the references cited therein.

Further publications

DIE REAKTION VON CHLORATOMEN MIT SAUERSTOFF - EINE
UNTERSUCHUNG DURCH BLITZSPEKTROSCOPIE.
G. Porter and F. J. Wright, *Zeit. Elektrochem.*, 1952, **56**, 782.

SIMPLE LIGHT SOURCE OF ABOUT 10 mμs DURATION.
G. Porter and E. R. Wooding, *J. Sci. Instr.*, 1959, **36**, 147.

NANOSECOND LIGHT SOURCES.
G. Porter and E. R. Wooding, *J. Phot. Sci.*, 1961, **9**, 165.

ABSORPTION SPECTRA FROM A PULSED ELECTRICAL DISCHARGE.
G. Black and G. Porter, *Spectrochim. Acta*, 1960, **16**, 1442.

VACUUM ULTRA-VIOLET FLASH PHOTOLYSIS OF WATER VAPOUR.
G. Black and G. Porter, *Proc. Roy. Soc.*, 1962, **A266**, 185.

FLASH PHOTOLYSIS.
G. Porter, "Technique of Organic Chemistry" Volume VIII, Part II,
Interscience Publishers, London, 1963, p. 1055.

FLASH PHOTOLYSIS AND PRIMARY PROCESSES IN THE EXCITED STATE.
G. Porter, "Nobel Symposium 5 - Fast Reactions and Primary Processes in Chemical
Kinetics", Interscience Publishers, London, 1967, p. 141.

STUDIES OF TRIPLET CHLOROPHYLL BY MICROBEAM
FLASH PHOTOLYSIS.
G. Porter and G. Strauss, *Proc. Roy. Soc.*, 1966, **A295**, 1. 113–148.

CHEMISTRY IN MICROTIME.
G. Porter and M. A. West, "Highlights of British Science",
Royal Society, London, 1978, pp. 155–174.

PICOSECOND STUDY OF STERN–VOLMER QUENCHING OF
THIONINE BY FERROUS IONS.
M. D. Archer, M. I. C. Ferreira, G. Porter and C. J. Tredwell,
Nouveau Journal de Chimie, 1976, **1**, (No. 1), 9.

CHEMISTRY IN MICROTIME.
George Porter.
Pauling Symposium of 1991. The Chemical Bond. Academic Press. 1992.

Chapter 2

ADIABATIC FLASH PHOTOLYSIS. COMBUSTION AND CARBON FORMATION

The first experiment designed to produce free radicals by flash photolysis of organic molecules showed that intensity of the source and yields of photolysis were not going to be a problem. The result of photolysis of acetone with a 4,000 J flash was total destruction of the acetone and formation of cobweb-like structures of nearly pure carbon. It was soon established that this was to be attributed to the very high temperature rise which resulted from the absorption process and the adiabatic radical-radical reactions which followed. Thermal diffusion to the walls of the vessel is relatively slow and temperature rises in the gas can attain thousands of degrees.

For the study of the spectra of radicals and the primary reactions of intermediates, nearly isothermal conditions are required and the large temperature rises can be a serious problem. Fortunately they can be reduced and almost eliminated by working with gas mixtures having a large excess of inert gas (or solvent); this is the procedure used for most (isothermal) flash photolysis studies. Some early work was also carried out in adiabatic conditions with three purposes in mind

(1) detection of the absorption spectra of some free radicals not formed under isothermal conditions such as CH, C_2 and C_3,

(2) studies by R.G.W. Norrish of explosion and related phenomena, such as knock,

(3) studies of the mechanism of carbon formation.

The intermediate stages in carbon formation were examined by analysis of products after rapid quenching and by time resolved spectroscopy. The formation of carbon by pyrolysis of hydrocarbons is accounted for by a succession of eliminations of hydrogen as far as acetylene, followed by Diels-Alder type reactions.

Reprinted from the Proceedings of the Royal Society, A, volume 216, p. 165, 1953

Studies of the explosive combustion of hydrocarbons by kinetic spectroscopy

I. Free radical absorption spectra in acetylene combustion

By R. G. W. Norrish, F.R.S., G. Porter and B. A. Thrush

(*Received* 16 *August* 1952)

[Plate 1]

The explosive oxidation of acetylene, initiated homogeneously by the flash photolysis of a small quantity of nitrogen dioxide, has been investigated by flash spectroscopy. The absorption spectra of OH, CH, C_2 (singlet and triplet), C_3, CN and NH, a number of which have not previously been observed, are described, and the relative concentrations, at all times throughout the explosion, are given. Four stages have been distinguished in the explosive reaction:

1. An initial period during which only OH appears.
2. A rapid chain branching involving all the diatomic radicals.
3. Further reaction, occurring only when oxygen is present in excess of equimolecular proportions, during which the OH concentration rises exponentially and the other radicals are totally consumed.
4. A relatively slow exponential decay of the excess radical concentration remaining after completion of stages 2 and 3.

The duration of stage 1 is 0 to 3 ms. In an equimolecular mixture at 20 mm total pressure, containing 1·5 mm NO_2, the durations of both stage 2 and stage 3 are approximately 10^{-4} s and the half-life of OH in stage 4 is 0·28 ms. A preliminary interpretation of these changes and of the radical reactions is given.

The light emission accompanying the high-temperature combustion of hydrocarbons consists mainly of the characteristic band spectra of the free radicals OH, CH, C_2 and, in the presence of nitrogen, CN and NH as well as the flame bands attributed to CHO. The numerous investigations of the occurrence of these spectra in flames and explosions, and the contribution which such observations have made to theories of combustion, have been summarized in the useful monograph by Gaydon (1948). While they have led to a better understanding of the types of combustion which occur under different conditions and the existence or otherwise of temperature equilibrium, the information which such studies have provided about the chemical kinetics of combustion has been rather disappointing. It is usually found that all these spectra appear together in the reaction zone of flames,

166   R. G. W. Norrish, G. Porter and B. A. Thrush

and the differences observed between one hydrocarbon and another are remarkably slight. Because of the narrowness of the reaction zone it is virtually impossible to differentiate between the spectra at the beginning and at the end of the reaction, although Gaydon & Wolfhard (1947) have had some success in this respect by using low-pressure flames. Our lack of knowledge of the concentration and reactions of the free radicals responsible for these spectra is reflected in the fact that, with the exception of OH, they are absent from most of the mechanisms of combustion which have been proposed so frequently during the last two decades.

Most investigators in the field of combustion spectroscopy would probably agree that the difficulties in the interpretation of the part played by these free radicals are to a large extent due to the fact that the spectra are observed almost exclusively in emission. As a result, the intensity of a band system is determined by the temperature and conditions of chemiluminescent excitation as much as by the concentration of the radical itself and is furthermore a summation over different stages of the reaction. On the other hand, the obvious advantages of absorption spectroscopy are subject to the well-known experimental limitations of this technique and, as far as we are aware, of the radicals mentioned above, only OH and, more recently, NH have been observed in absorption during combustion. This paper describes the observation of all the possible diatomic radicals in absorption as well as spectra of more complex molecules, and the recording of these spectra throughout the course of the combustion.

The experimental method of flash photolysis and spectroscopy by which these spectra were obtained has been fully described in previous communications (Porter 1950a, b). Kinetic studies of the OH radical in hydrogen/oxygen explosions have also been reported (Norrish & Porter 1952), as well as a preliminary account of the absorption spectra of the CN and NH radicals in explosions of ethylene with oxygen-nitrogen dioxide mixtures (Porter 1952). Our main purpose in this work was to investigate the occurrence of these three radicals in hydrocarbon combustion and to attempt the observation of other radicals, such as C_2 and CH, about which little is known. For this purpose acetylene was chosen as the most suitable hydrocarbon, because it shows C_2 and CH very strongly in emission, and the photochemical initiation was effected, as in the hydrogen-oxygen work, by means of a small addition of nitrogen dioxide to the acetylene-oxygen mixture.

Experimental

The principle of the method is to decompose the nitrogen dioxide photochemically into oxygen atoms and nitric oxide by means of a flash alongside the reaction vessel. This also results in the liberation of heat, by degradation of the absorbed radiant energy, and explosive reaction occurs homogeneously throughout the entire volume. A second flash, giving a continuous spectrum, is used to record the absorption spectra at any required time interval after initiation. Reference may be made to the original papers for experimental details, and all that will be necessary here is an account of the modifications introduced for the present problem.

The earlier work on flash-initiated explosions had shown that the total duration of the explosion, including the induction period, was frequently less than 1 ms, and the necessity therefore arose of adapting the apparatus for shorter times. The modifications introduced were as follows:

Flash-lamps

The duration of the photolysis flash was reduced in three ways. First, the length was reduced from 100 to 50 cm; secondly, the voltage was increased to 6000V, which made it possible to reduce the capacity, and hence the duration, without reducing the energy; thirdly, a further reduction in capacity was possible, owing to the fact that a lower energy flash was sufficient to initiate explosion in acetylene-oxygen-nitrogen dioxide mixtures. Throughout these investigations the capacity used was 35μF, giving an energy of 630 J/flash and an effective duration of 5×10^{-5} s. The spectroscopic flash-lamp eventually used was identical with those previously described, attempts to improve upon this design by operation at 20 000V having met with only limited success, owing to frequent explosion and unreliable firing characteristics.

Trigger and timing circuit

Operation of a mechanical trigger wheel at short times involves inconveniently high speeds of revolution. Furthermore, it has been found that with careful alinement it is possible to eliminate virtually all the scattered light from the photolysis flash without the necessity of a shutter, which was the main purpose of this device. It was therefore more convenient to use electronic methods, the photolysis flash being made to fire the spectroflash via a photocell, delay unit and thyratron. A comparison of the two methods shows that there is little difference in reproducibility, but the electronic method is more suitable for the shorter times for the reasons given. The circuit diagram of the timing unit is shown in figure 1.

The various time intervals were obtained by adjustment of the pre-set controls in the delay unit, but the actual measurement of the interval was in every case

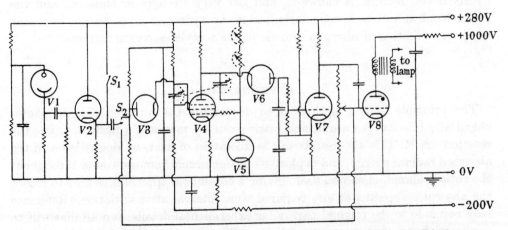

FIGURE 1. Circuit diagram of timing unit. $V1$, VA16; $V2$, 6J5; $V3$, EA50; $V4$, 6F32; $V5$, EB34; $V6$, EA50; $V7$, EC52; $V8$, MT57.

made on a cathode-ray oscillograph using a photocell placed near to both lamps. This was found to be essential at the shorter times, otherwise a considerable scatter in the delay time resulted, owing to a variable delay between the application of the trigger current and the main discharge. As before, the experiments were carried out in an arbitrary order, so that any changes observed are a function of the time interval only.

Acetylene was taken from a cylinder and redistilled *in vacuo*. Oxygen and nitrogen dioxide were prepared and purified as previously described. The mixtures were made up in a 2 l. spherical bulb and a minimum of 30 min was allowed for complete mixing. The spectrograph was a Littrow type instrument with interchangeable glass and quartz optics (Hilger E 478). Ilford H.P. 3, Selochrome, and Q. 3 plates were used in the visible, ultra-violet and far ultra-violet regions respectively, and a slit width of 0·02 mm was used throughout.

Intensity measurements

For extinction measurements of the continuous spectra, plate densities were determined by a microphotometer and compared with measurements on known densities from plate calibrations in the usual way. This method is unsuitable for complex line spectra because, owing to imperfect resolution, such measurements would not be a true indication of relative concentrations and also because the apparent intensity of a single line is very dependent on temperature. It was found, however, that the intensity distribution in a band was the same within the accuracy of measurement, in any two spectra, taken in different mixtures which had the same overall intensity. This is caused by the fact that over nearly the whole range of conditions under which the radicals appear the temperatures are sufficiently high so that small temperature variations from one mixture to another have little effect on the rotational level population. By the use of an optical comparator, it was possible to match equal intensities quite accurately, using the whole of the $v'' = 0$ sequence for this purpose.

In order that the intensities should be a true measure of concentration it was necessary to obtain spectra which bore a known concentration relationship. This problem was solved by a method which should be universally applicable to partly resolved band spectra. Series of spectra are taken under identical conditions using absorption paths of different lengths. Now for spectra of equal intensity at a given wave-length

$$\log I_0/I = \epsilon_1 c_1 l_1 = \epsilon_2 c_2 l_2,$$

where ϵ, c and l are the extinction coefficients, concentrations and path lengths for the two exposures. It is to be noted that, if $\epsilon_1 = \epsilon_2$, this applies even when, owing to incompletely resolved line contours, Lambert's and Beer's laws are not separately applicable, for, as long as the true line contour is the same in both cases, the apparent contour is also the same, being determined only by the number of molecules in the absorbing path. In this case, for equal intensities at all wave-lengths $c_1/c_2 = l_2/l_1$, and a true scale of concentrations is simply constructed.

In our experiments the method is not rigidly applicable, owing to the fact that the changes in concentration are accompanied by temperature and composition variations which affect the line contour and therefore the value of ϵ apparent. Nevertheless, owing to the fact already mentioned, that over a small range of conditions the relative temperature changes are insignificant, the line contour, and therefore the values of ϵ, are almost constant, and the method may be used with reasonable accuracy. Two vessels 50 and 25 cm in length were used, and comparisons of intensities were made during the radical decay where temperature changes are relatively small. By successive comparisons an intensity scale was obtained which was directly proportional to the radical concentration, and all other spectra were referred to this by means of a comparator.

Results

Total pressures of 1 or 2 cm of mercury were used, the explosion with higher pressures being violent enough to break the quartz end plates of the reaction vessel. Initial periods were found during which only the NO bands and a weak continuum were present, and which increased as the relative amount of nitrogen dioxide was reduced. The maximum initial periods observed were of the order of 3 ms; if the nitrogen dioxide pressure was then further reduced, no explosion occurred, no absorption spectra appeared and there was very little pressure change. Only a trace of nitrogen dioxide remained after the flash, even when no explosion took place, and its spectrum was entirely absent during the explosion.

The time interval over which free radical spectra were observed was 1 or 2 ms, but most of this was occupied by the decay, the initial rise between first detection and the time at which the intensity maximum was reached being usually less than 10^{-4} s. It will be seen that in some cases it has been possible to obtain spectra at intervals of 2×10^{-5} s during this rise which, although taken in different explosions, show little scatter and appear in a regular order. When an excess of acetylene was present and carbon appeared as a product, the scatter became more noticeable, but it was still possible, by taking a larger number of spectra, to measure the form of the intensity-time curve with reasonable accuracy. It was necessary to clean the reaction vessel between each explosion in the region of carbon deposition, even when the carbon was hardly visible, as otherwise the reduced transparency of the reaction vessel resulted in increased induction periods and a complete irreproducibility in the time-intensity measurements.

Preliminary investigations showed that the appearance of the band spectra was very dependent on the relative proportions of acetylene and oxygen, and that a complete change of the whole spectrum occurred in the region of equimolecularity, if nitrogen dioxide was calculated in terms of an equivalent amount of oxygen. When oxygen was in excess the spectrum of OH appeared at high intensity and some CN was also present. On the acetylene-rich side OH was barely detectable but C_2 and CN appeared very strongly and CH was also present. The NH radical was observed only in mixtures very near to equimolecular. The change from one type of spectrum to the other was very marked when the mixture ratio was changed by 1 or 2 % on either side of equimolecular proportions. In addition to the five diatomic radicals mentioned a weak band at 4051 Å was observed, as well as absorption from upper vibrational levels of NO, and a number of continuous spectra. As most of the absorption spectra are new, reproductions are given in figure 9, plate 1. In general, owing to the high temperatures, the appearance of the band systems of the diatomic molecules is very similar to that of their well-known emission spectra. Wave-length measurements of the band heads reported agreed within the accuracy of measurement with those given in the literature for the emission bands (Pearse & Gaydon 1950) and will not, therefore, be tabulated.

Description of spectra

OH : $^2\Sigma$–$^2\Pi$ system

The 0, 0, 1, 0, 2, 0 and 3, 0 sequences were observed, but not the 0, 1 sequence. The intensity distribution between rotational and vibrational levels was very similar, at a given overall intensity, to that observed in our related experiments on hydrogen-oxygen explosions (Norrish & Porter 1952). Furthermore, the total intensity in oxygen-rich mixtures was the same as that in hydrogen-oxygen explosions at approximately the same pressure, whereas, in emission from the inner cone of low-pressure flames, the intensity of OH is 500 times greater in acetylene than in hydrogen flames (Gaydon 1948). This is strong evidence that nearly all the OH radiation from such flames is of non-thermal origin.

C$_2$ Swan bands

This familiar system has not hitherto been detected in absorption in combustion processes. It was reported by Klemenc, Wechsberg & Wagner (1934) in absorption during the thermal decomposition of carbon suboxide, the 1, 0 band at 4737 Å being observed, but in spite of many attempts these authors were unable to repeat the observation (private communication).

The following band heads were identified in our spectra: the 2, 0, 3, 1, 4, 2; the 1, 0, 2, 1, 3, 2, 4, 3, 5, 4; the 0, 0, 1, 1, 2, 2; and the 0, 1, 1, 2, 2, 3. A careful search was made for the high-pressure bands of C$_2$ which have been interpreted as arising from an inverse predissociation from the $v' = 6$ level of the Swan band system (Herzberg 1946). If this is the case, one might expect an increase in intensity in absorption, owing to the effect of increased line width on incompletely resolved spectra, followed by an abrupt disappearance of higher bands, a phenomenon observed under similar conditions, for example, in the spectrum of SH (Porter 1950b). No such effect was observed, and the intensities of the band heads decreased quite regularly up to the last observed head, which was that of the 5, 4 band, but owing to extended rotational stucture the higher band heads were rather difficult to locate, and this cannot be taken as conclusive evidence against such a predissociation.

C$_2$ Mulliken bands

The absorption spectrum of this transition, which arises from the $^1\Sigma$ state, has not previously been observed, and there are very few examples known of absorption from upper electronic states. The spectrum appeared quite strongly, the lines of the P and R branches being considerably more intense than in the spectrum of CH, but it was difficult to photograph, owing to the very strong continuous absorption near 2300 Å where this spectrum appears.

The appearance of both singlet (Mulliken) and triplet (Swan) systems in absorption at similar intensity shows that the C_2 molecule and diradical have very similar energies and raises the question as to which is the ground state. Herzberg & Sutton (1940) have predicted that the $^1\Sigma$ state lies 5600 cm^{-1} above the $^3\Pi_u$ state; the temperature in our system is of the order of $3000°$ K, and, allowing for the lower wave-length of the singlet system, a separation of this order is in accordance with the observed intensities, if the f value of the singlet transition is near unity. It should be possible to obtain confirmation of this point by taking the spectra at different temperatures and using the intensity distribution in the higher vibrational levels of the Swan bands as a comparison.

C_3 *band at* 4051 Å

A single diffuse line at 4051 Å was observed very weakly in rich mixtures, along with the intense continuum in this region. This is probably identical with the 4050 Å group discovered in emission from comets by Swings, Elvey & Babcock (1941), and later in the laboratory by Herzberg (1942), and attributed by the latter to CH_2. On the basis of convincing evidence obtained by using the ^{13}C isotope, it has recently been attributed to the C_3 molecule (Douglas 1951), and the conditions of its occurrence in our experiments support the assignment to a polyatomic carbon molecule.

Doublet systems of CH

Three systems of this radical are known, all of which occur in emission in flames and none of which has been observed in absorption in the laboratory, though the 4315 and 3900 Å systems are well known in stellar sources. The only band to appear strongly in our experiments was the 3143 Å band, the other two being only just detectable. This is the opposite of flame-emission spectra, where the 4315 Å system is much the strongest and the 3143 Å band is only observed in the hottest flames. The close packing of lines in the Q head of the 3143 Å band increases the sensitivity of the absorption method, but the individual lines of the P and R branches of this band were also appreciably stronger than those of the other two systems.

All three systems have the same lower $^2\Pi$ level, and it has been suggested, on the grounds of an expected high concentration of CH in flames and the failure to detect it in absorption, that the $^2\Pi$ state is not the ground level but lies above an unobserved $^4\Sigma$ state (Gaydon 1948, 1951). Now that all these systems have been observed in absorption the original reason for this suggestion disappears, and it can be said that the $^2\Pi$ state cannot lie more than a few thousand wave numbers above the ground state, the f values of these transitions being low (Herzberg 1951). Other considerations indicate that the $^2\Pi$ is, in fact, the lowest level (Porter 1951).

CN *and* NH

The violet $^2\Sigma$–$^2\Sigma$ system of CN was the strongest and most persistent feature of all our spectra. The bands identified were the 0,1, 1,2, 2,3, 3,4, 4,5; the 0,0, 1,1, 2,2, 3,3, 4,4; and the 1,0, 2,1 and 3,2. The 0,0 band of this system has been observed in absorption by Kistiakowsky & Gershinowitz (1933) and by White (1940) in thermally dissociated cyanogen.

172   R. G. W. Norrish, G. Porter and B. A. Thrush

The line-like Q branches of NH at 3360 and 3370 Å and the extended triplets of the P and R branches appeared in mixtures near to equimolecular, the intensity being similar to that of the 3143 Å band of CH.

Continuous spectra

The only other discrete spectra observed were the bands of the hot NO molecule which were always present during the induction period. In particular, the spectra of formaldehyde and of aromatic molecules such as benzene were absent at all times. In acetylene-rich mixtures a strong continuous absorption was present, which was associated, in part at least, with the deposition of solid carbon. The spectrum is complex, the wave-length-intensity distribution being a function of time, and consists of at least two parts with the following characteristics:

(*a*) A continuum, with a maximum at 3800 Å, whose intensity varies with time in the same way as the carbon radicals.

(*b*) A continuum whose intensity increases from 3000 Å to shorter wave-lengths and shows an approximate proportionality to λ^{-4}, and which reaches maximum intensity after 3 ms and decays for a period of several minutes.

Detailed measurements of these spectra, and their interpretation, will be given in later parts.

Dependence of radical concentration on mixture ratio

A series of mixtures were investigated in which the acetylene pressure was varied, the partial pressures of nitrogen dioxide and oxygen being constant at 1·5 and 10 mm respectively. The intensities of the various absorption spectra were measured in the manner described throughout the explosion and the maximum

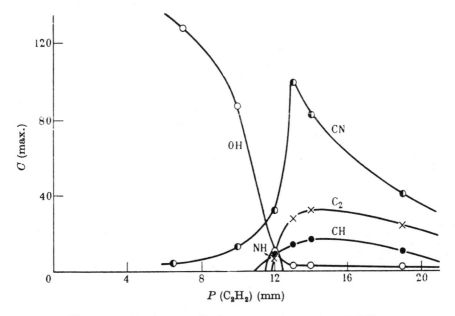

FIGURE 2. Maximum radical concentration against $P(C_2H_2)$;
$P(O_2) = 10$ mm, $P(NO_2) = 1\cdot5$ mm.

intensity of each radical obtained in this way for each mixture. The results are plotted in figure 2, the acetylene partial pressures investigated being 7, 10, 12, 13, 14 and 19 mm Hg.

No significance should be attached to the relative concentrations between one radical and another, except as an approximate indication of intensity. The complete change in radical concentration which occurs about equimolecular proportions of acetylene and oxygen is very striking. The decrease in the concentration of CN, C_2 and CH as the acetylene pressure is increased beyond 14 mm Hg was accompanied by a greatly increased continuous absorption and the deposition of visible quantities of solid carbon.

Dependence of radical concentration on time

The OH and CN radicals were observed at some point during the explosion of every mixture investigated, and the concentrations of these two radicals throughout the explosions are plotted on a logarithmic scale in figures 3 and 4 respectively. When excess oxygen was present, and no carbon was deposited, the reproducibility was good, and it has been possible to measure the OH concentration at intervals of 2×10^{-5} s throughout the development of the explosion, the induction period in

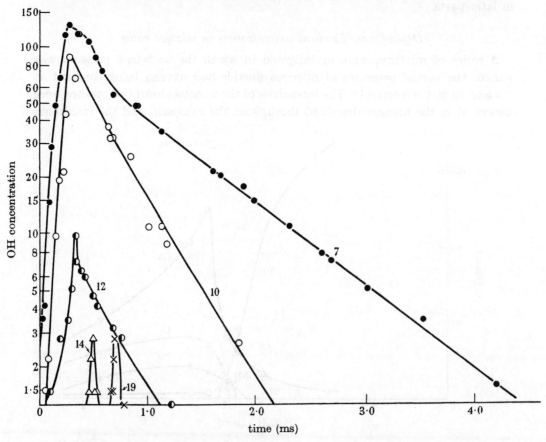

FIGURE 3. OH concentration (logarithmic scale) against time: effect of $P(C_2H_2)$; $P(O_2) = 10$ mm, $P(NO_2) = 1.5$ mm. Numbers on curves show pressure in mm.

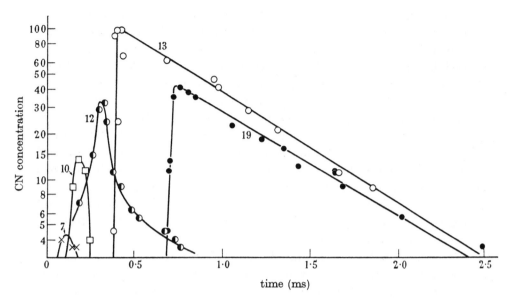

FIGURE 4. CN concentration (logarithmic scale) against time: effect of $P(C_2H_2)$; $P(O_2) = 10$ mm, $P(NO_2) = 1\cdot5$ mm. Numbers on curves show pressure in mm.

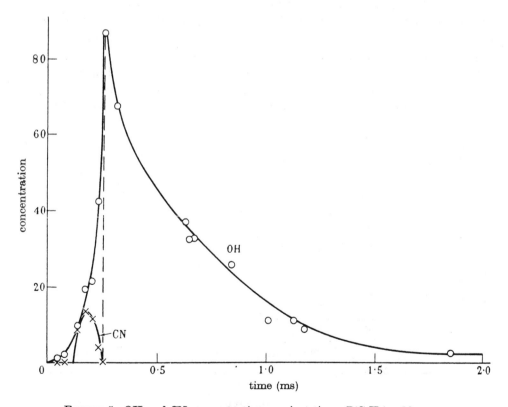

FIGURE 5. OH and CN concentrations against time; $P(C_2H_2) = 10$ mm, $P(O_2) = 10$ mm, $P(NO_2) = 1\cdot5$ mm.

the mixture containing 7 mm acetylene being less than $30\,\mu$s. The rate of decay of the two radicals at a given concentration towards the end of the reaction depends markedly and in opposite senses on the mixture ratio, the decay rate of OH being increased by increasing acetylene/oxygen ratios, and the CN decay rate being simultaneously decreased. The final decay is exponential in both cases. A comparison of the curves for the two radicals reveals two other important facts, which are shown more clearly in figure 5, where the intensities of OH and CN during the explosion of the mixture containing 10 mm C_2H_2 are plotted. First, the OH radical appears before the CN radical is detected. This point is fully confirmed in the later results where the OH concentration does not rise to such high values, and the phenomenon was encountered in every case studied. Secondly, the maximum CN concentration is reached 10^{-4} s before the OH maximum, and the CN decay is complete before the OH decay begins, the same being true in the 7 mm acetylene mixture (figures 3 and 4).

The C_2 and CH radicals were only detectable in the richer mixtures, and in all cases their intensity-time curves were very similar. The occurrence of these radicals in the mixtures containing 13 and 19 mm acetylene is shown in figures 6 and 7 respectively. Owing to the rapid rise and the carbon deposition, there is some scatter of the points, but as each set of intensities for a given time is recorded on the same spectrum the points are more significant than is indicated by the curves themselves, and, by comparison of corresponding points, it is possible to say quite definitely that the peak concentrations of CN, C_2 and CH, and also of C_3 when present, are reached at the same time and that the radicals are formed concurrently. For the same reason, one can say that the OH radical appears earlier and the intensity falls to zero before the decay of the carbon radicals begins. In all other mixtures the same conclusions apply, and no distinction whatever could

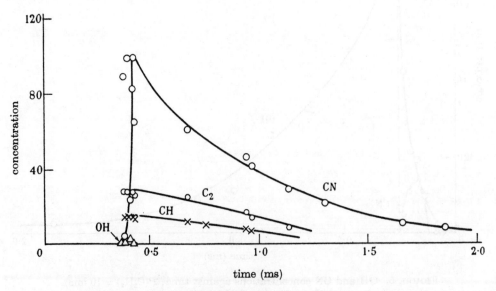

FIGURE 6. Radical concentrations against time; $P(C_2H_2) = 13$ mm,
$P(O_2) = 10$ mm, $P(NO_2) = 1.5$ mm.

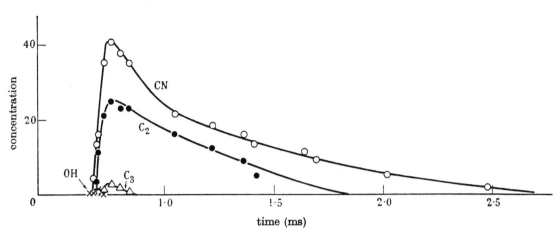

FIGURE 7. Radical concentrations against time; $P(C_2H_2) = 19$ mm,
$P(O_2) = 10$ mm, $P(NO_2) = 1.5$ mm.

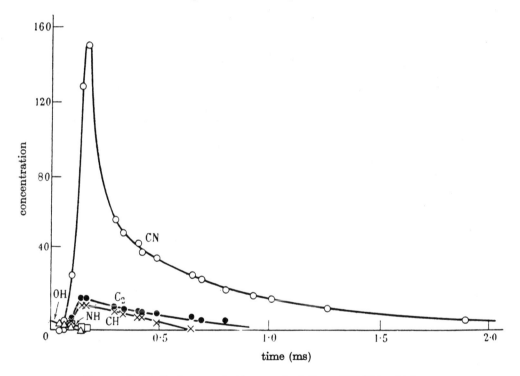

FIGURE 8. Radical concentrations against time; $P(C_2H_2) = 14.5$ mm,
$P(O_2) = 10$ mm, $P(NO_2) = 3$ mm.

be made between the times of appearance and rates of growth of the carbon radicals.

Effect of nitrogen dioxide pressure

Although the relationship of CN to C_2 and CH in time was very close there was a strong dependence of the intensities of these radicals on nitrogen dioxide pressure, as is shown in figure 8. Conditions here are identical with figure 6 except that

1·5 mm of the O_2 has been replaced by an equivalent amount of NO_2. The first effect of this change is greatly to reduce the initial period. Secondly, although the carbon radicals all appear together, the rate of growth and the peak intensities reached are increased for CN and considerably reduced for C_2 and CH. The effect of doubling the NO present is to increase the peak CN concentration by 50 % and to reduce peak C_2 and CH concentrations by 40 and 13 % respectively. The OH radical appears at the same intensity, but in the mixture containing 3 mm NO_2 its early appearance is particularly noticeable, and it appears at maximum intensity before any other radical is observed and rapidly falls to zero before the carbon radical maximum is reached. The NH radical, which was absent in figure 6, now appears and reaches maximum intensity after the OH but before CN. Here, as in all other cases, the NH radical only appears when both CN and OH are present.

Effect of nitrogen, water and carbon dioxide

The addition of nitrogen to the explosive mixture had the inert-gas effects, observed in hydrogen/oxygen explosions, of increasing the induction period and reducing the maximum concentration of radicals observed. For example, in the mixture containing acetylene, oxygen and nitrogen dioxide at pressures 14, 10 and 3 mm, the addition of 17 mm of nitrogen increased the induction period from 0·15 to 1·6 ms and decreased the maximum radical concentration to about one-half.

The influence of water on hydrogen/oxygen mixtures was the same as that of an inert gas, but in acetylene explosions small additions of water had a very specific and striking effect on the appearance of the radicals. This is illustrated by the data in table 1, which refer to the peak radical concentrations observed in the 13 mm acetylene, 10 mm oxygen, 1·5 mm nitrogen dioxide mixture, to which varying amounts of D_2O have been added. The intensities refer to the lighter isotopic species only.

TABLE 1. EFFECT OF ADDED D_2O ON MAXIMUM RADICAL INTENSITY

D_2O (mm)	OH	CN	C_2	CH	NH	time to max. rate (sec)
0	3	98	28	15	0	$4·8 \times 10^{-4}$
2·0	3·5	62	17	13	2	3×10^{-4}
3·5	37	24	5	9	12	2×10^{-4}
5·0	100	11	0	0	0	5×10^{-5}

Comparison with figure 2 shows that the addition of water has very nearly the same effect as the addition of an equivalent amount of oxygen in displacing the position at which the carbon radicals appear. More exact comparison shows that, in small amounts, it is equivalent to rather less than $\frac{1}{2}O_2$. Water has the additional effect of reducing the initial period, the times given in table 1 being those between the beginning of the flash and the time of maximum rate. Water therefore plays an important chemical role in the initial reactions, for its physical effect would be to increase, slightly, the thermal capacity of the system and therefore the initial period.

The addition of carbon dioxide to the above mixture had the same effect as the addition of water in replacing the carbon radicals by OH, and quantitatively it, also, was equivalent to slightly less than $\frac{1}{2}O_2$. On the other hand, it resulted in a greatly increased initial period, in the same way as nitrogen.

DISCUSSION

These investigations, incomplete as they must be at this stage, form a beginning to a complete kinetic analysis of hydrocarbon explosions. Our immediate interest will be to discover the relationship of the diatomic radicals observed to the combustion process.

The concentration of any radical R at a time t during the explosion will be given by an expression of the form

$$[R]_t = \int_0^t \sum_1 k_1(A)^a(B)^b \dots \mathrm{d}t - \int_0^t \sum_2 k_2(R)^r(C)^c(D)^d \dots \mathrm{d}t = \int_0^t K_1 \, \mathrm{d}t - \int_0^t K_2 \, \mathrm{d}t,$$

where K_1 and K_2 are the sums of the rate terms of all reactions by which R is formed and destroyed, respectively, (A), (B), etc., being the concentrations of the other molecules taking part. These concentrations are all functions of time and the rate constants, k, being functions of temperature, are also functions of time. The evaluation of this expression is clearly not possible at present, and we must begin by seeking qualitative relationships which may lead to some simplification.

The general form of the curves is explained as follows. The flash liberates oxygen atoms and a certain amount of heat and immediately afterwards the system is at a temperature T_i. As long as T_i is above a certain critical value, further exothermic reactions will occur resulting in heat liberation at a rate greater than heat dissipation to the wall, and consequently in a further temperature rise. For a given radical, which is to be observed, terms K_1 must exceed terms K_2, and the radical concentration then grows rapidly. In general, the radical growth will be accompanied by chain branching and further heat liberation resulting in explosive reaction. Eventually the reactants A, B, etc., will be consumed and terms K_1 will become smaller in spite of the high temperature. When $K_1 = K_2$ the peak radical concentration is reached. K_2 now becomes greater than K_1 and the radical concentration falls eventually to zero.

In most cases the decay of the radical is many times slower than the initial rise in concentration, and this must be interpreted as meaning that the observed decay rate is in fact that of the terms K_2, terms K_1 being negligible owing to complete removal of one reactant. The latter parts of the decay curves of CN and OH give very satisfactory unimolecular plots, the half-lives being 0·38 ms for CN in the 13 mm acetylene mixture and 0·63 ms for OH in the 7 mm mixture. It is probable that this implies an approximate constancy of composition and of temperature during this period and a zero value of terms K_1, owing to complete consumption of one reactant.

The interpretation of the maximum radical concentrations in figure 2 is now possible. It was shown by Bone (1932) that the explosion of acetylene and oxygen in equimolecular proportions occurred according to the overall equation

$$C_2H_2 + O_2 = 2CO + H_2,$$

and that excess oxygen resulted in water formation whilst excess acetylene gave carbon deposition. A naïve interpretation of the results of figure 2 would be the statement that the OH radical is formed only during the combustion of H_2 in excess oxygen and the radicals C_2, CH and CN result only from the cracking of excess acetylene and carbon formation, after consumption of the oxygen according to the above equation. This statement is, however, in discord with many of the facts recorded in the concentration—time curves and ignores the reactions by which the radicals are removed. These curves show that in every case a low maximum radical concentration is accompanied by a proportionately high rate of removal, indicating that the low concentration is to a great extent, if not entirely, due to increased K_2 terms rather than decreased K_1 terms. Anticipating part of the later discussion, the observed peak concentrations are to be interpreted as follows. In mixtures of all proportions investigated, both OH and the carbon radicals are formed to some extent during the chain-branching reactions which occur as the explosion develops. They react rapidly together and with the original reactants until one reactant is consumed and the other remains in excess. The concentration of carbon radicals or OH may then increase further by the cracking of any remaining acetylene or by the subsequent reaction between hydrogen and oxygen, though the former is clearly not of great importance if acetylene is increased much above equimolecular proportions, for the concentration of carbon radicals then falls again, owing to the lower temperature. When water or carbon dioxide are present in the original mixture, they are rapidly reduced by the carbon radicals or by the material from which these radicals are formed in the same way as oxygen or OH.

Time relationship of the radical concentrations

We may now inquire whether the chain-branching processes in which the radicals take part are so coupled that all the radical concentrations increase proportionately. It has already been noted that this is the case for the radicals C_2, CH and CN, and the reactions by which they are formed must therefore proceed from a common source or, alternatively, the formation of one results in the rapid reaction to form the others. So close is this correspondence that we shall frequently refer to these radicals together as the carbon radicals, and if, as is often the case, only CN is observed, owing to its high extinction and also to reasons of reactivity, which will be understood later, we shall infer that the other carbon radicals are probably being formed at a proportionate rate. The justifications for this assumption, which has important consequences, are as follows:

(*a*) The concentrations of C_2, CH and CN in figure 2 fall in proportion as the acetylene is decreased from 13 to 12 mm, and the absence of C_2 and CH in weaker mixtures is therefore expected, for a further decrease, in proportion to CN, results in a concentration of C_2 and CH below the limit of detectability.

(*b*) In all cases when CH and C_2 are observed, their rate of formation is proportional to that of CN.

(*c*) The rate of decay of C_2 and CH increases as the proportion of oxygen is increased in the same way as CN, indicating that the decreased concentration is, in part at least, caused by increased rate of removal.

The fact that C_2 and CH are formed concurrently rather than consecutively is important in view of the reports of a number of investigators that one appears before the other in emission in the extended reaction zones of flames. Gaydon (1948) has pointed out that C_2 usually appears lower in the flame than CH and has inferred that this is the order of formation, possibly owing to carbon breakdown to C_2 followed by reaction with OH. Although diffusion processes may alter the concentration distribution in a flame, it seems to us that a more probable explanation is to be found in conditions of excitation in different parts of the reaction zone, for there can be little doubt that the difference in times of formation during the explosion reaction is insufficient to explain the spatial resolution of these two radicals in flames.

The appearance of the OH radical is quite distinct from the carbon radicals. It invariably occurs before all other radicals at the end of the induction period, and there is no question of this always being caused by different limits of detectability, for in some cases (e.g. figure 8) its intensity has fallen considerably before any other radical spectrum appears. We must conclude that in all mixture ratios the early part of the reaction proceeds via a chain-branching process, which involves OH formation, but does not produce the carbon radicals in detectable amounts. The reaction therefore has at least two stages, and further examination of figure 5 shows that a third stage is involved when excess oxygen is present. The complete decay of CN, and, by inference, of the other carbon radicals also, is followed by further chain branching involving the formation of OH. Now the early disappearance of the carbon radicals implies, by our previous discussion, that the reactant from which they are formed has been consumed, and this must be acetylene, carbon or some other carbon compound. As there is an excess of oxygen, some of it must react with hydrogen, which will now be present if carbon has been consumed, and the third phase of the reaction, in which OH alone is observed and reaches a very high concentration, is therefore to be identified with the final hydrogen-oxygen reaction.

The explosive reaction between acetylene and oxygen, in the presence of a small amount of NO, can therefore be divided into the following four stages, the figures given in brackets being the approximate duration of each stage in the mixture containing 10 mm C_2H_2, 10 mm O_2 and 1·5 mm NO_2:

Stage 1. An initial period towards the end of which the OH radical concentration increases rapidly (10^{-4} s).

Stage 2. A rapid reaction involving C_2, CH, OH and, in the presence of NO. also CN and NH, the semi-stationary concentration of these radicals remaining low until the consumption of one reactant is nearly complete (10^{-4} s).

Stage 3. (Which only occurs when oxygen is present in excess of equimolecular proportions.) A further reaction during which the OH concentration rises exponentially and the carbon radicals are totally consumed (10^{-4} s).

Stage 4. A relatively slow exponential decay of the excess radical concentration remaining after the completion of stages 2 and 3 (half-life of OH = 0·28 ms).

The early part of stage 1 may be similar in mechanism to the slow isothermal reaction, but the rotational temperature of OH is very high, even when it first appears, and the reactions in the latter part probably involve a different mechanism. Formaldehyde and glyoxal, the products of slow oxidation, were never observed. The marked influence of water vapour on the induction period may be due to chain initiation by the dissociation of water at temperatures above about 1000° C. In the slow oxidation of acetylene the influence of water on the induction period does not seem to have been studied, but a heterogeneous reaction between acetylene and water to form acetaldehyde occurs at 300° C (Bone & Andrew 1905). Water has a negligible effect on the induction period during ethylene oxidation (Bone, Haffner & Rance 1933).

Stage 3 is undoubtedly the reaction of the hydrogen produced in stages 1 and 2 with the remaining oxygen, and there is a close correspondence between the intensity-time curves of this stage with those obtained with equivalent quantities of hydrogen. The maximum intensity of OH in the mixture containing 7 mm acetylene is approximately double that which would be obtained from 7 mm of hydrogen and a slight excess of oxygen under the same conditions. The decay of OH in stage 4 is to be attributed to its reaction with hydrogen and carbon monoxide, and the details of the OH reactions in stages 3 and 4 are best arrived at by a study of these reactions individually (Norrish & Porter 1952). The decay of the carbon radicals in stage 4 will be discussed in connexion with the continuous spectra and carbon formation.

The essentially new part of our mechanism is stage 2, and we shall, therefore, examine the reactions occurring during this process in a little more detail.

The C_2 and CH radicals

The proportionality in rates of formation of these two radicals shows either that one is formed from the other very rapidly or that they arise from a common source. There can be little doubt that the connecting reaction is

$$C_2 + OH = CH + CO, \tag{1}$$

for the results show that the radicals C_2 and OH react together so rapidly that they are never present in significant amounts at the same time. The rapid disappearance of both C_2 and CH in mixtures containing oxygen is then explained if CH reacts with O_2 or OH.

The carbon radicals could be formed either:

(1) by decomposition of acetylene, or

(2) by breakdown of small carbon particles formed from acetylene.

This important question is closely connected with carbon formation and will be discussed in this context.

The CN and NH radicals

At the very earliest part of the initial period which can be observed the decomposition of NO_2 into NO and oxygen is complete, and the radicals CN and NH must therefore be formed by the subsequent reactions of NO. The close correspondence in time between the formation and disappearance of CN and C_2 has been noted, but there are several significant differences in the effect of concentration on their relative intensities, the most useful being given by a comparison of

182   R. G. W. Norrish, G. Porter and B. A. Thrush

figures 6 and 8. As expected, more CN is formed in the mixture containing more NO_2, but the amount of C_2 formed is reduced by 40 %. This large reduction cannot be caused by removal of acetylene, for even if all the NO reacts with acetylene to form CN, this only corresponds to the removal of 10 % of the acetylene. We conclude that either:

(1) C_2 is very rapidly removed by the reaction

$$C_2 + NO = CN + CO, \tag{2}$$

until the NO is consumed; or

(2) the C_2 and CN are formed from a common source X, and the reaction of X with NO occurs more readily than its reaction to form C_2.

It is clear that the formation of C_2 from CN is not the important mechanism.

The NH radical is always associated with the presence of both OH and CN. This is shown as a function of mixture ratio in figure 2, of time in figure 8, and of added water in table 1. This is very strong evidence for the reaction

$$CN + OH = CO + NH. \tag{3}$$

As in our hydrogen-oxygen work, the formation of NH by direct reaction of NO with H_2, H or OH does not seem to occur. Unlike the other radicals, the disappearance of NH is rapid in mixtures of all proportions, showing that it reacts, not only with O_2 or OH, but also with the carbon radicals, presumably by

$$NH + C_2 = CN + CH \tag{4}$$

and

$$NH + CH = CN + H_2. \tag{5}$$

For simplicity we have written the above reactions in terms of the observed molecules, but it must be remembered that a high concentration of atoms is also probably present, and the reactions as written indicate the overall changes, which may proceed via atoms. It is now a simple matter to write a complete reaction scheme for stage 2 by the addition to the above of several equations of equal thermodynamical probability, but the discussion has been confined to reactions for which there is direct experimental evidence. The outstanding question which remains is what proportion of the oxidation mechanism proceeds via stage 2. The high intensities of the absorption spectra suggest that this may be by no means insignificant, but the quantitative answer cannot be given until absolute concentration determinations of the radicals concerned become available.

We are indebted to the Government Grants Committee of the Royal Society for the loan of the Littrow spectrograph used in this work. We are also indebted to the Anglo-Iranian Oil Company for financial support.

REFERENCES

Bone, W. A. 1932 Proc. Roy. Soc. A, 137, 243.
Bone, W. A. & Andrew, G. W. 1905 J. Chem. Soc. 87, 1232.
Bone, W. A., Haffner, A. E. & Rance, H. F. 1933 Proc. Roy. Soc. A, 143, 16.
Douglas, A. E. 1951 Astrophys. J. 114, 466.
Gaydon, A. G. 1948 Spectroscopy and combustion theory. London: Chapman and Hall.

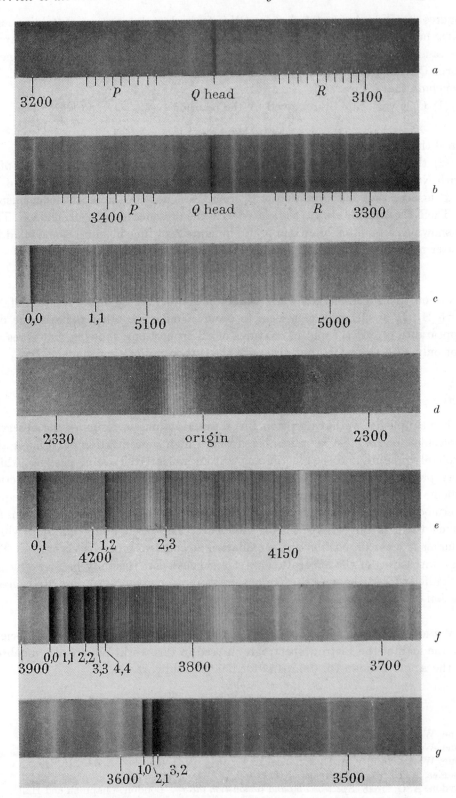

FIGURE 9. (*a*) CH $^2\Sigma^+-^2\Pi$ system. (*b*) NH $^3\Pi-^3\Sigma$ system. (*c*) C$_2$ 'Swan' bands $^3\Pi-^3\Pi$ (0,0) progression. (*d*) C$_2$ 'Mulliken' bands $^1\Sigma-^1\Sigma$. (*e*) CN violet system $^2\Sigma-^2\Sigma$ (0,1) progression. (*f*) CN violet system $^2\Sigma-^2\Sigma$ (0,0) progression. (*g*) CN violet system $^2\Sigma-^2\Sigma$ (1,0) progression.

Studies of the explosive combustion of hydrocarbons. I 183

Gaydon, A. G. 1951 *Disc. Faraday Soc.*, Hydrocarbons, **10**, 108.

Gaydon, A. G. & Wolfhard, H. G. 1947 *Disc. Faraday Soc.*, The labile molecule, **2**, 161.

Herzberg, G. 1942 *Astrophys. J.* **96**, 314.

Herzberg, G. 1946 *Phys. Rev.* **70**, 762.

Herzberg, G. 1951 *Molecular spectra and molecular structure*, **1**, 386. New York: D. Van Nostrand.

Herzberg, G. & Sutton, R. B. 1940 *Canad. J. Res.* A, **18**, 74.

Kistiakowsky, G. B. & Gershinowitz, H. 1933 *J. Chem. Phys.* **1**, 432.

Klemenc, A., Wechsberg, R. & Wagner, G. 1934 *Z. phys. Chem.* A, **170**, 97.

Norrish, R. G. W. & Porter, G. 1952 *Proc. Roy. Soc.* A, **210**, 439.

Pearse, R. W. B. & Gaydon, A. G. 1950 *The identification of molecular spectra*. London: Chapman and Hall.

Porter, G. 1950a *Proc. Roy. Soc.* A, **200**, 284.

Porter, G. 1950b *Disc. Faraday Soc.*, Spectroscopy and molecular structure, **9**, 60.

Porter, G. 1951 *Disc. Faraday Soc.*, Hydrocarbons, **10**, 108.

Porter, G. 1952 *Boll. Sci. Chim. Industr. Bologna*, **1**.

Swings, P., Elvey, C. T. & Babcock, H. W. 1941 *Astrophys. J.* **94**, 320.

White, J. U. 1940 *J. Chem. Phys.* **8**, 79, 459.

Reprinted from The Mechanism of Carbon Formation, Scheveningen Netherlands Conference 1954, Agard Memorandum, AG13/M9, p. 1

THE MECHANISM OF CARBON FORMATION

George Porter *

*University of Cambridge ***

An excellent review of the processes leading to carbon formation in flames, which covers most of the work up to 1952, has been given by Gaydon and Wolfhard (1). The present account will be limited mainly to the views of the author which have been summarised briefly elsewhere (2). In addition the results of investigations by the flash photolysis technique (3) which have a bearing on the problem are summarised.

It will be convenient to study the chemical processes leading to carbon formation in the following systems in turn :

1. Pyrolysis of pure hydrocarbons.
2. Pyrolysis of acetylene.
3. Diffusion flames.
4. Explosions in closed vessels
5. Premixed flames.

The pyrolysis of pure hydrocarbons.

The primary result of pyrolysis of saturated hydrocarbons is thermal decomposition. With unsaturated compounds polymerisation, as well as decomposition, may occur and the relative extent of these two processes depends on the experimental conditions. The interpretation of the vast literature on this subject is by no means simple, partly because many of the results refer to heterogeneous reactions, although often interpreted otherwise, and partly because the amount of conversion is frequently so great that secondary reactions of the products of primary pyrolysis have become important. If these complications are excluded however it is found, as might be expected on the basis of general chemical experience, that decomposition predominates over polymerisation at high temperatures. There are several methods of showing that this must be the case, of which the following is probably the most instructive.

Let us represent the processes of polymerisation and decomposition by the following equations :
$$AB = A + B \qquad (1)$$
$$2AB = (AB)_2 \qquad (2)$$
Now the rate constant, k, of either process, is given by the expression [4]

$$k = \frac{KT}{h} \quad \exp{-\frac{\Delta H^{\neq}}{RT}} \exp{\frac{\Delta S^{\neq}}{R}}$$

where $\Delta H^{\neq}$ and $\Delta S^{\neq}$ are the enthalpy and entropy differences between the activated complex and the reactants in their standard states and the other symbols have their usual significance. The relative rate constants of the two reactions are therefore given by

$$k_1/k_2 = \exp\frac{\Delta H_2^{\neq}-\Delta H_1^{\neq}}{RT} \exp\frac{\Delta S_1^{\neq}-\Delta S_2^{\neq}}{R} .$$

The decomposition of a hydrocarbon is usually a very endothermic process and the value of $\Delta H_1^{\neq}$, is high. $\Delta H_2^{\neq}$, on the other hand is relatively small. Typical values for the activation energies, which are very nearly equal to $\Delta H_1^{\neq}$, are as follows :

* Professor of Physical Chemistry

* Cambridge, England.

Decompositions.	E(k.cal/mole)	Ref.	Addition reactions.	E(k.cal/mole)	Ref.
CH_4	79	5	Dimerisation of		
C_2H_6	69,8	6	C_2H_4, C_3H_6, C_4H_8		
C_3H_8	63.3	7	and C_5H_{10}	38	7
C_4H_{10}	58.7	7	$2C_4H_6 = C_8H_{12}$ etc.	24	4
C_3H_6	72	8	$C_2H_4 + C_4H_6 = C_6H_{10}$		
			and other Diels-Alder additions	20	4
			$2C_2H_2 = C_4H_4$	40	9

A similar generalisation can be made about the entropy terms. The entropy of the activated complex in a polymerisation reaction will be near to that of the polymer molecule so that such reactions have a high negative value of $\Delta S^{\neq}$. Thus, for the dimerisation of ethylene and of butadiene, the values of $\Delta S_p^{\neq}$, the entropy change for a standard state of 1 atmosphere pressure, are -30.1 and -38.5 cal/degree/ mole respectively, at $300°K^{(4)}$. For dissociation reactions however the value of $\Delta S^{\neq}$ is usually near zero, that is to say they usually have a frequency factor within an order or two of 10^{14} sec^{-1}.

We can therefore state, with some generality, that for the reactions represented by equations 1 and 2

$$\Delta H_2^{\neq} < \Delta H_1^{\neq} \qquad \text{and} \qquad \Delta S_2^{\neq} < \Delta S_1^{\neq}$$

It follows that, as the temperature is increased, the concentration of AB being constant, the rate of reaction 1 relative to 2 will increase until above a certain temperature, decomposition becomes predominant. It is also true that, above a certain temperature, depolymerisation will occur more rapidly than polymerisation, this being the particular case when reaction 1 is the reverse of reaction 2.

Absolute rates.

It is clear from the above considerations that if polymerisation is to occur as a precurser to carbon formation, it must do so at the lower temperatures before decomposition becomes predominant. We must therefore enquire firstly as to the magnitude of this temperature and secondly as to the rate of polymerisation at lower temperatures.

A study of the above figures, and also the experimental rate data on hydrocarbon decomposition and polymerisation shows that the temperature at which the rate of decomposition becomes the greater at normal pressures, is usually about 1000°C or less. Calculation also shows that, at the temperature where the rate constants k_1 and k_2 become equal for 1 atmosphere pressure, the absolute value of k_2 rarely exceeds $10^2 atm^{-1}sec.^{-1}$. The low activation energies of polymerisation are accompanied by low frequency factors. Thus for butadiene the frequency factor is $9.2 \times 10^9 cc.mole^{-1}sec.^{-1}$ giving a rate of polymerisation at 1300°K of 40 sec^{-1} at 1 atmosphere.

Any small amount of polymer which is formed during the preheating period will normally be decomposed again when the higher temperatures are encountered, like any other hydrocarbon. The only exception to this statement would be the case where the hydrocarbon remained in the preheating zone, at temperatures below 1000°C, long enough for polymers of very high molecular weight to be formed. It is possible that such molecules, of colloidal or microscopic size, might form carbon at higher temperatures by a different mechanism involving "charring", that is a solid phase dehydrogenation. This mechanism has been supported by Wolfhard and Parker (10), who obtained evidence of such particles during experiments on hydrocarbon pyrolysis. Their temperatures were always 1000°C and their contact times were about 10 seconds. Under these conditions the formation of high polymers might be possible, but the experiments have little relevance to the mechanism of carbon formation in most flames, where the time spent in the preheating zone is of the order of milliseconds or less. In view of the rate considerations already discussed it is improbable that polymers of this size could be formed in a time of less than a few seconds.

Although accurate rate data ate not available for the high temperature reactions of all hydrocarbons there is little doubt that the above considerations apply to most fuels normally encountered. It is found, in practice, that, at normal pressures, decomposition of olefins and acetylenes predominates over polymerisation at temperatures above 1000° C(7). Polyenes and polyacetylenes are rather labile substances, even at low temperatures and long chain saturated hydracarbons decompose quite as readily as the paraffins which we have discussed. With aromatic molecules condensation occurs to a decreasing extent, relative to decomposition, at temperatures above 750°C (11).

Conclusion 1. *When the time for half reaction is of the order of one second or less, the pyrolysis of hydrocarbons does not result in polymerisation but in decomposition to smaller molecules, owing to the correspondingly high temperatures.*

It should perhaps be mentionde that there is one apparent exception to this rule, since solid carbon is itself a polymer and is certainly formed at very high temperatures. This is of course due to the unusually high heat of decomposition of solid carbon which is probably between 130 and 170 k.cal/gm. atom.

Mode of decomposition of hydrocarbons.

The very extensive data on the decomposition products of hydrocarbons is complicated, in many cases, by heterogeneous reactions, low temperature polymerisation during preheating, and by secondary reactions of the primary products. When these difficulties are eliminated the following scheme describes the primary reactions very satisfactorily :

$$2CH_4 = C_2H_6 + H_2 \qquad (12)$$

$$C_2H_6 = C_2H_4 + H_2 \qquad (13)$$

Higher paraffins = Lower paraffins + olefins   (14)
Higher olefins = Lower olefins + paraffins + acetylene   (15)

$$C_2H_4 = C_2H_2 + H_2 \qquad (16,17)$$

$$\text{Aromatics} = Carbon + CH_4 + C_2H_2 + H_2 \quad (15)$$

$$C_2H_2 = Carbon + H_2 \qquad (15,17)$$

Thus, with the possible exception of aromatic molecules, the decomposition of any hydrocarbon leads eventually to the formation of acetylene and hydrogen. In many cases, and in particular with aromatic molecules, where the temperatures of decomposition are very high, very little acetylene is actually found in the products but this is due to the rapid decomposition of acetylene to carbon and hydrogen at the temperatures necessary for pytolysis of the original hydrocarbon.

Conclusion 2. *The thermal decomposition of hydrocarbons results in dehydrogenation and cracking to smaller molecules and the last stable hydrocarbon to be observed before carbon formation is acetylene.*

These conclusions may appear to be a rather trivial result of our discussion but, if accepted, they greatly simplify the problem since we need now only consider the last stage of carbon formation from acetylene which will be a common mechanism for all fuels. Furthermore they rule out nearly all the mechanisms of carbon formation which have hitherto been proposed in any detail, since these require the intermediate formation of polymeric hydrocarbons.

The pyrolysis of acetylene

Acetylene decomposes to carbon and hydrogen very readily and, at normal pressures, the reaction is explosive above $850°C$[18]. The process is of great technical importance and is now of particular interest since it holds the key to the mechanism of carbon formation from other hydrocarbons as well as from acetylene itself.

Owing to the explosive nature of the decomposition the study of acetylene pyrolysis tells us little about mechanism. The final products are carbon and hydrogen and many workers have isolated quantities of tar and aromatics from the products of slow pyrolysis. These can usually be traced to heterogeneous reactions or low temperature polymerisation but the evidence is by no means clear. The important point to be settled is whether any stable intermediates, in particular aromatic hydrocarbons, are formed as precursors to carbon deposition. The following experiments provide fairly clear evidence on this point.

Results of flash pyrolysis experiments.
In order to eliminate the two uncertainties mentioned above it is necessary to raise the temperature of the hydrogen very rapidly well over $1000°C$ and heating should be in the body of the gas, rather than by the wall, in order to reduce heterogeneous effects. In addition, the products should be cooled very rapidly so that it is possible to quench the reaction and thereby stabilise any intermediate products formed between acetylene and carbon. These conditions are satisfied by using the technique of flash pyrolysis.

The method is as follows. A molecule which absorbs light is irradiated for a fraction of a second (usually about one millisecond) by an intense light flash. This decomposes a fraction of the molecules by photochemical action and also raises the temperature of the system to a very high value owing to the degradation of the excess thermal energy of the photolytic fragments and of the electronic energy of the molecules which do not decompose. The free radicals formed photolytically in this way recombine very rapidly and subsequent reaction occurs by a purely pyrolitic mechanism. The products found are entirely in accordance with this interpretation. The temperature effects can be eliminated by the addition of inert gas which increases the thermal capacity of the system, and in the presence of sufficient inert gas the products revert to the normal products of isothermal photolysis.

The substance which illustrate the processes leading to carbon formation most clearly are ketene, acetene and diacetyl. The overall mechanism of normal photolysis of these compounds is as follows :

$$2CH_2CO = C_2H_4 + 2CO$$

$$CH_3COCH_3 = C_2H_6 + CO$$

$$CH_3COCOCH_3 = C_2H_6 + 2CO$$

The carbon monoxide takes no further part in the reaction but, if the temperature is allowed to rise i.e. if no inert gas is added, the ethylene and ethane undergo pyrolysis and large quantities of carbon are formed. Analysis of the products showed that acetone and diacetyl formed ethylene and acetylene under these conditions and ketene formed acetylene. In all cases the amount of unsaturation increased as the temperature was increased by reducing the amount of inert gas present. The most detailed data are available for ketene [17] and the results of a typical series of analysis are shown in Fig.1. The original pressure of ketene was 10mm/Hg and the products, expressed as percentage of CO, are plotted against the pressure of inert gas, which is **inversely** proportional to the k temperature rise. (Fig. 1).

As the temperature is increased the amount of ethylene falls and is replaced by acetylene and hydrogen. At still higher temperatures carbon is formed and the acetylene concentration falls again. Apart from small quantities of methane, ethane and propylene no other products were detected. In particular, aromatic hydrocarbons such as benzene, which have discrete spectra and would have, discrete spectra and would have been detected, if present, at concentrations of 0.1% of the total, were absent. There is no doubt that if such compounds were intermediates between acetylene and carbon a dectectable quantity would have been quenched out in the products since a considerable fraction of the ketene remained undecomposed, and similar experiments on the flash photolysis of mixtures containing benzene invariably showed detectable quantities of benzene remaining after photolysis. The solid product was typical "acetylene carbon" with a mean particle diameter of 500A.

The above experiments were concerned entirely with the stable products remaining after photolysis, but studies have also been carried out during the course of the reaction itself. The method used was to record the absorption spectra at various times after the reaction had been initiated by the photolytic flash. For this purpose a second source, the spectroscopic flash, is used, which is able to record the absorption spectrum in a period of 50μ seconds at any desired time interval after photolysis. The experimental arrangement for this type of experiment is illustrated in Fig. 2.

Investigations by this method showed that absorption by carbon particles was present at the shortest time which could be studied after the flash photolysis of acetone, ketone and diacetyl. This absorption attained a steady value after about 3 milliseconds and then decayed slowly over a period of several minutes as the carbon particles coagulated and settled. No discrete spectra of any kind were present at any time during this process of carbon deposition although the spectra of C_2 and of benzene and other aromatic molecules would have been readily detected.

These results show that the pyrolysis of acetylene leads to the formation of carbon and hydrogen without the intermediate appearance of stable hydrocarbon polymers, a fact which is in accordance with our previous discussion about the pyrolysis of hydrocarbons in general.

The mechanism of acetylene decomposition.

We now have to consider the possible mechanisms by which acetylene might decompose to solid carbon and hydrogen without initially polymerising to polyacetylenes or to aromatic hydrocarbons. This first possibility is a further breakdown to carbon and hydrogen, followed by condensation of the carbon atoms or radicals. This might occur via C_2 or carbon atoms thus :

$$(1) \quad C_2H_2 = C_2 + H_2 \qquad nC_2 = C_{2n}$$

$$(2) \quad C_2H_2 = 2C + H_2 \qquad nC = C_n$$

No single experimental investigation is at present able to give a rigourous argument for the exclusion of reaction 1, but taken as a whole the evidence against it is overwhelming. Gaydon and Wolfhard have shown that the concentration of C_2 in premixed flames is too low to account for the rate of carbon formation (19). Parker and Wolfhard's experiments on diffusion flames (20), mentioned in the next section, show that C_2 does not appear during the process of carbon formation but only during its combustion. Finally our investigations of the C_2 spectrum in absorption during combustion in closed vessels show no correlation between C_2 concentration and carbon formation whereas a consideration of any reasonable mechanism of C_2 formation and polymerisation to carbon predicts such a correlation if the main part of the carbon is formed from C_2. We believe that the C_2 observed at high concentration at the prevailing temperature and that C_2 is formed from solid carbon rather than the reverse. This explains why no C_2 radicals are observed in very rich mixtures or during flash pyrolysis, when copious quantities of carbon are formed, since the temperature under these conditions is relatively low.

Another very strong argument against both mechanisms 1 and 2 is their improbability on energetic grounds. Norrish has suggested the carbon atom mechanism since the atomic spectrum cannot be observed and there is no experimental evidence to discount it. It can be ruled out by the following consideration. Although the latent heat of sublimation of carbon is uncertain, most chemists agree that it lies between 125 and 170k.cal/gm.atom. It follows that reaction 2 is endothermic by an amount between 196 and 286k.cal/gm.mole of acetylene. Since the activation energy of the reaction 2 must be at least as great as this, the exponential factor in the rate expression must have a value less than 10^{-14}. Even with a high frequency factor of 10^{-14} this would result in an average life of one second for the decomposition of an acetylene molecule at 3000° K. This is of course far longer than the observed reaction times in flames as well as in our pyrolysis experiments where the temperatures are much lower.

Even reaction 1 can be ruled out on energetic grounds when it is remembered that carbon deposition occurs at a measurable rate when acetylene is heated to 600° C, and explosively above 800° C.

Again the heats of reaction are uncertain but it would be difficult to support a value much less than 100k.cal/mole for the value of $-\Delta H_1$. At 800° C, and with a normal frequency factor, the half time of reaction would then be 10^6 seconds. It might be argued that, although impossible 800° C, this mechanism occurs at much higher temperatures, but since we have to find another mechanism anyway there is little to be said for retaining reaction 1.

We have now shown that hydrogen is not eliminated after polymerisation of acetylene nor is it eliminated before polyremisation and we are driven to the conclusion that the processes occur simultaneously.

Conclusion 3. *Solid carbon is formed from acetylene by a process involving a simultaneous condensation and dehydrogenation.*

This is perhaps not unexpected, since it is energetically economic to use the heat of polymerisation for the endothermic processes of dehydrogenation. No further experimental evidence is available concerning this last step and we must be content to show that reasonable mechanisms can be written. The first of these [2] involves the removal of an H_2 molecule at the same time as the acetylene reacts with the growing carbon chain somewhat as follows

$$\tilde{C} - + C_2H_2 \rightarrow \tilde{C} - C = C \rightarrow \tilde{C} = C = C -$$

$$\begin{array}{cccc} | & & & | \\ H & H\ H\ H & +H_2 & H \end{array}$$

This chain would be strained and might be expected to react readily with more acetylene to give a conjugated structure:

$$\tilde{C} = C = C = C - \rightarrow \tilde{C} = C = \dot{C} = C -$$

Once such a chain has been formed the acetylene may build up a ring structure by Diels-Alder type additions, as suggested by Dr. Thrush (21).

The second possibility is a chain reaction as follows :

$$H + C_2H_2 = C_2H + H_2$$

$$C_n + C_2H = C_{n+2} + H$$

The first reaction should be nearly thermoneutral and the latter will be exothermic. The spectrum of the C_2H radical is unknown and we are not therefore able to investigate these two possibilities further at present.

Diffusion Flames.
If, in a diffusion flame, carbon is formed before the fuel reaches the oxygen zone and if the time during which the fuel exists in the preheating zone is less than the time required to form high

polymers viz less than about 1 second, then the above conclusions relating to carbon formation by pyrolysis must apply to the mechanism in the diffusion flame as well.

Recent experiments by Wolfhard and **Parker** (20) have shown that the above conditions are satisfied, at least in the diffusion flames which they studied. These workers used a flat diffusion flame in which the fuel and oxygen were nicely separated and their results show that :
(a) Carbon is formed and reaches its maximum concentration, judged from its continuous spectrum, before the zones containing oxygen or OH are reached.
(b) Acetylene could be followed by means of its absorption spectrum, as far as the beginning of the continuous absorption, at which point the measured temperature was 2000° C.

It follows, as these authors concluded, that carbon is formed by a purely pyrolytic mechanism. **It also follows that the fuel survives the low temperature region below 1000°C and cannot therefore** polymerise to higher hydrocarbons. This is also to be expected from the fact that the total time taken for the fuel to pass through the visible flame was less than 1 second. In other diffusion flames conditions are very similar. It would be unsafe to calculate diffusion times across the reaction zone since it might be argued that diffusion was blocked by the solid carbon, but one can always set an upper limit to the available time in the preheating zone from the total time that the fuel spends in the flame. This is rarely greater than a few seconds.

Conclusion 4. Carbon formation in a diffusion flame occurs by the pyrolytic mechanism of 1,2 and 3 unless the fuel is maintained at temperatures near to, but less than, 1000° C for times of the order of seconds. In all diffusion flames normally encountered the preheating times are much less than this.

Explosions in closed vessels.
We shall now consider carbon formation in systems containing oxygen. The presence of oxygen might affect the reaction in two ways. Firstly the amount of oxygen present will determine whether or not carbon is formed at all, since the oxidation processes will compete with those of **pyrolysis. This is a matter of stoichiometry and will be considered later. Secondly, when condi-**tions are such that carbon is formed, the mechanism may be modified by the presence of oxygen.

From a theoretical standpoint we should not expect oxygen to change the mechanism of carbon formation which we have adopted. Carbon will not normally be formed until the oxygen is consumed and we shall then have present, in addition to hydrocarbons, the products of oxidation which may include partial oxidation products such as aldehydes, alcohols, peroxides, ethers and ketones as well as carbon monoxide. All the former substances decompose thermally to carbon monoxide and hydrocarbons which will then undergo pyrolysis in the usual manner. Even at low temperatures there is no evidence that oxygen increases the rate of polymerisation relative to that of thermal decomposition.

There are two methods of studying the explosive reaction between hydrocarbons and oxygen as a function of time. Firstly, the reaction may be made into a flow system, of which the laminar flame is the simplest case. The reaction may then be studied as a function of time by investigating the flame as a function of distance along the dimension of propagation. The disadvantages of this system are twofold. Firstly the reaction zone, except at very low pressures, is very small in extent. Secondly the interpretation of the system is complicated by the fact that diffusion from one part of the flame to another is an important factor, and this is especially true at low pressures.

The second method is to study a homogeneous reaction in a closed vessel as a function of time. This eliminates the complications of diffusion and also makes possible the use of a very **long reaction path which enables intermediates which cannot be observed in flames to be detected** by absorption spectroscopy and similar methods. The method is one with considerable experimental difficulties however, the main ones being those associated with recording phenomena in a time short compared with that of the explosion and of producing an explosion which is homogeneous throughout the reaction vessel. The application of flash photolysis techniques provides a partial solution to these problems. By initiating the reaction photochemically throughout the body of the gas the explosion can be made to proceed reasonably homogeneously. The reaction may be followed in time by means of flash spectroscopy which has already been' **referred to,** and which makes possible a time resolution of about 50μ seconds. Alternatively photoelectric methods may be used which, although the **sensitivity** of detection of a discrete spectrum is less, have a response time of the order of one microsecond.

Norrish, Porter and Thrush (22,23) have used this technique to study explosions of various hydrocarbons with oxygen, the explosion being initiated by the photochemical decomposition of small amounts of nitrogen dioxide. The present author also carried out a preliminary investigation of explosions of acetone and ketene with oxygen. Most of this work has been described in detail elsewhere and only an outline of the main observations relating to carbon formation will therefore be given.

Owing to the long reaction path it is possible to observe the radicals C_2, CH, and, in the presence of nitrogen, NH and CN, in absorption. The only other discrete spectrum involving carbon which was observed was the 4050A group which is now assigned with some certainty to C_3. A further continuous spectrum was observed, with a characteristic maximum at 3900A, as well as a quite different continuum appearing at later times which was due to absorption by carbon itself. This latter continuum decayed slowly over several minutes as the carbon settled whilst the first continuum was only present for about 1 millisecond and had a time dependence similar to the radicals C_2 and C_3.

A study of these spectra as a function of time and mixture ratio leads to the following conclusions :

1. The observed concentrations of radicals C_2, C_3 and the 3900A spectrum (which is probably a higher carbon radical) are mainly those in equilibrium with solid carbon. It is found that their concentration time curves follow the temperature-time curves very closely and that their appearance, as a function of fuel or mixture ratio, shows little correlation with the amount of carbon formed but a close correlation with temperature. Thus C_2 appears at higher concentration in acetylene explosions but is always below the limit of detection in the cooler explosions of methane. It is quite impossible to reconcile our results with a mechanism of formation of carbon from C_2.

2. At a pressure of 20mm acetylene the absorption by carbon particles reaches a constant value 1.5 milliseconds after the beginning of observable reaction and the final form of the carbon is a roughly spherical particle 500A in diameter.

3. Small traces of oxygen added to ketene have no observable effect on the rate or extent of carbon formation. Larger amounts gradually decrease the amount of carbon formed in the same way as occurs in the explosions of hydrocarbons.

All our results are in accordance with the theory that carbon formation in explosions of hydrocarbons with oxygen occurs by pyrolysis of the hydrocarbon in the same way as in the absence of oxygen. The main function of the oxygen is to increase the temperature so that pyrolysis of excess hydrocarbons occurs very rapidly once the oxygen has been consumed. That pyrolysis occurs by the reaction of acetylene with the growing carbon chain is a necessary conclusion if it can be shown that the alternative mechanism via C_2 or intermediate polymers are not possible.

Exactly the same arguments apply to the elimination of these mechanisms in the presence of oxygen as in pure hydrocarbons. Thus carbon is formed in the above explosions when the total time that the gas exists at temperatures between 300 and 2500°K is less than 10^{-4} seconds and its formation is complete in 1.5 milliseconds. As previously shown, polymerisation of hydrocarbons in such a time, and at temperatures below the decomposition temperature is negligible. The inapplicability of the C_2 mechanism has already been referred to.

Stoichiometry of Carbon Formation.

An important point to be considered at this stage is the stoichiometry of hydrocarbon-oxygen mixtures which show carbon formation. We have the following overall **reactions** to consider :

8

$$\text{Hydrocarbon} = \text{Carbon} + \text{Hydrogen} \qquad (1)$$

$$\text{Oxygen} + \text{Hydrocarbon} = CO + H_2 + H_2O \qquad (2)$$

$$\text{Oxygen} + \text{Carbon} = CO \qquad (3)$$

$$H_2O + \text{Carbon} = CO + H_2 \qquad (4)$$

All these reactions are known to occur, and their relative rates in the explosion process and in the period available for equilibration before cooling will determine the products.

The relative rates of 1 and 2 will be determined mainly by the oxygen concentration. In our explosions no trace of absorption by carbon particles is present in oxygen rich mixtures so that, if reaction 1 occurs, the carbon particles are burned before they grow to detectable concentrations. The fact that carbon radicals are detected at low concentration in oxygen rich mixtures of all proportions may indicate that a small amount of reaction 1 occurs in all mixtures, the incipient growth of the carbon particles by reaction with acetylene being arrested by oxidation processes 3 and 4.

The question of stoichoimetry during the explosive process has also been investigated by Norrish Porter and Thrush [23]. The stoichiometric mixture corresponding to complete oxidation of carbon to CO during the actual explosion process is very clearly indicated by the radical concentration observed during the explosion. The peak radical concentrations appearing in explosions of varying mixture ratios of acetylene and oxygen, each mixture being sensitised by 1.5mm nitrogen dioxide, are shown in Fig. 3.

A very marked change in the concentration of all radicals occurs at about equimolecular proportions of acetylene and oxygen. In mixtures on the fuel rich side of this mixture ratio the carbon radicals rise rapidly in concentration, which indicates that carbon is being formed, as is in fact found. On the lean side OH rises rapidly in concentration owing to reaction between hydrogen and oxygen to give water. This characteristic change in radical concentration therefore gives the composition of the mixture in which the oxygen is just consumed and carbon formation begins. In the case of acetylene the mixture ratio corresponds to the stoichiometric equation:

$$C_2H_2 + O_2 = 2CO + H_2 \qquad (1)$$

It is well known that equimolecular mixtures of acetylene and oxygen do in fact give these products, and no carbon formation, but the fact that this ratio is also derived from measurements of radical concentration at the peak of the explosive reaction shows that this result is not merely due to the subsequent attainment of equilibrium in the hot products after the chain branching reaction is complete.

Further experiments were carried out in which part of the oxygen was replaced by water and it was found that water reacted almost exactly as its equivalent amount of oxygen i.e. $\frac{1}{2}O_2$. It follows that in acetylene explosions, any carbon formed is rapidly oxidised by reactions 3 and 4, during the explosion itself and that carbon formation only occurs when acetylene is in excess of the amount required by equation 1.

These facts suggest that the stoichiometry may be either a matter of preferential combustion of carbon or one of simple attainment of equilibrium. Neither of these theories is valid, however, as is shown immediately other hydrocarbons are investigated. The marked change in radical concentration as the mixture ratio is altered by a small amount is still observed in explosions of ethylene, ethane and methane but the mixture composition corresponding to this change is no longer that required for combustion to CO and H_2. The amount of oxygen increases with increasing saturation of the hydrocarbon until, in methane, nearly twice as much oxygen is present as would be required by this reaction, suggesting the equation

$$CH_4 + O_2 = CO + H_2O + H_2. \qquad (2)$$

It follows that, in mixtures slightly on the fuel rich side of this ratio, where both water and carbon are formed, the reaction between these two products does not occur. This was confirmed by addition of water to methane-oxygen mixtures which had no effect on the radical concentrations in complete contrast to acetylene explosions. For details of the exact determination of the stoichiometric ratios, and a discussion of other differences between the various hydrocarbons the original papers by Norrish, Porter and Thrush may be consulted (22,23).

These results recall the work of Bone (24) which was carried out many years ago and has never been adequately explained, Bone's own interpretations being based on the hydroxylation theory of combustion which has since lost favour. Bone found that, whilst explosions of methane and ethane containing oxygen slightly in excess of the quantity required to burn all the carbon to CO deposited carbon and formed water, the corresponding mixtures of ethylene and acetylene formed no carbon. Again it was found that the mixture $C_2H_2 + 2H_2 + O_2$ exploded to give only CO and H_2, whilst the exactly equivalent mixture $C_2H_6 + O_2$ gave copious quantities of carbon and water. Our results show that these rather unexpected results are not entirely a matter of approach to equilibrium of the hot gases after the explosion since almost identical phenomena are observed when the stoichiometry is estimated from the products present at the peak temperature. The equation 2 given above for methane is in fact identical with that given by Bone on the basis of analyses of the final products.

It is now apparent that the rates of reactions 1 to 4 are not so different that one can generalise about all hydrocarbons, indeed great differences are found in the four simple hydrocarbons referred to here. At present only the following indications of these differences are possible.

1. The more water is present in the final products of chain branching the leaner the mixtures in which carbon formation will occur.

2. Water formation will probably occur more readily in the saturated hydrocarbons, since oxygen cannot at first attack the carbon directly.

3. The amount of water formed during the explosive reaction is not the only factor involved since added water is removed during acetylene explosions, but not during methane explosions. This may be due to the higher temperature of the former. In particular the carbon radical concentrations are much higher in explosions of the unsaturated hydrocarbons and it is possible that the rapid reaction between carbon and water proceeds in the gas phase via carbon radicals.

We may try to summarise the conclusions of this section as follows : Carbon formation in explosions of hydrocarbons, premixed with oxygen, occurs by the same pyrolytic mechanism, as in the absence of oxygen. The relative rate of this mechanism compared with oxidation both of the hydrocarbon and of carbon itself determines the amount of carbon finally formed. In hot flames, and in explosions where cooling of the products is slow, carbon will be formed when oxygen is less than the amount required to burn the fuel to CO and H_2. In other cases carbon may be formed in leaner mixtures, depending on the rate of water formation and of its subsequent reduction.

Premixed flames.

Little will be said here on this subject since these flames have been discussed most thoroughly by Gaydon and Wolfhard. We shall simply note that there is no reason to regard our homogeneous explosions as being chemically different from the premixed flame, the latter being the flow system and the former the static system. If this interpretation is correct, the above conclusions must apply also to premixed flames. Two minor qualifications should perhaps be made :

1. The rate of cooling of an explosion in a closed vessel is very rapid compared with that of the burned gases in a flame, so that differences may occur due to a greater degree of establishment of equilibrium in the flame gases.

2. Much confusion is caused by the fact that many so called premixed flames involve a considerable amount of diffusion. Thus the carbon formed in a Bunsen flame burning a rich mixture is eventually oxidised by the entrained air. A less obvious source of difficulty is that complications

may arise owing to changes in mixture strengh near the reaction zone owing to differing rates of diffusion of the different species.

Critique of other proposed mechanisms.

Most other theories of carbon formation have been discussed in developing the arguments for the acetylene theory; these being in particular the C_2 polymerisation theory and the hydrocarbon polymer theory. There is one other mechanism which has been proposed which we must consider, since its author has supported it not as an alternative to the theory given here, but as a necessary addition. This theory, proposed by Behrens (25) states that carbon in flames is formed by the Boudouard equilibrium.

$$2CO = C + CO_2 + 40Kcal \quad (1)$$

which, he suggests, proceeds by the mechanism

$$CO + OH = CO_2 + H \qquad (2)$$

$$CO + \ H = C_{solid} + OH \quad (3)$$

We may immediately discount the detailed scheme involving reaction 3 on energetic grounds. Reaction 3 is meaningless unless it implies that a carbon atom is formed, which subsequently polymerises, and the formation of such an atom by reaction 3 would be endothermic by 120K. cal/mole or more, depending on the value adopted for D (CO). At 3000° K this leads to an exponential rate factor less than 10^{-9} and a reaction time of the order of seconds in flames at 1 atmosphere pressure.

If we consider reaction 1 without enquiring into the detailed mechanism we are faced with the difficulty that the equilibrium mixture consists almost entirely of carbon monoxide unless the temperature is less than 1400° K. The equilibrium constants, defined by $K_p = (CO)^2/(CO_2)$, are as follows :

T° K	K_p (atm.)
700	7.4×10^{-8}
1000	3,65
1400	$3,9 \times 10^{5}$
2000	1.7×10^{9}

Not only is carbon formed at much higher temperatures than this, but it is usually found from analysis of the products, that the whole reaction is quenched at temperatures around 1200° C. (1) Any detailed mechanism for reaction 1 would have to proceed via C atoms and would again be far too endothermic to occur at these low temperatures. Furthermore, if this equilibrium is to lead to large quantities of carbon at low temperatures it is difficult to see why carbon is not formed in flame gases consisting almost exclusively of CO and hydrogen.

More recently Behrens (26) has accepted the mechanism of the present author for the initiation of carbon formation but retains the Boudouard mechanism for the subsequent particle growth. One might get over the equilibrium difficulties by supposing that the CO_2 concentration was kept very low by rapid reaction between fuel molecules and any carbon dioxide formed by reaction 1 ; the amount of carbon present would not of course affect the equilibrium. Reaction 1 is then retained only as a detailed mechanism rather than a thermodynamical necessity and the theory now states that carbon formation from hydrocarbons proceeds via a collision between two carbon monoxide molecules, which can only from a carbom atom. There seems to be no evidence for this improbable view; it certainly cannot apply in pyrolysis or diffusion flames and the carbon particles formed under such conditions are remarkably similar in structure to those formed in premixed flames and explosions. One of Behren's arguments against the acetylene theory is based on the position of the carbon zone in different flames. In the present author's view such observations which are based on luminosity, only become meaningful when the particle size, its emissivity,the temperature and particularly the exact stoichiometry in each part of the flame have been considered.

Particle size and structure.

The carbon particles formed during combustion and homogeneous gas phase pyrolysis are very similar. They consist of agglomerates of smaller particles which are roughly spherical and between 100 and 1000A in diameter. This alone is strong evidence that the same mechanism is common to all these processes, indeed it is remarkable that a greater variation in particle size is not found in view of the widely different conditions of concentration and temperature. The particle size, on the basis of the acetylene theory, will be determined by the rates of initiation and propagation of the chains and by the concentration of acetylene. The latter will itself depend on the rates of the various stages of pyrolysis. Each will depend on the detailed temperature structure of the flame zones where pyrolysis occurs. There is therefore no difficulty in accounting for minor differences of particle structure, but a detailed investigation of the effect of these various parameters, on particle size, would be a valuable means of obtaining a better understanding of the polymerisation process.

It is with pleasure that I express my indebtedness to Professor R.G.W.Norrish, Dr.B.A.Thrush, and Dr.K.Knox in collaboration with whom most of the experimental work described in this paper has been carried out for many useful discussions on the problems here discussed.

REFERENCES

1. Gaydon A.G. and Wolfhard H.G. Flames. Chapman and Hall. 1953.
2. Porter G.   4th. Combustion Symposium. Williams and Wilkins 1952.p248.
3. Porter G.   Proc.Roy.Soc. 1950.A200 284.
4. Glasstone S. Laidler.K.J. and Eyring H. Theory of Rate Processes. McGraw-Hill. 1941.
5. Kassel L.S. J.Am.Chem.Soc. 1932.54. 3949.
6. Sachsse H.Z.physic.Chem. 1935.B31.87
7. Steacie E.W.R. Atomic and Free Radical Reaction. Reinhold. 1946.
8. Szwarc M.   J.Chem.Phys. 1949.17.284.
9. Taylor H.A. and van Hook.A. J.Phys.Chem. 1935.39.811.
10. Parker W.G. and Wolfhard H.G. J.Chem.Soc. 1950.2038.
11. Zanetti J.E. and Egloff G. Ind.Eng.Chem. 1917.9.350.
12. Storch H.H. J.Am.Chem.Soc. 1932.54.4185.
13. Tropsch H and Egloff G. Ind.Eng.Chem. 1935.27.1063.
14. Stubbs F.J. and Hinshelwood C.N. Discussions Faraday Soc. 1951.10.129.
15. Egloff G.   Reactions of Pure Hydrocarbons. A.C.S.Monograph 1937.
16. Storch H.H. J.Am.Chem.Soc. 1932.54.4185.
17. Knox   K.   Norrish.R.G.W. and Porter.G. J.Chem.Soc. 1952.1477.
18. Bone   W.A. and Coward H.F. J.Chem.Soc. 1908.93.1197.
19. Gaydon A.G. and Wolfhard H.G. Proc.Roy.Soc. 1950.A201.570.
20. Wolfhard H.G. and Parker W.G. Proc.Phys.Soc. 1952.65.2.
21. Thrush B.A. Thesis. Cambridge University. 1953.
22. Norrish R.G.W. Porter.G. and Thrush.B.A. Proc.Roy.Soc. 1953.A216.165.
23. Norrish R.G.W. Porter G. and Thrush.B.A. Proc.Roy.Soc. In course of publication.
24. Bone   W.A. and Townend D.T.A. Flame and Combustion in Gases, Longmans, Green & Co.1927.
25. Behrens H. 4th. Combustion Symposium. Wiliams and Wilkins. 1952.p.538.
26. Behrens H. ibid. p.252.

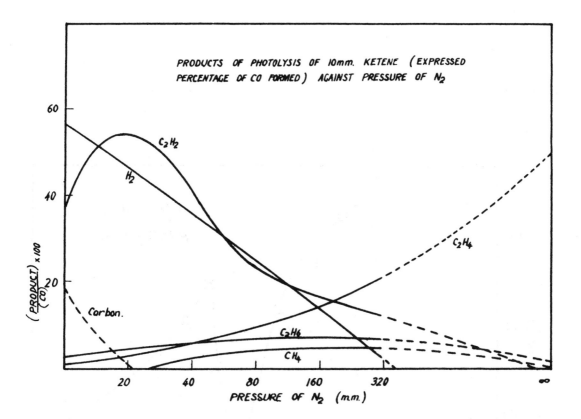

Fig. 1.

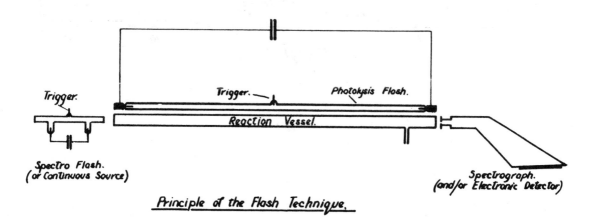

Fig. 2.

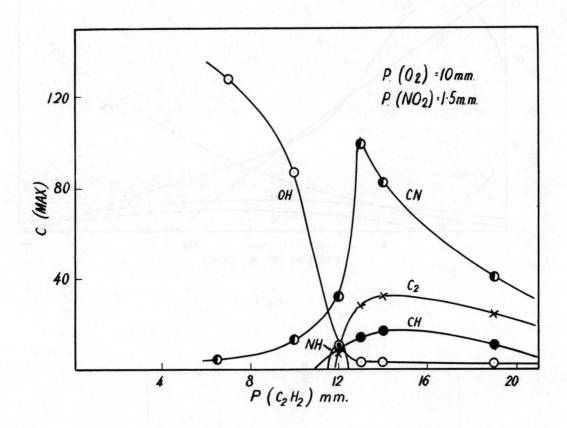

Fig. 3

Further publications

DETECTION OF DIATOMIC RADICAL ABSORPTION SPECTRA
DURING COMBUSTION.
R. G. W. Norrish, G. Porter and B. A. Thrush, *Nature*, 1952, **169**, 582.

LIMITING PRODUCTS IN THE PHOTOLYSIS OF
ACETALDEHYDE AT HIGH INTENSITY.
M. A. Khan, R. G. W. Norrish and G. Porter, *Nature*, 1953, **171**, 513.

THE PHOTOLYSIS OF ACETALDEHYDE, DIACETYL AND ACETONE
AT HIGH INTENSITY.
M. A. Khan, R. G. W. Norrish and G. Porter, *Proc. Roy. Soc.*, 1953, **A219**, 312.

KINETIC STUDIES OF GASEOUS EXPLOSIONS.
R. G. W. Norrish, G. Porter and B. A. Thrush, Fifth International Symposium on
Combustion, 1954, p. 651.

STUDIES OF THE EXPLOSIVE COMBUSTION OF
HYDROCARBONS BY KINETIC SPECTROSCOPY
II. COMPARATIVE INVESTIGATIONS OF HYDROCARBONS AND
A STUDY OF THE CONTINUOUS ABSORPTION SPECTRA.
R. G. W. Norrish, G. Porter and B. A. Thrush, *Proc. Roy. Soc.*, 1955, **A227**, 423.

SPECTROSCOPIC STUDIES OF THE HYDROGEN-OXYGEN
EXPLOSION INITIATED BY THE FLASH PHOTOLYSIS OF
NITROGEN DIOXIDE.
R. G. W. Norrish and G. Porter, *Proc. Roy. Soc.*, 1952, **A210**, 439.

THE PHOTOCHEMICAL DECOMPOSITION OF KETEN
BY MEANS OF LIGHT OF VERY HIGH INTENSITY.
K. Knox, R. G. W. Norrish and G. Porter, *J. Chem. Soc.*, 1952, 1477.
R. G. W. Norrish and G. Porter, *Proc. Roy. Soc.*, 1952, **A210**, 439.

CARBON FORMATION IN THE COMBUSTION WAVE.
G. Porter, Fourth Symposium (International) on Combustion, M.I.T.,
The Williams and Wilkins Company, Baltimore, 1953, p. 248.

Chapter 3

ATOM RECOMBINATION

In the first flash photolysis studies it was observed that chlorine gas, as well the other halogens, were bleached reversibly by the flash. It was known that the recombination of atoms must require a third body if the product was to be stabilised, although this expected termolecular reaction had not been observed directly. Extensive studies of iodine vapour in the presence of other (inert) gases fully confirmed expectations but also revealed some less predicable features of these reactions. The third order recombination rate constants varied by a factor of more than a thousand between different third body (chaperone) molecules, one of the most effective chaperones being the iodine molecule itself. The temperature coefficients of recombination were measured and found to be negative (corresponding to negative "activation energies") in all cases and there was a fairly close proportionality between this negative temperature coefficient and the logarithm of the third-order rate constant. All these observations are interpreted in terms of the involvement of an inter-mediate complex formed from, and in equilibrium with, one iodine atom and the third body.

Extracted from the Proceedings of the Royal Society, A, *volume* 216, pp. 152, 153, 156–165, 1952

The recombination of atoms
I. Iodine atoms in the rare gases

By M. I. Christie, R. G. W. Norrish, F.R.S., and G. Porter

(*Received* 16 *August* 1952)

The recombination of iodine atoms, in the presence of five rare gases, has been studied directly by means of the flash technique. The termolecular rate law

$$-d[I]/dt = k[I]^2[M]$$

is accurately obeyed and the absolute rate constants k, at 20° C, are 1·73, 1·86, 2·42, 3·41 and $3·44 \times 10^{-32}$ ml.² molecules^{-2} s^{-1} for $M =$ He, Ne, A, Kr and Xe respectively.

These data are briefly discussed in terms of the various theories of termolecular reactions.

It is to be expected on energetic grounds (Polanyi 1920), and it now seems well established experimentally, that a third body is necessary to stabilize the collision complex formed by the union of two atoms. Similar restrictions may also apply to the recombination of simple free radicals, though the experimental data are at present very uncertain.

The existing information on atomic recombination reactions has been obtained by a variety of means. The recombination of hydrogen atoms, produced by the electrical discharge, has been studied by flow methods (Smallwood 1929; Senftleben & Riechemeier 1930; Steiner 1932; Amdur & Robinson 1933; Farkas & Sachsse 1934). The rate of bromine atom recombination has been deduced from the kinetics and the overall rate of the hydrogen-bromine reaction (Bodenstein & Lütkemeyer 1925; Jost & Jung 1929; Ritchie 1934; Hilferding & Steiner 1935). The most complete data are those obtained by Rabinowitch & Lehmann (1935) and Rabinowitch & Wood (1936*a, b*) in their classical work on the rate of recombination of bromine and iodine atoms in a number of different foreign gases. Their method involved a photo-electric determination of the concentration of halogen molecules in a photochemical stationary state, from which, knowing the rate of light absorption and the quantum yield, and after making some corrections for thermal and heterogeneous effects, they were able to calculate the absolute rate constants of the homogeneous recombination. Although a number of assumptions were involved in this work, their results are probably the most reliable at present available.

Theoretical discussions of the termolecular recombination process have been given by Wigner (1937, 1939), Rabinowitch (1937), Rice (1941) and Careri (1949, 1950), all of whom considered the experimental values of the rate constants obtained by Rabinowitch and his co-workers. As far as absolute magnitudes are concerned, the errors in the experimental values are probably less than those involved in any calculations which can be made at present. In considering the relative values between one third body gas and another, however, better agreement between experimental and calculated values might be expected. This does not prove to be the case, the calculated differences being smaller than those found experimentally. Moreover, the relative values determined by the different experimental methods vary widely, Rabinowitch, for example, finding that helium was half as effective as argon in the recombination of bromine atoms, whilst Hilferding & Steiner found that helium was four times as effective as argon.

Even in the simplest case, namely, that of the series of inert gases, the factors which determine the relative efficiency of different gases as third bodies are at present unknown, and more accurate data on the rate constants of these termolecular reactions, over a wide range of conditions and for a large number of gases, are required before a correlation with the other physical properties of the gases can be obtained. Such measurements seem to be a promising way of elucidating the mechanism of energy transfer between two colliding gas molecules.

The flash technique developed in this laboratory is suitable for the direct investigation of rapid physical and chemical changes of this type. In addition to the advantages of easy interpretation and the elimination of uncertainties of reaction mechanism, heterogeneous reactions, etc., it will be shown to lead to a more accurate evaluation of absolute rate constants. The investigations described in this paper deal with the simplest systems for theoretical treatment, three atom complexes. The recombination of iodine atoms in the rare gases is studied, iodine being chosen first because of its high molecular extinction and low atomic reactivity. Subsequent parts will describe atomic recombination in polyatomic gases, temperature dependence, etc., and similar work is in progress on free radical reactions (see, for example, Porter & Wright 1953).

M. I. Christie, R. G. W. Norrish and G. Porter 156

Typical oscillograph traces are shown in figure 3. It was not possible, with the present optical arrangement, completely to eliminate scattered light from the flash, and the first trace shows the contribution from this source alone. Measurements of concentration were taken after the time at which the scan returns to zero in the 'blank'. The other traces in figure 3 show clearly the acceleration of atomic recombination with increasing pressure of argon.

If the rate of the recombination reaction

$$I + I + M = I_2 + M^x$$

is of the second order with respect to iodine atom concentration and directly proportional to inert gas concentration

$$-d[I]/dt = k[I]^2[M]$$

and $$1/[I]_t - 1/[I]_0 = k[M]t,$$

157   *The recombination of atoms. I*

where $[I]_0$ is the initial concentration of iodine atoms, $[I]_t$ is the concentration at time t, $[M]$ is the concentration of inert gas, $[I_2]$ and $[I]$ being relatively negligible as third bodies, and k is the termolecular rate constant.

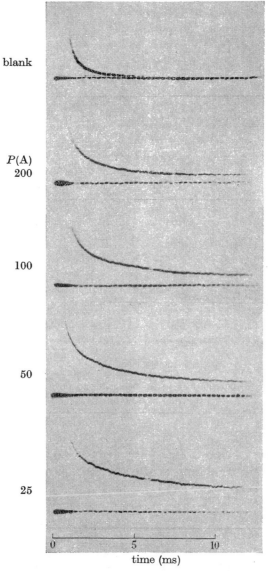

FIGURE 3. Oscillographic traces showing iodine recombination in argon.
$P(A) =$ argon pressure in mm Hg.

In figure 4 the relationship between $1/[I]_t$ and t is shown for five different pressures of argon, and the linearity of these plots shows that the second-order dependence on $[I]$ is accurately obeyed for all values of M between 25 and 200 mm Hg. The values of k, calculated from the gradients of figure 4, are as follows:

argon pressure (mm Hg)	25	50	100	150	200
$10^{32} k$ (ml.^{2} molecules^{-2} s^{-1})	2·45	2·42	2·35	2·37	2·42

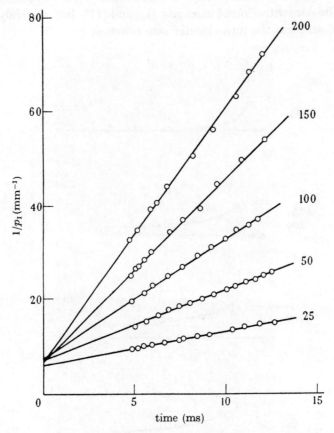

FIGURE 4. Bimolecular plots of iodine recombination in argon at pressure
from 25 to 200 mm Hg.

TABLE 1. RATE CONSTANTS, $k \times 10^{32}$, AT $20°$ C (CONCENTRATIONS
EXPRESSED IN MOLECULES/ML.)

gas	helium	neon	argon		krypton	xenon
pressure (mm Hg)			931A cell	IP 28 cell		
25	—	—	2·45	2·37	—	—
50	—	1·84	2·42	2·54	3·37	3·46
75	1·73	1·95	—	2·52	3·45	3·52
100	1·79	1·90	2·35	2·37	3·51	3·35
125	1·65	1·84	—	2·43	3·32	3·49
150	1·74	1·85	2·37	2·44	3·37	3·49
200	1·75	1·79	2·42	2·35	3·42	3·33
400	1·71	—	—	—	—	—
mean	1·73	1·86	2·42		3·41	3·44
standard deviation	0·04	0·05	0·06		0·06	0·07
coefficient of variation (%)	2·3	2·7	2·5		1·8	2·0

159   *The recombination of atoms. I*

The values of k are constant, with a coefficient of variation of 2 %, confirming the direct proportionality of rate with inert gas concentration.

The exact value of $1/I_0$, which is not required for our purpose, depends on the output of each separate flash. The average value corresponds to about 50 % decomposition into atoms with an initial pressure of I_2 of 0·16 mm Hg.

The influence of the nature of the inert gas on the rate is illustrated in figure 5, which compares the rate of recombination in the five rare gases, the pressure being 100 mm Hg in each case. A series of pressures was investigated for all five gases, and our complete results are summarized in table 1. At least four separate recordings were used in the rate determinations at each pressure. At all pressures for which rate constants are tabulated linear bimolecular plots were obtained throughout the recombination reaction. The two series of results for argon were obtained with two photocells of types IP 28 and 931 A respectively.

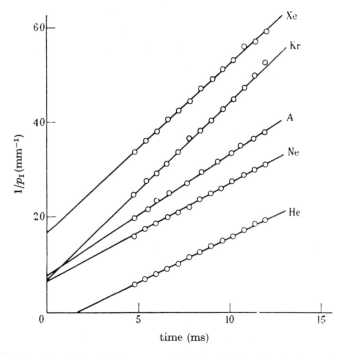

FIGURE 5. Bimolecular plots of iodine recombination in the five inert gases at 100 mm Hg pressure. (The $1/p_I$ ordinates for He and Xe are displaced by -10 and $+10$ mm^{-1} respectively.)

Assessment of errors

The possibilities of error were as follows:

(1) *Iodine molecule concentration and trace calibration*

Gillespie & Fraser (1936) claim an accuracy to $\pm 0·14$ % for their vapour-pressure formula. Our thermometer, calibrated against an N.P.L. standard, was read to $\pm 0·02°$ C. The possible error in the measurement of iodine concentration was therefore ± 1 %. The random error between similar determinations of trace deflexion in terms of iodine concentration was $\pm 2·5$ %, which includes the foregoing 1 %. There is also the possibility of a systematic error arising from deviations from Beer's law. The accuracy to which the validity of this law was confirmed experimentally leads to a possible maximum error in all the values of k of $\pm 5 \%$.

(2) *Time calibration*

Estimated error $\pm \frac{1}{2} \%$.

(3) *Inert gas*

No significant error arises from the pressure measurement, and the purity of the gases was high. The possibility of errors due to traces of air in the rare gases was considered and the rate constant of iodine atom recombination in laboratory air was therefore measured. It was found to be $4 \cdot 5 \times 10^{-32}$ ml.^{2} molecules^{-2}s^{-1}, and we conclude that no error from this source is possible.

(4) *Effects of inhomogeneity, heterogeneous recombination and temperature variation*

The construction of the lamp, and the use of a highly efficient reflector, assured that the illumination was constant throughout the gas volume, the light absorbed in a single diametrical traverse across the cell being only 2 %. A contribution from wall recombination and several complications owing to heat liberation are to be expected at low gas pressures. In helium, small deviations from these causes were observed towards the end of the recombination with pressures below 50 mm Hg. No such deviations occurred at any of the pressures used for the data of table 1, and as such effects are negligible at the highest pressures, and only dependent on total pressure, the constancy of the rate values is confirmation that such effects were insignificant.

(5) *Determination of gradient*

Estimated error $\pm 1 \%$.

The observed deviations in the values of k are commensurate with the random errors arising from (1) and (5). The possible systematic error of $\pm 5 \%$ arising from (1), and the $\pm \frac{1}{2} \%$ from (2), should be nearly constant for all the five gases.

DISCUSSION

Our results offer direct proof of the validity of the termolecular rate law for the recombination of iodine atoms in the rare gases over the whole range of pressures investigated. The only existing data on the absolute values of these rate constants are those of Rabinowitch & Wood (1936 a, b, c) for both helium and argon as third bodies, and a value for argon, published while our work was in progress, by Davidson, Marshall, Larsh & Carrington (1951). The latter authors have not yet given details of their method, but they used the flash technique and an apparatus which was clearly very similar to that described here. The data from these sources are compared in table 2. The identity, within the stated error, of the two values obtained by the flash method is reassuring, especially as the American workers used higher gas pressures and only 4 % decomposition. The general agreement with the results of Rabinowitch & Wood is very satisfactory, since the method of these earlier investigators required the accurate measurement of a relatively small change Their value for helium is expected to be the more accurate owing to smaller thermal effects.

It has been stated (Eyring & Rollefson 1932), on the basis of calculations by the transition state method, that the trihalogen molecules, such as I_3, have considerable stability, and therefore the iodine molecule should have a very specific effect in facilitating the recombination of iodine atoms. No anomalies such as would arise from this cause were detected, and an upper limit for the rate constant of the reaction

$$I + I + I_2 = I_2 + I_2 \qquad (1)$$

can be obtained from our results as follows. At the lowest inert gas pressure (25 mm Hg) the ratio of iodine to total pressure was greater than 1/200, and a deviation of 4 % from the rate constants observed at higher pressures would have been significant. Using the results for argon, we find that an upper limit for k_1 is 20×10^{-32} ml.^{2} molecules^{-2} s^{-1}. This limiting value does not indicate any very great specific action of the iodine molecule.

TABLE 2. COMPARISON OF EXPERIMENTAL VALUES OF k, IN
ML.^{2} MOLECULES^{-2} S$^{-1} \times 10^{32}$ FOR HELIUM AND ARGON

	He	A
Rabinowitch & Wood	$1 \cdot 8 \ \pm 0 \cdot 1$	$3 \cdot 6 \ \pm 0 \cdot 3$
Davidson and co-workers	—	$2 \cdot 5 \ \pm 0 \cdot 3$
this work	$1 \cdot 73 \pm 0 \cdot 04$	$2 \cdot 42 \pm 0 \cdot 06$

Comparison with theory

For the purpose of discussion we shall adopt the gas kinetic collision diameters given in table 3 (Partington 1952). In accordance with the usual practice, the value for the iodine atom is taken to be slightly greater than for the next rare gas. In the second column the values are given of the reciprocal collision efficiency $\alpha_{exp.}$ of the process

$$I + I = I_2$$

in the five rare gases at 1 atm pressure calculated by the kinetic theory expression for bimolecular collisions (Maxwell 1860). These values may first be compared with those obtained by use of the simple expression suggested by Bodenstein (1922) for calculating the ratio of the number of bimolecular collisions $Z_{bi.}$ to termolecular collisions $Z_{ter.}$, viz.

$$Z_{bi.}/Z_{ter.} = \sigma/\lambda = \sqrt(2)\,\pi N_M \sigma^3,$$

where λ is the mean free path. The appropriate σ to be used here is $\sigma_{IM}\left(= \dfrac{\sigma_I}{2} + \dfrac{\sigma_M}{2}\right)$

and if every termolecular collision leads to recombination then $Z_{bi.}/Z_{ter.} = \alpha_{calc.}$. The calculated and experimental values for 1 atm pressure are compared in table 3. The absolute values calculated are greater than the experimental ones by a factor of about 4. This is not serious, for even if the expression gives the true collision ratio, which is doubtful in view of its approximate nature, the discrepancy is not large when the probable steric and energy transfer requirements are taken into account. The fact that the experimental and calculated values are in good relative agreement for the five gases is more significant and implies an approximate proportionality of rate to σ_{IM}^3, which is shown more clearly by the ratios σ_{IM}^3/k given in line 4.

TABLE 3

	He	Ne	A	Kr	Xe	I
σ_{MM}	2·18	2·59	3·64	4·16	4·85	5·2
$\alpha_{exp.}$ (1 atm)	613	570	438	311	308	—
$\alpha_{calc.} = \dfrac{\sigma_{IM}}{\lambda}$ (1 atm)	178	152	104	87	71	—
$\dfrac{\sigma_{IM}^3}{k} \times 10^{-8}$	29	32	36	30	37	—

We may now consider the consecutive collisions

$$I + I = I_2^*,$$
$$I_2^* + M = I_2 + M^x,$$

where M is the inert gas atom. For this process Steiner (1932), following Herzfeld (1922), calculated the number, $Z_{ter.}$, of termolecular collisions from the bimolecular collision frequency of I_2^* with M, the concentration of I_2^* being τZ_{II}, where τ is the average lifetime of the I_2^* complex and Z_{II} the bimolecular collision frequency between iodine atoms. This leads to the expression

$$Z_{ter.} = 4\pi \sqrt{2} \, [N_I]^2 \, [N_M] \tau \sigma_{II}^2 \, \sigma_{I_2^* M}^2 RT \sqrt{\frac{M_M + M_{I_2^*}}{M_I M_M M_{I_2^*}}}, \qquad (A)$$

where $\sigma_{I_2^* M}$ is the collision diameter for an encounter between I_2^* and M. It is not clear what is the appropriate value of $\sigma_{I_2^*}$ to be used in this calculation, and Rabinowitch (1937) avoided this difficulty by considering, for the second stage, the collision between M and the two iodine atoms in the complex which results in the following slightly different expression

$$Z_{ter.} = 4\pi \sqrt{2} \, [N_I]^2 \, [N_M] \sigma_{II}^2 \, \sigma_{IM}^2 \frac{RT}{M_I} \sqrt{\left(\frac{M_I + M_M}{M_M}\right)} 2\tau. \qquad (B)$$

On the basis of these theories τ should be a constant independent of the rare gas, and, using the experimental rate constant in argon to determine τ, the rate constants may be calculated for the other gases. The results of this calculation, using expression (B), are given in line 2 of table 4, and it will be seen that, owing to the importance of the mass term, the calculated rate constants decrease with increasing atomic weight of the rare gas. The values of 2τ calculated from the experimental rate constants for each gas are given in line 3. These are comparable with the time of a molecular vibration, as would be expected, but they are not constant, and cannot therefore refer to the true lifetime of I_2^*, showing that, in this form, the theory is inadequate.

Rabinowitch has also considered the two types of triple collision (a) $(I' + I'') + M$, where the first double collision is that of two iodine atoms, and (b) $(I' + M) + I''$, where the first collision is that of an iodine and inert gas atom. The observed recombination coefficient k should be given by

$$k = 8\pi \sqrt{2} \, \sigma_{II}^2 \sigma_{IM}^2 \frac{RT}{M_I} \sqrt{\left(\frac{M_I + M_M}{M_M}\right)} (2\tau_{II} + \tau_{IM})\beta,$$

where τ_{II} is the collision period of the double collision in (a), τ_{IM} is the collision period of the double collision in (b), and β is the fraction of triple collisions which result in recombination. The experimental values of $(2\tau_{II} + \tau_{IM})\beta$ which are identical with the values of $2\tau(\text{exp.})$ calculated from formula (B) (table 4, line 3) are compared with the values of $2\tau_{II} + \tau_{IM}$ (table 4, line 5), estimated by a method suggested by Rabinowitch, using the Sutherland constant, obtained from the temperature coefficient of viscosity, as a measure of the 'softness' of a molecule and therefore of collision duration. It seems doubtful whether the value of diameter correction, $\Delta\sigma$, derived in this way is a quantitative measure of the path traversed during a collision, as suggested by Rabinowitch, but the values obtained by dividing $\Delta\sigma$ by the mean relative velocity should certainly be an indication of the *relative* collision duration for different gases.

TABLE 4

	He	Ne	A	Kr	Xe
$k_{\text{exp.}} \times 10^{32}$	1·73	1·86	2·42	3·41	3·44
$k_{\text{calc.}} \times 10^{32}$ (collision formula)	4·73	2·49	(2·42)	2·12	2·16
$2\tau_{\text{exp.}} \times 10^{13} = (2\tau_{II} + \tau_{IM})\beta$ (Rabinowitch)	1·21	2·47	3·30	5·35	5·26
$\tau_{IM} \times 10^{13}$ (calc. from Sutherland's constant)	1·30	2·69	4·53	6·43	8·70
$2\tau_{II} + \tau_{IM} \times 10^{13}$	19·24	20·63	22·47	24·37	26·64
$k_{\text{calc.}} \times 10^{32}$ (Wigner 1)	2·6	2·8	3·5	3·8	4·3
$k_{\text{calc.}} \times 10^{32}$ (Wigner 2)	7·0	7·7	9·4	10·3	11·5

Our results show that β defined in this way is not constant throughout the series of inert gases. A more general derivation would consider the possibility of different values of β for the two types of collision, leading to the replacement of 2τ by $2\tau_{II}\beta_1 + \tau_{IM}\beta_2$. In this case our results indicate that β_1 is considerably less than β_2, the best values, if the Sutherland constant calculations are adopted, being $\beta_1 = 0\cdot04$ and $\beta_2 = 0\cdot7$.

Rice (1941) calculated the effective collision diameters of iodine molecules with inert gas molecules from the experimental rate constants of the association reaction and thermodynamic data. Results of the right order are obtained after insertion of a somewhat arbitrary numerical factor, but the predicted relative effect of the different gases is automatically the same as that obtained by direct use of the kinetic theory expression for the association reaction (table 4, line 2).

Finally, our results may be compared with calculations of absolute rates by the method of Wigner (1937). Beginning with the general equations of motion, Wigner derived an expression for the probability of decrease of the relative energy of two atoms below zero, under the influence of a third body. In the case of two iodine atoms and a rare gas atom M his expression reduces to

$$k = -2\pi \left(\frac{\pi}{mkT}\right)^{\frac{1}{2}} \left[\int_{V_0 < 0} 2V_0 q_{II}^2 \sigma_{IM}^2 \, dq_{II} + \int_{V_0 < 0} V_0 q_{II}^3 \sigma_{IM} \, dq_{II} \right],$$

where V_0 is the relative potential energy of the two iodine atoms for infinite separation of M, m is the mass of the iodine atom, σ_{IM} is the collision diameter of an iodine atom with M, and q_{II} is the separation of two iodine atoms. The integrals were evaluated by Wigner from the Morse curves derived from spectroscopic data on the iodine molecule. The values of k calculated in this way, using the collision diameters of table 3, are given in table 4. The figures in line 6 are derived by considering only the ground state of the iodine molecule and dividing by a statistical weight factor of 16. The values in line 7 take account of three other attractive states of I_2 which can be formed from $^2P_{\frac{3}{2}}$ iodine atoms, each of statistical weight 2, the integrals for these states having also been given by Wigner (1939).

The main uncertainty in these calculations is in the choice of the appropriate statistical factor. In addition, the method leads only to an upper limit, and whilst the absolute values must be high, the relative values should be in good agreement with experiment for three atom problems. A comparison with our experimental values shows that the relative values are in fairly satisfactory accord with experiment.

We may summarize this comparison of the present experimental results with existing theories by the statement that, while no theory is sufficiently comprehensive to give more than an approximation to the absolute values of the rate constants, the Wigner theory and the simple expression of Bodenstein predict the correct relative effects of the different rare gases within the uncertainty of the gas kinetic collision diameters. The kinetic theory expression for termolecular collisions is inadequate in its simple form and requires a special interpretation of the meaning of collision duration.

One of us (M.I.C.) is grateful to the Carnegie Trust for the Universities of Scotland for the award of a Senior Scholarship.

REFERENCES

Amdur, I. & Robinson, A. L. 1933 *J. Amer. Chem. Soc.* 55, 1395, 2615.
Bodenstein, M. 1922 *Z. phys. Chem.* 100, 118.
Bodenstein, M. & Lütkemeyer, H. 1925 *Z. phys. Chem.* 114, 208.
Careri, G. 1949 *Nuovo Cim.* 6, 94.
Careri, G. 1950 *Nuovo Cim.* 7, 155.
Christie, M. I. & Porter, G. 1952 *Proc. Roy. Soc.* A, 212, 398.
Davidson, N., Marshall, R., Larsh, A. E. & Carrington, T. 1951 *J. Chem. Phys.* 19, 1311.
Eyring, H. & Rollefson, G. K. 1932 *J. Amer. Chem. Soc.* 54, 170.
Farkas, L. & Sachsse, H. 1934 *Z. phys. Chem.* B, 27, 111.
Gillespie. L. J. & Fraser, L. H. D. 1936 *J. Amer. Chem. Soc.* 58, 2260.
Herzfeld, K. F. 1922 *Z. Phys.* 8, 132.
Hilferding, K. & Steiner, W. 1935 *Z. phys. Chem.* B, 30, 399.
Jost, W. & Jung, G. 1929 *Z. phys. Chem.* B, 3, 83.
Maxwell, C. R. 1860 *Phil. Mag.* 19, 19.
Partington, J. R. 1952 *An advanced treatise on physical chemistry*, 1, 858. London: Longmans, Green and Co.
Polanyi, M. 1920 *Z. Phys.* 1, 337.
Porter, G. 1950a *Proc. Roy. Soc.* A, 200, 284.
Porter, G. 1950b *Disc. Faraday Soc.* 9, 60.
Porter, G. & Wright, F. J. 1953 *Disc. Faraday Soc.* no. 14 (in the Press).
Rabinowitch, E. 1937 *Trans. Faraday Soc.* 33, 283.
Rabinowitch, E. & Lehmann, H. L. 1935 *Trans. Faraday Soc.* 31, 689.
Rabinowitch, E. & Wood, W. C. 1936a *Trans. Faraday Soc.* 32, 907.
Rabinowitch, E. & Wood, W. C. 1936b *J. Chem. Phys.* 4, 497.

The recombination of atoms. I

Rabinowitch, E. & Wood, W. C. 1936c *Trans. Faraday Soc.* **32**, 540.

Rice, O. K. 1941 *J. Chem. Phys.* **9**, 258.

Ritchie, M. 1934 *Proc. Roy. Soc.* A, **146**, 828.

Senftleben, H. & Riechemeier, O. 1930 *Ann. Phys., Lpz.,* **6**, 105.

Smallwood, H. M. 1929 *J. Amer. Chem. Soc.* **51**, 1985.

Steiner, W. 1932 *Z. phys. Chem.* B, **15**, 249.

Wigner, E. P. 1937 *J. Chem. Phys.* **5**, 720.

Wigner, E. P. 1939 *J. Chem. Phys.* **7**, 646.

Reprinted from the Proceedings of the Royal Society, A, *volume* 261, pp. 28–37, 1961

The recombination of atoms

III. Temperature coefficients of iodine atom recombination

By G. Porter, F.R.S. and J. A. Smith

Department of Chemistry, University of Sheffield

(*Received* 11 *November* 1960)

The temperature coefficients of iodine atom recombination rates have been measured in eight gases having widely different efficiencies as third bodies or chaperons. If the rate constant is expressed in Arrhenius form, the activation energies, which are negative, decrease as the chaperon efficiency increases and it is the exponential term which principally determines the chaperon efficiency.

A priori statistical calculations based on intermediate complex formation predict values of the pre-exponential term which are in reasonably good absolute and relative agreement with experiment. The heat of formation of the complex is much greater than that which could result from normal van der Waals type interactions and the recombination is interpreted in terms of charge transfer complex formation between chaperon and iodine atom.

Allowance for the two effects described in part II (Christie, Harrison, Norrish & Porter 1955) namely, the high efficiency of I_2 as a third body and the disturbances caused by temperature inhomogeneity, have reconciled the differences between rate constants determined by different schools and there is now excellent agreement on the values for iodine atom recombination in the presence of the inert gases. Further theories of atom recombination have been proposed (Keck 1958; Bunker & Davidson 1958; Husain & Pritchard 1959; Jepson & Hirschfelder 1959; Keck 1960; Bunker 1960) most of which predict absolute rate constants in reasonable agreement with those measured experimentally for recombination in the presence of inert gases.

Two of the most significant features of iodine atom recombination have, however, received little further experimental attention and have not, in our view, been given a satisfactory theoretical interpretation. These are the observations that:

(*a*) Temperature coefficients are negative.

(*b*) The efficiency of different gases as third bodies or chaperons (Adam 1956) vary over a range far greater than can be understood in terms of differences in collision diameter or energy transfer efficiencies.

Theories of Husain & Pritchard (1959) and of Bunker & Davidson (1958) give possible qualitative explanations of negative temperature coefficients, but give poor quantitative agreement with the magnitude of these coefficients measured experimentally and no satisfactory explanation of the varying efficiency of the different gases as chaperons.

In part I (Christie, Norrish & Porter 1953) it was shown that the observed rate constants were better interpreted in terms of the scheme first suggested by Rabinowitch (1937)

$$I + M \rightarrow IM \qquad (k_1),$$

$$IM + I \rightarrow I_2 + M \qquad (k_3),$$

than of the alternative scheme

$$I + I \rightarrow I_2^* \qquad (k_4),$$

$$I_2^* + M \rightarrow I_2 + M \quad (k_5).$$

If, following Rice (1941), we assume that the reverse of reaction 1 is rapid so that the equilibrium

$$I + M \rightleftharpoons IM \quad (K)$$

is maintained, we have an immediate explanation both of the negative temperature coefficient and of the widely different efficiency of different chaperons. There are, however, two apparent objections to this theory:

(*a*) Van der Waals type interactions in the complex IM, estimated from second virial coefficients, are less than the binding energy necessary to give the observed magnitude of negative temperature coefficients.

(*b*) Chaperons giving higher recombination rates should also have more negative temperature coefficients. The study of temperature coefficients in a number of gases by Russell & Simons (1953) indicated that, contrary to this expectation, the coefficients were very nearly the same in all gases studied.

In this paper we describe an investigation of the temperature coefficients of iodine atom recombination in a number of gases of widely differing efficiency as chaperons and, on the basis of the results obtained, we propose a theory of recombination based on the intermediate formation of charge transfer complexes.

Experimental

Flash-photolysis measurements of recombination rates were carried out in the manner previously described. In order to reduce temperature inhomogeneities and the contribution of I_2 as a chaperon the partial pressure of I_2 was always very small, being typically one five-thousandth part of the total gas pressure.

The temperature of the reaction vessel was controlled by an outer jacket filled with silicone oil. A heating element enabled the temperature of this jacket to be varied between 20 and 250 °C, and thermocouple measurements at different parts of the vessel showed that the temperature was uniform to $\pm \frac{1}{2}$ °C. Owing to the low pressure of iodine and the consequently small optical density changes and also because of absorption of the iodine on the taps, difficulties were experienced with absolute measurements of extinctions. The method eventually used was as follows. A break-seal device was fitted to the cell and, after being filled, the cell was sealed off and transferred to the flash apparatus where it was connected to a vacuum line. Recombination rates were measured at a number of temperatures in a random order and finally, without disturbing the optical arrangement, the seal was broken and the optical density difference between the full and empty cell was determined by rapid evacuation.

Results

Good linear third-order plots were obtained in all experiments, the average of several traces being utilized for each rate determination. Except for ethyl iodide and the aromatic hydrocarbons, the temperature could be raised and lowered and the mixture flashed many times without any effect on the results. The rate constants for each gas were treated in two ways corresponding to the two rate equations

$$k = A_1 e^{-E_1/RT}, \tag{I}$$

and
$$k = TA_2 e^{-E_2/RT}, \tag{II}$$

the values of A and E being determined from plots of $\ln k$ against $1/T$ in the first case and of $\ln k/T$ against $1/T$ in the second case. Both types of plot were linear within the random errors of the separate determinations and the slopes were determined by least-squares treatment. A plot of the second type, for which the added gas was argon, is shown in figure 1. Good linear plots were also obtained for oxygen

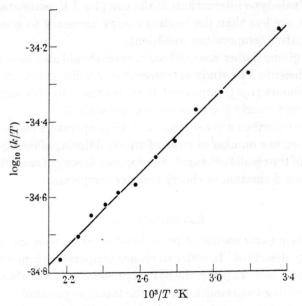

FIGURE 1. Plot showing the temperature dependence of iodine rates in argon atom recombination.

and carbon dioxide as well as for helium, although in this latter case the effect of temperature on rate was so small that a rise in temperature of 200 °C only reduced the rate by 30 %.

Determinations of rate constants in the presence of ethyl iodide, benzene, toluene and mesitylene were carried out in the presence of an excess of helium. This was necessary because the vapour pressures of the organic molecules were not high enough to ensure the absence of complications due to thermal effects, and helium was chosen owing to its very low efficiency as a chaperon and the small temperature coefficient of the associated recombination reaction. In calculating the separate rate constants from experiments on mixed gases M_1 and M_2, the linear relation

$$-dI/dt = (k_{M_1}[M_1] + k_{M_2}[M_2])\,[I]^2$$

was assumed to be valid.

Preliminary experiments on the hydrocarbons, particularly mesitylene, gave poor reproducibility and it was apparent that some permanent change was occurring in

the reaction mixture after flashing. Systematic investigations on mesitylene +
iodine mixtures showed that, in the ordinary Pyrex vessel, prolonged flashing at
room temperature resulted in the removal of some of the iodine, but that when the
light reaching the reaction mixture was filtered through a soda-glass tube no change
took place at room temperature. Even with the soda-glass filter in place, however,
some elimination of the iodine occurred after prolonged flashing at 150 °C. These
results indicate that it is absorption of the light by mesitylene which results in the
removal of iodine. In view of this complication a soda-glass filter was used for the
mesitylene measurements; the room-temperature measurements were carried out
first and the rate constant was determined from measurements at two temperatures
only, 22 and 154 °C.

There were some indications of similar effects with ethyl iodide, benzene and
toluene, but in these cases the reproducibility after several flashes was sufficiently
good to enable measurements to be made at several temperatures and reasonably
good linear Arrhenius plots were obtained.

TABLE 1. EXPERIMENTAL VALUES OF RATE CONSTANTS AND ARRHENIUS FACTORS

chaperon	$10^{32} k_{20}$ (ml.^{2} mol.$^{-2}$ s^{-1})	$10^{32} A_1$ (ml.^{2} mol.$^{-2}$ s^{-1})	$-E_1$ (kcal/mole)	$10^{36} A_2$ (ml.^{2} mol.$^{-2}$ s^{-1})	$-E_2$ (kcal/mole)
helium	0·84	0·42	0·4	2·6	1·4
argon	1·64	0·17	1·3	1·8	2·0
oxygen	3·72	0·28	1·5	2·9	2·2
carbon dioxide	7·41	0·37	1·75	4·1	2·4
benzene	43·9	2·38	1·7	24	2·4
toluene	107	1·03	2·7	11	3·4
ethyl iodide	144	2·33	2·4	24	3·1
mesitylene	223	0·19	4·1	2·0	4·8
iodine	760	—	—	2·9	(5·3)

The results are summarized in table 1 which gives the rate constants at 20 °C, and
the Arrhenius factors A_1, E_1, A_2 and E_2 defined by equations (I) and (II). For the
purpose of discussion we have also included in table 1 the values of recombination
rate constants and Arrhenius factors for iodine as chaperon. The rate constant is the
mean of the values obtained by Christie *et al.* (1955) and by Bunker & Davidson
(1958). The Arrhenius factors are from the results of Bunker & Davidson and are
not exactly comparable since these authors derived the activation energy assuming
the relation
$$k = A T^{\frac{1}{2}} e^{-E/RT}.$$

Owing to the high temperature coefficient of this recombination, activation energies
derived on the basis of equations (I) or (II) would not be very different.

Our values for absolute rate constants at 20 °C are in excellent accord with those
of other recent work. Thus the four most recent determinations of the rate constant
in argon are as follows

Christie *et al.* (1955)	$1·84 \pm 0·2 \times 10^{-32}$ ml.^{2} molecule^{-2} s^{-1}
Strong *et al.* (1957)	$1·67 \pm 0·07 \times 10^{-32}$
Bunker & Davidson (1958)	$1·60 \pm 0·16 \times 10^{-32}$
this work	$1·63 \pm 0·09 \times 10^{-32}$

Our rate constants for recombination in the other gases are in remarkably good agreement with those of Russell & Simons (1953).

Apart from the measurements of Russell & Simons, temperature coefficients in the gases studied here have previously been determined only for recombination in argon and again the agreement between different investigators is satisfactory. The values obtained for the activation energy, derived on the basis of the simple Arrhenius equation with a temperature-independent pre-exponential factor are

Russell & Simons (1953)	$-2 \cdot 1 \pm 0 \cdot 25$ kcal/mole
Strong *et al.* (1957)	$-1 \cdot 4$
Bunker & Davidson (1958)	$-1 \cdot 13$
this work	$-1 \cdot 3 \pm 0 \cdot 1$

Apart from the value of Russell & Simons, agreement is within the estimated experimental error.

DISCUSSION

Our results show that, contrary to the findings of Russell & Simons, the temperature coefficient of recombination decreases with increasing chaperon efficiency.

The scheme of recombination based on the equations

$$I + M \rightleftharpoons IM \quad (K),$$

$$I + IM \rightarrow I_2 + M \quad (k_3)$$

leads directly to the conclusion that the overall rate constant of the thermolecular recombination, k_0, defined by

$$-d[I]/dt = k_0[I]^2[M]$$

should also be given by

$$k_0 = 2k_3 K.$$

Calculation of absolute rate constants of recombination is therefore possible if k_3 and K can be evaluated. Since the strength of the IM bond is small we shall assume that reaction 3 has negligible activation energy and that its rate constant can therefore be calculated on the basis of collision theory. The equilibrium constant K can, in principle, be calculated *a priori* by statistical mechanical methods provided the heat of reaction is known. This we have determined experimentally from the temperature coefficient of reaction.

Calculation of k_3

The metathetical reaction 3 requires that the free iodine atom shall interact with the iodine atom in the complex. The appropriate collision diameter is therefore σ_{II}, the diameter of the iodine atom being assumed equal to that of xenon (Rabinowitch 1937) but the appropriate mass of the complexed atom is the mass of the whole complex.

The calculation, using the standard kinetic theory formulae given in part I, is straightforward except for the uncertainty as to what statistical factor should be assumed. We have taken the statistical factor equal to $\frac{1}{16}$, which implies that of the 16 possible combinations of two $^2P_{\frac{3}{2}}$ iodine atoms only one leads to recombination.

The values of k_3 calculated in this way are given in table 2 and it will be seen that, owing to the large relative mass of the iodine atom, they do not differ greatly for different chaperon molecules.

TABLE 2. CALCULATED VALUES OF RATE CONSTANTS AND
OTHER QUANTITIES AT 20 °C

chaperon	$10^{11} k_3$ (ml. mol.$^{-1}$s^{-1})	$10^8 \sigma_{1M}$ (cm)	$10^{32} A_{\text{calc.}}$ (ml.^{2}mol.$^{-2}$s^{-1})	$10^{32} k_{\text{calc.}}$ (1) (ml.^{2}mol.$^{-2}$s^{-1})	$10^{32} k_{\text{calc.}}$ (2) (ml.^{2}mol.$^{-2}$s^{-1})
helium	1·65	3·70	0·38	0·76	3·0
argon	1·56	4·42	0·26	2·4	8·1
oxygen	1·57	4·41	0·25	3·3	11
carbon dioxide	1·55	4·90	0·28	5·7	17
benzene	1·49	6·29	0·19	3·5	12
toluene	1·47	6·50	0·25	25	86
ethyl iodide	1·42	5·42	0·26	16	54
mesitylene	1·44	7·28	0·39	450	1500
iodine	1·35	6·21	0·23	—	2100

Calculation of K

The quantum statistical evaluation of K can be carried out in several ways. The method which we have preferred is based on the partition functions for a diatomic molecule, since it has the advantage that the factors which are important in controlling the rate of reaction are determined separately and the sources of inaccuracy in the calculations are more readily apparent. When the chaperon is polyatomic some assumptions must be made about the structure of the complex and in order to preserve generality, and also because little can be predicted about such structures, we have based our calculation on the following assumptions:

(*a*) The distance between the centres of mass of the iodine atom and chaperon molecule is equal to the gas kinetic collision diameter σ_{IM}.

(*b*) The internal vibrations of the chaperon molecule are unchanged in the complex.

(*c*) The chaperon molecule is free to rotate within the complex.

With these assumptions the partition functions for internal vibration and rotation of the chaperon disappear and we obtain the following simple expression for the equilibrium constant K:

$$K = \frac{\{[2\pi(m_I + m_M)\,kT]\}^{\frac{3}{2}}/h^3\,(8\pi^2 I_{IM}\,kT)/h^2 S\,[1 - e^{(-h\nu/kT)}]^{-1}}{\{(2\pi m_I kT)^{\frac{3}{2}}\}/h^3\,\{(2\pi m_M kT)^{\frac{3}{2}}\}/h^3}\,e^{-\Delta E/RT}$$

$$\equiv B\,e^{-\Delta E/RT}.$$

The moment of inertia I_{IM} is given by

$$I_{IM} = \frac{(m_I m_M)}{(m_I + m_M)}\,\sigma_{IM^2},$$

where $$\sigma_{IM} = \tfrac{1}{2}(\sigma_I + \sigma_M),$$

the values adopted for σ_{IM} being given in table 2. The symmetry factor S was taken as unity for all IM complexes except I_3 which was assumed to be linear with $S = 2$.

The principal unknown in this expression is the stretching frequency ν of the IM bond (other modes have disappeared as a result of assumption c). We have estimated this frequency from the expression

$$\nu_{IM} = a \left(\frac{D}{2\mu_{IM}} \right)^{\frac{1}{2}}$$

derived from the Morse equation assuming simple harmonic oscillation. Here D is the bond energy determined experimentally from our temperature coefficients of rate, μ is the reduced mass and a is the constant which appears in the Morse equation.

The value of a calculated for the IA complex, assuming a Lennard-Jones potential, was 2×10^8 cm^{-1} and since this was close to the values of a given by Partington (1949) for a variety of diatomic molecules it was adopted for all the IM complexes. This is perhaps the most doubtful part of our calculation but some assumption of this kind is unavoidable, and since this vibrational partition function is small and is not very different in the various IM complexes the error is not likely to be serious. ΔE, the heat of reaction at constant volume, is obtained from the experimentally determined temperature coefficient of reaction rate.

Comparison of calculated and experimental results

Since $k_0 = 2k_3 K$

and $K = B e^{-\Delta E/RT}$,

then $k_0 = 2k_3 B e^{-\Delta E/RT} = A_{\text{calc.}} e^{-\Delta E/RT}$.

Values of $A_{\text{calc.}} = 2k_3 B$ at 20 °C, calculated by the procedure described in the last section are given in table 2.

The correct value to be used for ΔE depends on the temperature coefficient of $2k_3 B$. Two cases may be distinguished:

(1) $h\nu \gg kT$

Then $k_3 \propto T^{\frac{1}{2}}$,

$B \propto T^{-\frac{1}{2}}$,

and $2k_3 B \propto T^0$.

Since $k_0 = A_1 e^{-E_1/RT}$,

then $\Delta E = E_1$, $A_{\text{calc.}} = 2k_3 B = A_1$

and $k_{\text{calc.}}(1) = A_{\text{calc.}} e^{-E_1/RT}$.

(2) $h\nu \ll kT$

Then $k_3 \propto T^{\frac{1}{2}}$,

$B \propto T^{\frac{1}{2}}$,

and $2k_3 B \propto T$.

Since $k_0 = T A_2 e^{-E_2/RT}$,

then $\Delta E = E_2$, $A_{\text{calc.}} = 2k_3 B = A_2 T$

and $k_{\text{calc.}}(2) = A_{\text{calc.}} e^{-E_2/RT}$.

Since $h\nu \approx kT$ the true value of k should lie between these extremes and the temperature coefficient of A should be between zero and unity. It is interesting to note that a negative temperature coefficient cannot arise from the pre-exponential term alone. The values of $k_{\mathrm{calc.}}(1)$ and $k_{\mathrm{calc.}}(2)$ at 20 °C are given in table 2.

The calculations predict correctly the order of magnitude of the rate constants and reproduce, reasonably well, the relative differences between various chaperon molecules. The agreement of absolute values is to some extent fortuitous owing to uncertainty as to the correct value of the statistical factor. The calculated rate constants are low for benzene and ethyl iodide and it is possible that the error lies in our experimental determination of the temperature coefficient to which the calculations are very sensitive.

Calculation shows that the A factor should be nearly constant for all chaperons and that the rate is principally determined by the exponential term. The most significant feature of our results is that they are in accordance with this prediction and that, although there are considerable variations in A_1 and A_2, there is no systematic variation of these factors with rate constant. This lends strong support to the theory of recombination on which the calculations have been based.

Nature of the IM bond

The theory so far outlined gives a reasonably satisfactory account of the principal features of iodine atom recombination kinetics, and fair quantitative agreement with absolute rate constants provided the experimentally determined temperature coefficient is used to derive the energy of the IM bond. The only additional requirement for an absolute calculation of recombination rates is the calculation of this bond energy. It will not be surprising, however, if this proves to be impracticable since the absolute calculation of the bond energies of stable diatomic molecules, with the exception of H_2, is also not feasible. It should, however, be possible to indicate the nature of the bonding forces which are involved and to predict the trend of these forces between complexes of different chaperon molecules.

The most obvious assumption, and that which has been made by other workers in this field, is that the forces between the iodine atom and the chaperon molecule are of the normal van der Waals type. This conflicts with our measurements of the IM bond energy in two respects. First, the magnitude of such forces calculated from the London dispersion formula, or derived experimentally from the second virial coefficients is less than those derived from temperature coefficients of recombination by a factor of 4 for helium and more for the other gases. Secondly, the calculated values of these forces are relatively constant for the different gases except helium whilst a considerable variation is found in practice. In making this calculation the usual combining rule

$$E_{IM} = (E_I E_M)^{\frac{1}{2}},$$

where E_{IM}, E_I and E_M are the bonding energies for the IM, II and MM complexes is used, and the properties of the iodine atom are taken to be identical with those of xenon. These are doubtful assumptions but unavoidable in the absence of any data about the properties of iodine atom gas. It is unlikely that they introduce any great error in the calculation of unspecific forces of the dispersion type. This method of calculation was used by Bunker & Davidson who were satisfied with the agreement between the calculated and observed rate constants for recombination in argon, hydrogen and n-butane. But, as already pointed out, this agreement was largely fortuitous depending on the values which they chose for the various molecular parameters, and agreement between theory and experiment in the much more significant matters of temperature coefficients and relative efficiences of different gases was poor.

We are therefore led to suppose that some more specific type of bonding is operative in the complex formed between an iodine atom and the chaperon molecule, and we believe that the explanation is to be found in terms of a charge-transfer type of complex (Porter & Smith 1959). Apart from the fact that the experimental findings leave no reasonable alternative, there are good reasons for supposing that such complexes exist and will have bonding energies of the required magnitude.

Since the discovery of charge-transfer spectra in iodine solutions by Benesi & Hildebrand (1949), and their interpretation as such by Mulliken (1950), the charge-transfer type of complex has been found in many types of molecule. Since, according to the theory of Mulliken, the principal requirements are a low ionization potential of the donor and a high electron affinity of the acceptor it is to be expected that the bond energy of the complex will be greater for iodine atoms as acceptors (electron affinity = 3·33 eV) than for iodine molecules. The binding energy of the benzene-molecular iodine complex in solution is 1·55 kcal/mole so that the energy which we have measured for the iodine atom + benzene complex of between 1·7 and 2·4 kcal seems entirely reasonable. Furthermore, the efficiency of various chaperons as third bodies and the magnitude of the negative temperature coefficient increase roughly as the ionization potential decreases, a fact which was first pointed out by Russell & Simons.

The absolute calculation of recombination rates now becomes a matter of calculating the energies of charge transfer complexes. Unfortunately the theory of such complexes is, as yet, imperfectly developed. The formula derived by Hastings, Franklin, Schiller & Matson (1953) predicts values for the IM complexes which, although greater than those observed by a factor of approximately 2, are relatively in rather good agreement with the values determined experimentally from temperature coefficients.

Further progress in developing this theory of recombination can be made in two ways. First, development of the theory of charge-transfer complexes would enable more exact calculations to be made for comparison with experimentally determined complex-bond energies. Secondly, one might hope to observe the charge transfer complexes of the iodine atom directly in flash-photolysis experiments.

We have attempted to detect the absorption of the charge-transfer complex between iodine atoms and benzene or ethyl iodide directly in the gas phase without success. On the other hand, very recently, transient absorption spectra have been observed in the flash photolysis of iodine solutions by Rand & Strong (1960), by Bridge (1960) and by Gover & Porter (1960). These are almost certainly to be attributed to the charge transfer complex of the iodine atom predicted by the gas

37   *The recombination of atoms. III*

phase work, and direct studies of the properties of these complexes which are now possible should provide clear evidence concerning their role in atom recombination.

We are grateful to the European Office of the United States Army for financial support in this work.

REFERENCES

Adam, N. K. 1956 *Physical chemistry*, p. 460. Oxford.

Benesi, H. A. & Hildebrand, J. H. 1949 *J. Amer. Chem. Soc.* **71**, 2703.

Bridge, N. K. 1960 *J. Chem. Phys.* **32**, 945.

Bunker, D. L. & Davidson, N. 1958 *J. Amer. Chem. Soc.* **80**, 5085, 5090.

Bunker, D. L. 1960 *J. Chem. Phys.* **32**, 1001.

Christie, M. I., Harrison, A. J., Norrish, R. G. W. & Porter, G. 1955 *Proc. Roy. Soc.* A, **231**, 446 (part II).

Christie, M. I., Norrish, R. G. W. & Porter, G. 1953 *Proc. Roy. Soc.* A, **216**, 152 (part I).

Gover, T. A. & Porter, G. 1961 (part IV of this series).

Hastings, S. H., Franklin, J. L., Schiller, J. C. & Matson, F. A. 1953 *J. Amer. Chem. Soc.* **75**, 2900.

Husain, D. & Pritchard, H. O. 1959 *J. Chem. Phys.* **30**, 1101.

Jepson, D. W. & Hirschfelder, J. O. 1959 *J. Chem. Phys.* **30**, 1032.

Keck, J. C. 1958 *J. Chem. Phys.* **29**, 410.

Keck, J. C. 1960 *J. Chem. Phys.* **32**, 1035.

Mulliken, R. S. 1950 *J. Amer. Chem. Soc.* **72**, 600.

Partington, J. R. 1949 *An advanced treatise on physical chemistry*, vol. 1. London: Longmans, Green and Co.

Porter, G. & Smith, J. A. 1959 *Nature, Lond.* **184**, suppl. 7, 446.

Rabinowitch, E. 1937 *Trans. Faraday Soc.* **33**, 283.

Rand, S. J. & Strong, R. L. 1960 *J. Amer. Chem. Soc.* **82**, 5.

Rice, O. K. 1941 *J. Chem. Phys.* **9**, 258.

Russell, K. E. & Simons, J. 1953 *Proc. Roy. Soc.* A, **217**, 271.

Strong, R. L., Chien, J. C. W., Graf, P. E. & Willard, J. E. 1957 *J. Chem. Phys.* **26**, 1287.

Reprinted from the *Discussions of the Faraday Society*, 1962, No. 33

Mechanism of Third-Order Recombination Reactions

By G. Porter

Dept. of Chemistry, The University, Sheffield 10

Received 19th January, 1962

Two theories of third-order recombination and second-order dissociation give a satisfactory explanation of observed negative temperature coefficients and the wide differences in efficiency of chaperon molecules. Recent measurements of iodine atom recombination temperature coefficients in the presence of twelve chaperons are used to test the theories and it is shown that whilst the radical-molecule complex theory predicts both the absolute rates and their dependence on chaperon the energy transfer theory is less successful.

The recombination of two atoms or molecules R, in the presence of a chaperon M, occurs in two steps since, if we define precisely the separation at which two species are in collision, formation of a bimolecular collision complex must precede formation of the termolecular one. The recombination reaction may therefore proceed in two ways depending on whether R_2^* or RM* is the first bimolecular complex to be formed. If the complex RM* can be stabilized to a species RM by collision with a second M, a third mechanism of recombination is possible. The complete reaction scheme is therefore as follows:

$$R + R \underset{2}{\overset{1}{\rightleftharpoons}} R_2^* \qquad K_1$$

$$R + M \underset{4}{\overset{3}{\rightleftharpoons}} RM^* \qquad K_3$$

$$R_2^* + M \underset{6}{\overset{5}{\rightleftharpoons}} R_2 + M \qquad K_5$$

$$RM^* + M \underset{8}{\overset{7}{\rightleftharpoons}} RM + M \qquad K_7$$

$$RM^* + R \underset{10}{\overset{9}{\rightleftharpoons}} R_2 + M \qquad K_9$$

$$RM + R \underset{12}{\overset{11}{\rightleftharpoons}} R_2 + M \qquad K_{11}$$

If dissimilar species R and R′ are involved 36 equations are necessary but we can confine discussion to the above case without loss of generality.

If the system is at equilibrium

$$d[R_2]/dt = 0 = [k_5 K_1 + k_9 K_3 + k_{11} K_3 K_7][R]^2[M] - [k_6 + k_{10} + k_{12}][R_2][M].$$

If, as is usually the case, the system is not at equilibrium, the stationary-state treatment is applicable to the recombination reaction provided $[R_2^*]$, [RM*] and [RM] are much smaller than [R], and that the reaction has proceeded to steady state conditions. Then the termolecular recombination rate constant k_r, when the second term in the above equation is much smaller than the first, is given by

$$k_r = \frac{k_1 k_5}{k_2 + k_5[M]} + \frac{k_3 k_9(k_8 + k_{11}\theta)}{(k_4 + k_9\theta[M])(k_8 + k_{11}\theta) + k_7 k_{11}\theta[M]} +$$

$$\frac{k_3 k_7 k_{11}}{(k_4 + k_9\theta[M])(k_8 + k_{11}\theta) + k_7 k_{11}\theta[M]}$$

where $\theta = [R]/[M]$.

If the reaction is third order, $k_5[M] \ll k_2$ and $k_7 k_{11}\theta[M] \ll (k_4 + k_9\theta[M])(k_8 + k_{11}\theta)$ corresponding to a short lifetime of R_2^* and a weak complex RM. Then

$$k_r = \frac{k_1 k_5}{k_2} + \frac{k_3 k_9}{k_4 + k_9\theta[M]} + \frac{k_3 k_7 k_{11}}{(k_4 + k_9\theta[M])(k_8 + k_{11}\theta)}$$

If [R] is small so that $k_9\theta[M] \ll k_4$ and $k_{11}\theta \ll k_8$,

$$k_r = \frac{k_1 k_5}{k_2} + \frac{k_3 k_9}{k_4} + \frac{k_3 k_7 k_{11}}{k_4 k_8}$$

$$= k_5 K_1 + k_9 K_3 + k_{11} K_3 K_7 \tag{1}$$

which is equal to k_r at equilibrium.

Since $[RM^*]/([RM^*] + [RM]) = \exp -\Delta E/RT$, where ΔE is the heat of formation of RM from R and M at constant volume, $K_7 = \exp(\Delta E/RT) - 1$. Since $k_9 = k_{11}$ when ΔE is small and $K_7 \gg 1$ when ΔE is large we obtain

$$k_r = k_5 K_1 + k_{11} K_3 \exp(\Delta E/RT). \tag{2}$$

In atom-recombination reactions the principal experimental facts to be accounted for are as follows:

(1) rate constants involving atomic chaperons [1] are about $3 \times 10^9 1.^2$ mole^{-2} sec^{-1};

(2) the efficiency of different chaperons [2] varies over a factor of at least 10^3;

(3) temperature coefficients are negative and become increasingly so as the efficiency of the chaperon increases.[3]

Theories of third-order recombination, and the microscopically reversible theories of second-order dissociation, differ widely in the values which they derive for the individual constants in eqn. 2, though all are capable of giving reasonable agreement with the observed rate constant for three-atom processes. Most theories assume $\Delta E = 0$ in eqn. (2) and calculate the efficiency of three-body collisions either of the " billiard ball " type (see ref. (4) for discussion of early theories) or with the introduction of energy transfer restrictions, so that k_5 and k_{11} are less than the corresponding collision numbers Z_5 and Z_{11}. Thus, Husain and Pritchard [5] postulate that deactivation of R_2^* may only occur when the kinetic energy of R_2^* and M is less than a critical amount E^*, in which case

$$k_5 = Z_5[1 - \exp(-E^*/RT)].$$

This equation gives a good account of the observed temperature dependence of rate but predicts absolute values which are too low and nearly independent of the chaperon used. The theories of Keck,[6, 7] Bunker [8] and of Jepson and Hirshfelder [9] predict neither the observed temperature dependence nor the dependence on chaperon molecule, though Keck's variational theory predicts the absolute rates in the five inert gases very well.

There are two theories of atom recombination at present in use which give a fairly satisfactory account of all the observations, including temperature coefficients and chaperon effects. We shall first describe the two theories briefly and then attempt to evaluate them in terms of the recent data of Porter and Smith [3] on iodine-atom recombination temperature coefficients.

ENERGY TRANSFER THEORY [10]

ΔE is assumed to be zero and no distinction is made between the terms $k_5 K_1$ and $k_{11} K_3$. The value of k_5 is calculated via the reverse rate constant k_6 and the overall equilibrium constant K of

$$I_2 = 2I, \quad K.$$

The collision (energy transfer) theory expression used to evaluate k_6 is [11]

$$k_6 = k_d = Z \left(\frac{E_0}{RT} \right)^n \frac{1}{n!} \exp \left(-\frac{E_0}{RT} \right),$$

where $2n+2$ is the number of square terms in the collision complex which can contribute energy for dissociation, E_0 is the dissociation energy of I_2 and Z is the rate constant for $I_2 + M$ collisions. The recombination constant

$$k_r = k_5 K_1 = \frac{k_d}{K} = \frac{Z \left(\dfrac{E_0}{RT} \right)^n \dfrac{1}{n!}}{BT^{\frac{1}{2}}[1 - \exp(-310/T)]},$$

where B is independent of temperature. Since Z varies as $T^{\frac{1}{2}}$,

$$k_r = \frac{C}{n!} \left(\frac{E_0}{RT} \right)^n,$$

where C is independent of temperature at low temperatures and proportional to the first power of temperature at high temperatures. The constant B is independent of chaperon and Z varies by little more than a factor of two over all the chaperons which we have investigated so that, since we are concerned with variations in the rate constant over a factor of 10^3 we may treat C as a constant. For iodine-atom recombination we calculate that

$$k_r = 7 \times 10^7 \left(\frac{E_0}{RT} \right)^n \frac{1}{n!} 1.^2 \text{ mole}^{-2} \text{ sec}^{-1} \text{ at } 300°\text{K}.$$

If the experimental results are plotted in Arrhenius form the apparent negative activation energy $E_a = nRT$ at low temperatures and $(n-1)RT$ at high temperatures.

RADICAL-MOLECULE COMPLEX THEORY [12,3,13]

The fundamental assumption here is that $\Delta E > RT$ so that

$$k_r = k_{11} K_3 \exp (\Delta E / RT).$$

Since the product R_2 of reaction (11) has energy $E_0 - \Delta E$ it cannot dissociate and no energy transfer restrictions are imposed so that $k_{11} = Z_{11}$. Statistical evaluation of the constant K_3 shows that the product $k_{11} K_3$ is insensitive to the nature of M and is independent of temperature at low temperatures and proportional to T at high temperatures. Again, within a factor of two we can regard the pre-exponential factor as constant and use the average of the calculated values of Porter and Smith [3] to give

$$k_r = 5 \times 10^8 \exp\left(E_a/RT\right) \text{l.}^2 \text{ mole}^{-2} \text{ sec}^{-1},$$

where $E_a = \Delta E$ at low temperatures and $E_a = \Delta E - RT$ at high temperatures.

COMPARISON OF THE TWO THEORIES

The only data which are sufficiently accurate and extensive for this comparison are flash-photolysis studies of halogen-atom recombination, and even these do not allow a distinction to be made between the two theories on the basis of the temperature dependence alone since both expressions give equally good accounts of the experimental results. The temperature dependence of reaction rates obtained from shock-tube experiments alone are rather uncertain though a combination of these data with flash-photolysis results is more reliable and it has been shown,[10] for bromine atom recombination in argon, that the energy transfer theory can account for the dissociation rate over a temperature range corresponding to a variation of rate by a factor of 10^{27}. The complex theory accounts for the data nearly as well though Givens and Willard [14] have shown that the energy transfer theory gives a somewhat better fit if the temperature dependence measured in shock-tube experiments is correct.

We have recently measured the temperature coefficients of recombination of iodine atoms in the presence of eight chaperons of widely different efficiencies and four chaperons have been studied by Bunker and Davidson [13] and by Engleman and Davidson.[15] All the available data are summarized in table 1, the rate constants being defined by eqn. (2) and therefore equal to one half of those tabulated by Porter and Smith. These results enable us to compare the theories in the following way.

TABLE 1

M	k_{27} (l.^{2} mole^{-2} sec^{-1}) $\times 10^{-9}$	E_a(kcal)	ΔE (kcal)	n	ref.
He	1·5	0·4	1·0	1·7	3
	1·4	0·66	1·26	2·1	15
A	3·0	1·3	1·9	3·2	3
	3·0	1·4	2·0	3·3	16
	2·9	1·13	1·73	2·9	13
H_2	5·7	1·22	1·82	3·0	15
O_2	6·8	1·5	2·1	3·5	3
CO_2	13·4	1·75	2·35	3·9	3
C_4H_{10}	36	1·65	2·25	3·7	13
C_6H_6	80	1·7	2·3	3·8	3
	105	1·97	2·57	4·3	15
CH_3I	160	2·55	3·15	5·2	15
$C_6H_5CH_3$	194	2·7	3·3	5·5	3
C_2H_5I	262	2·4	3·0	5·0	3
$C_6H_3(CH_3)_3$	405	4·1	4·7	7·8	3
I_2	1600	4·4	5·0	8·3	13

The value of E_a from the Arrhenius plot gives n and ΔE from which, by use of the expressions derived above, the rate constants can be calculated absolutely. Since, in each case, we have two observable quantities and only one adjustable parameter a test of the theories can be made. The results are shown in fig. 1 where the rate constants have been plotted on a logarithmic scale against n and ΔE for twelve chaperons. The points are experimental values and the two lines are calculated from the two theories. The high-temperature approximation, which should be the more correct in the temperature region investigated, was used in both cases.

The agreement between experimental values and the complex theory calculation is within the estimated uncertainties of calculation and experiment, and, since the calculations are absolute, except for the use of the experimental E_a, the good fit, both of gradient and absolute values is significant. On the other hand, the energy transfer theory predicts higher values and a more rapid increase in rate with n than is observed. This reflects what has already been noticed—that in order to account for the large negative temperature coefficients this theory predicts very small collision efficiencies.

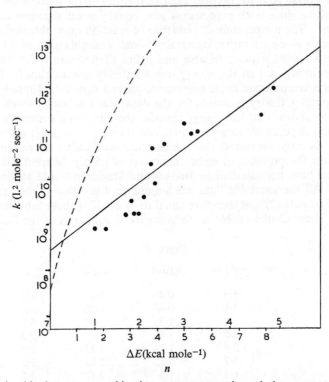

FIG. 1.—Relationship between recombination rate constant k_r and the parameters ΔE and n. The points are experimental; the full curve is calculated from the complex theory and the broken curve is calculated from the energy transfer theory.

This is not surprising since it is difficult to see how theory can be correct in assuming that energy flow from all available degrees of freedom can occur during the time of one collision, i.e., about 10^{-13} sec. A further objection is that the number of square terms required, $2(n+1)$, seems to be impossibly large in many cases and to show a poor correlation with molecular complexity. It seems most improbable that I_2 and Br_2 would contribute more degrees of freedom as chaperons than hydrocarbons and alkyl iodides.

The main objection which can be raised to the complex theory concerns the nature of the RM complex. Clearly, the interaction is unspecific since even iodine itself, the most efficient chaperon, fits the correlation with boiling point, ionization potential and similar properties found by Russell and Simons.[2] Attempts to interpret recombination rates in terms of Van der Waals forces [13] were not, however, successful since these forces are too small, and show too little variation between chaperon molecules, to account for either temperature coefficients or relative rates. This difficulty is removed in the charge-transfer complex theory of Porter and Smith [3] and excellent support for this theory has now been provided by the direct observation of the absorption spectra of the charge-transfer complexes themselves in flash-photolysis experiments.[17,18]

The question now arises as to what extent the charge-transfer complex theory of recombination will be applicable to systems other than iodine. Similar flash-photolysis experiments on bromine-atom recombination,[14] although more limited, have established that the rate coefficients have negative temperature coefficients similar to iodine as well as a similar dependence on chaperon. Recent work on the reactions $H + OH + M$ and $OH + OH + M$ also show a marked variation with chaperon molecule and a good correlation with halogen-atom results.[17] Since the electron affinity of OH is at least comparable with that of halogen atoms, charge-transfer complexes are probably involved here as well. It would not be surprising if interactions of a few kcal/mole were found quite commonly between free radicals and molecules for the following reason. The highest filled orbital of the chaperon M will combine with the odd electron orbital of R to form two new molecular orbitals, one bonding and one antibonding, the separation depending on the difference in energy of the original orbitals. Two electrons will occupy the lowest bonding orbital in the complex and one the antibonding orbital so that, by the usual molecular orbital argument, bonding will result. Whether this type of bonding is generally to be termed a charge-transfer complex is largely a matter of definition. Perhaps the less definitive term radical-molecule complex would be more useful until more has been learned about the existence of such species.[19]

It is unlikely that the radical-molecule complex mechanism will be important in the recombination (or the inverse molecular dissociation) involving larger radicals. The relative importance of the three mechanisms of recombination represented by eqn. (1)-(12) can be estimated in the following way. The rate constants k_1, k_3, k_5 and k_{11} are, by collision theory, of the order of magnitude 10^{11} l. mole^{-1} sec^{-1} at 300°K. For this temperature, eqn. (2) therefore becomes

$$k_r \approx 10^{22}\tau_{R_2}^* + 10^{22}\tau_{RM}^* \exp(\Delta E/RT) \text{ l.}^2 \text{ mole}^{-2} \text{ sec}^{-1},$$

where $\tau_{R_2}^*$ and τ_{RM}^* are the lifetimes of the collision complexes R_2^* and RM*. Three cases may be distinguished.

(a) Three atoms. The lifetimes τ will be about 10^{-13} sec and

$$k_r = 10^9 + 10^9 \exp(\Delta E/RT)$$

so that if $\Delta E = 0$ the two processes contribute equally and if $\Delta E > 0$ the complex mechanism predominates.

(b) R are atoms and M is polyatomic. Then $\tau_{R_2} \sim 10^{-13}$ sec, but τ_{RM}^* may be greater owing to the increased number of degrees of freedom if there is any attraction between the atoms and molecules. Again, unless $\Delta E = 0$, the complex mechanism predominates.

(c) R are radicals. In this case, $\tau_{R_2}^*$ will be increased by the increased number of degrees of freedom. If the values of ΔE are, like those found with iodine, about 1-3 kcal for typical chaperon molecules we may expect that the complex mechanism will still predominate for recombination of diatomic radicals but probably not for triatomic radicals. What little evidence is available supports this; for example, the rate constants for combination of small radicals such as OH are similar to those for atoms [20] whilst the third-order rate constant for methyl radical recombination [21]

THIRD-ORDER RECOMBINATION REACTIONS

is about 10^5 times greater implying a lifetime of $C_2H_6^*$ of 10^{-8} sec and the utilization of approximately 14 degrees of freedom.

It would be helpful to have data on the effects of different chaperons on a wider variety of radical recombination reactions; those with few atoms, proceeding by the complex mechanism, should show a considerable variation with chaperon, similar to that of the halogen atoms, whilst those like methyl, which become termolecular only at low pressures, should show a dependence on the chaperon determined only by the energy transfer rate constant k_5.

1 Christie, Harrison, Norrish and Porter, *Proc. Roy. Soc. A*, 1955, **216**, 446.
2 Russell and Simons, *Proc. Roy. Soc. A*, 1953, **217**, 271.
3 Porter and Smith, *Nature*, 1959, **184**, suppl. 7, 446; *Proc. Roy. Soc. A*, 1961, **261**, 28.
4 Christie, Norrish and Porter, *Proc. Roy. Soc. A*, 1953, **216**, 152.
5 Husain and Pritchard, *J. Chem. Physics*, 1959, **30**, 1101.
6 Keck, *J. Chem. Physics*, 1958, **29**, 410.
7 Keck, *J. Chem. Physics*, 1960, **32**, 1035.
8 Bunker, *J. Chem. Physics*, 1960, **32**, 1001.
9 Jepson and Hirshfelder, *J. Chem. Physics*, 1959, **30**, 1032.
10 Palmer and Hornig, *J. Chem. Physics*, 1957, **26**, 98.
11 Fowler and Guggenheim, *Statistical Thermodynamics* (Oxford, 1952), p. 497.
12 Rice, *J. Chem. Physics*, 1941, **9**, 258.
13 Bunker and Davidson, *J. Amer. Chem. Soc.*, 1958, **80**, 5085, 5090.
14 Givens and Willard, *J. Amer. Chem. Soc.*, 1959, **81**, 4773.
15 Engleman and Davidson, *J. Amer. Chem. Soc.*, 1960, **82**, 4770.
16 Strong, Chien, Graf and Willard, *J. Chem. Physics*, 1957, **26**, 1387.
17 Rand and Strong, *J. Amer. Chem. Soc.*, 1960, **82**, 5.
18 Gover and Porter, *Proc. Roy. Soc. A*, 1961, **262**, 476.
19 Hausser and Murrell, *J. Chem. Physics*, 1957, **27**, 500.
20 Black and Porter, *Proc. Roy. Soc. A*, in course of publication.
21 Dodd and Steacie, *Proc. Roy. Soc. A*, 1954, **223**, 283.

Further publications

THE RECOMBINATION OF ATOMS. II. CAUSES OF VARIATION IN
THE OBSERVED RATE CONSTANT FOR IODINE ATOMS.
M. I. Christie, A. J. Harrison, R. G. W. Norrish and G. Porter, *Proc. Roy. Soc.*, 1955,
A231, 446.

MECHANISM OF ATOM RECOMBINATION.
G. Porter and J. A. Smith, *Nature*, 1959, **184**, 446.

THE RECOMBINATION OF ATOMS.
IV. IODINE ATOM COMPLEXES.
T. A. Glover and G. Porter, *Proc. Roy. Soc.*, 1961, **A262**, 476.

THE RECOMBINATION OF ATOMS.
V. IODINE ATOM RECOMBINATION IN NITRIC OXIDE.
G. Porter, Z. G. Szabo and M. G. Townsend, *Proc. Roy. Soc.*, 1962, **A270**, 493.

Chapter 4

THE TRIPLET STATE

Before the introduction of the laser, the transient species detected and studied by flash photolysis were atoms and free radicals in Boltzmann equilibrium and in their ground electronic states. Most allowed electronic transitions occur in times much less than a microsecond and were therefore undetectable in the early flash photolysis studies. There was however on important class of spin forbidden transitions, the radiative transitions between singlet and triplet states, which typically occur in times of microseconds. Since the radiationless intersystem crossing occurs more rapidly than this it can compete with fluorescence so that the efficiency of triplet formation can be high.

The identification of phosphorescence of organic molecules in rigid solvents with emission from the triplet state was made in the 1940s by Lewis, Kasha, Terenin and their associates. Absorption under the same conditions was observed by McClure in 1950. The first observation of triplet states of organic molecules in ordinary solutions at room temperature and also in the gas phase, with lifetimes in the microsecond range, are described below. This made possible the extensive studies of what were really new chemical species, with their own chemical and physical properties.

In this chapter the new triplet absorption spectra are described along with the measurement of their extinction coefficients. The determinations of physico-chemical properties of the triplet states, such as acidity constants, and the kinetics of quenching and energy transfer are described. Mechanisms of radiationless conversion are discussed.

Reprinted from THE JOURNAL OF CHEMICAL PHYSICS, Vol. 21, No. 11, 2088, November, 1953

Triplet States in Solution

GEORGE PORTER AND MAURICE W. WINDSOR
*Department of Physical Chemistry, University of Cambridge,
Cambridge, England*
(Received August 19, 1953)

PHOTOCHEMICAL and fluorescence studies of aromatic molecules in solution indicate that metastable biradicals in the lowest triplet level, formed by a radiationless transition from the excited singlet state, may be intermediates in the chemical change. At present the observed kinetic laws of the over-all reaction do not allow a choice to be made between the several possible mechanisms and direct observations on the triplet molecule itself are lacking. While absorption from triplet levels can be observed by using steady cross illumination in glasses at very low temperatures,[1,2] under which conditions the lifetime approaches the true radiative lifetime, the rapid deactivation or reaction makes this method inapplicable in solution. The high efficiency of photochemical oxidation in solution[3] coupled with the observations of McClure[2] suggest, however, that if time intervals of the order of 10^{-5} second could be resolved and a high intensity source used, the direct study of triplet-state molecule reactions in solution might become possible.

We have applied the method of flash photolysis and flash spectroscopy[4] to the study of the photolysis of aromatic molecules in solution and in all polynuclear compounds so far investigated we have detected transient absorption spectra which are attributed to absorption from the lowest triplet level. The example in the accompanying figure was obtained with a 10^{-5} molar solution of anthracene in hexane from which dissolved gases had been carefully removed. Each spectrum was recorded by a flash lasting 25 μsec. The fluorescence resulting from the photolysis flash alone is shown in the first spectrum. The second spectrum, taken before the photolysis flash, shows the absorption of normal anthracene in the ground singlet state and subsequent spectra were taken in arbitrary order at increasing time intervals after photolysis. Immediately after photolysis there is a decrease in the absorption of the singlet state accompanied by the appearance of new absorption bands at 4203 and 3980A and a continuous absorption beginning at 2800A which increases to lower wavelengths until it overlaps the second singlet system. The bands are attributed to absorption from the lowest triplet level to two vibrational levels of an excited triplet state, the continuum arising from the transition to a second electronically excited triplet state. The triplet spectra disappear, and the singlet absorption returns to its original intensity, with a half time of about 100 μsec. The conversion time is unchanged in ethanol solution but is increased by a factor of ten in the more viscous solvent, glycerol. Similar spectra have been obtained with solutions of naphthalene, phenanthrene, naphthacene, triphenylene, chrysene, pyrene, 3,4-benzphenanthrene, 1,2-benzanthracene, 1,2,5,6-dibenzanthracene, perylene, rubrene, coronene, β-bromonaphthalene, α-chloroanthracene and also diacetyl. All have lifetimes in hexane of 100 μsec or less and there is no obvious correlation of lifetime with molecular structure.

It is clear that the formation of triplet states by photolysis in solution is a general phenomenon. It is also one of major importance, as is shown by the fact that in some cases about 50 percent conversion to the triplet state is produced by a single flash. These high conversions make possible the direct measurement of absolute concentrations and hence of extinction coefficients and transition probabilities of the triplet transitions. Of most promise to the chemist is the fact that the reactions of triplet molecules in solution become amenable to direct kinetic investigation.

We are grateful to the British Rayon Research Association for financial support and to Dr. E. J. Bowen and Professor J. W. Cook for samples of hydrocarbons.

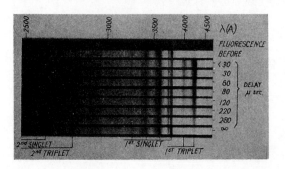

FIG. 1. Flash photolysis of a 10^{-5} molar solution of anthracene in **hexane**.

[1] Lewis, Lipkin, and Magel, J. Am. Chem. Soc. **63**, 3005 (1941).
[2] D. S. McClure, J. Chem. Phys. **19**, 670 (1951).
[3] E. J. Bowen, Faraday Soc. Discussions **14**, 143 (1953).
[4] G. Porter, Proc. Roy. Soc. (London) **A200**, 284 (1950).

Reprinted from the *Transactions of the Faraday Society*,
No. 393, Vol. 51, Part 9, September, 1955

PRIMARY PHOTOCHEMICAL PROCESSES IN AROMATIC MOLECULES

PART 2.*—OBSERVATIONS ON THE TRIPLET STATE IN AROMATIC VAPOURS

By George Porter and Franklin J. Wright
Physical Chemistry Dept., University of Cambridge

Received 28th February, 1955

Conversion to the triplet state has been observed following the optical excitation of aromatic molecules in the vapour phase. The absorption spectra of the triplet states are tabulated for nine molecules. The lifetime of the triplet state is less in the vapour phase than in solution or in rigid glasses and the rates of intersystem crossings, both from upper singlet to triplet, and from triplet to ground state, are independent of collision rate with an inert gas, such as carbon dioxide, over a wide range of pressures. Our observations emphasize the important differences between these two formally identical processes.

The phosphorescence of aromatic molecules, which can be observed in rigid or very viscous media, results from a transition between a metastable level of the molecule and the ground state. It is now generally accepted that this metastable level is the lowest triplet level of the molecule, and it will be referred to as. the triplet state throughout this paper. The state has its own distinct absorption spectrum, with high extinction coefficients, which must arise from allowed transitions to higher levels of the same multiplicity. The decay time of this absorption, in rigid glasses at low temperatures, is identical with the decay time of the phosphorescent emission [1] and it follows that both arise from the same metastable state.

In ordinary fluid solvents and in the gas phase, phosphorescence of long duration is never observed, but it has been shown in part 1 that the triplet phosphorescent state is nevertheless formed with high efficiency in solution after absorption of

* *Discussions Faraday Soc.*, 1954, **17**, 178, forms part 1 of this series.

radiation which initially elevates the molecule to an upper singlet level.   The absence of phosphorescence in fluid solvents is due to a more rapid radiationless deactivation of the triplet level and the rate of this process increases with decreasing viscosity of the medium.   In attempting to gain a better understanding of this deactivation process, and also of the process by which the upper singlet level is converted to the triplet level, it is natural to enquire what is the probability of these processes in the isolated molecule or in the nearest practical equivalent —the gas phase at low pressures.   Even the knowledge that the triplet state is formed under such conditions would be of assistance in eliminating certain possible mechanisms, and kinetic studies should be particularly valuable since our understanding of the kinetics of collisional processes is far more advanced for the gaseous than for condensed phases.   We have therefore used the flash photolysis technique, which was successful for the study of the triplet state in fluid solvents, to investigate the triplet state in the vapour phase.

EXPERIMENTAL

The flash photolysis apparatus has been previously described.[2]   For the present work we used a quartz reaction vessel, one metre in length with a photolysis flash lamp of the same length, both surrounded by a reflector of magnesium oxide.   The photolysis flash was operated with a condenser of 56 μF charged to 7000 V and had a half-peak duration of 70 μsec.   Spectra were recorded by means of a spectroscopic flash of 30 μsec duration and a large Littrow (Hilger E1) spectrograph.   The spectral region between 2100 and 5600 Å was investigated, though the lower wavelength limit was usually fixed by the singlet absorption of the substance being studied.

In order to obtain a sufficient vapour pressure, the reflector assembly was enclosed in a furnace.   To prevent condensation the reaction vessel had double windows with an evacuated space between them.   All experiments were conducted in the presence of a considerable excess pressure of inert gas, carbon dioxide or nitrogen being used for this purpose.   The former was prepared by vacuum fractionation of dry ice and the latter from sodium azide.   Anthracene and naphthalene were of spectroscopic purity.   We are grateful to Prof. J. W. Cook for gifts of fluorene, 1 : 2-benzanthracene, pyrene and triphenylene and to Dr. E. J. Bowen for the naphthacene and phenanthrene.   All other materials were the purest grade available commercially and were purified immediately before use by fractional distillation or by a single recrystallization.   Whenever possible several samples from different sources were investigated.   The temperature was chosen to give a vapour pressure of the order of 1 mm Hg, so that the concentrations were comparable with those which had been used in fluid solvents.   Inert gas was added to a total pressure of 700 mm Hg in order that no appreciable temperature rise should accompany the change.

RESULTS

All the polycyclic aromatic hydrocarbons investigated showed transient spectra during photolysis which disappeared immediately irradiation ceased, i.e. after about 100 μsec. As in condensed phases, negative results were obtained with benzene and its derivatives (but see part 3).   The wavelengths of all transient spectra which were observed are given in table 1 along with the frequencies of triplet state spectra recorded in hexane at room temperature and in E.P.A. glass at 77° K.   Unless otherwise stated, measurements refer to centres of bands of similar width to those of anthracene and naphthalene (fig. 1 and 2).

ANTHRACENE

This substance was of particular interest since it has been the most thoroughly studied in fluid solvents.   The absorption spectra before, during and after photolysis are shown in fig. 1, the pressure of anthracene being 0·2 mm and that of carbon dioxide 700 mm Hg. The two new bands which appear during photolysis are identical in appearance with those of the published triplet spectra obtained in liquid hexane (part 1), but are shifted to shorter wavelengths.   This shift is a characteristic difference between spectra in condensed and gaseous phases and the singlet spectrum shows a similar shift.   In spite of the high resolving power of the spectrograph the bands showed no fine structure, nor even a head, but were quite as diffuse as those observed in solution.   This is not entirely caused by the rather

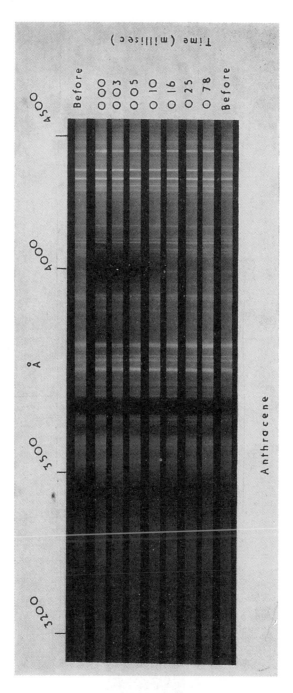

Fig. 1.

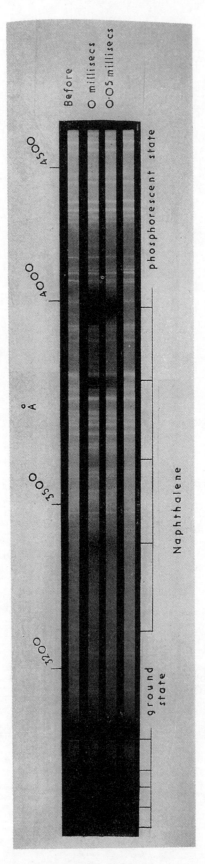

Fig. 2.

high temperature (120° C) since the singlet spectrum showed quite sharp bands, this being particularly evident with naphthalene (fig. 2). The diffuseness seems to be a characteristic of triplet-triplet spectra and we have found it quite useful as a preliminary diagnosis of such transitions.

NAPHTHALENE

The triplet spectrum of naphthalene at 0·7 mm pressure admixed with 700 mm of carbon dioxide is shown in fig. 2. It consists of five bands, the first three of which correspond to those of the triplet spectrum of naphthalene observed in solution, as reference to table 1 will show. The appearance of the two further bands at shorter wavelengths is rather puzzling since, although both are stronger than the third band at 3581 Å, they

TABLE 1

molecule	temp. (° C)	gas (CO_2)		$\Delta\nu$ (gas-hexane)	hexane (cm^{-1})	$\Delta\nu$ (hexane—E.P.A.)	E.P.A. (cm^{-1})
		λ (Å)	ν (cm^{-1})				
naphthalene	50	3971	25,180	970	24,210	210	24,000
		3766	26,550	910	25,640	175	25,465
		3581	27,930	720	27,210	260	26,950
		3408	29,340				
		3236	30,900				
α-methyl naphthalene	40	4045	24,720	1020	23,700	—	—
		3787	26,410	1280	25,130		
anthracene	120	4031	24,810	1020	23,790	370	23,420
		3812	26,230	1100	25,130	380	24,750
phenanthrene	200	4571	21,880	1090	20,790	510	20,280
		4301	23,250	1170	22,080	350	21,730
		4066	24,590	1060	23,530	330	23,200
naphthacene	160	4231	23,630	—	20,400	—	—
		4001	24,990		to 25,000		
1 : 2-benz-anthracene	200	wide band from 21,740 to singlet with max. at 22,120		—	20,620 22,200 23,260	—	20,450 21,100 21,700 23,050
pyrene	100	3945	25,350	1020	24,330	260	24,070
					25,840	—	—
					27,100	200	26,900
fluorene	95	4900	20,410	—	wide band 25,000 to 29,400	—	23,750 25,300 25,800
		wide band from 26,300 to 31,250 with max. at 28,360 and 30,230					
triphenylene	200	24,390 to singlet		—	22,500 to 33,500	—	23,200

Description of triplet absorption bands in the vapour phase and comparison with observations in hexane (part 1 and ref. (3)) and E.P.A. solutions (ref. (1) and (4)).

were not observed in solution. Impurities were at first suspected but it was found that the five bands were observed with the same relative intensity from different samples of naphthalene and, more significantly, the same samples investigated in solution by Mr. Windsor showed the first three bands only. The two additional bands which are observed in the gas phase appear to belong to the same progression and if this is the case the relative intensities of the five bands is also unusual. These intensities were as follows :

ν (cm^{-1})	25,183	25,546	27,925	29,345	30,903
$\epsilon_{relative}$	1·00	0·60	0·09	0·40	0·14

OTHER HYDROCARBONS

In cases where clear bands are observed both in the gas phase and in the condensed phases the same correlation between the observed spectra exists as with anthracene and naphthalene, i.e. the same bands are observed but show a shift of about 1000 cm⁻¹ to lower wavelengths in the gas phase as compared with solutions in hexane at room temperature. This applies to phenanthrene, pyrene and α-methyl naphthalene where there can be little doubt as to the identity of the vapour phase spectra with those obtained in solution. With 1 : 2-benzanthracene and triphenylene the spectra are more diffuse in the gas phase, probably owing to the elevated temperature, while naphthacene, which showed a fairly sharp band spectrum in our vapour phase experiments, has only a diffuse region of absorption in solution. Apart from these differences of structure and sharpness, the absorption regions are similar to the spectra attributed to triplet states in solution, and may in all cases be fairly definitely attributed to the same species. The only exception is the 4900 Å band of fluorene which may be the band of a transient dissociation product (see part 3).

DEPENDENCE OF TRIPLET CONCENTRATION ON PRESSURE

It had been hoped that if the triplet state were observed it would be possible to investigate the kinetics of its deactivation directly in the same manner as was applied to the study of its reaction in solution. Unfortunately this proved to be impossible with the present techniques since the lifetimes of the spectra were less than the duration of the photolysis flash. Since the concentrations of triplet molecules were similar to those observed in solution it is unlikely that the life-time is very much shorter and it may be possible, with improved time resolution, to carry out such direct measurements. At present, however, we must resort to a less powerful approach.

In attempting to measure the decay time of the triplet state, its extinction was measured as a function of time and compared with the oscillographic recording of light intensity. It was found that the extinction, and therefore, if Beer's law is valid, the concentration of the triplet, was proportional to the light intensity at all times within the accuracy of our measurement of the two quantities. We may therefore apply stationary state kinetics to the problem and, in order to reduce the possibility of error due to departure from steady state conditions, the relative concentrations were measured at the peak of the flash when the rate of change of intensity is momentarily zero. Measurements were made at the centre of the strongest band and, in view of the low pressures of aromatic vapour and the absence of any fine structure, the use of Beer's law is almost certainly justified.

The relative concentrations of triplet, expressed as optical density of the absorption maximum in a 1-m path, were measured as a function of the pressure of carbon dioxide, all other factors, including flash intensity, concentration of aromatic vapour and temperature, being held constant. The results for anthracene and naphthalene were as follows.

NAPHTHALENE

Temp. 50° C, pressure 0·65 mm

pressure of CO_2 (mm Hg)	500	100	50	20	5	0
optical density of triplet max.	0·79	0·75	0·80	0·60	0·50	0·29

ANTHRACENE

Temp. 120° C, pressure 0·22 mm

pressure of CO_2 (mm Hg)	500	100	20	10	5	2	0
optical density of triplet max.	0·81	0·80	0·81	0·82	0·78	0·66	0·25

Each density is the mean of four independent measurements whose mean deviation was 5 %.

The stationary state concentration of triplet molecules is constant, and independent of pressure, at pressures of carbon dioxide higher than 10 mm for anthracene and at pressures higher than 50 mm for naphthalene.

DISCUSSION

The observation of photo-excitation to the phosphorescent state in gases has considerable bearing on questions concerning the nature of this state and the mechanisms of its formation and deactivation. The results of this work, along with those of part 1, fully confirm the contention of Lewis and Kasha [5] that formation of the triplet state should be general to all phases. The possibility of explaining phosphorescent phenomena in terms of ionic species was already exceedingly remote, but can now definitely be ruled out since the magnitude of the energy changes in the gas phase is known. Thus the ionization potential of naphthalene vapour is 8·3 eV,[6] whereas the energy of the quantum of highest frequency which could have been absorbed in our experiments was only 6 eV, and it will be shown later that quanta of much lower energy are also effective in forming the triplet state.

The mechanism of conversion to and from the triplet state must now be considered in connection with our observations on the pressure dependence of the triplet stationary state concentration. The question of particular interest is whether the rate-determining process of conversion between states of different multiplicity is a collisional one. We shall therefore consider the following processes:

$$A + h\nu = A^* \qquad k_1$$
$$A^* = A + h\nu \qquad k_2$$
$$A^* = A_T \qquad k_3$$
$$A^* + M = A_T + M \qquad k_{3a}$$
$$A_T = A \qquad k_4$$
$$A_T + M = A + M \qquad k_{4a},$$

where A is the molecule in its ground state, A^* is the excited molecule in an upper singlet level, A_T represents the lowest triplet level and M is a third body. These are probably all the reactions which need be considered under the present conditions. Quenching of anthracene fluorescence by inert gases is negligible and self-quenching has been shown to be small at the pressures used here.[7] Anthracene has no effect on the rate of deactivation of the triplet state at relatively high concentrations in solution. In any case, since the pressure of aromatic vapour was constant, self-quenching reactions would merely affect the magnitude of the constants k_2 and k_4 without altering the arguments to follow. Wall reactions are excluded since the times were too short for diffusion across the vessel to be significant.

By the usual steady state treatment the concentration of triplet molecules obtained by the various combinations of these reactions is as follows:

(A) 1, 2, 3 and 4. $\{A_T\} = k_1 k_3 I_{abs}/k_4(k_2 + k_3)$,

(B) 1, 2, 3a and 4a. $\{A_T\} = k_1 k_{3a} I_{abs}/k_{4a}(k_2 + k_{3a}M)$,

(C) 1, 2, 3a and 4. $\{A_T\} = k_1 k_{3a} M I_{abs}/k_4(k_2 + k_{3a}M)$,

(D) 1, 2, 3 and 4a. $\{A_T\} = k_1 k_3 I_{abs}/k_{4a}M(k_2 + k_3)$,

where $M \equiv [M]$.

Except at low pressures we have found that $\{A_T\}$ is independent of pressure and therefore only scheme A is applicable. At pressures greater than about 10 mm and 50 mm for anthracene and naphthalene respectively it follows that neither conversion process is rate-determined by collisions with inert gas. It should be mentioned that although schemes B and C become independent of gas pressure when k_2 and $k_{3a}M$ respectively become large, neither of these cases can apply since they correspond to complete absence either of fluorescence or of triplet state formation, whilst it is known that the fluorescence yield of anthracene in the gas phase is nearly unity [8] and triplet formation is, of course, the quantity measured.

The low pressure results have been left for separate discussion since their significance is less certain. When the pressure of carbon dioxide is reduced to a few millimetres a considerable temperature rise is to be expected and this is

TRIPLET STATE IN AROMATIC VAPOURS

confirmed by a broadening of the band structure immediately after photolysis at low pressures. However, the marked decrease in triplet concentration suggests that there may be a real pressure dependence at lower pressures, and if this is accepted reaction (3), but not reaction (4), must become a collisional process. In this case we should replace reaction (3) by

$$A^* = A_T' \qquad k_5$$
$$A_T' = A^* \qquad k_6$$
$$A_T' + M = A_T + M \qquad k_7,$$

where A_T' is the molecule in its lowest triplet level before removal of excess vibrational energy by collision. This type of reverse process must of course eventually become significant for both reactions (3) and (4) at very low pressures.

Reactions (1), (2), (4), (5), (6) and (7) lead to the relation

$$\{A_T\} = k_1 k_5 k_7 M I_{abs}/k_4[k_2 k_6 + k_7 M(k_2 + k_5)],$$

which is qualitatively in accord with our results over the whole pressure range. The predicted linear relation between $1/A_T$ and $1/M$ is only very approximately followed, but this is not surprising in view of the fact that at low pressures the anthracene concentration becomes a significant factor and the reaction is no longer isothermal. Therefore, whilst there is some indication that reaction (3) becomes pressure dependent at lower pressures, we shall limit our conclusions to the more reliable high-pressure range where the system does not depart significantly from isothermal conditions. In this region, at pressures of carbon dioxide greater than 10 mm for anthracene and 50 mm for naphthalene, both the rate of conversion of the excited singlet state to the triplet and of the triplet to the ground state are constant over a wide pressure range of the inert gas (50-fold for anthracene).

The independence of pressure of both these processes, especially the triplet state deactivation, would have been a somewhat surprising result if experiments in other media had not prepared us to expect something of the kind. Since, however, the triplet state deactivation is slower in solution than in the gas phase, it would have been even more surprising if it had been found to be kinetically dependent on collision rate. Since the viscosity of a gas is independent of pressure over the range which we have studied, and is less than that of a liquid, the gas-phase behaviour is formally in accordance with the findings of part 1 on the viscosity dependence of this conversion. There was already evidence to suggest that the singlet-triplet conversion also cannot be rate determined by collisions. In the few cases where data are available it is found that the fluorescent yield is similar in the gas, liquid and solid phases,[9] and since this radiative process is competitive with conversion to the triplet state we should expect the latter process also to be independent of the phase, and therefore independent of gas pressure over the pressure range at which these fluorescence determinations were made. We have not yet been able to measure quantum yields of triplet state formation, but a comparison of triplet concentrations formed in the gas phase and in solution under otherwise identical conditions indicates that the yields cannot be greatly different in the two phases.

Our knowledge of the two conversion processes may be summarized as follows :

UPPER SINGLET → TRIPLET

The rate is independent of inert gas (CO_2) pressure over a wide range but may become pressure dependent at about 10 mm pressure for anthracene and 50 mm for naphthalene. The rate is also not very different in gas, liquid and solid phases.

The rate constant of conversion must be comparable with that of fluorescence in those molecules where both fluorescence and phosphorescence are observed, i.e. k_3 usually lies in the range 10^6 to 10^9 sec^{-1} for aromatic molecules.

TRIPLET → GROUND STATE

The rate is independent of inert gas (CO_2) pressure above 10 mm pressure for anthracene and 50 mm for naphthalene and probably also at still lower pressures. The rate is, however, very dependent on the phase. It decreases with increasing viscosity of the solvent until the limiting rate of the radiative conversion is reached.

The rate constant in the gas phase, for all molecules so far studied, is $> 10^4$ sec^{-1} and probably $< 10^5$ sec^{-1}. Its minimum value, in rigid media, is less, and probably very much less, than the rate constant of the radiative process which, for the majority of aromatic molecules, lies between 10^3 and 10^{-1} sec^{-1}.

These results confirm and emphasize the differences between these two formally identical processes which were indicated in part 1.

[1] Craig and Ross, *J. Chem. Soc.*, 1954, 1589.
[2] Porter and Wright, *Faraday Soc. Discussions*, 1953, **14**, 23.
[3] Porter and Windsor, *Molecular Spectroscopy* (Institute of Petroleum Conference, 1954).
[4] McClure, *J. Chem. Physic*, 1951, **19**, 670.
[5] Lewis and Kasha, *J. Amer. Chem. Soc.*, 1944, **66**, 2100.
[6] Sugden, private communication.
[7] Bowen, *Trans. Faraday Soc.*, 1954, **50**, 97.
[8] Stevens, private communication.
[9] Gilmore, Gibson and McClure, *J. Chem. Physics.*, 1952, **20**, 829.

Reprinted from Proceedings of the Chemical Society, October, 1959, page 291

TILDEN LECTURE*

The Triplet State in Chemistry

By GEORGE PORTER
(UNIVERSITY OF SHEFFIELD)

THE concept of a biradical, or a molecule having two separate unsatisfied valencies, has been frequently used in chemistry. The spectroscopist has been equally familiar with the term "triplet state", used to

* Delivered before the Chemical Society at Burlington House, London, on December 11th, 1958; at King's College, Newcastle upon Tyne, on January 20th, 1959; at The University, Southampton, on January 30th; at The University Leicester, on February 16th; and at Marischal College, Aberdeen, on March 10th.

describe an atom or molecule having two electrons with parallel spin. Spectroscopic observations of the triplet state were, until recently, confined to atoms and diatoms whilst the most characteristic examples of biradicals are found in more complex organic molecules. One of the most important recent advances of molecular spectroscopy has been the observation of the triplet states of a wide variety of organic molecules and the determination of their energy levels. Equally important to the chemist is the development of methods which make possible the direct observation of triplet states in solution and in the gas phase, and the study of their physical stability and chemical reactivity.

It will be appropriate to consider first how the triplet state arises and why it is of importance in chemistry. Molecules having an even number of electrons, *i.e.*, nearly all chemically stable molecules, have odd multiplicity and usually have a singlet ground state, but excited configurations give rise to both singlet and triplet states. If the spins of electrons occur in antiparallel pairs the resultant spin angular momentum is zero, giving a single or "singlet" level. If the spins of two electrons are parallel the resultant spin angular momentum is $\frac{1}{2} + \frac{1}{2} = 1$, in atomic units, and this vector has three components, $+1$, 0, and -1, which results in a "triplet" of energy levels in an internal molecular or external field. The degree of splitting of the three components of the triplet, in the absence of an external field, is determined by the coupling between spin and orbital motions which in turn depends on the atomic masses. In light atoms, such as those of which most organic molecules are composed, the coupling is small and it will be sufficient to regard the triplet state in such molecules as a single energy level.

According to the Pauli principle a single non-degenerate atomic or molecular orbital can accom-

configuration of lowest energy, will usually be a singlet:

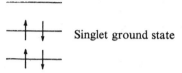

Singlet ground state

but each singly excited configuration may be singlet or triplet; *e.g.*, the lowest excited configuration:

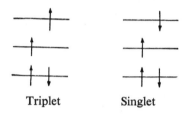

Triplet   Singlet

The singlet and the triplet state arising from the same configuration have different energies, and the triplet state lies the lower. This has long been known for atoms in the empirical form of Hund's rule.

The difference in energy between singlet and triplet levels of the same configuration does not arise from interactions between the spins of the electrons, which are relatively quite weak, but from the different spatial electron distributions which are dictated by the spin functions in accordance with the Pauli principle. This principle may be stated in the more general form: "The total wave function of an atom or molecule is antisymmetric to the exchange of any two electrons." Allowing for indistinguishability of the two electrons we obtain the following satisfactory wave functions for the ground and first excited configurations:

Singlet $\quad \Psi_s^* = \{\psi_a(1)\psi_b(2) + \psi_a(2)\psi_b(1)\}\{\alpha(1)\beta(2) - \alpha(2)\beta(1)\}$

Triplet $\quad \Psi_T^* = \{\psi_a(1)\psi_b(2) - \psi_a(2)\psi_b(1)\}\left\{\begin{array}{c} \alpha(1)\ \alpha(2) \\ \alpha(1)\ \beta(2) + \alpha(2)\ \beta(1) \\ \beta(1)\ \beta(2) \end{array}\right\}$

Singlet $\quad \Psi_0 = \{\psi_a(1)\psi_a(2)\}\{\alpha(1)\ \beta(2) - \alpha(2)\ \beta(1)\}$

modate only one electron of each spin quantum number. The maximum number of electrons in any orbital is therefore two and these must have antiparallel spins so that the ground state, having the

Here ψ_a and ψ_b are the co-ordinate wave functions for the orbitals a and b, and α and β the spin wave functions corresponding to spin $\frac{1}{2}$ and $-\frac{1}{2}$. Other combinations of these spin and co-ordinate functions would not change sign on interchanging electrons 1 and 2 and would be contrary to the Pauli principle. The spin functions therefore determine the co-ordinate functions which are acceptable.

When we used the "single electron" molecular orbitals the singlet and triplet states arising from the same configuration had exactly the same energy but coulombic repulsion between the two electrons was completely ignored. This repulsion will increase the energy by an amount which depends on the averaged relative position of the two electrons. Now the probability that electron 1 is in a volume $d\tau_1$ and at the same time electron 2 is in a volume element $d\tau_2$ is proportional to $\Psi\Psi^*d\tau_1 d\tau_2$ where, for the excited singlet state

$$\Psi_s = \psi_a(1)\,\psi_b(2) + \psi_a(2)\,\psi_b(1)$$
and for the triplet
$$\Psi_T = \psi_a(1)\,\psi_b(2) - \psi_a(2)\,\psi_b(1)$$

The two functions therefore lead to different probability density distributions and it has been shown by Dickens and Linnett, for several typical cases, that if the probability is plotted as a function of the co-ordinates the chance of the two electrons' being at the same point is finite when the singlet wave function is used but is zero with the triplet wave function. Quite irrespectively of electron repulsion the electrons are further apart in the triplet than in the singlet state.

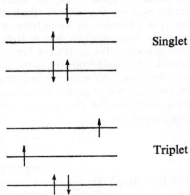

Singlet

Triplet

When electron repulsion is considered the electrons will be further separated in both states but will remain further apart in the triplet than in the singlet state. The coulombic repulsion will therefore be less in the triplet which will be of lower energy than the singlet. A second consequence of the different electron distributions is that, in so far as a "biradical" has two *separate* unpaired electrons, the triplet state is more of a biradical than the corresponding singlet. A true biradical, in which the two electrons would be so far separated that no exchange occurred, would of course have degenerate singlet and triplet levels and be better regarded as two independent doublets.

If two orbitals are degenerate or have nearly equal energies the *lowest* singlet and triplet states will have equal energies in the single electron approximation and, after allowance for electron interaction, the triplet state will be the lower.

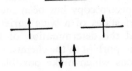

Triplet ground state

This is the case in oxygen, and also in some highly conjugated organic molecules where a near-degeneracy of orbital energies results in a triplet ground state.

Triplet Levels in Atoms and Diatomic Molecules

To illustrate how this works out in practice we may refer to some typical atomic and molecular energy levels. The levels of three atoms, He, Ca, and Hg are shown in Fig. 1.

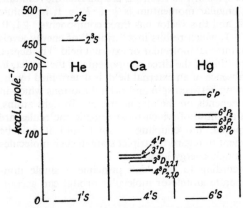

FIG. 1. *Lowest excited levels of some atoms.*

In a heavy atom, such as mercury, spin–orbit interaction results in a considerable energy difference between the three components of the triplet and also in strong intercombination lines.

The levels of diatomic molecules are quite similar, as shown in Fig. 2, for H_2, CO, I_2, and GaCl. Again in the heavy molecule, iodine, the singlet–triplet transition is quite strong and is, indeed, responsible for its colour. All the states shown on this diagram are stable except for the triplet level of H_2 which is entirely repulsive and is the state which arises from the antisymmetric combination of co-ordinate wave functions in the Heitler–London treatment. In molecular-orbital language, one of the electrons is promoted from a bonding to an antibonding orbital so that the energy is greater than that of the two

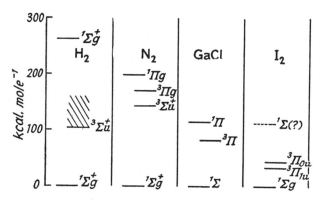

FIG. 2. *Lowest excited levels of some diatomic molecules*

normal 1*s* hydrogen atoms with which the state correlates. The corresponding singlet has an even higher energy with respect to two 1*s* hydrogen atoms but correlates with a 1*s* and a 2*p* atom and is therefore stable with respect to its dissociation products.

But the instability of the first excited triplet state of hydrogen is by no means general. Some of the higher excited triplet states of H_2 have deeper potential minima than the singlet states, and the lowest triplet states of most diatomic molecules are quite stable. The calculations of Heitler and London have only been carried out for *s* states and apply, of course, only to two electrons, other electrons being ignored. It is probable that the hydrogen situation also occurs in other molecules in which all electrons, apart from those in very low energy levels, are used in forming single bonds, *e.g.*, in saturated paraffins. On the other hand the excitation of non-bonding or antibonding electrons will not necessarily reduce the bond strength whilst the excitation of electrons in a multiple bond, although it may reduce the bond energy, will not usually result in an unstable state. Since few molecules, apart from the paraffins, have all outer electrons in single bonding orbitals, we shall expect to find that most lower triplet states are stable.

To summarise, the lowest excited state of nearly all stable molecules is a triplet state and, except in molecules whose electrons are all occupied in single-bond formation, the lowest triplet state is probably stable. This state has two separate unpaired electrons and, in so far as excited states play a part in chemistry, we shall expect that the lowest triplet level will be of prime importance.

Triplet States in Polyatomic Molecules

The positions of the triplet states of simple polyatomic molecules such as H_2O, CO_2, C_2H_2, and C_2H_4 are not yet known. We may be quite certain

that such states exist and it is probable that they are stable. In some of these molecules the triplet may lie near enough to the ground state to be significant in the determination of thermodynamic properties, especially at high temperatures, such as the heat capacities of water and carbon dioxide in combustion processes.

The dearth of our knowledge about triplet states of polyatomic molecules arises from the fact that radiative transitions between states of different multiplicity, *e.g.*, singlet to triplet, are forbidden, and because the probability of radiationless processes of energy dissipation is high. Two special methods are now available for the direct study of the triplet states of polyatomic molecules. The first is observation of the phosphorescence of rigid solutions and the second is the study of absorption spectra of the triplet state after flash activation in solids, liquids, or gases.

Phosphorescence of Rigid Solutions.—The phosphorescence of solid solutions of aromatic molecules and dyes has been known for more than half a century. In 1935 Jablonski suggested that the phenomenon involved a transition from a metastable level of the molecule which was reached by radiationless crossing from the first excited state as shown in Fig. 3.

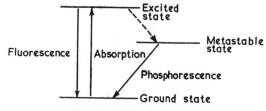

FIG. 3. *Jablonski diagram*

In view of the previous discussion it is now natural to conclude that this metastable level is the lowest triplet level of the molecule, although this interpretation was not found until a decade later. In the meantime attempts were made to explain phosphorescence and the Jablonski metastable state in terms of tautomeric or isomeric forms of the molecule. Interpretation in terms of the triplet state was first suggested by Lewis, Lipkin, and Magel, and also by Terenin, and was put in a clear convincing form by Lewis and Kasha in 1944, since when extensive

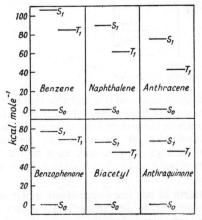

FIG. 4. *Lowest singlet and triplet levels of some organic compounds*

states in solution by means of their absorption spectra. Strong absorption spectra to higher triplet levels were known to exist and had been detected in rigid media even before the triplet state theory was developed. It was not clear, however, whether the absence of phosphorescence in ordinary solutions meant that the triplet state was not generally formed or that it was deactivated, in fluid media after formation, by some radiationless process which was very rapid compared with the radiative process of phosphorescence.

The principle of our method is shown in Fig. 5. The photolysis flash cannot excite the triplet state directly to any significant extent but it populates upper singlet states which may pass, by radiationless conversion, into the lowest triplet state as they are known to do in the rigid media. After a short time interval has elapsed a second spectroscopic monitoring flash is triggered; light from this passes through the solution and records the absorption spectrum of the lowest triplet level as it passes to an excited triplet state.

The experiment was immediately successful and the first record of triplet–triplet absorption spectra in solution is shown in Plate 1 (facing p. 310). It was obtained from a 10^{-5}M-solution of anthracene in hexane which had been thoroughly outgassed.

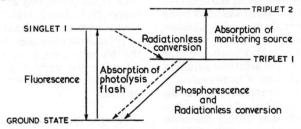

FIG. 5. *Transitions involved in flash-photolysis studies of the triplet state*

work has established the lowest triplet energy levels and radiative lifetime of a large number of substances. Some typical energy levels of polyatomic molecules are shown in Fig. 4.

Observation of Triplet States by Flash Photolysis of Solutions and Vapours.—Important as phosphorescence studies have been in the development of triplet-state theory and the measurement of the lowest triplet energy levels, they are not normally applicable to the study of triplet states in solution or in the gas phase which are, of course, of most interest to the chemist.

In 1952, Mr. Windsor and I decided to apply the method of flash photolysis, which had been successfully used in the detection of transient free radicals in the gas phase, to attempt the observation of triplet

The life of the triplet state is seen to be about 200 μsec., which is short enough to explain why phosphorescence is not observed under these conditions but long enough to be studied kinetically by flash techniques.

Absorption spectra of many other triplet states were observed in this way in a variety of solvents and shortly after these experiments Dr. F. J. Wright, using a longer reaction vessel, surrounded by a furnace to give the required vapour pressure, was able to record the triplet–triplet absorption spectra of several of the same molecules in the gas phase.

One may ask how these absorption spectra are known to be those of triplet states; indeed other spectra are obtained under identical conditions which we have assigned to free-radical dissociation products. The assignments are based on several types of evidence. First, the study of a series of related molecules usually makes possible the identification of spectra which arise from free radicals and similar transient products. Secondly, the spectra assigned to triplet states are also observed in rigid solvents where they decay with the same rate as the phosphorescence whilst free radicals and similar products, if formed at all in rigid solvents, are usually trapped indefinitely.

Recently, theoretical calculations and absolute measurements of extinction coefficients have provided a third type of evidence. In many of our flash-photolysis experiments the proportion of molecules

triplet transitions. Semiempirical molecular-orbital calculations, particularly those of Pariser, have predicted the positions of these levels for the linear polyacenes, and a comparison of our measurements with Pariser's calculations is shown in Figs. 6 and 7. The combined errors of the experimental estimations and the calculations are quite high, but the general picture leaves little doubt as to the identity of the theoretically predicted triplet levels and those which we have measured.

Triplet State Kinetics in Gases and Solutions

These observations opened up the possibility of direct studies of the triplet state in chemical reactions.

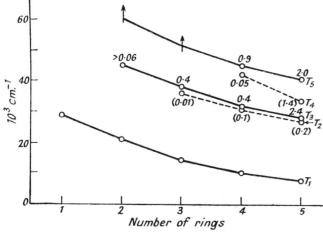

FIG. 6. *Experimental triplet levels and oscillator strength of the linear polyacenes*

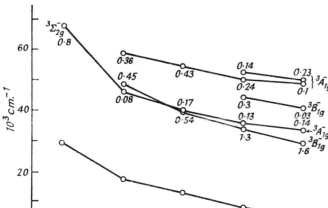

FIG. 7. *Calculated triplet levels and oscillator strengths for the linear polyacenes (Pariser)*

converted into the triplet state is so high that the singlet state is depopulated to a significant extent, and it has therefore been possible to measure the absolute concentrations and extinction curves of the

But before chemical behaviour is investigated the more fundamental question of triplet state deactivation in the absence of chemical change must be better understood. The outstanding problem was simply this: "Why is phosphorescence of long duration observed only in solid or rigid media?" It is clear, since we have found that the triplet state is readily formed in gases and liquids, that the absence of phosphorescence must be caused by a rapid radiationless process of some kind. What is this process and why does it not occur in rigid media?

Our first experiments showed that the decay was predominantly of the first order and that the rate constant decreased markedly with increasing viscosity. The viscosity effect suggested some form of diffusion control and it therefore seemed obvious that we must look for collisional deactivation processes. Several of these were found, e.g., oxygen deactivated the triplet state with nearly unit encounter efficiency, but in the absence of any known quenching molecules the decay was essentially of first order in triplet concentration and the lifetime was still very much shorter in solution than the radiation lifetime found in rigid media.

Flash-spectroscopic methods involve measurement of photographic plate density, which is not a very accurate procedure, and subsequent kinetic work has been carried out mainly by photoelectric methods. Here a continuous monitoring source is used and the transient absorption is recorded at one wavelength throughout its formation and decay. Typical records of this kind, obtained by Mr. M. R. Wright for triplet naphthalene in solution are shown in Plate 2. Measurement of these traces leads to very accurate kinetic data and we found that the decay was, in

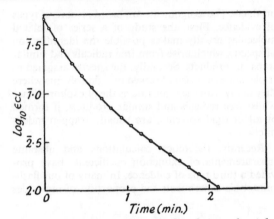

FIG. 8. *First-order plot of the decay of triplet naphthalene in hexane*

energy-transfer from an excited state which is apparently independent of collisions, indeed the results in the gas phase indicate that the decay rate is independent of gas pressure. The only reasonable explanation of these facts seems to be as follows:

(1) The rate-determining process of triplet-state deactivation in solution and in the gas phase, in the absence of specific quenching by other molecules, is radiationless conversion to an isoenergetic level of the ground state.

(2) The transition is inhibited or prevented in viscous or rigid media for the following reason. If a

TABLE 1. *First- and second-order rate constants of triplet naphthalene decay in several solvents.*
(*Concentration of naphthalene* $= 10^{-4}$M)

Solvent	Viscosity (cp)	$10^{-3} k_1$ (sec.$^{-1}$)	$10^{-9} k_2$ (1. mole^{-1} sec.$^{-1}$)
n-Hexane	0·3	12·1 $\pm$ 1·3	2·1 $\pm$ 0·6
Water	1·1	7·5 $\pm$ 0·6	4·1 $\pm$ 1·2
Ethylene glycol	21·1	0·97 $\pm$ 0·1	0·22 $\pm$ 0·03
Liquid paraffin I	33·0	1·5 $\pm$ 0·1	0·39 $\pm$ 0·02
Liquid paraffin II	167	0·31 $\pm$ 0·03	0·08 $\pm$ 0·01

fact, partly of the second order, i.e., there was some contribution to the triplet decay from triplet–triplet encounters. A first-order plot from the data of Plate 2 is shown in Fig. 8. The conclusion that the decay was predominantly of the first order was, however, confirmed and it was possible to measure both second- and first-order decay constants and to show that not only the second-order decay, but also the first-order decay was viscosity-dependent.

The rate constants for naphthalene in various solvents are shown in Table 1.

These findings are rather strange. We have a spontaneous unimolecular process which is apparently diffusion-controlled. We are also dealing with

molecule is rigidly held in one particular nuclear configuration it will oscillate back and forth between the two electronic states but will not be able to lose energy since no conversion into kinetic energy can take place. It will therefore remain in this configuration until energy is lost by radiation.

It is interesting that the appearance of phosphorescence is then dependent, not only on the spin-forbidden nature of the radiative transition, but also on the configurational forbiddeness of the radiationless transition in rigid media. The early explanation of the metastability of this level in terms of an isomeric form is therefore in some respects reintroduced.

Although the rate of radiationless conversion is not generally very dependent on the properties of the solvent apart from its viscosity, some molecules are able to quench the triplet state with high efficiency. Apart from those which quench by chemical reaction, two other principal types of quenching molecule have been distinguished.

(*a*) *Paramagnetic molecules.* Windsor showed that oxygen and nitric oxide quench the triplet state of anthracene and other aromatic molecules with nearly unit encounter efficiency. More recently we have found that paramagnetic ions of the first transition and rare-earth series also quench it but that the quenching efficiency is not related to the magnetic susceptibility as it is in the somewhat similar process of nuclear spin conversion in ortho- and para-hydrogen. The process probably involves complex-formation, the function of the paramagnetic molecule being that the complex formed between triplet and paramagnetic quencher can dissociate into the ground singlet and unchanged quencher without violation of the spin conservation rules.

(*b*) *Molecules with lower energy triplet states.* The transfer of electronic energy by the process:

A* (triplet) + B (singlet) → A (singlet) and B* (triplet)

is allowed by the spin-conservation rules and is possible provided the triplet of B lies lower than that of A. Terenin showed, by phosphorescence measurements in rigid glasses, that transfer of this type does take place although with very low efficiency, the rate constants being about 10 l. mole^{-1} sec.$^{-1}$. Mr. Wilkinson has recently shown that in ordinary solutions this type of process can be very efficient, rate constants greater than 10^8 l. mole^{-1} sec.$^{-1}$ being found. This high efficiency of transfer coupled with its long lifetime suggests an important role for the triplet state in radiation chemical and biological systems. The flash photolysis of chlorophyll solutions, for example, shows that extremely efficient conversion into a metastable—probably triplet—state occurs and it will be surprising if this state does not play an important part in the primary processes of photosynthesis.

The Triplet State in Photochemistry

We are now in a position to consider those collisional processes of the triplet state which result in chemical change. We have every reason to expect that the triplet state will play an important, if not a predominant, part in many photochemical reactions. It is formed in high yield from many molecules as exemplified by the following quantum efficiencies:

Benzene 18%	Triphenylene 36%
Anthracene 13%	Benzophenone 55 %

Its electronic structure leads us to expect a high reactivity typical of a biradical. The excited singlet state, which may also be regarded as a biradical in some respects, has a shorter intrinsic lifetime than the 10^{-4} sec. which we have found to be typical of most triplet states in solution so that the triplet has a greater probability of reaction.

I have chosen three examples which have played an important part in the development of photochemistry. These are anthracene, the quinones, and the aldehydes and ketones, all of which are readily written in biradical form:

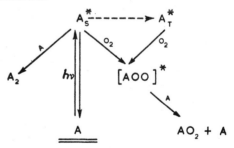

Anthracene.—Anthracene solutions, on irradiation, yield dianthracene and, if oxygen is present, the transannular anthracene peroxide. The fact that the two reactions proceed independently and apparently require the postulate of two distinct excited states had already been indicated by the fluorescence and photochemical investigations by Bowen.

Kinetic flash-photolysis studies showed us that the anthracene triplet state is deactivated by oxygen at nearly every encounter and that encounters with normal anthracene have no effect, suggesting that the dimer is formed from the singlet state and the peroxide from the triplet state. Consideration of all the evidence from fluorescence and flash studies shows this view to be correct although peroxide is also formed *via* the singlet state at high oxygen concentrations. The scheme of reactions is shown in Fig. 9. The oxidation reaction is really a photo-sensitised oxidation of anthracene by anthracene and a similar mechanism applies to photosensitised oxidation of other molecules by anthracene, as well as the photo-oxidation of anthracene by other molecules.

$$A_S^* \dashrightarrow A_T^*$$

FIG. 9. *Scheme of photo chemical reactions for anthracene*

Here we have a rather well-established case of triplet-state participation and of the fact that singlet and triplet states may have quite distinct chemical behaviour.

Quinones.—Quinones and dyes of related structure are also able to photosensitise the oxidation of other molecules, but the mechanism of photosensitisation is quite different from that of anthracene, and indeed these two examples illustrate the two principal modes of photosensitised oxidation. In anthracene the primary reaction of the excited state was with oxygen; in the quinones the primary reaction is with the substrate. The whole scheme of reactions, both in the presence and in the absence of oxygen, is shown in Fig. 10 without distinguishing for the moment between the importance of singlet and triplet states. Oxidation of the substrate occurs even in the absence of oxygen, the quinone being simultaneously reduced to the semiquinone and subsequently to the quinol. That oxidation of alcohols involves hydrogen-abstraction by the excited state has been shown by the observation that the semiquinone neutral radical appears first even when the equilibrium form is the radical-ion. In other cases, *e.g.*, the oxidation of ferrous ion, the primary process is probably electron-transfer.

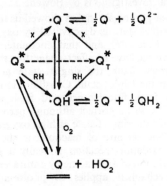

FIG. 10. *Scheme of photochemical reactions of quinones.*

The question we are concerned with is whether the triplet or the singlet state is responsible for the oxidation reaction and the results on the two systems so far studied, *viz.*, duroquinone and Methylene Blue, are somewhat surprising. Dr. Bridge studied the flash photolysis of duroquinone in alcohol solutions and was able to detect three separate short-lived transients which were assigned to the triplet, the radical, and the semiquinone radical ion. The spectra of triplet and radical in viscous paraffin solution are shown in Plate 3. It was then possible to follow the triplet and the radical concentrations by recording their extinctions simultaneously. If the radical R were formed from triplet T then the relation

$$d[R]/dt = k[T]$$

would be valid, the decay of the radical being relatively slow. That this is not the case is readily seen from the oscilloscope traces of the two species in Plate 4 where the rate of formation of radical is nearly zero at the time when the triplet concentration is a maximum. Detailed analysis shows that the rate of radical formation is proportional to light intensity, F, and we can conclude that the radical is formed from the singlet state and independently of the triplet. On the other hand a preliminary study of the Methylene Blue–ferrous ion system by Parker suggests that the opposite is true in this case and that photo-oxidation occurs predominantly *via* the triplet state. I may be forgiven if I do not attempt to generalise from two examples which give opposite results.

Ketones.—More attention has been paid to the photochemistry of ketones and aldehydes than to that of any other class of compounds. The greater part of this work has been concerned with photodissociation which is of relatively little importance in solution where the excited carbonyl compounds react as typical biradicals. Indeed this was pointed out

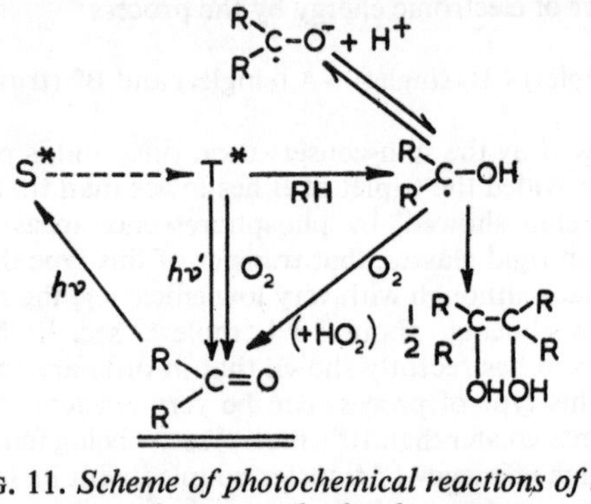

FIG. 11. *Scheme of photochemical reactions of benzophenone and other ketones.*

in the early work of Backström before the triplet-state theory had been developed. Essentially the situation is very similar to that found for the quinones, but hydrogen-abstraction now gives ketyl radicals which dimerise as shown in Fig. 11 for the particular case of benzophenone. The ketyl radicals which are formed have recently been detected, in both their neutral and their ionised forms, by Mr. Wilkinson in flash-photolysis experiments very similar to those of Bridge on the quinones. Wilkinson has also been able to show that the abstraction of hydrogen by benzophenone to yield the ketyl radical occurs by reaction of the triplet state. In this case absorption by the triplet state is not observed and a more subtle method had to be used. As already mentioned, transfer of energy can occur from an excited triplet to a second molecule with a lower energy triplet level. Benzophenone has a higher triplet state than naphthalene but its excited singlet state lies above the lowest excited singlet of naphthalene so that, if benzophenone is excited in the presence of naphthalene, transfer between triplet states can occur but transfer between singlet states is energetically impossible. Three solutions were examined by flash photolysis, using wavelengths absorbed only by benzophenone. This is shown in Plate 5. The first (a) was benzophenone alone which showed the ketyl radical, the second (b) was naphthalene alone which showed no transient spectra since no light was absorbed, and the third solution (c) contained naphthalene and benzophenone at the same concentrations as in experiments (a) and (b) and showed the triplet of *naphthalene* and complete absence of the ketyl radical. This transfer of energy from the triplet of benzophenone has produced the triplet of naphthalene and has quenched the triplet of benzophenone before it reacted to give the ketyl radical, showing that the triplet state was responsible for all ketyl radicals originally formed.*

These few examples are sufficient to show that the triplet state is sometimes of predominant importance in photochemical reactions whilst in other cases the singlet state only is responsible even when triplet is present.

There are several reasons why the first excited singlet and triplet states have different chemical properties. In the first place, even where they correspond to the same type of electronic transition, the electron distribution in the two states is different, as we have already seen. Secondly, in some molecules, it may happen that owing to different singlet–triplet splitting of different types of transitions, the lowest excited states of the triplet and singlet manifolds may arise from different types of electron transition, resulting in excited states having quite distinct electron-distributions and chemical reactivities. The important development of the last few years is that this distinction has been recognised and it is now an essential part of the study of photochemistry to find the degree of conversion into singlet and triplet states and to separate their reactions.

The Triplet State in Thermal Reactions

Whilst it is clear that excited electronic states are of primary importance in photochemical and radiation chemical reactions their participation in thermal reactions is less apparent. Thermal population of the triplet state will only occur to a measurable extent at normal temperatures when the excitation energy is less than about 15 kcal. mole^{-1}. This is quite rare. It is also rare to find bond-dissociation energies as small as this, but bonds are broken nevertheless. In fact, energetically the fission of single bonds and the formation of triplet states are quite comparable as is shown by the typical energies in the annexed Table.

	Triplet energy (kcal. mole^{-1})
Benzene	85
Naphthalene	61
Anthracene	42
Pentacene	22
Naphthol	60
Anthraquinone	56
Benzophenone	68
Acetone	70
Biacetyl	55
	Bond energy (kcal. mole^{-1})
C–H	80—100
C–C	60—85

Some reactions which follow dissociation into free radicals are compared with analogous reactions following the formation of the biradical state in Table 2. The reactions may be divided into three classes which are distinguished by the energies required to dissociate a bond or to form the biradical state.

TABLE 2. *Comparison of radical and biradical reactions.*

Radical	Biradical
	Formation and equilibrium
$\overset{h\nu,\ heat}{AB \longrightarrow \dot{A} + \dot{B}}$	$\overset{h\nu,\ heat}{A=B \longrightarrow \dot{A} - \dot{B}}$
	Transfer reactions
$\dot{A} + RX \rightarrow AX + \dot{R}$	$\dot{A} - \dot{B} + RX \rightarrow \dot{A} - BX + \dot{R}$
	Addition
$\dot{A} + B{=}C \rightarrow AB - \dot{C}$	$\dot{A} - \dot{B} + C{=}D \rightarrow \dot{A} - B - C - \dot{D}$
	Isomerisation
	or ABCD
$AB(1) \rightarrow \dot{A} + \dot{B} \rightarrow AB(2)$	$A{=}B(1) \rightarrow \dot{A} - \dot{B} \rightarrow A{=}B(2)$
	Participation in the transition state
$\dot{X} + AB \rightarrow X .. A .. B \rightarrow XA + \dot{B}$	$\dot{X} + A{=}B \rightarrow \dot{A} \overset{\cdot X \cdot}{\cdots} \dot{B} \rightarrow XA - \dot{B}$

* I am grateful to Mr. Wilkinson for these unpublished results.

182

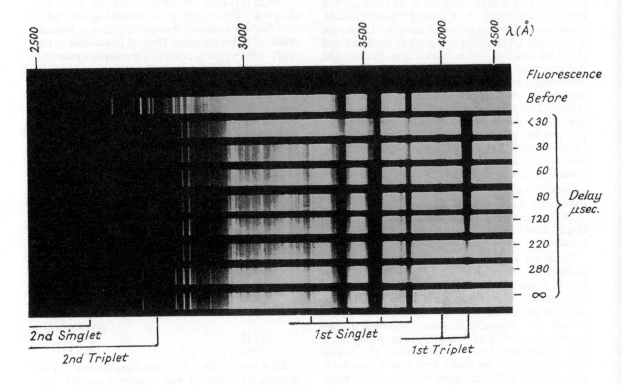

PLATE 1.

*Flash-spectroscopic record of the triplet state
of anthracene in hexane solution.*

[G. PORTER: The Triplet State in Chemistry.]
(See page 295.)

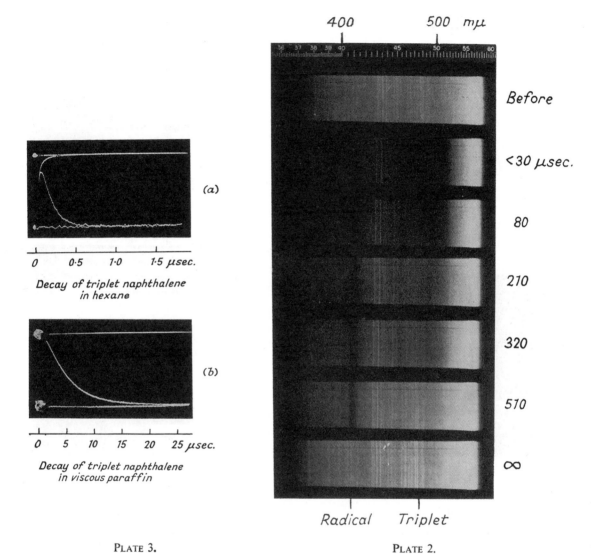

(a)

Decay of triplet naphthalene
in hexane

(b)

Decay of triplet naphthalene
in viscous paraffin

400   500 mμ

Before

<30 μsec.

80

210

320

570

∞

Radical   Triplet

PLATE 3.

Duroquinone transients in viscous paraffin.

PLATE 2.

Decay of triplet naphthalene in (a) *hexane,*
(b) *viscous paraffin.*

[G. PORTER: The Triplet State in Chemistry.]
(See pages 297 and 299.)

184

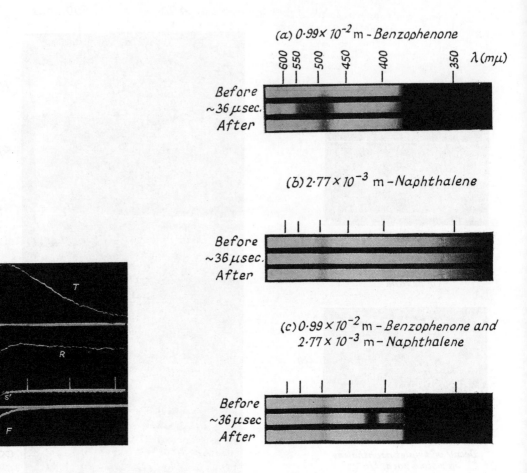

(a) 0·99 × 10⁻² m - Benzophenone

$\lambda (m\mu)$
600 550 500 450 400 350

Before
~36 μsec.
After

(b) 2·77 × 10⁻³ m - Naphthalene

Before
~36 μsec.
After

(c) 0·99 × 10⁻² m - Benzophenone and
2·77 × 10⁻³ m - Naphthalene

Before
~36 μsec
After

PLATE 4.

Photoelectric record of the decay of triplet
and radical from duroquinone.

PLATE 5.

Demonstration of energy transfer from triplet
benzophenone to naphthalene in benzene
solution.

[G. PORTER: The Triplet State in Chemistry.]
(See pages 299 and 300.)

(i) *Energies of* 15 *kcal. or less.* Here we may compare molecules such as hexaphenylethane and Chichibabin's hydrocarbon (I) which, at room temperature, are in equilibrium with a measurable proportion of radical and triplet state respectively. In the extreme case we have diphenylpicrylhydrazyl, which is a free radical even in the solid state, and Schlenck's hydrocarbon (II) which has a triplet ground state.

(ii) *Energies between* 15 *kcal. and* 40 *kcal.* In this range the equilibrium concentration of radical or biradical is too small to be observed directly but dissociation and activation occur at a significant rate at normal temperatures. Molecules with weak bonds, such as the acyl peroxides, yield free radicals and molecules with low triplet-state excitation energies, such as the linear polyacenes, react in a manner characteristic of biradicals. In high-temperature pyrolysis, bond fission and triplet excitation may be important at energies above this range and the triplet state may contribute to the complexities of some of these decompositions and the apparent occurrence of two separate transition states which has recently been found by Hinshelwood and his collaborators.

(iii) *Energies greater than* 40 *kcal.* The limit of 40 kcal. is chosen rather arbitrarily to represent an energy which is greater than the activation energy of reaction, so that direct thermal dissociation or excitation is precluded. But just as bond energies still play a part in determining the rate of radical reactions such as hydrogen abstraction, the triplet state may be expected to be involved in a similar way in the transition state, *e.g.*, for radical addition at a multiple bond. Some evidence that this is so has been provided by Szwarc who found that the logarithm of the "methyl affinity" of aromatic hydrocarbons was closely related to their triplet-state energy and explained this finding in exactly the same way as Polanyi explained the relation between activation energy and bond energy in metathetical reactions.

Direct evidence of the participation of triplet states in thermal reactions is not available, as it is for photochemical reactions, but this is not surprising since evidence for free radicals in thermal reactions is also mainly indirect. In photochemistry triplet-state formation is now known to be comparable in importance with dissociation as a primary process, and it will not be surprising if, in thermal reactions as well, the energy levels and reactivities of the triplet state become as much a part of chemistry as bond-dissociation energies and the reactions of free radicals.

Reprinted from the *Faraday Society Discussions*, 1959, No. 27

INTRAMOLECULAR AND INTERMOLECULAR ENERGY CONVERSION INVOLVING CHANGE OF MULTIPLICITY

By George Porter and M. R. Wright

Dept. of Chemistry, The University, Sheffield 10

Received 16*th January*, 1959

Radiationless transitions between states of different multiplicity are considered with special reference to conversion between triplet and singlet energy levels. Rates of this conversion have been measured for naphthalene and anthracene and rate constants are given for the first-order intramolecular process, the second-order process involving two triplet molecules and also of quenching by other species, particularly paramagnetic ions. It is established that the first-order rate is strongly viscosity dependent and this is attributed to a structural difference between the two states.

Conversion from the triplet state is induced by paramagnetic molecules and ions but the quenching rate constant shows no correlation with magnetic susceptibility. A general theory of " paramagnetic quenching " is proposed in which the function of the quenching molecule is one of overall spin conservation.

All processes, including energy transfer, which involve a change in total electron spin momentum have a low probability compared with the corresponding processes in which spin momentum is conserved. Nevertheless it is now quite clear that conversion between electronic states of different multiplicity is often of primary importance in changes which involve excited molecules. The reason for these apparently conflicting statements lies in the high probability of radiationless conversion between electronic states which may, in the absence of spin and other restrictions, have a rate constant exceeding 10^{11} sec^{-1}. If a change of multiplicity is involved the rate may be reduced by a factor of 10^4 but is still high enough to compete effectively with other modes of deactivation of the excited state.

Since the ground state of most molecules is a singlet, the multiplet of most interest is the triplet and, in particular, the triplet state of lowest energy. This has the following properties.

(i) It is the lowest excited electronic state of the molecule.

(ii) It has a lifetime, even in fluid solvents, which is typically of the order of 10^{-4} sec, i.e. several orders of magnitude greater than that of the excited singlet states.

(iii) Its chemical behaviour is usually characteristic of a biradical.

The triplet state is now frequently postulated as an intermediate in energy transfer processes and in chemical and biochemical change but only in a very few cases has its role been established.

Most work on the triplet state has been concerned with the radiative process of phosphorescence in rigid media and only recently have data become available concerning its properties in the more common fluid solvents. By means of the flash photolysis technique it is possible to observe the triplet state directly by means of its absorption spectrum and to follow its concentration as a function of time after irradiation. In this way, fairly extensive data have now been

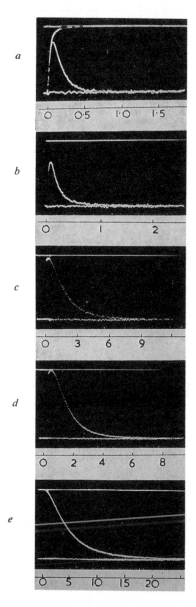

Fig. 1.—Oscillographic records of the decay of triplet naphthalene in (*a*) *n*-hexane, (*b*) water, (*c*) ethylene glycol, (*d*) paraffin $\eta = 33$ cp, (*e*) paraffin $\eta = 167$ cp. Time units : msec.

accumulated on the triplet states of a variety of molecules.   Here, two fundamental processes involving change of multiplicity will be discussed.   The first is the intra-molecular process of radiationless conversion between triplet and singlet states and the second is the deactivation of the triplet state by other molecules, particularly those which themselves have multiplicities higher than singlet.   The experimental findings on each of these processes are somewhat unusual, and have not received a satisfactory explanation.

RESULTS

The data to be presented were obtained by a combination of flash photolysis and spectrophotometric recording.   The apparatus has already been briefly described [1] and further experimental details will be given elsewhere.   In this paper we are concerned mainly with first-order decay constants of the triplet state in various solvents and the constants of quenching by paramagnetic substances.

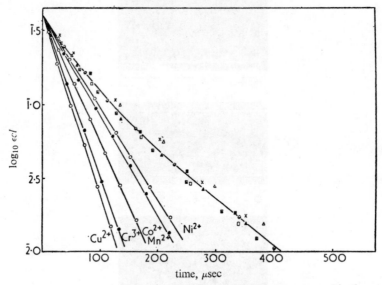

FIG. 2.—First-order plots of the decay of triplet naphthalene in water and in the presence of various ions.   Concentration of Zn^{2+} and Ga^{3+} was 5×10^{-4} M and of all other ions 2.5×10^{-4} M.

◑ No ion ;   △ $[Cu(CN)_2]^-$;   ▧ Ga^{3+} ;   ◳ Zn^{2+} ;
× Ce^{3+} ;   ▲ Nd^{3+} ;   ☐ Gd^{3+}.

The type of record obtained, as well as the two principal effects with which we are concerned, are illustrated by fig. 1 and fig. 2.   The former shows oscillographic records of the absorption by the triplet state of naphthalene in solvents of varying viscosity, as a function of time.   Fig. 2 shows first-order decay plots derived from this type of trace which illustrate the effect of ionic solutes on the lifetime of triplet naphthalene in water.

The first-order decay curves are not exactly linear and it is found that there is a small contribution from second-order (triplet-triplet) quenching which becomes quite considerable at high triplet concentrations.   The method of analysis of these curves which is used to derive first- and second-order decay constants has been described.[1]   As a result of the second-order contribution the first-order rate constants in earlier work were consistently higher than those reported here.   In the flash photographic work of Porter and Windsor,[2, 3] rather high triplet concentrations were measured and owing to the lower accuracy of the method, separation of first- and second-order processes was not possible.   Our rate constants are also slightly lower than those of Livingston and Tanner [4] who used the present method but found no second-order contribution.

20   CHANGE OF MULTIPLICITY

The occurrence of a second-order process immediately suggested a possible explanation of the viscosity dependence of triplet decay and it was therefore necessary to study the effect of viscosity on first- and second-order rates separately. The second-order rate was viscosity dependent as expected but the data in tables 1 and 2 show that the rate of first-order decay is also largely controlled by solvent viscosity. There are evidently some constitutional effects as well, which is not surprising since macroscopic viscosity is only approximately related to diffusion coefficients and to effects on a molecular scale. At the higher viscosities, when the encounter rate is truly diffusion controlled, the rate constant in related solvents, e.g. paraffins 1 and 2, is approximately inversely proportional to viscosity. The limiting values in rigid solvents refer to the radiationless process and are derived from phosphorescence lifetimes.

TABLE 1.—FIRST-ORDER (k_1) AND SECOND-ORDER (k_2) DECAY CONSTANTS OF TRIPLET NAPHTHALENE IN VARIOUS SOLVENTS AT 20°C. SECOND-ORDER CONSTANTS ARE BASED ON THE LIMITING VALUE $\epsilon = 10{,}000$ AND c IS A CONSTANT WHICH IS LESS THAN UNITY

solvent	viscosity (cp)	k_1(sec^{-1})	ck_2 (l. mole^{-1} sec^{-1})
n-hexane	0·3	$1·2 \times 10^4$	$2·1 \times 10^9$
water	1·1	$7·5 \times 10^3$	$4·1 \times 10^9$
ethylene glycol	21·1	$9·7 \times 10^2$	$2·2 \times 10^8$
paraffin 1	33·0	$1·5 \times 10^3$	$3·9 \times 10^8$
paraffin 2	167	$3·1 \times 10^2$	$8·0 \times 10^7$
rigid glass	very high	< 1	$< 10^6$

TABLE 2.—FIRST-ORDER (k_1) AND SECOND-ORDER (k_2) DECAY CONSTANTS OF TRIPLET ANTHRACENE IN VARIOUS SOLVENTS AT 20°C

solvent	viscosity (cp)	k_1 (sec^{-1})	k_2 (l. mole^{-1} sec^{-1})
n-hexane	0·3	$1·3 \times 10^3$	$1·6 \times 10^{10}$
tetrahydrofurane	0·5	$2·3 \times 10^3$	$1·0 \times 10^{10}$
ethylene glycol	21·1	$2·8 \times 10^2$	$8·8 \times 10^8$
paraffin 1	33·0	$5·6 \times 10^2$	$1·2 \times 10^9$
paraffin 2	167	$1·6 \times 10^2$	$2·5 \times 10^8$
rigid glass	very high	< 10	$< 10^6$

The rate constants of quenching of triplet naphthalene by various ions, determined in the same way, are given in table 3. Of the ions investigated, diamagnetic ions have no effect (or at high concentrations possibly a small negative effect) whilst paramagnetic ions all quench the triplet state, though the efficiency varies over a wide range and is apparently quite unrelated to the magnetic susceptibility, provided the ion is paramagnetic.

TABLE 3.—RATE CONSTANTS OF QUENCHING OF TRIPLET NAPHTHALENE BY IONS IN WATER AND ETHYLENE GLYCOL

ion	k_Q(l. mole^{-1} sec^{-1} × 10^{-7}) in water	in ethylene glycol	no. of unpaired electrons	normal state of ion	paramagnetic susceptibility (Bohr magnetons)
K$^+$	$0·00 \pm 0·01$	$0 \pm 0·1$	0	$^1S_0$	diamagnetic
Zn^{2+}	$0·0 \pm 0·2$	$0 \pm 0·2$	0	$^1S_0$	diamagnetic
Ga^{3+}	$0·0 \pm 0·1$	$0 \pm 0·2$	0	$^1S_0$	diamagnetic
Cu(CN)$_2^-$	$0·0 \pm 0·1$	—	0	$^1S_0$	diamagnetic
Cu^{2+}	$7·5 \pm 0·7$	$7·3 \pm 0·7$	1	$^2D_{5/2}$	1·93
Ni^{2+}	$2·3 \pm 0·4$	$2·4 \pm 0·3$	2	$^3F_4$	3·21
Co^{2+}	$5·0 \pm 0·6$	$4·4 \pm 0·4$	3	$^4F_{9/2}$	5·01
Cr^{3+}	$6·9 \pm 0·6$	—	3	$^4F_{3/2}$	3·82
Fe^{2+}	—	$3·8 \pm 0·3$	4	$^5D_4$	5·30
Fe^{3+}	$2·9 \pm 0·4$	—	5	$^6S_{5/2}$	5·85
Mn^{2+}	$2·8 \pm 0·4$	$1·6 \pm 0·2$	5	$^6S_{5/2}$	5·81
Nd^{3+}	—	$0·04 \pm 0·005$	3	$^4J_{9/2}$	3·60
Gd^{3+}	—	$0·007 \pm 0·002$	7	$^8S_{7/2}$	8·01

Preliminary data indicate that the same is true of anthracene triplet quenching, but that the quenching rate constants of a particular ion depend both on the triplet molecule and on the solvent.

The only other paramagnetic molecules, for which quenching constants are available, are given in table 4. The rate constants of quenching by O_2 and NO are taken from the results of Porter and Windsor.[3] The effects are certainly general to other molecules, as well as anthracene, but rate constants have not yet been accurately determined.

TABLE 4.—RATE CONSTANTS OF QUENCHING OF TRIPLET ANTHRACENE BY PARAMAGNETIC MOLECULES IN HEXANE SOLUTION

quenching molecule	no. of unpaired electrons	k_Q(l. mole^{-1} sec^{-1})
O_2	2	4×10^9
NO	1	4×10^9
triplet anthracene	2	$1 \cdot 6 \times 10^{10}$

DISCUSSION

INTRAMOLECULAR RADIATIONLESS CONVERSION BETWEEN TRIPLET AND SINGLET STATES

It is not always appreciated that the appearance of phosphorescence in a wide variety of molecules and its absence in fluid solvents is largely unexplained. In molecules such as the aromatic hydrocarbons, radiationless crossing from the upper singlet state S_1 to the triplet T_1 occurs with a rate constant which is typically of the order of 10^8 sec^{-1} in rigid media, whilst the apparently similar conversion from T_1 to the ground state S_0 is not observed in rigid media and must have a rate less than 10^{-1} sec^{-1} in benzene and other molecules with similar radiative triplet lifetimes. If the rates of the two radiationless conversions were of comparable magnitude, no phosphorescence would be observed, and this is the reason for the absence of phosphorescence in ordinary solutions and gases.

Studies of the kinetics of triplet state decay in different solvents have shown that a number of bimolecular processes may occur, e.g. quenching by oxygen or a second triplet, but that, in the absence of such quenchers, the triplet state has a natural lifetime which is generally much less than the radiative life and which must be attributed to the process of radiationless conversion to the ground state.[2] The second-order decay would of course be diffusion controlled in view of its high rate and a viscosity dependence is to be expected and is, in fact, found. The results in table 1 and 2 establish that the first-order decay, which is the predominant process at low intensities, is also a function of solvent viscosity.

It is clearly unprofitable to consider the triplet decay as a process which is rate-determined by energy transfer to the solvent, particularly since the rate attains its maximum value in the gas phase.[5] It has also been established that the first-order process being considered occurs without the intervention of any molecules other than those of the solvent. We must therefore conclude that the rate constants measured are those of the intramolecular radiationless conversion process from the triplet state T_1 to the ground state S_0. Energy transfer to solvent occurs after crossing and is not rate-determining. The outstanding problem is why the radiationless crossing T_1-S_0 should be viscosity dependent to such an extent that it is totally inhibited in rigid media.

The radiationless transition probability between two states i and j is proportional to the square of the matrix element W_{ij} of the perturbation function W which in turn is given by

$$W_{ij} = \int \psi_i W \psi_j \, d\tau$$

where ψ_i and ψ_j are the eigenfunctions of the two states i and j. The eigenfunctions can be separated into a product of electronic-rotational and vibrational functions which are independent to a first approximation and the electronic function can again be separated into a product of spin and co-ordinate functions. The spin functions of two electronic states of different multiplicity are orthogonal so that, in so far as the separation of the functions is a good approximation, the transition probability is zero, but it becomes finite in the presence of spin-orbit interaction.

The electronic eigenfunctions will be modified only slightly by the solvent and the dependence of transition probability on viscosity cannot be explained in any general way by a difference in the electronic terms of the matrix element. The position with regard to rotation is less clear. The selection rule for perturbations in isolated diatomic molecules is that both states must have the same total angular momentum, i.e. $\Delta J = 0$. In solution, free rotation is inhibited but the conservation of angular momentum is still at least as probable as in the isolated molecule. Further theoretical work on this point would be helpful but explanations of the viscosity effect in terms of inhibited rotation are made very unlikely by the experimental fact that predissociations and internal conversions are observed, even in rigid media, with high probability, e.g. crossings from state S_1 to T_1. A general selection rule based on conservation of total angular momentum would apply to all such transitions and it is probable that rotation can be neglected in discussing the transition probability as it can in discussing the Franck-Condon principle for radiative transitions.[6]

The part of the matrix element depending on the vibrational eigenfunction is

$$W_{ij}^v = \int \psi_i^v W^v \psi_j^v \mathrm{d}\tau,$$

where W^v is the part of the interaction energy depending on nuclear co-ordinates and ψ_i^v and ψ_j^v are the vibrational eigenfunctions. The transition probability therefore depends on the overlap of vibrational eigenfunctions and the Franck-Condon principle is valid just as in the more familiar case of radiative transitions in diatomic molecules. A high transition probability will be found only when, classically, the system can pass from one state to the other without a large alteration of position or momentum.

In a polyatomic molecule and in a viscous medium this restriction may be very important if the equilibrium configuration of the molecule in the two states is different. Recent work on simple molecules has shown that such is often the case.[7] The structures of molecules in their triplet states, and particularly aromatic ones in which we are most interested, are not known and evidence obtained from the weak absorption spectra or diffuse emission spectra in rigid media is at present uncertain. We believe, however, that the low radiationless transition probability from the triplet to the ground state and its viscosity dependence in a wide variety of molecules must be interpreted as showing that the triplet state has an equilibrium nuclear configuration which is considerably different from that of the ground state and that the configurational change which is necessary to attain a position with low reverse-crossing probability is inhibited by clamping of the distorted structure by the viscous or rigid solvent.

There are theoretical reasons for believing that the structure of the triplet state even of a molecule as simple as benzene, may differ significantly from that of the ground state. Because of the Pauli principle the probability distributions of two electrons in different orbitals with respect to each other are different in singlet and triplet states. For two electrons in a circle, Dickens and Linnett[8] have shown that

$$\psi_S \psi_S^* = (\tfrac{1}{2}\pi)^2 \{1 + \cos[(m-n)(\phi_2 - \phi_1)]\},$$
$$\psi_T \psi_T^* = (\tfrac{1}{2}\pi)^2 \{1 - \cos[(m-n)(\phi_2 - \phi_1)]\},$$

where ψ_S and ψ_T are the wave functions of the singlet and triplet systems, m and n are the quantum numbers of the two orbitals, and $\phi_2 - \phi_1$ is the angular separation of the electrons. The products $\psi\psi^*$ are proportional to the probability that the electrons have the angular separation $\phi_2 - \phi_1$. It is seen that, for the singlet state, the probability is a maximum when the electrons are coincident whilst in the triplet state the electrons have zero probability of being in the same place. Electron repulsion will change the situation quantitatively but the difference remains as is shown by the large splitting of singlet and triplet levels in aromatic molecules. In benzene the instantaneous electron distributions in the three states of interest may be schematically represented as follows:

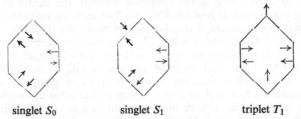

singlet S_0 singlet S_1 triplet T_1

The most stable nuclear configuration of a single Dewar form of T_1 would probably be folded about the vertical axis, a structure suggested on quite different grounds by Lewis and Kasha.[9]

Experimental evidence, which may indicate a different equilibrium configuration in the triplet and singlet ground states comes from the results of Craig, Hollas and King,[10] following work of Evans.[11] It is found that the radiative lifetime of the triplet state of benzene calculated from the integrated absorption coefficients is greater than 700 sec, earlier work being incorrect owing to the presence of oxygen. On the other hand, the measured lifetime from phosphorescence decay is 7 sec. These findings would be in accordance with our conclusion that transition from the triplet state occurs to a ground state molecule with a configuration very different from the equilibrium one.

If the structural difference between triplet and ground singlet states of benzene is accepted there is little difficulty in extending the arguments to most other aromatic molecules. It is interesting to note that structural isomeric differences between these states were originally proposed as an explanation of the phosphorescent state [12] but were later discarded in favour of the triplet state theory. In our view both spin and structural restrictions are necessary to the appearance of phosphorescence of long duration.

INTERMOLECULAR PROCESSES OF TRIPLET STATE DEACTIVATION

All energy transfer processes from excited states must involve other molecules but we have seen that, when only an inert solvent is present, energy transfer is not itself the rate-determining process. In the presence of certain molecules, however, the rate of deactivation of the triplet state is greatly enhanced. There are undoubtedly several mechanisms of quenching by which this may occur, some of which involve electron transfer, hydrogen atom transfer or other chemical reactions. Since most chemical reactions result in the formation of an addition compound or of two doublet radicals so that spin conservation is always possible, we shall not be concerned with such processes here.

A quenching molecule may induce a change of multiplicity in a second molecule without chemical change in at least three ways:

(i) It may induce perturbations, and particularly increase spin-orbit coupling, so that the spin selection rule is partially broken down.

(ii) A transfer of electronic energy may take place, spin momentum being conserved by excitation of the quencher to a state of different multiplicity.

(iii) Conservation of total spin momentum may be made possible during the encounter without, necessarily, any multiplicity change of, or energy transfer to, the quenching molecule.

24   CHANGE OF MULTIPLICITY

We shall be concerned mainly with the third of these processes which gives a new interpretation of so-called " paramagnetic quenching ". Since the quenching molecule is unchanged it will be appropriate to describe the process as " catalyzed spin conservation ".

Perturbations resulting in a breakdown of the spin selection rule may be brought about, in principle, by heavy atoms or by a magnetic field including the magnetic field of a neighbouring molecule. No effect which can be attributed to heavy atom catalysis of radiationless conversion between triplet and singlet states has been found. Table 3 shows that diamagnetic ions such as Zn^{2+} and Ga^{3+} have no measurable quenching effect on the triplet state of naphthalene in water. Livingston and Tanner [4] found no marked effect of carbon disulphide or of substituted benzenes, e.g. bromobenzene, on the triplet state lifetime of anthracene. Furthermore approximate measurements of triplet yields in solution indicate that heavy atoms in the solvent have no measurable effect on the S_1-T_1 crossing either.

On the other hand, paramagnetic molecules have a considerable effect on triplet state lifetimes and this has been attributed by some workers to magnetic perturbations, and breakdown of the spin selection rule. The data given in table 3 makes this interpretation very unlikely. Not only is there a wide variation in efficiency but in a related group, such as the ions of the first transition series, there is no correlation whatever with magnetic susceptibility except that all paramagnetic ions quench to some extent and diamagnetic ions have no effect. Quenching by magnetic perturbation would be expected to be a general and rather unspecified effect of all paramagnetic molecules which was closely related to magnetic susceptibility as it is in the process of nuclear spin change in the para-ortho hydrogen conversion induced by these same ions.

ENERGY TRANSFER WITH SPIN CONSERVATION

In a collisional process it is only necessary that the total spin momentum of the whole system be conserved rather than that of each separate partner. One process by which this may be achieved is by " spin transfer " from the excited molecule A to the quencher Q. This is automatically accompanied by energy transfer since a change of electronic state of both partners is involved. If the quenching molecule Q is a singlet this is the only process (apart from chemical reaction) by which a change in multiplicity of A can be induced whilst conserving spin momentum. In the case of a triplet state of A the process becomes

$$A^* \text{ (triplet)} + Q \text{ (singlet)} \rightarrow A \text{ (singlet)} + Q^* \text{ (triplet)}.$$

It is therefore a necessary condition for this type of quenching that the quencher Q shall have a triplet level lower than that of A. Such cases have been found by Terenin and Ermolaev [13] for transfer from the triplet state of benzophenone and a number of similar molecules to naphthalene and its derivatives. The rate constants are very low in rigid solvents and are not yet known in ordinary solutions.

CATALYZED SPIN CONSERVATION

All paramagnetic molecules which we have investigated quench the triplet state to some extent and we have given reasons why this is not to be attributed to the magnetic field alone. In many of the cases given in table 3, chemical reaction, particularly electron transfer, might reasonably be expected but it is highly improbable that this is the general explanation of the quenching action of paramagnetic molecules. There is no correlation with the oxidation-reduction potentials of the quenchers and, since both oxidizing and reducing ions are effective, it would be necessary to postulate formation of both the negative and positive radical ions of the hydrocarbons.

Most of the paramagnetic molecules studied have low-lying energy levels and quenching might therefore be interpreted simply as an electronic-energy transfer process. This may indeed contribute to the quenching mechanism when the energy levels are favourably situated. But there are many difficulties in accepting this as the general mechanism of "paramagnetic quenching". It is improbable that nitric oxide has a quartet level lower than the triplet level of anthracene and no correlation with the excited levels of the paramagnetic molecules and their quenching efficiency can be found.

An alternative explanation of the quenching effect of paramagnetic molecules on the triplet state will now be given.

The collisional process

$$A \text{ (triplet)} + Q \text{ (singlet)} \to A \text{ (singlet)} + Q \text{ (singlet)}$$

is forbidden by spin conservation rules.[14] On the other hand the process is allowed if the multiplicity of Q is higher than singlet.

More generally, in the process

$$A^* (S = x) + Q(S = y) \to A(S = x - 1) + Q(S = y),$$

where S is the spin quantum number and x and y are integral or half integral numbers with $x \geqslant 1$, the change is allowed by spin conservation rules for the overall system provided $y > 0$. We immediately obtain a common property of paramagnetic molecules which distinguishes them from singlet molecules. The otherwise forbidden spin change of A becomes allowed in the presence of a paramagnetic Q without necessarily involving energy transfer or other change in Q. Molecules of solvent must of course be present to remove the excess energy of A after the transition, just as occurs in the intramolecular conversion.

This quenching mechanism is quite distinct from magnetic perturbation effects and indeed its rate will be shown to be independent of magnetic susceptibility provided the quencher is not a singlet. It is useful to consider the process in two steps: (i) the formation of the collision complex AQ, and (ii) the dissociation of AQ to A and Q. The quenching rate constant will depend on the following factors:

    (i) spin-spin coupling between A and Q in the complex AQ,

    (ii) the lifetime of AQ,

    (iii) a spin statistical factor.

Consider first the spin statistical factor. If $x \leqslant y$, the possible spin quantum numbers of the complex AQ are

$$y + x, y + x - 1, \ldots y - x,$$

nd he total statistical weight g is given by

$$g = 2(y + x) + 1 + \ldots 2(y - x) + 1.$$

Of these possible states the only ones which can give the required products with spin $x - 1$ and y are

$$y + x - 1, \ldots y - x + 1,$$

the statistical weight of these states being

$$g_q = 2(y + x - 1) + 1 + \ldots 2(y - x + 1) + 1.$$

The probability that AQ will have a spin momentum which correlates with the required products is g_q/g, the general expressions for which are

$$\frac{g_q}{g} = \frac{2x - 1}{2x + 1} \quad \text{for } y > x - \tfrac{1}{2},$$

and

$$\frac{g_q}{g} = \frac{4xy}{(2x + 1)(2y + 1)} \quad \text{for } y < x - \tfrac{1}{2}.$$

26   CHANGE OF MULTIPLICITY

Values of g_q/g for values of x and y of interest are given in table 5.

TABLE 5.—STATISTICAL FACTORS FOR THE PROCESS

$$A(S = x) + Q(S = y) \rightarrow A(S = x - 1) + Q(S = y)$$

x	y	$\dfrac{g_q}{g}$	$\dfrac{g_p}{g_t}$	$\dfrac{g_q g_p}{g g_t}$
1	0	0	$\frac{1}{4}$	0
1	$> \frac{1}{2}$	$\frac{1}{3}$	$\frac{1}{4}$	$\frac{1}{12}$
$1\frac{1}{2}$	0	0	$\frac{1}{3}$	0
$1\frac{1}{2}$	$\frac{1}{2}$	$\frac{3}{8}$	$\frac{1}{3}$	$\frac{1}{8}$
$1\frac{1}{2}$	> 1	$\frac{1}{2}$	$\frac{1}{3}$	$\frac{1}{6}$
2	0	0	$\frac{3}{8}$	0
2	$\frac{1}{2}$	$\frac{2}{5}$	$\frac{3}{8}$	$\frac{3}{20}$
2	1	$\frac{8}{15}$	$\frac{3}{8}$	$\frac{1}{5}$
2	$> 1\frac{1}{2}$	$\frac{3}{5}$	$\frac{3}{8}$	$\frac{9}{40}$

If we confine ourselves to the process given, where the multiplicity of Q is unchanged and that of A must decrease, the only possibilities for the dissociation of AQ are to the products or to the original reactants unless $x > 2$. Two possibilities may now be distinguished.

CASE 1

The complex AQ is stable with respect to dissociation to the original reactants. In this case the probability of dissociation to products with altered spin is unity and the statistical probability of the overall process is given directly by g_q/g.

CASE 2

The energy of AQ with respect to original reactants is small compared with kT. In this case the probability of formation of products with changed spin will be given by

probability that AQ will dissociate to $A(S = x - 1) =$

$$\frac{\text{statistical weight of } A(S = x - 1)}{\text{sum of statistical weights of all allowed states of A}} = \frac{g_p}{g_t}$$

and values of g_p/g_t are given in table 5. The overall probability of conversion of A in this case is given by the product $g_q g_p / g g_t$ in the final column. In the last two rows, referring to quintet quenching, the factor $g_p/g_t = \frac{3}{8}$ does not include the small probability of singlet formation.

The interesting conclusion emerges from the figures of table 5 that, on purely statistical grounds, all paramagnetic molecules have an equal probability of inducing the transition between a triplet and a singlet state. Differences between quenching rate constants will therefore arise as a result of the other two factors mentioned above and, since these are not related in any direct way to the magnetic susceptibility, the experimental finding that molecules with unpaired electrons are quenchers but that their efficiency is not related to the multiplicity, becomes comprehensible.

The large difference in quenching rates between different molecules is now to be considered in terms of spin-spin interaction and the lifetime of the collision complex. These two factors are closely related since both depend on the overlap of the orbitals of the unpaired electrons in A and Q. If there is no interaction between the spin moments, each molecule must individually conserve spin momentum and no change will occur. If interaction is small quenching will occur with a probability less than that calculated on purely statistical grounds.

The rates in tables 3 and 4 fall broadly into three groups:

GROUP 1 $k \approx 10^{10}$ l. mole^{-1} sec^{-1} O_2, NO, aromatic triplet
GROUP 2 $k \approx 5 \times 10^7$ l. mole^{-1} sec^{-1} metal ions of first transition series
GROUP 3 $k \approx 2 \times 10^5$ l. mole^{-1} sec^{-1} ions of lanthanide rare earths.

On passing from group 1 to group 3 the orbitals of the unpaired electrons become increasingly deep-seated as we go from p to d to f electrons. The d electrons of the transition metal ions overlap readily with orbitals of other molecules but they form complexes in aqueous solution which shields the unpaired electrons and reduces overlap with electrons of the triplet. The f electrons of the rare earths are known from many different lines of evidence, e.g. magnetic susceptibility theory [15] to have relatively little interaction with the solvent or other environment.

The three examples of group 1 on the other hand are typical radicals or biradicals and the collision complex formed with a triplet state probably has a stability of at least several kcal. Spin-spin interaction will therefore be strong and the complex may have a considerable life. In view of the strong interaction, case 1 is almost certainly applicable here and the statistical factor is therefore g_q/g. The product of the reciprocal of this factor and the quenching rate constant is in close agreement with the calculated diffusion-controlled encounter rate.

It is to be expected that the radiationless transition probability is increased in the presence of an efficient paramagnetic quencher only by an amount corresponding to the difference between a spin forbidden and a spin allowed transition, i.e. by a factor of about 10^4. Now the lifetime of triplet anthracene in n-hexane in the absence of quenchers is 10^{-3} sec so that its lifetime when the spin restriction is removed should be about 10^{-7} sec. The average lifetime of the collision complex between triplet anthracene and oxygen, nitric oxide or a second triplet should therefore also be about 10^{-7} sec which is much longer than the duration of an encounter not involving chemical interaction and is in accordance with kinetic studies of anthracene photosensitized oxidation.[16]

We are grateful to the Royal Society for the loan of a monochromator and to the Geophysics research directorate, Air Force Cambridge research centre of A.R.D.C., USAF through its European Office for support of part of this work.

[1] Porter and Wright, M. R., *J. Chim. Physique*, 1958, **55**, 705.
[2] Porter and Windsor, *Faraday Soc. Discussions*, 1954, **17**, 178.
[3] Porter and Windsor, *Proc. Roy. Soc. A*, 1958, **245**, 238.
[4] Livingston and Tanner, *Trans. Faraday Soc.*, 1958, **54**, 765.
[5] Porter and Wright, F. J., *Trans. Faraday Soc.*, 1955, **51**, 1205.
[6] Herzberg, *Spectra of diatomic molecules* (Van Nostrand, 1950).
[7] Ramsay, *Ann. N.Y. Acad. Sci.*, 1957, **67**, 485.
[8] Dickens and Linnett, *Quart. Rev.*, 1957, **11**, 291.
[9] Lewis and Kasha, *J. Amer. Chem. Soc.*, 1944, **66**, 2100.
[10] Craig, Hollas and King, *J. Chem. Physics*, 1958, **29**, 976.
[11] Evans, *J. Chem. Soc.*, 1957, 1351, 3885.
[12] Lewis, Lipkin and Magel, *J. Amer. Chem. Soc.*, 1941, **63**, 3005.
[13] Terenin and Ermolaev, *Trans. Faraday Soc.*, 1956, **52**, 1042.
[14] Wigner, *Göttinger Nachrichten*, 1927, 375.
[15] van Vleck, *Theory of Electric and Magnetic Susceptibilities* (Oxford, 1932).
[16] Livingston, *J. Chim. Physique*, 1958, **55**, 887.

Reprinted from the Proceedings of the Royal Society, A, *volume* 260, pp. 13–30, 1961

Acidity constants in the triplet state

By G. Jackson and G. Porter, F.R.S.

Department of Chemistry, University of Sheffield

(*Received* 2 *August* 1960)

Acidity constants of the lowest triplet state have been determined for seven aromatic molecules by two independent methods. In the first method the triplet state is populated by flash photolysis and the acid-base equilibrium is studied spectrophotometrically. In the second method the difference between acidity constants in the triplet and ground states is calculated from energy levels of the acid and base derived from phosphorescence spectra.

For five of the molecules investigated—2-naphthol, 1- and 2-naphthoic acids, 2-naphthylamine and acridine—acidity constants are now known for three electronic states. It is found that acidity constants of the triplet and ground states are comparable while that of the first excited singlet state differs by a factor of about 10^6 in each case. The electron density distribution in the three states is briefly discussed.

The acidity constant is one of the most useful indications of the electron distribution within a molecule and differences in the proton affinities of organic molecules can be satisfactorily accounted for in terms of electronic structure (Ingold 1953). When a molecule is electronically excited the electron density distribution changes and the acidity constant in the excited state should again provide an experimental approach to the determination of this distribution. Such information is of interest both to the photochemist, who wishes to understand the differences in reactivity between molecules in their ground and in their excited states, and to the theoretical chemist who is concerned with the calculation of the electronic structure of the excited state and the nature of the transitions involved.

Pioneering work in this field has been carried out by Förster (1950) and Weller (1952) following an early observation of Weber (1931). It was shown that in several aromatic molecules the change in fluorescence spectrum, corresponding to the change in protonation of the first excited singlet state, occurred at a pH different from that at which the ground state changed its state of protonation. In some cases, acid-base equilibrium in the excited state was established before fluorescence occurred and acidity constants in the excited singlet state could therefore be determined. Independent confirmation of these acidity constants was also possible from spectroscopic measurements of the energy levels, provided assumptions were made about the entropy changes involved.

It would be useful to have similar data about the acidity constants of the lowest triplet state, particularly for those cases where comparison can be made with acidity constants in the ground and first excited singlet states. Unfortunately the method used by Förster and Weller is not applicable to triplet states since phosphorescence is not normally observed in fluid solvents and equilibrium will not be rapidly established in rigid media. We have been able to determine these constants, however, by direct observation of the absorption spectrum of the triplet state base and its conjugate acid. The triplet state has been populated in aqueous solution by

14   G. Jackson and G. Porter

flash methods in the manner first used by Porter & Windsor (1954) and studies of the relative concentration of the two forms as a function of pH has made possible the measurement of acidity constants. Confirmation of these measurements has been obtained from phosphorescence spectra of the acid and base in rigid solvents.

EXPERIMENTAL

Flash photolysis measurements were made in outgassed solutions contained in a quartz reaction vessel 21 cm long and 1·6 cm in diameter. The photolysis flash was derived from three flash lamps arranged symmetrically around the reaction vessel. Each lamp was connected to a condenser of 3 μF capacity working up to 20 kV and the three lamps discharged through a common external hold-off gap. When lower energies were required the number of lamps was reduced.

The spectra were recorded photographically by means of a vertical capillary tube flash lamp. Spectra were recorded in a single flash on a medium Hilger quartz spectrograph using Ilford HP 3 plates. Each plate was calibrated by means of a seven-step neutral density wedge and densities were measured by means of a double-beam recording microdensitometer. Triplet spectra were obtained by subtracting microdensitometer traces of the solution after flashing from the blank before photolysis. Both traces were linear in optical density but were not corrected for change of plate sensitivity with wavelength.

The delay between photolysis and recording flashes was set approximately by an electronic delay unit and was measured exactly for each separate exposure by photographically recording the oscillograph of the light flashes.

Triplet state emission spectra were recorded on a phosphoroscope of conventional design (Lewis & Kasha 1944). A rotating can exposed the sample for one-third of a rotation and rotated at 1330 rev/min. The light source was a high-pressure mercury arc with a heat filter of 4 cm water in a quartz cell. The spectra were recorded on a small Hilger glass prism spectrograph on Ilford HPS plates, exposure times being between 1 and 4 h with a slit width of 0·25 mm.

Solvents were required which would form clear glasses at liquid-nitrogen temperatures, which would not show significant phosphorescence and which could be used for the study of strongly acidic and strongly basic solutions. Water is unsuitable since it cracks badly at low temperatures and alcoholic solvents were eventually used, the principal ones being

    (a) ethanol + potassium hydroxide;

    (b) ether (5 parts), isopentane (5 parts), ethanol (2 parts by volume);

    (c) sulphuric acid (1 part), n-propanol (10 parts), ethanol (2 parts).

Solvents were purified by distillation and passage over activated silica gel. Background phosphorescence of solvents a and b was negligible but c gave a broad phosphorescence band of moderate intensity. This was due to an impurity in the n-propanol which was not completely removed by distillation. Fortunately the diffuse nature of this emission made it possible to subtract the contribution of the solvent to the emission of the solution with fair accuracy.

Analysis of a spectrum with two overlapping components

If an absorption spectrum is obtained which consists of two triplet spectra superimposed then, provided the spectrum of each separate triplet is known, the relative concentrations of each in the compound spectrum can be estimated. In the present work, only the shapes of the separate spectra were known, not the extinction coefficients and, before the ratio of the concentrations can be calculated, assumptions must be made about the extinction coefficients.

Suppose that we have a protonated triplet a with maximum absorption at wavelength 1, and the corresponding unprotonated triplet b with maximum at wavelength 2, and a trace of optical density against wavelength is available for each species. A plate is obtained from which the optical densities at the two triplet maxima, are found to be D^1 and D^2 due to unknown concentrations c_a and c_b; the ratio of these concentrations is required.

Now
$$D^2 = \epsilon_a^2 c_a l + \epsilon_b^2 c_b l$$
and
$$D^1 = \epsilon_a^1 c_a l + \epsilon_b^1 c_b l,$$

where l is the path length, ϵ is the extinction coefficient, and superscripts and subscripts refer to wavelength and species, respectively, so that

$$\frac{D^2}{D^1} = \frac{\epsilon_a^2 c_a + \epsilon_b^2 c_b}{\epsilon_a^1 c_a + \epsilon_b^1 c_b},$$

and since
$$\epsilon_a^2 = (d_a^2/d_a^1)\,\epsilon_a^1 \quad \text{and} \quad \epsilon_b^2 = (d_b^2/d_b^1)\,\epsilon_b^1,$$

where d_a and d_b are optical densities of the pure species a and b, then

$$\frac{D^2}{D^1} = \frac{c_a(d_a^2/d_a^1)\,\epsilon_a^1 + c_b(d_b^2/d_b^1)\,\epsilon_b^1}{\epsilon_a^1 c_a + \epsilon_b^1 c_b}.$$

By rearrangement this becomes

$$\frac{c_a}{c_b} = \frac{\epsilon_b^1}{\epsilon_a^1}\left[\frac{d_b^2/d_b^1 - D^2/D^1}{D^2/D^1 - d_a^2/d_a^1}\right].$$

The ratios in the square bracket are all known; the ratio of extinction coefficients must be estimated before c_a/c_b can be computed.

Calculation of pK values

The pK for an acid dissociation in aqueous solution corresponds to that value of the pH at which the concentration c_a of the protonated form equals that of the unprotonated form c_b. When the values of c_a/c_b are available, a graph of c_a/c_b or $c_a/(c_a+c_b)$ against pH yields the pK as that pH value where

$$\frac{c_a}{c_b} = 1 \quad \text{or} \quad \frac{c_a}{c_a+c_b} = 0{\cdot}5.$$

This applies whether the c's are ground, excited singlet, or triplet concentrations.

For some of the triplets observed, the triplet absorption of either the acid or base is almost negligible, and the pK can be estimated by watching the disappearance of the stronger band. Suppose the triplet absorption of the acid form is much stronger than that of the alkaline form and, at a particular pH, $[T]$ is the known optical density measured at the wavelength of the acid band. If $[T_{\mathrm{max.}}]$ is the optical density at the extreme pH where only triplet a can be assumed present, then

$$[T_{\mathrm{max.}}] = \epsilon_a c l$$

and at the intermediate pH

$$[T] = \epsilon_a c_a l + \epsilon_b c_b l.$$

If

$$c_a + c_b = c$$

then

$$\frac{[T]}{[T_{\mathrm{max.}}]} = \frac{\epsilon_a c_a + \epsilon_b (c - c_a)}{\epsilon_a c}. \tag{1}$$

When c_a is zero, (1) becomes

$$\frac{[T]}{[T_{\mathrm{max.}}]} = \frac{\epsilon_b c}{\epsilon_a c} = \frac{\epsilon_b}{\epsilon_a}. \tag{2}$$

At the pK when $c_a = c_b$,

$$\frac{[T]}{[T_{\mathrm{max.}}]} = \frac{\epsilon_a c_a + \epsilon_b c_a}{2 \epsilon_a c_a} = \frac{1 + \epsilon_b/\epsilon_a}{2}. \tag{3}$$

In a plot of $[T]/[T_{\mathrm{max.}}]$ against pH, ϵ_b/ϵ_a can be read off directly from (2) and substitution in (3) gives the value of $[T]/[T_{\mathrm{max.}}]$ at the pH equivalent to the pK. In this method no assumption is made about extinction coefficients but the total amount of triplet formed is taken to be constant over the range investigated.

Calculation of acidity constants from phosphorescence spectra

As already mentioned, an independent estimate of acidity constants in the excited state was made by Förster and Weller from spectroscopically determined energy levels. In order to apply the same method to the triplet state the energies of the triplet states of both acidic and basic forms of the molecule above the ground state must be determined from phosphorescence spectra. These energy levels are shown in figure 1.

The energy changes ΔE are directly determined from the short wavelength limit of the phosphorescence spectra of the two forms. They are related to the heat content changes ΔH and ΔH^* as follows:

$$\Delta E_{HA} + \Delta H^* = \Delta E_A + \Delta H.$$

The standard free energy change ΔG in the reaction is related to this heat change, the entropy change ΔS and the equilibrium constant K by

$$\Delta G = \Delta H - T \Delta S = - RT \ln K.$$

If it is now assumed that the entropy changes for the ground and excited state reactions are the same

$$\Delta G - \Delta G^* = \Delta H - \Delta H^* = \Delta E_{HA} - \Delta E_A$$

and

$$\mathrm{p}K - \mathrm{p}K^* = \frac{\Delta E_{HA} - \Delta E_A}{RT}.$$

If the ground state constant pK_G is known, the constant pK_T for the triplet state can therefore be calculated from spectral data. Here and throughout this paper all pK's are derived from *acidity* constants, K_a.

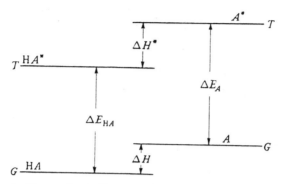

FIGURE 1. Energy level diagram of an acid and its conjugate base.

RESULTS

The principal compounds investigated were quinoline, acridine and five derivatives of naphthalene. These were chosen since some data on the acidity constants of their singlet states were available and because they have strong characteristic triplet absorption spectra in solution. Preliminary flash photolysis examinations of 1-naphthol, 1-naphthylamine, diphenylamine, triphenylamine and fluorescein showed these compounds to be unsuitable for pK measurements by this method, owing either to poorly characterized triplet spectra in aqueous solution at flash energies insufficient to cause decomposition, or to too small a difference between the acid and base triplet absorption spectra.

2-naphthol

Flash photolysis of a 5×10^{-5} molar solution of 2-naphthol in liquid paraffin, using a flash energy of 450 J, resulted in the appearance of a strong transient absorption spectrum which is shown in figure 2. The sharp band at 4650 Å was observable for 1 ms and therefore resulted from a different species from the other three bands with maxima at 4330, 4110 and 3815 Å whose lifetimes were about half that of the 4650 band. The long wavelength band will be shown to be due to the 2-naphthoxyl radical and the other bands, by analogy with naphthalene and other derivatives, is clearly the absorption spectrum of the triplet state. In the non-polar solvent, liquid paraffin, it is to be expected that only the protonated form of the triplet will be observed. In aqueous solutions all spectra were less characteristic and they became much sharper in aqueous ethylene glycol solutions. Unfortunately the significance of pH measurements in such solutions is questionable and aqueous solutions were therefore used in spite of the rather diffuse spectra which were obtained.

The absorption spectrum in strong acid solutions (N and N/10 sulphuric acid) had a sharp maximum at 4320 Å and was replaced, in N/10 sodium hydroxide solution, by a spectrum with a broader maximum displaced at 4600 Å. These are

18   G. Jackson and G. Porter

to be attributed respectively to the acidic and basic forms of the triplet molecule
and the relative intensity of the two forms was next investigated as a function of
pH. A few preliminary experiments showed that the pH of unbuffered solutions
changed by more than one unit on outgassing and flashing and all subsequent work
was carried out on buffered solutions. The buffers used were:

(1) $H_3BO_3 + KCl + NaOH$ from pH 8 to 10,

(2) $KH_2PO_4 + NaOH$ from pH 6 to 8,

(3) $CH_3COOH + CH_3COONa$ below pH 6.

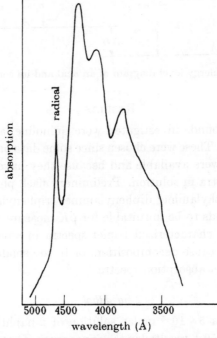

FIGURE 2. Absorption spectra of the triplet of 2-naphthol and the
2-naphthoxyl radical in viscous paraffin.

The spectra of the acidic and basic forms in buffered acidic and basic solutions are
shown in figure 3. Flash energies of only 100 J were used for this work and spectrophotometric determinations after flashing showed that the decomposition was less
than $\frac{1}{2}$ % per flash.

Measurements of optical densities were made at the shortest delay and the pK
in the triplet state was determined by the treatment already given assuming that
the optical density of the acid at 4320 Å was equal to that of the base at 4600 Å.
The resulting graph of concentration ratio versus pH is shown in figure 4 from which
we find that the pK of the triplet state of 2-naphthol

$$pK_T = 8.3 \text{ at } 20 \,°C$$

which is to be compared with $pK_G = 9.5$ for the ground state and $pK_S = 3.1$ for
the first excited singlet.

The assumption of equal extinction coefficients for the maxima of the two states was arbitrary and the effect of this assumption on the result must now be considered. If it is assumed that the maximum extinction coefficient of one form is twice that of the other the value of pK_T is shifted by 0·3 unit. It is unlikely that the extinction coefficients differ by much more than this. In the ground state the

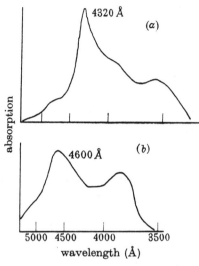

FIGURE 3. Absorption spectra of the triplets of (*a*) 2-naphthol (pH 4·6) and (*b*) 2-naphtholate ion (pH 10·9) in water.

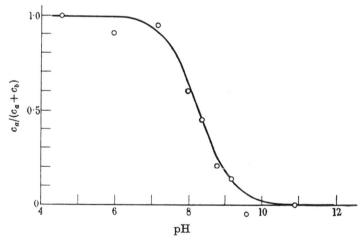

FIGURE 4. Plot for the determination of pK_T for 2-naphthol, assuming equal extinction coefficients.

maxima of the acidic and basic forms are at 3250 and 3450 Å with extinction coefficients of 1470 and 2290 Å respectively. Furthermore, it is unlikely that the population of the triplet state will be very dependent on pH, in view of the similarity of the spectra. If the efficiency of population of the two forms is taken to be identical, and this is probably the case below pH 9, the relative intensities of the two spectra lead to the value $pK_T = 8·1$.

20 G. Jackson and G. Porter

It is worth noting that, in order that the value of pK_T should equal that of the ground state, the ratio of extinction coefficients would be 0·06. Such a ratio is very unlikely and we may conclude that the triplet is a slightly stronger acid than the ground state but a very much weaker acid then the first excited singlet.

The photographic plates automatically recorded both the absorption spectra of the singlet and the fluorescence, so that pK_G and pK_S could also be determined. The fluorescence of both forms was noted and varied with pH as described by Förster and Weller.

The 2-naphthoxyl radical

The sharp peak at 4650 Å in liquid paraffin has been assigned to the 2-naphthoxyl radical for the following reasons:

(1) Its lifetime shows that it is a different species from the triplet state and analogy with similar investigations on phenols and hydroquinones suggests the naphthoxyl radical.

(2) Flash photolysis of methyl 2-naphthyl ether in liquid paraffin showed the same band at 4650 Å while the triplet bands were shifted to 4350, 4130, 3860 and 3520 Å.

(3) Further work by Porter & Strachan (1958) on the lines already described has shown that photolysis of a 5×10^{-5} molar solution of 2-naphthol in *iso*pentane-methyl *cyclo*hexane glass at 77 °K results in the rapid appearance of a sharp band at 4638 Å which remains as long as the glass is rigid, and disappears on warming. This must be attributed to a free radical and again analogy with similar work on the phenols leaves little doubt that it is to be identified with the 2-naphthoxyl radical.

Phosphorescence spectra by 2-naphthol

Phosphorescence spectra of 2-naphthol in ethanol and in ethanol + KOH solutions at 77 °K are shown in figures 5(a) and (b), respectively. The spectrum in figure 5(a) arises from the protonated molecule and the long-wavelength limit at 4735 Å agrees well with that reported by Lewis & Kasha (1944).

The phosphorescence spectrum of the base is similar to that of the acid at longer wavelengths but the shorter wavelength bands E and F do not appear, and doubling the exposure failed to reveal any further structure. Identification of the position of the origin is therefore difficult but a shift of about 200 Å is indicated both from the bands A to D at longer wavelengths and from the extrapolated short-wavelength limits. This separation leads to the following values:

$$E_{HA} = 6 \cdot 05 \, \text{kcal/mole}, \quad E_A = 58 \cdot 1 \, \text{kcal/mole}$$

and
$$pK_T = 7 \cdot 7.$$

2-naphthoic acid

The triplet absorption spectrum obtained by flash photolysis of 2-naphthoic acid (5×10^{-5} M) in liquid paraffin has maxima at 4300 and 4120 Å. In alkaline solution (N/10 NaOH) an intense absorption was observed with maximum at 4140 Å, while in N/10 HCl the absorption was weaker with maximum at 4280 Å. The relative intensity of these two maxima was studied as a function of pH in buffered solutions of pH between 1·25 and 7 and, with the same method as for 2-naphthol and the assumption of equal extinction coefficients, the value $pK_T = 3·7$ was obtained. Since the intensity of the spectrum of the base was approximately twice that of the acid it is more probable that the extinction coefficients are also in this ratio and this leads to the value $pK_T = 4·0$.

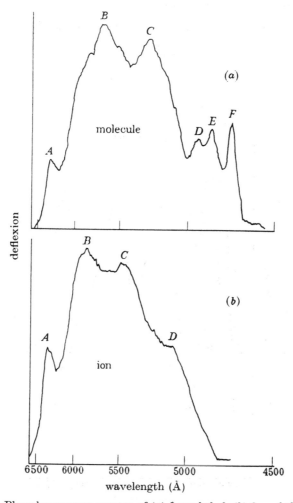

FIGURE 5. Phosphorescence spectra of (*a*) 2-naphthol, (*b*) 2-naptholate ion.

Both acidic and basic forms gave phosphorescence spectra with sharp maxima and short-wavelength limits which were identical within the accuracy of measurement, and very similar to the spectra of undissociated 1-naphthoic acid (see later section). From the limits $E_{HA} = E_A = 57·5\,\text{kcal/mole}$ and $pK_T = 4·2$.

1-*naphthoic acid*

Triplet absorption maxima in N/10 NaOH and N-HCl occurred at 4390 and 4460 Å, respectively. The relative intensity was studied in buffered solutions but only an approximate measurement of the pK was possible owing to the closeness of the maxima, the diffuseness of the spectra and strong interference from the fluorescence of the 1-naphthoic acid between 4700 and 3450Å. The value of pK_T was estimated as $3 \cdot 8 \pm 0 \cdot 5$.

The phosphorescence spectra of 1-naphthoic acid in EPA and alkaline ethanol glasses are shown in figure 6. The maxima and short-wavelength limits are very well characterized and lead to the values:

$$E_{HA} = 58 \cdot 0 \, \text{kcal/mole}, \quad E_A = 59 \cdot 3 \, \text{kcal/mole},$$

$$pK_T = 4 \cdot 6.$$

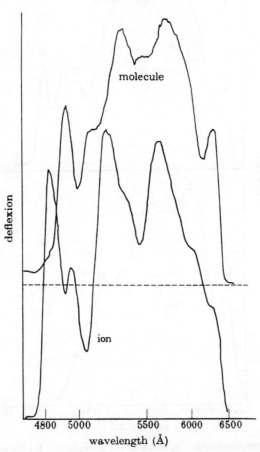

FIGURE 6. Phosphorescence spectra of 1-naphthoic acid and 1-naphtholate ion.

Acridine

Acidic solutions of acridine showed negligible transient absorption but alkaline solutions showed a strong transient with maximum at 4430Å and a weaker maximum at 5200Å. The acridinium cation therefore apparently has no observable triplet absorption spectrum and the pK of the acridine triplet was estimated from the relative concentrations of the acridine molecule triplet absorption. A plot of $T/T_{\text{max.}}$, referred to $T = T_{\text{max.}}$ at pH 7·4, is shown in figure 7 together with the calculated variation of the ratio $[A]/[A]+[AH^+]$ for acridine in the ground state.

The curves are closely similar for the two states and the value obtained for the pK in the triplet state is p$K_T = 5\cdot6$.

Independent confirmation of this value was not possible from phosphorescence data since both the acridinium ion and the unprotonated molecule showed a broad diffuse phosphorescence band whose maximum could not be estimated.

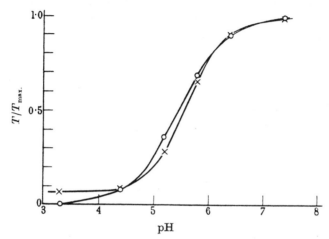

FIGURE 7. Plot for the determination of pK_G and pK_T of acridine. ○, Ground state; ×, triplet.

Quinoline

Absorption of both the acidic and basic forms of quinoline were observed with maxima at 4650 and 4180Å, respectively, but, as for acridine, the absorption of the unprotonated molecule was much stronger than that of the acid cation. An additional transient was observed which had a diffuse maximum at about 4900Å and whose intensity reached a maximum at pH 5. The origin of this band is unknown. In view of this complication, and the weakness of the acid triplet absorption the acidity constant was determined from a plot of $T/T_{\mathrm{max.}}$ against pH, which led to the value p$K_T = 6\cdot0$.

Phosphorescence spectra of the acid and base are shown in figure 8. Though very different in form, the spectra give fairly well characterized short wavelength maxima which lead to the values

$$E_{\mathrm{H}A} = 61\cdot6\,\mathrm{kcal/mole}, \quad E_A = 62\cdot6\,\mathrm{kcal/mole},$$

$$\mathrm{p}K_T = 5\cdot8.$$

2-naphthylamine

Flash photolysis of 2-naphthylamine in liquid paraffin resulted in the appearance of a transient absorption with a principal broad maximum at 4570 and a second maximum at 3770Å. In neutral and alkaline aqueous solutions a similar spectrum was obtained (figure 9) and in addition a rather sharp band at 5260Å was present at pH less than 7. In strongly acidic solutions the 5260Å band was still observed but the other bands disappeared and were replaced by a spectrum with maxima

24   G. Jackson and G. Porter

at 4140 and 3870 Å. The species responsible for the absorption at 5260 Å is possibly an amino radical but this is not yet established. The other bands are attributed to the triplet absorption spectra of the acid and base of 2-naphthylamine.

The acidity constant of the triplet state was estimated from a plot of $c_a/c_a + c_b$ based on the relative intensities of the maxima of acid and base at 4140 and 4570 Å, respectively. Assumption of equal extinction coefficients at the two maxima leads to the value $pK_T = 3 \cdot 5$. In fact the acid absorption is approximately double the intensity of that of the base and a better assumption is probably that $\epsilon_a^{4140} = 2\epsilon_b^{4570}$ which gives $pK_T = 3 \cdot 3$.

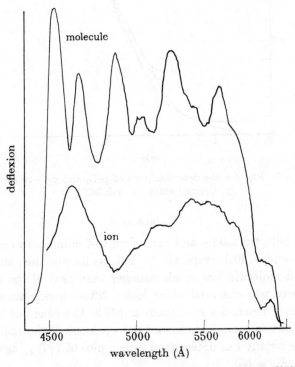

FIGURE 8. Phosphorescence spectra of quinoline and quinolinium ion.

Phosphorescence of the neutral basic molecule showed three broad bands with maxima at 4800, 5200 and 5700 Å. The protonated molecule showed a similar, though weaker, phosphorescence with the three maxima shifted about 100 Å to shorter wavelengths and with a short wavelength band at 4690 Å. These spectra lead to the following values:

$$E_{HA} = 60 \cdot 9 \, \text{kcal/mole}, \quad E_A = 59 \cdot 6 \, \text{kcal/mole},$$

$$pK_T = 3 \cdot 1.$$

N.N-*dimethyl 2-naphthylamine*

The pK of this amine in its ground state was not known and was therefore estimated spectrophotometrically from the absorption maxima of the base and acid at 2950 and 2800 Å, respectively. The value found was $pK_G = 4 \cdot 9$.

Flash photolysis in strongly acid solution showed two sharp transient bands at 4090 and 3880 Å. In neutral or alkaline solution these bands disappeared and were replaced by a weaker and more diffuse absorption extending from 6000 to 3600 Å. The acidity constant was therefore determined from the variation with pH of $T/T_{\mathrm{max.}}$ for the acid and the result obtained from measurements at nine different acidities with pH between -1.5 and 13 was $pK_T = 2.7$.

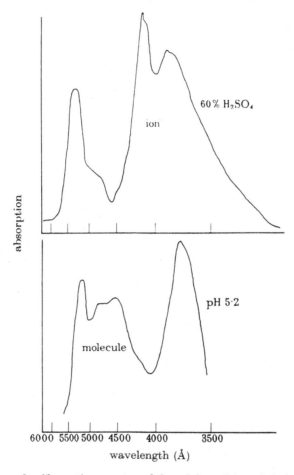

FIGURE 9. Absorption spectra of the triplets of 2-naphthylamine and 2-naphthylaminium ion.

Phosphorescence spectra of the acidic and basic molecules were of quite different form. The short-wavelength limit of the acid was clearly defined at 4710 Å, but the base showed only a broad maximum at 5700 Å which tailed off to shorter wavelengths. The extrapolated limit was 4900 Å which agrees well with that estimated by Lewis & Kasha for this molecule. This leads to the values

$$E_{\mathrm{H}A} = 60.7 \,\mathrm{kcal/mole}, \quad E_A = 58 \,\mathrm{kcal/mole},$$

$$pK_T = 2.9.$$

26   G. Jackson and G. Porter

The results of these investigations are summarized in table 1 along with the pK values for the ground state and for the first excited singlet state where these are known. Agreement between pK_T values obtained by the two methods is quite good. It is immediately apparent that the acidity constants of the triplet states of the molecules studied are quite close to those of the ground state while the acidity of the excited singlet state is very different. Before discussing the results further it is therefore necessary to consider the possibility that our measurements do not represent true equilibrium values and that an apparent similarity with the ground state results because the ground level determines the state of protonation of the triplet in the flash photolysis experiments.

TABLE 1

molecule	pK_G	pK_S	pK_T (flash photolysis)	pK_T (phos- phorescence)
2-naphthol	9·5	3·1	8·1	7·7
2-naphthoic acid	4·2	10–12	4·0	4·2
1-naphthoic acid	3·7	10–12	3·8	4·6
acridine	5·5	10·6	5·6	—
quinoline	5·1	—	6·0	5·8
2-naphthylamine	4·1	−2	3·3	3·1
$N.N$-dimethyl 1-naphthylamine	4·9	—	2·7	2·9

Most measurements were made at the shortest delays and, although no changes in relative intensity of the two spectra were ever observed at later times, we will consider that the time for equilibration in the excited state is approximately equal to the resolving time of our apparatus, i.e. 10^{-5} s. (In future work of this kind it would be better to use a slightly longer delay so that the time is more accurately specified.)

Consider first the flash photolysis experiments on 2-naphthol, and the rate processes in the singlet state first reached by light absorption. The rate constants for the forward and reverse reactions of the equilibrium

$$(ROH)^* + H_2O \rightleftharpoons (RO)^{*-} + H_3O^+,$$

where $(ROH)^*$ and $(RO)^{*-}$ are 2-naphthol and the 2-naphtholate ion in the first excited singlet state, have been determined by Weller (1955) as

$$\vec{k} = 4·1 \times 10^7 \, \text{s}^{-1},$$

and $$\overleftarrow{k} = 5·1 \times 10^{10} \, \text{l. mole}^{-1} \text{s}^{-1} \text{ at } 20\,^{\circ}\text{C},$$

while the natural radiative lifetimes of naphthol and naphtholate ion were $1·1 \times 10^{-8}$ and $8·1 \times 10^{-9}$ s, respectively.

Thus for the dissociation reaction in the excited singlet, the time to $1/e$th of complete reaction

$$\tau_{ROH} = 1/\vec{k} = 2·4 \times 10^{-8} \, \text{s}.$$

The association reaction has a lifetime of

$$\tau_{RO^-} = 1/\overleftarrow{k}[H^+]$$

which gives the values

pH	5	7	8	9	11
τ_{RO^-}	2×10^{-6}	10^{-4}	10^{-3}	10^{-2}	2 s

For the base and acid catalysis of this reaction

$$(ROH)^* + B^- \underset{k_a}{\overset{k_b}{\rightleftharpoons}} (RO)^* + HB$$

in solutions containing different buffer materials, the rate constants, determined by Weller (1959) are given in table 2. In the present work, $H_3BO_3 + KCl + NaOH$ and $KH_2PO_4 + NaOH$ buffers were used, mainly the former. The rates do not vary

TABLE 2

forward reaction B^-	k_b (l.mole^{-1}s^{-1})
$H.COO^-$	2.4×10^9
$CH_3.COO^-$	2.9×10^9
$C_2H_5.COO^-$	2.86×10^9
$C_3H_7.COO^-$	2.76×10^9
$H_2PO_4^-$	6.0×10^8

back reaction HB	k_a
$H.COOH$	2.8×10^8
$CH_3.COOH$	0.33×10^8
$H_3.PO_4$	2.91×10^9

greatly with different buffers and if it is assumed that the forward rate constant is the same for boric as for phosphoric acid, then $k_b = 6 \times 10^8$. The half-life of the dissociation reaction in the excited singlet

$$\tau_{ROH} = 1/\overrightarrow{k}_b[B^-]$$

and since the buffer concentration was typically 0.1 M, the life to be expected is about

$$\tau_{ROH} = 1/(6 \times 10^8 \times 0.1) = 1.7 \times 10^{-8}\,s.$$

Thus the forward rate is probably not greatly affected by the presence of buffer, at the most a factor of 10 increase is likely.

The back reaction, however, is affected much more strongly by the buffer

$$\tau_{RO^-} = 1/k_a[HB],$$

and again taking the phosphoric acid figure gives

$$\tau_{RO^-} = 1/(2.9 \times 10^9 \times 0.1) = 3.4 \times 10^{-9}\,s.$$

Comparing this value with the values given above for different pH's in the non-buffered solution, we see that in the region of the measured pK_T, the association reaction becomes very much faster with buffer present.

The mode of population of the molecule and ion triplets over the pH range covered may now be seen. For pH's < 9.5, the naphthol is mainly undissociated in solution, and the photoflash excites $ROH(S)$ which dissociates with a life of about

2×10^{-8} s, while decaying naturally in $1 \cdot 1 \times 10^{-8}$ s. The two processes will compete in the inverse ratio of their half-lives, less dissociating than are deactivated. Some population of $RO^-(T)$ via the dissociation to $RO^-(S)$ is to be expected and this will be independent of the pH. The back association when catalyzed is also fast and can probably compete successfully with the natural decay, the rate again being independent of pH (until strongly acid solutions are considered and the proton takes over the role of the acid catalyst). Since $pK_S = 3 \cdot 093$ there will be a shift of the ground equilibrium to the right on excitation, the proportion of $RO^-(S)$ depending on the exact values of k_b. Population of the triplet level would probably follow this trend, a bigger fraction of triplet ion being formed than was originally present in the ground state.

If the base and acid catalyses of the triplet reaction are the same as for the excited singlet, and their velocity constants are similar, then in 10^{-5} s equilibrium will easily be established. Taking extreme values of k_a and k_b, there is still a wide time margin for the changes to occur. It appears then that the triplet pK measured experimentally corresponds to a true equilibrium within the triplet level, and this is confirmed by the general agreement with acidity constants determined from phosphorescence measurements which are independent of any assumptions about the establishment of equilibrium. (It must be noted that, had buffers been absent, the back reaction would have been much too slow to progress before the spectra were recorded and the pK_T value calculated would have been merely a consequence of the singlet and ground pK's, and would bear no relation to the triplet equilibrium.) Similar considerations for the other molecules studied shows that in every case equilibrium in the triplet state should be established under the experimental conditions in a time less than 10^{-5} s and hence the measured acidity constants correspond to true equilibrium in the triplet state. In the case of 1-naphthoic acid, and 2-naphthoic acid this statement would have been true even in the absence of buffers; for the other molecules the presence of buffers was essential to the attainment of equilibrium in the time available.

COMPARISON OF ACIDITY CONSTANTS IN THE THREE STATES

The acidic and basic properties of aromatic molecules in their normal ground states have been rather well accounted for in a qualitative fashion by considering the resonance of the normally accepted structure with other canonical forms. Thus, for the four principal structural types studied in this paper, the proton affinity can be understood in terms of a small contribution of structure (b) to the 'normal' structure (a) (table 3). There are, of course, other similar structures of both types.

If the excited singlet state is also to be described in terms of these canonical structures it is clear, from the measured acidity constants, that the contribution of type (b) structures is much greater in the excited singlet than in the ground state.

Although our data are limited to a few molecules there is a clear indication that the contribution of the polar structures of type (b) is much less in the triplet than in the lowest excited singlet level and a possible qualitative explanation of this is to be found in considerations of electron correlation in the two states. It has been shown (Dickens & Linnett 1957) that two electrons in singly occupied orbitals have a high probability of being in the same position in the excited singlet state but a low probability of coincidence in the triplet state. Structures of type (b), on this basis, are probable ones in the excited singlet, while the electron distribution in the triplet state would be better described by a contribution from structures of type (c), in which the two electrons are spacially separated.

TABLE 3

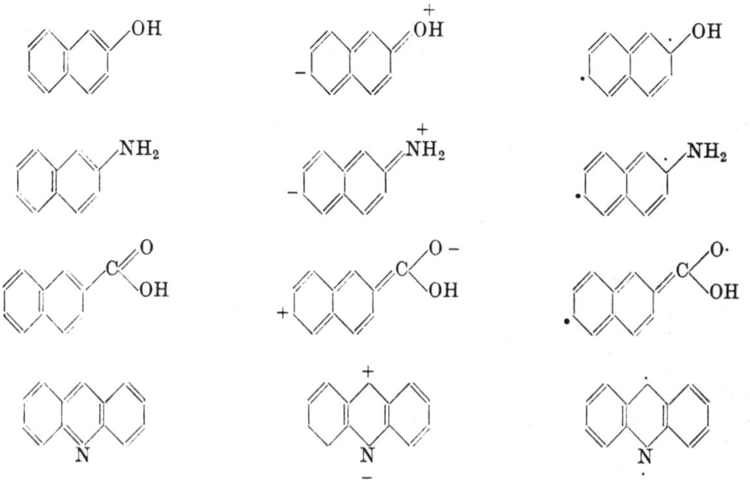

Useful as such considerations as these have been in organic chemistry, only complete calculations of electron density can give a quantitative account of acidity constants. Calculations of this type have largely substantiated the observations of acidity constants in the ground state and have been extended to include excited states. Electron densities and bond orders in the ground and excited states calculated by Coulson & Jacobs (1949) predict a migration of charge away from nitrogen on excitation of aniline but towards the nitrogen on excitation of pyridine which is just what is observed for the singlet states of the analogous naphthylamine and acridine molecules. In pyridine itself, however, to which the calculations apply, it seems that the acidity constant in the excited singlet state is nearly the same as in the ground state (Weller 1957). The contributions of charge-transfer states, in which the electron is transferred from the nitrogen into the lowest unoccupied orbital of the ring, have been calculated for aniline by Murrell (1955).

Calculations of this type have not been carried out for the molecules which we have studied and no detailed calculations have been carried out for the triplet states. Some recent preliminary calculations by Murrell (1960, private communication) on aniline have, however, shown that the charge transfer character is indeed much greater for the upper singlet than for the triplet state, the percentage charge transfer being 17 % in the singlet and only 4 % in the triplet. These calculations are being refined and extended.

G. Jackson and G. Porter

Although not applicable to all molecules studied, there is another possible explanation of pK differences between singlet and triplet states in molecules such as quinoline, acridine and the naphthoic acids. In these molecules both π-π and n-π transitions are possible and, since the singlet-triplet splitting of these two types of level are different, it is quite possible that the triplet and excited singlet states are of different types, the most probable case being an n-π singlet and π-π triplet because n-π singlet-triplet splitting is usually less than for π-π. Since π-π transitions might be expected to affect the charge density on the nitrogen or oxygen less than n-π transitions the acidity constant would be less affected by excitation to the triplet state than to the excited singlet.

References

Coulson, C. A. & Jacobs, J. 1949 *J. Chem. Soc.* p. 1984.
Dickens, P. G. & Linnett, J. W. 1957 *Quart. Rev. Chem. Soc.* 11, 291.
Förster, T. 1950 *Z. Electrochem.* 54, 42.
Ingold, C. K. 1953 *Structure and mechanism in organic chemistry*, p. 723.
　　　London: Bell.
Lewis, G. N. & Kasha, M. 1944 *J. Amer. Chem. Soc.* 63, 3005.
Murrell, J. N. 1955 *Proc. Phys. Soc.* A, 68, 969.
Porter, G. & Windsor, M. W. 1954 *Disc. Faraday Soc.* 17, 178.
Porter, G. & Strachan, E. 1958 *Trans. Faraday Soc.* 54, 1595.
Weber, K. 1931 *Z. Phys. Chem.* B, 15, 18.
Weller, A. 1952 *Z. Electrochem.* 56, 662.
Weller, A. 1955 *Z. Phys. Chem.* 3, 238.
Weller, A. 1957 *Z. Electrochem.* 61, 956.
Weller, A. 1959 *Disc. Faraday Soc.* 27, 28.

Reprinted from the Proceedings of the Royal Society, A, volume 264, pp. 1–18, 1961

Energy transfer from the triplet state

By G. Porter, F.R.S. and F. Wilkinson

Department of Chemistry, University of Sheffield

(*Received* 10 *April* 1961)

[Plates 1 and 2]

Energy transfer from the triplet level of a donor molecule resulting in quenching of the donor and elevation of the acceptor molecule from its singlet ground state to a triplet state has been observed between a number of donor-acceptor pairs in fluid solvents. In most cases the mechanism of the transfer has been unequivocally established by observation of the triplet state absorption spectra of both species. Energy transfer from excited singlet donors to the triplet state of the acceptor is not observed.

When the energy of the acceptor triplet is considerably lower than that of the donor, transfer is diffusion-controlled but there is no evidence for long-range resonance transfer of the kind found in the analogous singlet energy transfer processes. As the triplet energies become comparable the transfer probability is reduced and no quenching is observed by molecules with triplet levels higher than that of the donor. Transfer of triplet energy between pairs of aromatic hydrocarbons has been illustrated and it has been established that complex formation between donor and acceptor cannot be important under the conditions of these experiments.

The longer a molecule persists in an excited state the greater its chance of meeting another which can accept its excitation energy. In most molecules the ground state is a singlet and the lowest excited state is a triplet which has a lifetime several orders of magnitude greater than that of excited singlet states. It is therefore hardly surprising that the triplet state is frequently postulated as an intermediate in energy transfer processes, although the role played by the triplet state has been established for only a very few cases.

Quenching of the triplet state occurs both by chemical reaction and by physical processes. Chemical reactions include electron transfer, hydrogen atom abstraction from the solvent, dissociation and addition of oxygen. In these reactions spin conservation is always possible. Paramagnetic molecules, e.g. O_2, NO, aromatic triplets, transitional and inner transitional metal ions have been characterized as physical quenchers. The quenching efficiency is not proportional to magnetic susceptibility and occurs by the process

$$A^*(\text{triplet}) + Q(\text{multiplet}) \rightarrow A(\text{singlet}) + Q(\text{multiplet})$$

with overall spin conservation and without any change in the multiplicity of Q (Porter & Wright 1959).

When Q is a singlet state, spin conservation is only possible if there is a change in the multiplicity of Q. The energy transfer process represented as

$$A^*(\text{triplet}) + Q(\text{singlet}) \rightarrow A(\text{singlet}) + Q^*(\text{triplet})$$

is spin-allowed. Such a process has been postulated by Terenin & Ermolaev (1956) to explain the sensitized phosphorescence of naphthalene and its derivatives by light absorbed by benzophenone and similar compounds in rigid media at $-195\,^{\circ}\text{C}$.

2   G. Porter and F. Wilkinson

The concentrations required to bring about transfer in rigid media were very high and the work has been questioned because of the possibility of complex formation (McGlynn, Boggus & Elder 1960).

This paper deals with flash photolysis studies of the same process in solution at room temperature and provides unequivocal proof of the occurrence of this process in a variety of systems. During the course of this work Bäckström & Sandros (1958, 1960) by quite different methods have also provided clear evidence of this process.

Experimental

The flash photographic apparatus was the same as that used by Porter & Jackson (1961). Spectra were recorded on Ilford Selochrome and H.P.3 plates in a medium quartz Hilger spectrograph. Each plate was calibrated internally by means of a 7 step neutral filter of rhodium on quartz. Band positions of transient absorption peaks were subject to errors of ± 5 Å at 2500 Å which increased to ± 20 Å at 6000 Å.

The flash photoelectric apparatus was originally very similar to that described by Porter & Wright (1958). During the course of this work several modifications were made in order to allow measurements at shorter times after irradiation. Kinetic analysis of the oscillographic traces, which gave a direct measure of any transient during its formation and decay, was only made when the scattered light pulse from the photolysis lamps had decayed so as to give no measurable deflexion under working conditions, i.e. when the intensity of the scattered light at the particular wavelength being investigated was less than 1 % of the monitoring light intensity. The modifications, which included the use of a zirconium arc lamp as the monitoring light source, addition of nitrogen to the lamp fillings and improved circuit design to reduce the flash duration (Gover & Porter 1961) made accurate kinetic measurements possible 50 μs after the photolysis flash.

The reaction vessels were designed to fit both types of flash photolysis apparatus. They were constructed from quartz, Pyrex or soda-glass and were essentially similar to those described by Porter & Windsor (1954). Some of them had outer jackets so that 0·6 cm of filter solution could be placed around the reactants. The reaction vessels were thoroughly cleaned before use and grease traps were introduced below the tap to ensure that no grease could contaminate the solution.

Solutes

Napththalene (B.D.H. micro-analytical reagent) was used without further purification. 1-Bromonaphthalene was fractionated. 1-Iodonaphthalene was treated with sodium sulphite to remove iodine, washed, dried and fractionated. The fraction collected was kept over calcium in the dark. The anthracene was the same as that used by Porter & Windsor (1954). Phenanthrene was freed from anthracene by treatment with maleic anhydride (Kooyman & Farenhorst 1953) in purified benzene and then recrystallized from spectroscopic alcohol. Triphenylene, 1:2-benzanthracene and benzophenone were recrystallized from spectroscopic ethyl alcohol. Diacetyl was fractionally distilled in a nitrogen atmosphere under reduced pressure (100 mm), and stored in the dark. Iodine was re-sublimed and ethyl iodide was passed through a column of activated alumina and fractionated.

Hexane was spectroscopic grade and was not further purified. Analar benzene was extracted with concentrated sulphuric acid, washed, dried and fractionated. Hopkins and Williams G.P.R. ethylene glycol was filtered and used without further purification at first. Later supplies of this solvent were less pure, judged by ultra-violet transmission and were fractionally distilled. The fraction which was used still did not have as good transmission properties as the original samples and gave an increased decay rate of triplet states. This did not affect the accuracy of measurement of quenching rate constants since care was taken to ensure that solutions used to investigate any particular pair of compounds were made up from the same stock of solvent.

Thorough outgassing of all solutions was necessary because of the high efficiency of quenching of the triplet state by oxygen. A freeze, pump, melt, shake, freeze procedure was repeated until the pressure of air above the solution after freezing in liquid nitrogen was less than 0·001 mm. When ethylene glycol was used as solvent the bulk of the solution was not frozen but a small reservoir was cooled to liquid-nitrogen temperature to prevent loss of solute or solvent.

The absorption spectra of all solutions investigated were measured before and after flashing. In order to check that no change in concentration had occurred as a result of the outgassing procedure, spectra were recorded before and after outgassing.

Filter solutions

When mixtures of donor and acceptor compounds were flashed an attempt was made to filter the light from the photolysis flash so that light was only absorbed by the donor compound. Soda-glass or Pyrex reaction vessels were used in some cases. Frequently a very highly concentrated solution of the acceptor compound was used in the outer jacket of the reaction vessel. This filter interfered very little with the higher wavelength absorption of the donor compound. Checks were made to ensure that such filters were working efficiently by flashing the acceptor compound alone in a filtered reaction vessel when the amount of light absorbed was shown to be negligible.

RESULTS

Unless otherwise stated all the results refer to thoroughly outgassed solutions. No attempt was made to control thermostatically the reaction vessels but the temperature was always between 20 and 25 °C. Rate constants given are the average of at least two separate runs, four traces from each run being analyzed in most cases.

Flash photolysis of single compounds in solution

Each solute was studied first in the absence of other solutes. The triplet-triplet absorption spectra observed are given in table 1 and the first-order decay constants in table 2.

The band positions and relative intensities are in good agreement with those measured by Porter & Windsor (1958). No transient absorption was detected when

4   G. Porter and F. Wilkinson

solutions 10^{-5} to 10^{-3} M in 1-iodonaphthalene were flashed in either hexane or ethylene glycol. In the case of phenanthrene Porter & Windsor reported two additional bands at 5200 Å and 5085 Å. The present authors also observed these bands when they used the same sample of phenanthrene but they were not obtained from a sample of phenanthrene which had been extensively purified.

TABLE 1. BAND MAXIMA OF TRIPLET-TRIPLET ABSORPTION SPECTRA

compound	solvent	$\lambda_{max.}$(Å)	$\nu_{max.}$(cm^{-1})	relative intensity
naphthalene	hexane	4100	24390	1·07
		3890	25706	0·55
		3690	27100	0·15
	benzene	4150	24100	1·0
		3920	25510	0·60
		3720	26880	0·15
	ethylene glycol	4150	24100	1·0
		3920	25510	0·55
		3780	26880	0·10
1-bromonaphthalene	hexane	4190	23870	1·0
		3980	25130	0·55
		3750	26670	0·20
	ethylene glycol	4200	23800	1·00
		3980	25130	0·60
		3780	26460	0·25
1-iodonaphthalene	hexane ⎫ ethylene glycol ⎬	no detectable absorption		
anthracene	hexane	4650	21510	v. weak
		4230	23640	1·00
		4000	25000	0·30
	ethylene glycol	4690	21320	v. weak
		4250	23530	1·00
		4010	24940	0·30
phenanthrene	hexane	4800	20830	1·00
		4500	22220	0·60
		4280	23360	0·25
		3990	25060	0·10
	ethylene glycol	4850	20620	1·00
		4530	21980	0·60
		4280	23420	0·25
		4000	25000	0·10
triphenylene	hexane	4280	23360	1·00
		4050	24570	0·90
	benzene	4280	23360	1·00
		4070	24570	0·90
	ethylene glycol	4280	23360	1·00
		4070	24570	0·90
1:2-benzanthracene	benzene	5380	18590	0·10
		4880	20490	1·00
		4580	21830	0·65
		4350	22990	0·50
		4020	24880	0·30
pentacene	hexane	4920	20330	0·30
		4600	21740	0·20
		3850	25970	0·10
		3000	33330	1·00

No decomposition was detected of naphthalene, phenanthrene, triphenylene, or 1:2-benzanthracene. Slight decomposition occurred with 1-bromonaphthalene, 1-iodonaphthalene, anthracene and pentacene.

The values of k_1, the first-order rate constant for triplet state decay, were found to be independent of the initial concentration of the solute except for triphenylene. In this case, above about 10^{-3} M, the transient absorption spectrum changed from fairly sharp peaks to a broad general absorption upon which the normal triplet-triplet spectrum seemed to be superimposed. The singlet-singlet absorption spectrum was examined at concentrations above 10^{-3} M but no new bands nor any deviation from Beer's law were observed. Above 10^{-3} M the decay of the transient absorption measured at the triplet maxima was complex. The values given for triphenylene in tables 1 and 2 refer to solutions of concentration less than 10^{-3} M.

TABLE 2. THE FIRST-ORDER RATE CONSTANTS OF TRIPLET STATE DECAY

	$10^{-3}k_1(\text{s}^{-1})$	
	hexane	ethylene glycol
naphthalene	11 ± 1	$1 \cdot 0 \pm 0 \cdot 1$
1-bromonaphthalene	12 ± 2	$1 \cdot 2 \pm 0 \cdot 2$
anthracene	$1 \cdot 1 \pm 0 \cdot 1$	$0 \cdot 26 \pm 0 \cdot 01$
phenanthrene	$10 \cdot 7 \pm 0 \cdot 8$	$1 \cdot 1 \pm 0 \cdot 1$
triphenylene	18 ± 3	$1 \cdot 0 \pm 0 \cdot 1$
1-benzanthracene	$6 \cdot 3 \pm 0 \cdot 5$ (benzene)	—
pentacene	9 ± 1	—

Further outgassing, by the procedure outlined in the experimental section, had no effect on the measured rate constants. The values given in table 2 for naphthalene and anthracene are in good agreement with those obtained by Porter & Wright (1958). However, towards the end of this research a few runs were made using different outgassing procedures. In one run, for example, after the normal outgassing, the whole solution of naphthalene in hexane was distilled from one side arm to another and back a dozen times. The reaction vessel was pumped out after every distillation. A value of $2 \cdot 8 \times 10^3 \text{s}^{-1}$ was obtained for k_1 compared with $1 \cdot 1 \times 10^4 \text{s}^{-1}$ given in table 2 for naphthalene in hexane. The value of k_1 in this case increased with the number of times the solution was flashed.

Porter & Windsor (1958) assigned the transient they observed upon flashing benzophenone in liquid paraffin to the triplet state because of the similarity between its absorption spectrum and that observed by McClure & Hanst (1955) as a transient in a rigid solvent. A detailed examination by the present authors (Porter & Wilkinson 1961) indicates that the majority of the transient absorption in fluid media with bands at ~ 3300 and 5440 Å is due to the ketyl radical $\phi_2 \overset{.}{\text{C}}$—OH which is formed from the triplet state by hydrogen atom abstraction from the solvent.

Diacetyl phosphoresces in solution and also gives a transient absorption below 3300 Å when flashed. An estimate of the transient lifetime, from the photographic apparatus and a 10^{-2} M solution of diacetyl in benzene gave $\tau = 300 \pm 100$ μs. The phosphorescence lifetime was also measured for 10^{-2} M diacetyl in benzene on a flash photoelectric apparatus available in our laboratories. This gave a value for the

6   G. Porter and F. Wilkinson

mean life of the phosphorescence decay of $600 \pm 50\,\mu s$. The significant difference between these two values indicates that the absorption and the emission do not arise from the same species.

Scattered light from the photolysis flash was negligible and good fluorescence spectra were recorded photographically by a single flash from many of the solutions studied.

TABLE 3. QUENCHING RATE CONSTANTS

	donor	acceptor	solvent	k_Q(l. mole^{-1} s^{-1})
1	phenanthrene	naphthalene	hexane	$2 \cdot 9 \pm 0 \cdot 7 \times 10^6$
			ethylene glycol	$2 \cdot 3 \pm 0 \cdot 8 \times 10^6$
2	triphenylene	naphthalene	hexane	$1 \cdot 3 \pm 0 \cdot 8 \times 10^9$
3	phenanthrene	1-bromonaphthalene	hexane	$1 \cdot 5 \pm 0 \cdot 8 \times 10^8$
			ethylene glycol	$1 \cdot 5 \pm 0 \cdot 8 \times 10^7$
4	phenanthrene	1-iodonaphthalene	hexane	$7 \pm 2 \times 10^9$
			ethylene glycol	$2 \cdot 1 \pm 0 \cdot 2 \times 10^8$
5	naphthalene	1-iodonaphthalene	ethylene glycol	$2 \cdot 8 \pm 0 \cdot 3 \times 10^8$
6	1-bromonaphthalene	1-iodonaphthalene	ethylene glycol	$8 \pm 4 \times 10^7$
7	benzophenone	naphthalene	benzene	$1 \cdot 2 \times 10^9$
8	diacetyl	1:2-benzanthracene	benzene	$3 \pm 2 \times 10^9$
9	phenanthrene	iodine	hexane	$1 \cdot 4 \pm 0 \cdot 6 \times 10^{10}$
10	anthracene	iodine	hexane	$2 \cdot 4 \pm 0 \cdot 2 \times 10^9$

Systems showing no quenching, for which upper limits were obtained

	donor	acceptor	solvent	
1	naphthalene	phenanthrene	hexane	$\leqslant 2 \times 10^4$
			ethylene glycol	$\leqslant 1 \times 10^5$
2	naphthalene	triphenylene	hexane	$\leqslant 5 \times 10^4$
3	1-bromonaphthalene	phenanthrene	ethylene glycol	$\leqslant 5 \times 10^4$
4	naphthalene	benzophenone	benzene	$\leqslant 1 \times 10^4$
5	1:2-benzanthracene	diacetyl	benzene	$\leqslant 5 \times 10^4$
6	anthracene	phenanthrene	ethylene glycol	$\leqslant 5 \times 10^3$
7	anthracene	naphthalene	hexane	$\leqslant 4 \times 10^4$
8	anthracene	1-iodonaphthalene	ethylene glycol	$\leqslant 2 \times 10^4$
9	anthracene	ethyl iodide	hexane	$\leqslant 1 \cdot 6 \times 10^4$
10	phenanthrene	ethyl iodide	hexane	$\leqslant 3 \times 10^5$

Flash photolysis of mixed solutes

Previous workers who have studied triplet energy transfer have dealt almost exclusively with carbonyl compounds as energy donors. In order to establish the generality of this type of process it seemed important to investigate whether it occurred when both energy donor and acceptor were unsubstituted aromatic hydrocarbons. Having established that transfer of this type did occur the next step was to investigate the factors upon which the efficiency of the transfer process depends.

The results obtained for every pair of compounds studied are summarized in table 3. In the account which follows the donor is written first.

(i) *Phenanthrene and naphthalene*

This pair of compounds was studied in hexane, with a soda-glass reaction vessel as filter, and in ethylene glycol, with a concentrated solution of naphthalene in hexane in the outer jacket of a Pyrex reaction vessel as filter. At the highest concentrations of naphthalene used only a trace of triplet naphthalene was formed by direct excitation through the soda-glass filter. The concentrated solution of naphthalene in hexane was a very efficient filter and no triplet naphthalene was detected when solutions of naphthalene were flashed through this filter. Figure 5, plate 1 shows the effect of adding 6.9×10^{-3} M of naphthalene to a 2.3×10^{-3} M solution of phenanthrene in ethylene glycol. The triplet state absorption of phenanthrene has been completely suppressed and has been replaced by the absorption spectrum of triplet naphthalene.

Although the triplet-triplet absorption maxima of phenanthrene and naphthalene are quite separate, there is overlap of the weak tails of these bands which interferes with the kinetic analysis when both are present. The quenching rate constants can be estimated, though less accurately, from the series of spectra obtained from the flash photographic apparatus.

The rate of decay of the triplet state can be written as

$$-\mathrm{d}[T]/\mathrm{d}t = k_1[T] + k_2[T]^2 + k_Q[Q][T], \tag{1}$$

where k_1 and k_2 are the first- and second-order rate constants of decay of the triplet donor T, k_Q is the bimolecular quenching rate constant and $[Q]$ is the concentration of the acceptor or quencher.

At low triplet state concentrations the $k_2[T]^2$ term can be neglected and, if the concentration of the quencher which increases the rate of decay of the donor by a factor of 2 is called the half-quenching concentration then

$$k_Q[Q_{\frac{1}{2}}][T] = k_1[T] \quad \text{or} \quad k_Q = k_1/[Q_{\frac{1}{2}}]. \tag{2}$$

The half-quenching concentrations for the quenching of triplet phenanthrene by naphthalene were $4 \pm 1 \times 10^{-3}$ M and $1.0 \pm 0.25 \times 10^{-3}$ M in hexane and ethylene glycol, respectively, k_1 values for phenanthrene were $10.7 \pm 0.8 \times 10^3$ and $2.1 \pm 0.2 \times 10^3 \mathrm{s}^{-1}$, giving quenching rate constants of triplet phenanthrene by singlet naphthalene of $2.9 \pm 0.9 \times 10^6$ and $2.3 \pm 0.8 \times 10^6$ l. mole^{-1} s^{-1} in hexane and ethylene glycol, respectively.

The first-order rate constant for the decay of triplet naphthalene formed by energy transfer was identical with that formed by direct excitation in the presence or absence of phenanthrene.

In these experiments the concentration of naphthalene was comparable with that of phenanthrene so a check was made on the absorption spectra of separate components and that of the mixture to see if there was any indication of complex formation. No new bands were observed and, for all the mixtures studied, the absorption spectrum of the mixture was simply the sum of the separate absorption of the two components.

Fluorescence measurements were made under identical conditions, the same reaction vessel being used with a concentrated solution of naphthalene as filter. Special care was taken to develop the separate plates under similar conditions. The microdensitometer traces of phenanthrene fluorescence were compared for four runs each containing a fixed concentration of phenanthrene in glycol and (a) zero, (b) 1.25×10^{-3} M, (c) 2.5×10^{-3} M, and (d) 5.0×10^{-3} M of naphthalene.

8   G. Porter and F. Wilkinson

The intensity of fluorescence in these four solutions was the same to $\pm 5\%$ and strong fluorescence was observed from phenanthrene even when no triplet phenanthrene absorption was detectable (see, for example, figure 5).

(ii) *Triphenylene and naphthalene*

Although difficulties were experienced in the interpretation of the flash photolysis of triphenylene alone at concentrations above 10^{-3} mole/l., 1.01×10^{-3} M solutions of triphenylene in hexane containing varying amounts of naphthalene were investigated in a soda-glass reaction vessel. The production of the triplet-triplet absorption of naphthalene under these conditions showed that energy transfer was taking place. The effect of adding 1.15×10^{-4} M of naphthalene is illustrated in figure 1.

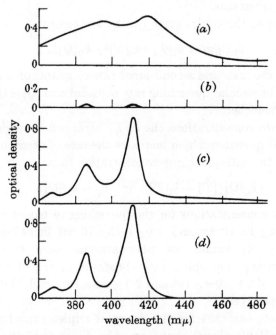

FIGURE 1. Transient absorption spectra following photolysis, in hexane solution of (a) 1.01×10^{-3} M triphenylene, (b) 1.15×10^{-4} M naphthalene, (c) 1.01×10^{-3} M triphenylene, and 1.15×10^{-4} M naphthalene. In (d) the solution was the same as in (b) but the filter was removed.

Because of the overlap of the triplet absorption spectra of triphenylene and naphthalene, photoelectric measurements were not useful and the quenching constant was estimated from the photographic records with the following results:

$$[Q_{\frac{1}{2}}] = 2 \pm 1 \times 10^{-5} \,\mathrm{M}, \quad k_1 = 1.8 \pm 0.3 \times 10^4 \,\mathrm{s^{-1}}$$

and therefore $k_Q = 1.3 \pm 0.8 \times 10^9$ l. mole^{-1} s^{-1}.

(iii) *Phenanthrene and 1-bromonaphthalene*

This pair of compounds was studied in hexane and in ethylene glycol with a concentrated solution of naphthalene used as filter. Energy transfer was confirmed

Porter & Wilkinson   *Proc. Roy. Soc. A, volume* 264, *plate* 1

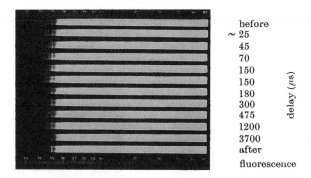

(*a*) 6.9×10^{-3}M naphthalene

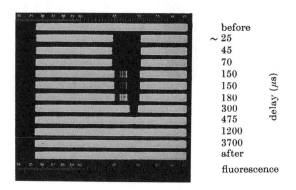

(*b*) 2.3×10^{-3}M phenanthrene

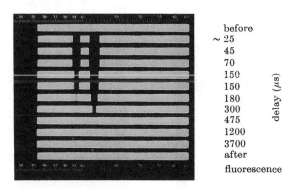

(*c*) 6.9×10^{-3}M naphthalene $+ 2.3 \times 10^{-3}$M phenanthrene.

FIGURE 5. Absorption spectra illustrating energy transfer from the triplet state of phenanthrene to naphthalene (ethanol solution).

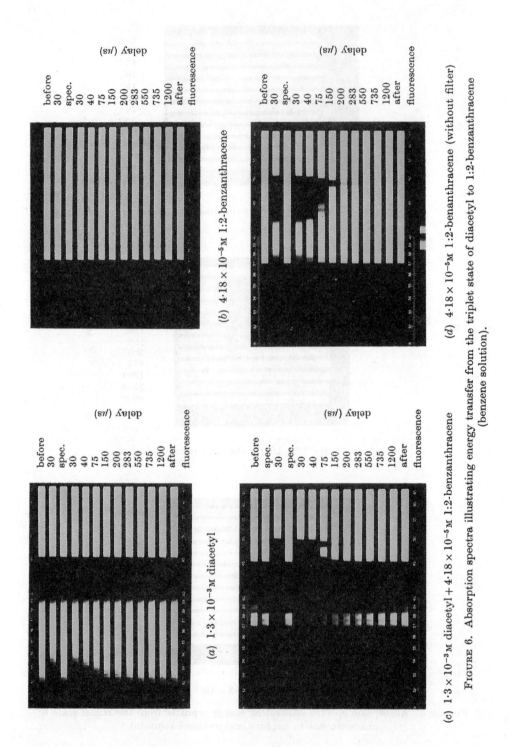

(a) 1.3×10^{-3} M diacetyl

(b) 4.18×10^{-5} M 1:2-benzanthracene

(c) 1.3×10^{-3} M diacetyl + 4.18×10^{-5} M 1:2-benzanthracene

(d) 4.18×10^{-5} M 1:2-benanthracene (without filter)

FIGURE 6. Absorption spectra illustrating energy transfer from the triplet state of diacetyl to 1:2-benzanthracene (benzene solution).

by the appearance of the triplet-triplet absorption of 1-bromonaphthalene. By the same method as before the following values were derived:

Hexane

$$[Q_{\frac{1}{2}}] = 1 \cdot 0 \pm 0 \cdot 5 \times 10^{-4}\,\text{M}, \quad k_1 = 10 \cdot 7 \pm 0 \cdot 8 \times 10^3\,\text{s}^{-1},$$

$$k_Q = 1 \cdot 5 \pm 0 \cdot 8 \times 10^8\,\text{l. mole}^{-1}\,\text{s}^{-1}.$$

Ethylene glycol

$$[Q_{\frac{1}{2}}] = 2 \pm 1 \times 10^{-4}\,\text{M}, \quad k_1 = 2 \cdot 1 \pm 0 \cdot 2 \times 10^3\,\text{s}^{-1}$$

$$k_Q = 1 \cdot 5 \pm 0 \cdot 8 \times 10^7\,\text{l. mole}^{-1}\,\text{s}^{-1}.$$

(iv) *Phenanthrene and 1-iodonaphthalene*

In the case of phenanthrene with naphthalene and 1-bromonaphthalene the fact that energy transfer had taken place was illustrated by the appearance of the triplet-triplet absorption spectra of the acceptors. However, 1-iodonaphthalene does not show a triplet-triplet absorption and this type of illustration was not possible. On the other hand, the effect of 1-iodonaphthalene on the decay of triplet phenanthrene can be studied accurately on the flash photoelectric apparatus since there is no interference from the acceptor triplet. This was done in the solvents hexane and ethylene glycol with a concentrated solution of naphthalene in hexane as filter.

Various concentrations of 1-iodonaphthalene were added to phenanthrene solutions and the rate of decay of triplet phenanthrene was measured from first-order plots which were accurately linear at low triplet concentrations. The first-order rate constant obtained from these plots (k_A) decreased each time the solution in hexane was flashed, indicating that the 1-iodonaphthalene was being destroyed and therefore only the values from the first flashes of each run were used to determine the bimolecular rate constant k_Q for energy transfer. This effect was not observed for phenanthrene and 1-iodonaphthalene in ethylene glycol where the concentrations of 1-iodonaphthalene were approximately 100 times greater.

The rate of decay of the triplet state is given by

$$-\mathrm{d}[T]/\mathrm{d}t = \{k_1 + k_Q[Q]\}\,[T] = k_A[T];$$

therefore,
$$k_A = k_1 + k_Q[Q]. \tag{3}$$

Plots of k_A against $[Q]$ were linear and from the slopes the values $k_Q = 7 \pm 2 \times 10^9$ and $2 \cdot 1 \pm 0 \cdot 2 \times 10^8$ l. mole^{-1}s^{-1} were obtained for the quenching rate constants in hexane and ethylene glycol, respectively. The plot of k_A against $[Q]$ for triplet phenanthrene with 1-iodonaphthalene as quencher in ethylene glycol is shown in figure 2.

Again the fluorescence yield of phenanthrene was unaffected by the presence of the 1-iodonaphthalene and the absorption spectrum of the mixture was the sum of the absorption spectra of the separate solutes.

(v) *Naphthalene and 1-iodonaphthalene*

The absorption spectra of naphthalene and 1-iodonaphthalene are so similar that it is not possible to filter the system selectively. Measurements were made with a quartz reaction vessel and ethylene glycol as solvent. The concentration of

G. Porter and F. Wilkinson

naphthalene was 10^{-3} M and the highest concentration of 1-iodonaphthalene used was 4×10^{-5} M so that most of the light was absorbed by the naphthalene. Figure 3 shows the plot of k_A against the concentration of 1-iodonaphthalene in glycol which leads to the value $k_Q = 2 \cdot 8 \pm 0 \cdot 3 \times 10^8$ l. mole^{-1} s^{-1}.

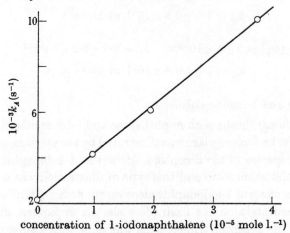

FIGURE 2. Plot of k_A, the decay constant of triplet phenanthrene, against $[Q]$, the concentration of 1-iodonaphthalene.

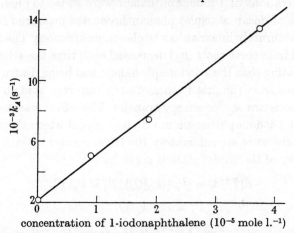

FIGURE 3. Plot of k_A, the decay constant of triplet naphthalene, against $[Q]$, the concentration of 1-iodonaphthalene.

(vi) 1-*Bromonaphthalene and* 1-*iodonaphthalene*

Here again selective filtration was not possible and a quartz reaction vessel was used. Two concentrations of 1-iodonaphthalene were studied and the concentration of 1-bromonaphthalene was always at least ten times greater than that of 1-iodonaphthalene. The quenching rate constant derived was $8 \pm 4 \times 10^7$ l. mole^{-1} s^{-1}.

(vii) *Benzophenone and naphthalene*

Studies on this system are described in detail elsewhere (Porter & Wilkinson 1961). The change in the transient absorption observed upon flashing 10^{-2} M benzophenone in benzene was studied as the amount of added naphthalene was varied in steps from

10^{-4} to 10^{-2} M. As naphthalene was added the amount of absorption by the ketyl radical decreased and absorption by triplet naphthalene appeared. To ensure that the naphthalene itself was not irradiated a soda-glass reaction vessel was used.

As the triplet state of the donor is not observed, direct measurements of quenching are not possible. The relevant reactions are as follows:

$$\phi_2CO(T) \rightarrow \phi_2CO(S), \tag{1}$$

$$\phi_2CO(T) + RH \rightarrow \phi_2C\text{—}OH + R^{\cdot}, \tag{2}$$

$$\phi_2CO(T) + Q(S) \rightarrow \phi_2CO(S) + Q(T), \tag{3}$$

where RH = solvent, Q = acceptor naphthalene, and T and S refer to triplet and ground states.

At the concentration of naphthalene which reduces the amount of ketyl formed by one half

$$k_3[Q_{\frac{1}{2}}] = k_1 + k_2[RH].$$

The half-quenching concentration was found to be 4×10^{-4} M. Using the value of Bäckström & Sandros for the lifetime τ_B of triplet benzophenone in benzene we obtain

$$\tau_B = \frac{1}{k_1 + k_2[RH]} = 1 \cdot 9 \times 10^{-6} \quad \text{and} \quad k_3 = k_Q = 1 \cdot 2 \times 10^9 \, \text{l. mole}^{-1} \text{s}^{-1}.$$

Measurements of ketyl concentrations were made $70 \, \mu$s after the flash, and this may make the above estimate of k_Q somewhat low.

(viii) *Diacetyl and 1:2-benzanthracene*

Solutions of diacetyl, 1:2-benzanthracene and mixtures of the two in benzene solution were flashed with a concentrated solution of 1:2-benzanthracene used as filter. Figure 6, plate 2 shows

(i) energy transfer to form the triplet state of 1:2-benzanthracene;

(ii) the absence of fluorescence when the triplet state of 1:2-benzanthracene is formed by energy transfer, proving that its formation bypasses the first excited singlet state.

The value of the quenching constant derived from these photographic results was $k_Q = 3 \pm 2 \times 10^9$ l. mole^{-1} s^{-1} in good agreement with the value $3 \cdot 8 \times 10^9$ l. mole^{-1} s^{-1} measured by a quite different method by Bäckström & Sandros (1958).

(ix) *Quenching by iodine*

The effect of three different concentrations of iodine, the highest being 10^{-5} M, on the rate of decay of triplet phenanthrene in hexane was studied on the flash photoelectric apparatus. The concentration of phenanthrene was $5 \cdot 0 \times 10^{-3}$ M throughout these experiments and a concentrated solution of iodine in hexane was used as a filter. The k_1 values for solutions which contained iodine decreased with the number of times the solution was flashed. This indicated that the iodine was being used up and therefore only the rate values obtained from the first flashes were used. The value of k_Q obtained in this way was $1\cdot4\pm0\cdot4\times10^{10}\,\text{l. mole}^{-1}\text{s}^{-1}$.

Similar measurements were carried out using anthracene as the donor, the concentration of anthracene being $1\cdot39\times10^{-4}\,\text{M}$. Three different concentrations of iodine were used, the highest being $5\times10^{-6}\,\text{M}$. Again iodine was consumed on flashing and only the first flash was used for rate measurements. A value of $k_Q = 2\cdot4\pm0\cdot2\times10^{9}\,\text{l. mole}^{-1}\text{s}^{-1}$ was obtained for the quenching rate constant of anthracene by iodine in hexane.

(x) *Systems which showed no quenching of the triplet state*

For a number of the pairs of compounds described so far, the lifetime of the triplet state of the acceptor compound, formed by energy transfer, generally in the presence of a high concentration of the donor compound, was found to be identical with that of the triplet state of the acceptor alone formed by direct excitation. Thus k_Q for triplet naphthalene with phenanthrene as quencher in ethylene glycol was $\leqslant 2\times10^{4}\,\text{l. mole}^{-1}\text{s}^{-1}$. Upper limits of other k_Q values obtained in this way are included in table 4.

No quenching was observed in the following systems:

(*a*) Anthracene and phenanthrene

The decay of triplet anthracene in ethylene glycol was unaffected by the presence of $1\cdot9\times10^{-3}\,\text{M}$ of phenanthrene.

Thus $k_Q \leqslant 5\times10^{3}\,\text{l. mole}^{-1}\text{s}^{-1}$.

(*b*) Anthracene and naphthalene

The decay of triplet anthracene in hexane was unaffected by $1\cdot2\times10^{-3}\,\text{M}$ of naphthalene.

Thus $k_Q \leqslant 4\times10^{4}\,\text{l. mole}^{-1}\text{s}^{-1}$.

(*c*) Anthracene and 1-iodonaphthalene

The decay of triplet anthracene in ethylene glycol was unaffected by $9\cdot3\times10^{-4}\,\text{M}$ of 1-iodonaphthalene.

Thus $k_Q \leqslant 2\times10^{4}\,\text{l. mole}^{-1}\text{s}^{-1}$.

(*d*) Anthracene and ethyl iodide

The decay of triplet anthracene in hexane showed only slight quenching in the presence of $1\cdot24\times10^{-2}\,\text{M}$ of ethyl iodide. A concentrated solution of naphthalene in hexane was used as filter for these experiments. The filter was not perfect and the ethyl iodide decomposed to give iodine which then quenched the triplet anthracene. This was clear since both the absorption of iodine and the amount of quenching increased as the solution was flashed.

The upper limit derived for the quenching rate constant was $1\cdot6\times10^{4}\,\text{l. mole}^{-1}\text{s}^{-1}$.

(*e*) Phenanthrene and ethyl iodide

Again a concentrated solution of naphthalene was used as a filter and the ethyl iodide slightly decomposed each time the solution was flashed. The rate of decay of triplet phenanthrene in hexane was only very slightly quenched in the presence of $1\cdot24\times10^{-2}\,\text{M}$ of ethyl iodide which gave the value derived for

$$k_Q \leqslant 3\times10^{5}\,\text{l. mole}^{-1}\text{s}^{-1}.$$

Discussion

This work has not been directly concerned with the mechanism of triplet state decay in pure fluid solvents but two points may be referred to in this connexion. First, the excellent agreement between our earlier values of k_1 and those of Porter & Wright (1958) is probably an illustration of the contention of Livingston, Jackson & Pugh (1960) that a constant level of impurity such as oxygen can be obtained with a high degree of reproducibility. Our later values of k_1 were significantly lower and this indicates that the earlier values related to pseudo first-order processes. The quenching efficiency of aromatic molecules with lower triplet levels found in this work suggests one possible source of quenching impurity.

Secondly, our finding that naphthalene, 1-chloronaphthalene and 1-bromonaphthalene have comparable triplet lifetimes whilst for 1-iodonaphthalene no triplet absorption could be detected seems very significant. In rigid media the phosphorescence lifetimes of the halogenated naphthalenes decrease regularly from 2·6 s for naphthalene to 0·0028 s for 1-iodonaphthalene and the rate constant for radiationless decay might be expected to show a similar heavy atom effect. A possible explanation of our findings is that the true first-order radiationless process is only predominant in 1-iodonaphthalene and the rate process in the other naphthalenes is mainly pseudo first order and therefore leads to approximately constant decay rates.

Our measurements of quenching constants are not subject to these uncertainties. k_Q values were determined from the slope of the linear plots of k_A against, $[Q]$ and any change in the value of k_1 affects the intercept but not the slope of this plot. The presence of impurities in the solvent will therefore not affect the value of k_Q provided the level of impurity is constant and this is shown to be the case by the constancy of k_1 values.

Evidence for energy transfer from the triplet state

Energy levels and other relevant data for the compounds investigated are given in table 4.

The energy level diagram appropriate to the pairs phenanthrene-naphthalene, triphenylene-naphthalene, phenanthrene-bromonaphthalene, phenanthrene-iodonaphthalene, and benzophenone-naphthalene is shown in figure 4.

Our results provide unequivocal proof of the occurrence, in these pairs of compounds, of the process

donor (triplet) + acceptor (singlet) → donor (singlet) + acceptor (triplet)

since both the quenching of the donor triplet and the formation of the acceptor triplet have been observed directly. The only energetically possible alternative

donor (upper singlet) + acceptor (singlet) → donor (singlet) + acceptor (triplet)

is not only improbable on grounds of spin conservation and the short lifetime of the upper singlet but is experimentally excluded by our observation that, in all cases where fluorescence was observed, the fluorescence yield of the donor was unchanged by the presence of acceptor.

14   G. Porter and F. Wilkinson

Only in the cases of iodonaphthalene and iodine as acceptors were the triplet states not observed spectroscopically. Iodonaphthalene quenched the triplet state of molecules whose triplet state lay higher (e.g. naphthalene and phenan-

TABLE 4. ENERGIES OF FIRST EXCITED SINGLET AND TRIPLET STATES, PHOSPHO-
RESCENCE LIFETIMES (τ), LIFETIMES IN SOLUTION (τ_{solution}) AND PHOSPHORES-
CENCE YIELDS (Φ)

compound	long-wavelength limit of singlet absorption (cm^{-1})	first triplet level (cm^{-1})	τ (s)	τ_{solution} (10^{-4} s)	Φ
benzophenone	27 800	24 400[2]	0·006[6]	1·6 × 10^{-2} (benzene)[8]	0·84[6]
triphenylene	28 200	23 500[3]	15·9[2]	0·56 (hexane)	0·6[6]
phenanthrene	28 900	21 600[3]	3·3[3]	0·93	0·23
naphthalene	31 200	21 300[2]	2·6[3]	0·91	0·09[6]
1-bromonaphthalene	31 200	20 700[2]	0·018[3]	0·83	0·55[4]
1-iodonaphthalene	31 200	20 500[4]	0·0025[3]	—	0·70[4]
diacetyl	21 750	19 700[2]	0·00225[3]	0·60 (benzene)	—
1:2-benzanthracene	20 600	16 500[3]	0·3[3]	1·59 (hexane)	0·001
anthracene	26 000	14 700[2]	0·09[7]	9·1 (benzene)	0·0001
iodine	33 774[1]	11 888[5]	—	—	—

References

(1) Elliot (1940)   (5) Brown (1931)
(2) Lewis & Kasha (1944)   (6) Gilmore, Gibson & McClure (1952)
(3) McClure (1949)   (7) McGlynn, Padbye & Kasha (1955)
(4) Terenin & Ermolaev (1958) (8) Bäckström & Sandros (1960)

The values of Φ for which references are not given have been calculated from the ϕ_p/ϕ_f ratios in reference (3) by assuming $\phi_p + \phi_f = 1$.

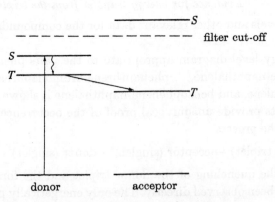

FIGURE 4. Energy levels of donor and acceptor.

threne) but not of molecules with lower triplet states (e.g. anthracene) and it clearly operates by the same mechanism. The mechanism of quenching by iodine is less certain since iodine quenched the triplet states of all molecules investigated. But the lowest triplet level of iodine is of lower energy than those of all other molecules investigated so that this behaviour is fully in accordance with a triplet energy transfer mechanism. Attempts to establish a negative result between pentacene and iodine were not successful owing to the very low solubility of pentacene and extensive decomposition. The mechanism of quenching by iodine is, however, complicated by the fact that chemical changes occur which result in the consumption of iodine. Whether this is a separate quenching mechanism, a reaction which follows triplet quenching or a process which is quite distinct from the quenching cannot yet be decided.

The possibility that quenching by iodine occurs as a result of heavy atom enhancement of spin-orbit coupling must be considered. Previous work on the effect of heavy ions such as Zn^{2+}, Ga^{3+} and Pb^{2+} (Porter & Wright 1958) and our observation, that ethyl iodide has a negligible effect on triplet lifetime, seem to exclude this possibility. Finally, the well-known tendency of iodine to form change transfer complexes with aromatic molecules suggests a further mechanism by which iodine might operate. However, the concentration of iodine in the mixtures studied was so low that, even if it were all complexed, it would have resulted in a reduction of the concentration of free donor molecules by less than 1 %.

TABLE 5. VISCOSITIES AND DIFFUSION CONTROLLED RATE CONSTANTS AT 25 °C IN THREE SOLVENTS

	$\eta(P)$	k_d(l. mole^{-1} s^{-1})
hexane	0·00326	$2·0 \times 10^{10}$
benzene	0·00647	$1·0 \times 10^{10}$
ethylene glycol	0·199	$3·3 \times 10^{8}$

Similar considerations rule out complex formation as a significant factor in the other systems studied. For example, 4×10^{-5} M iodonaphthalene increased the rate of disappearance of triplet phenanthrene in ethylene glycol by a factor of 5. The ground-state concentration of phenanthrene in this experiment was 2×10^{-3} M and hence, if all the iodonaphthalene were complexed, this would change the phenanthrene concentration by only 2 %. Even if the complex triplet had an absorption spectrum similar to that of triplet phenanthrene and decayed very rapidly it would have negligible effect on the observed decay rate of triplet phenanthrene.

Finally, the 'trivial process' of light emission and reabsorption can be excluded in our experiments since the fraction of triplet molecules which decay by a radiative process in fluid solvents is entirely negligible.

Mechanism of the energy transfer

The rate constants of energy transfer given in table 3 fall into three groups.

(a) When the triplet level of the donor is considerably greater than that of the acceptor the rate is quite close to the encounter rate calculated from the equation of Debye (1942)
$$k_d = 8RT/3000\,\eta \text{ l. mole}^{-1}\text{s}^{-1},$$
where k_d is the diffusion controlled rate constant and η is the solvent viscosity. This is illustrated by comparing the values of k_d in table 5, which have been calculated from this equation, with the rate constants in table 3, for systems where the triplet levels are well separated.

(b) When the triplet levels of donor and acceptor lie very close, but the energy of the acceptor is below that of the donor, quenching occurs with a rate considerably less than the encounter rate. This is most marked in the phenanthrene + naphthalene system, where the quenching rate is less than the encounter rate, even in the viscous solvent ethylene glycol. The quenching rate by bromonaphthalene, which has a slightly lower triplet level, is higher and quenching by iodonaphthalene is very nearly diffusion controlled.

(c) When the triplet level of the donor is below that of the acceptor no quenching is observed.

There are no indications that quenching efficiency decreases as the separation of the acceptor triplet below that of the donor triplet increases.

Energy transfer therefore occurs between molecules during an encounter at normal collisional separation. Under these conditions overlap of the orbitals of the two molecules occurs and an exchange transfer mechanism is possible. In the region of overlap the electrons are indistinguishable and the acceptor may emerge, by a 'collision of the second kind' in an electronically excited state.

According to Dexter (1953) not only has spin momentum to be conserved in such a collision but M_D^* must equal M_A^* and $M_D = M_A$, where M_D^* and M_A^* refer to the multiplicity of the donor and acceptor molecules in their excited states and M_D and M_A to multiplicities in their ground states. This condition is satisfied in the cases being considered here.

Dexter gives the probability of energy transfer by an exchange mechanism between two molecules as

$$P_{DA} = \frac{4\pi^2}{h} Z^2 \int f_D(E) \, F_A(E) \, dE,$$

where $\int f_D(E) \, F_A(E) \, dE$ is a type of overlap integral between the emission spectrum of the donor and the absorption spectrum of the acceptor. Z^2, which has the dimensions of energy squared, is a quantity that cannot be directly related to optical experiments. The separation and concentration dependences are hidden in Z^2 which varies approximately as $Y(e^4/K^2R_0^2) \exp(-2R/L)$ where R_0 is the effective average Bohr radius of the unexcited states of D and A, L is the effective average Bohr radius of the excited and unexcited states of D and A, e is the charge on the electron, K is the dielectric constant of the medium, R is the distance between the two molecules, and Y is a dimensionless quantity $\ll 1$ which takes account of cancellation as a result of sign changes in the wave functions.

Transfer times between neighbouring molecules can be very short but they increase rapidly with molecular separation—for example, in a typical case, the transfer time is increased by a factor of the order 10^2 when the molecular separation is increased by one molecular diameter. Exchange transfer would therefore lead to values of quenching constants not exceeding those calculated for diffusion controlled encounters between molecules having normal cross sections.

Energy transfer between molecules in singlet states often occurs over greater distances with correspondingly greater rate constants. For example, the rate constants of energy transfer from 1-chloroanthracene to perylene in various solvents lie in the range $(1\cdot4$ to $2\cdot5) \times 10^{11}$ l. mole^{-1} s^{-1} (Bowen & Livingston 1954), and similar transfer occurs in rigid solvents with a transfer distance of 34 Å (Bowen & Brockle-hurst 1955). These high transfer efficiencies are interpreted in terms of inductive resonance or dipole-dipole transfer and the relevance of this mechanism to triplet energy transfer deserves consideration.

Förster (1948) derived a formula from which quantitative predictions can be made provided energy transfer occurs after thermal equilibrium has been established.

$n_{D^* \to A^*}$, the transfer frequency, is given by

$$n_{D^* \to A^*} = \frac{9 \times 10^6 (\ln 10)^2 cx^2}{16\pi^4 n^2 N^2 R_{DA}} \int \epsilon_e^D(\bar{\nu}) \, \epsilon_a^A(\bar{\nu}) \, \frac{\mathrm{d}\bar{\nu}}{\bar{\nu}^2}, \tag{1}$$

where c is the velocity of light, x is an orientation factor, n is the refractive index, N is the Avogadro number, R_{DA} is the distance between the two molecules D and A, $\epsilon_a^A(\bar{\nu})$ is the molar extinction coefficient of the acceptor A and $\epsilon_e^D(\bar{\nu})$ represents the intensity of emission of D measured in the same units as the extinction coefficient.

If R^0 is defined as the critical transfer distance for which excitation transfer and spontaneous deactivation are of equal probability

$$n_{D^* \to A^*} = \frac{1}{\tau^D} \left(\frac{R_0}{R_{DA}} \right)^6, \tag{2}$$

where τ^D is the actual mean lifetime of the energy donor D. Substituting, and using numerical values of the constants, we obtain

$$R_0^6 = \frac{1\cdot69 \times 10^{-33}}{n^2} \frac{\tau^D}{\bar{\nu}_0^2} J(\bar{\nu}), \tag{3}$$

where $J(\bar{\nu}) = \int_0^\infty \epsilon_e^D(\bar{\nu}) \, \epsilon_e^A(\bar{\nu}) \, \mathrm{d}\bar{\nu}$ is called the overlap integral, and $\bar{\nu}^0$ is the wave number of the zero-point emission of the donor.

For typical cases of sensitized fluorescence R_0 values from 30 to 100 Å are calculated.

The values obtained for $J(\bar{\nu})$ will obviously be much smaller when singlet-triplet transitions are substituted in equation (3) because of the very low transition probabilities for singlet-triplet transitions. Calculations of R_0 values were made for the pairs of compounds which have been studied experimentally. The molar extinction coefficients were calculated from the following equation

$$\frac{1}{\tau} = \frac{8000\pi n^2 (\ln 10) c}{\Phi N} \int \frac{(2\bar{\nu}_0 - \bar{\nu})^3}{\bar{\nu}} \epsilon(\bar{\nu}) \, \mathrm{d}\bar{\nu}, \tag{4}$$

where τ is the phosphorescence lifetime, Φ is the quantum yield of phosphorescence, $\bar{\nu}_0$ is the position of the zero-zero band expressed in wave-numbers and is the median position between the phosphorescence and the singlet-triplet absorption maxima on a wave-number scale. The singlet-triplet spectra were generally drawn as the mirror image of the phosphorescence on a wave-number scale.

The data available to make these calculations are not very extensive, most of the information used being from the sources given in table 4, but the values given in table 6 show the magnitude of transfer distances predicted by this theory for triplet energy transfer.

Clearly these distances, which in most cases are much less than collision diameters, are so small that the theory used for their calculation is no longer valid. The calculations show, however, that long-range dipole-dipole transfer is not likely to play a significant role in triplet energy transfer processes of the type being considered, and this is in accordance with our experimental findings and with the conclusions of Terenin & Ermolaev (1958).

TABLE 6. CALCULATED VALUES OF R^0 (FÖRSTER THEORY)

donor	acceptor	solvent	R^0 (Å)
phenanthrene	naphthalene	hexane	0·11
triphenylene	naphthalene	hexane	0·11
phenanthrene	1-bromonaphthalene	hexane	0·40
	1-iodonaphthalene	hexane	0·62
naphthalene		hexane	0·52
1-bromonaphthalene		hexane	1·26
benzophenone	naphthalene	benzene	0·18
diacetyl	1:2-benzanthracene	benzene	0·19
phenanthrene	iodine	hexane	3·84

We conclude that triplet energy transfer to singlet molecules occurs only during encounters at normal collision distances and probably occurs by an exchange transfer mechanism. The efficiency of this process approaches unity as the triplet level of the acceptor falls below that of the donor.

REFERENCES

Bäckström, H. L. J. & Sandros, K. 1958 *Acta Chem. Scand.* **12**, 823.
Bäckström, H. L. J. & Sandros, K. 1960 *Acta Chem. Scand.* **14**, 48.
Bowen, E. J. & Livingston, R. 1954 *J. Amer. Chem. Soc.* **76**, 6300.
Bowen, E. J. & Brocklehurst, B. 1955 *Trans. Faraday Soc.* **51**, 774.
Brown, W. G. 1931 *Phys. Rev.* **38**, 1187.
Debye, P. J. W. 1942 *Trans. Electrochem. Soc.* **82**, 205.
Dexter, D. L. 1953 *J. Chem. Phys.* **21**, 836.
Elliot, A. 1940 *Proc. Roy. Soc.* A, **174**, 273.
Förster, Th. 1948 *Ann. Phys., Lpz.,* **2**, 55.
Gilmore, E. H., Gibson, G. E. & McClure, D. S. 1952 *J. Chem. Phys.* **20**, 829.
Gover, T. A. & Porter, G. 1961 In course of publication.
Kooyman, E. C. & Farenhorst, E. 1953 *Trans. Faraday Soc.* **49**, 58.
Lewis, G. W. & Kasha, M. 1944 *J. Amer. Chem. Soc.* **66**, 2100.
Livingston, R., Jackson, G. & Pugh, A. C. 1960 *Trans. Faraday Soc.* **56**, 1635.
McClure, D. S. 1949 *J. Chem. Phys.* **17**, 905.
McClure, D. S. & Hanst, P. L. 1955 *J. Chem. Phys.* **23**, 1772.
McGlynn, S. P., Boggus, J. D. & Elder, E. 1960 *J. Chem. Phys.* **32**, 357.
McGlynn, S. P., Padbye, M. E. & Kasha, M. 1955 *J. Chem. Phys.* **23**, 593.
Porter, G. & Jackson, G. 1961 *Proc. Roy. Soc.* A, **260**, 13.
Porter, G. & Wilkinson, F. 1961 In course of publication.
Porter, G. & Windsor, M. W. 1954 *Disc. Faraday Soc.* **17**, 178.
Porter, G. & Windsor, M. W. 1958 *Proc. Roy. Soc.* A, **245**, 238.
Porter, G. & Wright, M. R. 1958 *J. Chim. phys.* **55**, 705.
Porter, G. & Wright, M. R. 1959 *Disc. Faraday Soc.* **27**, 18.
Terenin, A. W. & Ermolaev, V. L. 1956 *Trans. Faraday Soc.* **52**, 1042.
Terenin, A. W. & Ermolaev, V. L. 1958 *J. Chim. phys.* **55**, 698.

Reprinted from the Proceedings of the Royal Society, A, *volume* 296, pp. 435–441, 1967

Quantum yields of triplet formation in solutions of chlorophyll

By P. G. Bowers and G. Porter, F.R.S.*

Department of Chemistry, The University of Sheffield

(*Received* 4 *July* 1966)

The method of flash photolysis has been used for direct measurement of the quantum yields of triplet formation in dilute chlorophyll solutions at 23 °C. In ether solution, the triplet yields were found to be 0·64 and 0·88 for chlorophyll a and b, respectively. In very dry hydrocarbon solution, the yields decrease by at least a factor of 5. It is suggested that the low yields of fluorescence and intersystem crossing in dry solvents are due to dissociation of excited singlet dimer chlorophyll into ground state monomer species.

Introduction

It is well known that the lowest excited singlet state of chlorophyll in solution exhibits both fluorescence and intersystem crossing to the triplet (Livingston 1960a) In polar ('wet') solvents, such as ether, in which the pigment exists as a monomer-solvate, the fluorescence quantum yields are 0·32 and 0·12 for chlorophyll a and b, respectively (Weber & Teale 1957). These yields decrease to less than 0·01 in non-polar ('dry') solvents, usually very pure hydrocarbons, where chlorophyll exists as a dimer (Katz, Pennington, Thomas & Strain 1963; Amster & Porter 1966).

Triplet chlorophyll has been characterized by its absorption spectrum following flash photolysis (Linschitz & Sarkanen 1958). It has also been reported that 3 to 4 times more triplet is formed in polar, than in non-polar solvents (Livingston 1960). If correct, this would imply that a third major primary process, such as internal conversion ($S_1 \to S_0$), is operative at least in non-polar solvents. The problem is of considerable interest because recent work on other molecules, notably aromatic hydrocarbons and their derivatives, has shown that internal conversion efficiencies are usually small or negligible (Medinger & Wilkinson 1965; Hammond & Lamola 1965). Furthermore, a knowledge of the way in which the primary photochemical behaviour is influenced by environment and state of aggregation is of obvious importance to studies of chlorophyll *in vivo*.

In this paper, we report direct measurements of the quantum yields of triplet formation for both chlorophyll a and b in 'wet' and 'dry' solution. In order to measure these quantities, a simple modification to the conventional technique for flash kinetic spectroscopy was developed, whereby transient optical density changes produced by flashing chlorophyll solutions were observed under conditions such that the total light absorbed could be precisely evaluated.

* Present address: The Royal Institution, 21 Albemarle Street, London W. 1.

436   P. G. Bowers and G. Porter

EXPERIMENTAL

The photolysis block used is shown in figure 1. The krypton-filled flash lamp was fitted with a cylindrical reflector of aluminium foil. Exciting light passed along the collimation channel, through the filters and into the Pyrex cuvette of square (1 cm) cross section. This cell was 5 cm long, and fitted with a side arm to facilitate outgassing. The filter consisted of Wratten 47B gelatin film, protected from over-heating by 2 cm water. The transmission of this filter did not undergo any transient or permanent changes on flashing. Maximum flash energies were 50 to 70 J.

Optical density changes in the chlorophyll solutions were monitored perpendicular to the direction of excitation, using the photoelectric method (Bridge & Porter

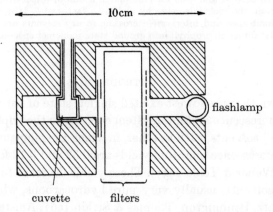

FIGURE 1. Photolysis block used for quantum yield determinations.

1958). Because of the high extinction coefficient of chlorophyll, most precision could be obtained by observing depletion of the ground state in the Soret absorption band, rather than triplet-triplet absorption at longer wavelengths. The relative lifetimes of the flash ($\sim 70\ \mu\mathrm{s}$), the triplet state ($\sim 500\ \mu\mathrm{s}$), and the first excited singlet state ($\sim 10^{-8}\,\mathrm{s}$), were such that when the decay curve was recorded at the end of a flash, all molecules were either in the ground state or the triplet state. The short extrapolation of this decay curve to zero time then gave the optical density change from which the initial triplet concentration could be deduced. The optical density changes were typically 0·02 to 0·07. Neutral filters were used to vary the incident flash intensity.

Extinction coefficients for triplet-triplet absorption in the region of the Soret band were required as a minor correction. These were found in separate experiments, using high energy unfiltered flashes to obtain near-complete conversion to the triplet.

Incident light intensities were measured with ferrioxalate actinometer solution (0·15M) in the cuvette. These intensities were less than the greatest at which this actinometer has been tested (Hatchard & Parker 1956).

Crystalline chlorophyll a and b was obtained from Sandoz Limited, Basle. Additional samples were very generously supplied by Dr P. J. McCartin (E. I. DuPont de Nemours), and by Dr W. D. Bellamy (General Electric Company). The concentration of solutions employed ($< 10^{-6}$M) was such that less than ten per cent of the exciting light was absorbed, thus avoiding greatly inhomogeneous triplet concentrations. A permanent decomposition amounting to a few parts per cent occurred after about twenty flashes.

Diethyl ether was Hopkins and Williams microanalytical reagent, and used without further purification. 3-Methylpentane was subjected to vigorous purification, as previously described (Godfrey & Porter 1966), and stored over calcium hydride.

The procedure of Amster & Porter was used in the preparation of dry chlorophyll solutions. This involves allowing the solvent to remain over a few grains of crystalline chlorophyll, in the presence of a small quantity of calcium hydride. In this work, the solution was allowed to form only after the solvent had been degassed and sealed off in the reservoir attached to the reaction cell. Dissolution was extremely slow, and the maximum concentrations attainable were around 10^{-6} M. The Soret extinction coefficients of the dry solutions were found by recording the reappearance of the wet spectrum on adding a drop of ether to a portion of the dry solution decanted from excess solid chlorophyll.

CALCULATION OF QUANTUM YIELDS

Extinction coefficients

The concentration c_3, of triplet chlorophyll produced, was calculated from the initial optical density change at the monitoring wavelength (the Soret band maximum in each case), by the use of the equation

$$\Delta \text{o.d.} = c_3(\epsilon_1 - \epsilon_3)\, 1,$$

where ϵ_1 and ϵ_3 are the ground and triplet state extinction coefficients. The literature values for ϵ_1 in ether vary considerably. We used what appeared to be the highest reliable values, $1 \cdot 21 \times 10^5$ and $1 \cdot 55 \times 10^5$ l. mole^{-1} cm^{-1} for chlorophyll a and

TABLE 1. EXTINCTION COEFFICIENTS FOR CHLOROPHYLL SOLUTIONS

	$10^{-5}\epsilon$ (l. mole^{-1} cm^{-1})			
	singlet		triplet	
	ether	hydrocarbon	ether	hydrocarbon
chlorophyll a	1·21 (428 nm)	0·94 (431 nm)	0·40 (430 nm)	0·40*
chlorophyll b	1·55 (452 nm)	1·13 (453 nm)	0·21 (450 nm)	0·26 (450 nm)

* Assumed.

b, respectively (Trurnit & Colmano 1959; Zcheile & Comar 1941; Kutyurin, Ulebekova & Artamkina 1962). All other extinction coefficients were based on the above-quoted figures, and are listed in table 1. The values for triplet-triplet absorption are close to those found by Linschitz & Sarkanen for chlorophyll in pyridine.

Light absorbed

In order to obtain optical density changes large enough to measure accurately, it was necessary to use the comparatively broad band of exciting light transmitted

438   P. G. Bowers and G. Porter

by the 47 B filter. This covered the whole of the Soret band. The overall fraction of
light absorbed per flash was found by integration

$$\frac{I_a}{I_0} = \frac{\int T_f(\nu)\,(1 - T_c(\nu))\,\mathrm{d}\nu}{\int T_f(\nu)\,\mathrm{d}\nu},$$

where T_f and T_c are the fractional transmission profiles of the filter and chlorophyll
solutions respectively. This procedure assumes a uniform output from the flash
lamp over the wavelength range transmitted by the filter. In fact the lamp spectrum

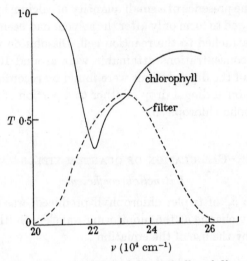

FIGURE 2. Fractional transmission profiles of chlorophyll and filter from which the light
absorbed was found by integration (vertical scale exaggerated).

in this region contains a number of sharp lines, although the overall output from
these is small compared with the underlying continuum. The adequacy of the
integration procedure was verified by using chlorophyll solutions more concentrated
than those employed in the quantum yield determinations; it was found that the
fraction of the flash absorbed, calculated as above, agreed well with that obtained
by direct photoelectric measurement of the incident and transmitted beams.

Corrections

The apparent triplet quantum yield $\phi_{\mathrm{app.}}$, found by the procedure described,
differs from the true value ϕ_T because the flash is of finite duration and converts a
considerable fraction of the total ground state chlorophyll to the triplet. In the later
stages of the flash there will be:

(a) recirculation of molecules which have fluoresced;

(b) absorption of exciting light by the triplet state;

(c) decreased absorption of exciting light by the depleted ground state.

These effects all lead to $\phi_{\mathrm{app.}} < \phi_T$, and have to be allowed for by making measurements at several intensities, and extrapolating $\phi_{\mathrm{app.}}$ to zero intensity. Fluorescence
reabsorption was unimportant at the concentrations used here.

Typical intensity plots are shown in figure 3. The fractional conversion is just the ratio of the number of ground state molecules initially present to the quanta absorbed. Understandably, the correction is more important for chlorophyll *a* than for chlorophyll *b*, because in the former the fluorescence yield is higher (more recirculation probability), and the triplet spectrum relatively more intense in the Soret region. Because of the scatter of points (each representing a single flash), and the difficulty of exact mathematical analysis, ϕ_T was found by linear mean square extrapolation, although the true curve is undoubtedly more complex.

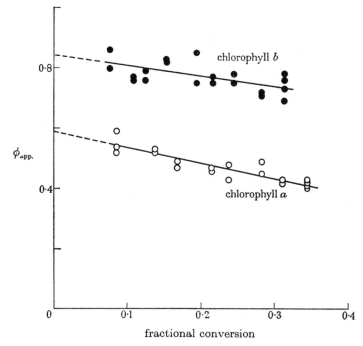

FIGURE 3. Apparent quantum yield for triplet formation as a function of the fractional conversion, for either solutions at 23 °C.

RESULTS AND DISCUSSION

Wet solutions

The ϕ_T values for several determinations for chlorophyll *b* in ether are shown in table 2. The reproducibility is moderately good for this type of measurement, and there is no obvious effect in varying the concentration or the source of chlorophyll.

Table 3 summarizes the mean yields found for both forms of chlorophyll in ether. The errors quoted include the uncertainty in the triplet extinction coefficients.

TABLE 2. REPRODUCIBILITY OF THE TRIPLET YIELD MEASUREMENTS
FOR CHLOROPHYLL *b* IN ETHER AT 23 °C

source	concentration (M)	ϕ_T
Du Pont	3.0×10^{-7}	0.80
Sandoz	4.3×10^{-7}	0.77
Du Pont	4.8×10^{-7}	0.96
Du Pont	5.1×10^{-7}	0.84
Du Pont	8.0×10^{-7}	1.03

P. G. Bowers and G. Porter

The sum of the fluorescence and triplet yields is unity within the experimental error in both cases. There can be little doubt that radiationless conversion from the excited singlet to the ground state, if it occurs at all, only plays a very minor part in the primary photochemistry of solvated chlorophyll.

TABLE 3. QUANTUM YIELDS FOR TRIPLET FORMATION AT 23 °C

	solvent	ϕ_T	ϕ_F	$\phi_T + \phi_F$
chlorophyll a	ether	0.64 ± 0.09	0.32	0.96
	3-Me pentane	< 0.10	< 0.01	< 0.1
chlorophyll b	ether	0.88 ± 0.12	0.12	1.00
	3-Me pentane	0.17 ± 0.03	< 0.01	< 0.2

Dry solutions

The spectra of the dry solutions were similar to those which have been previously published (Fernandez & Becker 1959; Amster and Porter 1966). The ratio of the two peaks at 659 and 641 nm in chlorophyll b was 0.83, slightly lower than the value of unity which is presumed to correspond to complete dimer formation.

The decay of triplet chlorophyll in general follows both first and second order kinetics. Under our conditions, the chlorophyll b triplet had a decay time of about $500\,\mu s$, and this did not change appreciably in going from ether to hydrocarbon solution, although detailed kinetic studies were not made. The triplet yield, however, decreased by more than a factor of 5 (table 3). The decay time of the chlorophyll a triplet, approximately $500\,\mu s$ in ether, decreased markedly to around $100\,\mu s$ or less in hydrocarbon solution, and in some experiments no transient could be observed at all. The short lifetime meant that accurate extrapolation of the decay curve back to zero time was not possible, and for this reason only an upper limit to ϕ_T could be calculated. Again, this is much lower than the yield in ether.

It is important to note that the triplet yields in dry solution were calculated assuming the chlorophyll to be completely dimerized. If it were assumed that only monomer was present, the yields would be twice as high.

Since the dry chlorophyll solutions are only very weakly fluorescent, the question arises as to the fate of the excited singlet dimers. Dissociation to give ground and excited state singlet monomers

$$^1A_2 \to {}^1A + A$$

can be ruled out, because the fluorescence yield is so low, and the fluorescence spectrum differs from that of the wet (monomer) solutions. Furthermore, the excited singlet monomer crosses to the triplet with high yield. Similarly, direct dissociation of $^1A_2$ into monomer triplets cannot be important in view of the low triplet yields. It is conceivable that the triplet yields which we observed in dry solutions are due to excitation of a small residual concentration of monomer.

The excited dimer $^1A_2$ must therefore either cross to a triplet state whose decay is too rapid to be observed (unlikely), or undergo radiationless deactivation to the ground state. Radiationless conversion from excited singlet to ground state is uncommon in molecules so far investigated, and it is probable that in the present case, the deactivation occurs via dissociation of the excited singlet dimer into two ground state monomer molecules.

The behaviour of the dry solutions is reminiscent of chlorophyll at high concentration, either in certain glasses at room temperature (Porter & Strauss 1966), in solution (Livingston 1960 b), or in the chloroplast (Müller, Rumberg & Witt 1963). In these cases, the fluorescence and triplet yields are reduced by self-quenching. It is possible that an excited singlet dimer is again the species through which the energy is dissipated,

$$^1A + A \rightarrow {}^1A_2.$$

Several methods have now been described for measuring intersystem crossing efficiencies. Medinger and Wilkinson have made use of the heavy atom effect in inducing crossover to the triplet and quenching of fluorescence. Hammond & Lamola have measured ϕ_T for a large number of compounds by studying the sensitized *cis*-trans isomerization of the olefins, while Parker has recently described a method based on sensitized delayed fluorescence (Parker & Joyce 1966). In the procedure we have described, there are no prior requirements concerning the donor-acceptor energy levels, nor are any assumptions made about energy transfer efficiencies. On the other hand, the direct method is probably not capable of the high precision claimed for the various energy transfer techniques. Unfortunately, there are very few compounds which have been studied under similar conditions using more than one method. We have, however, used the method described here to determine the quantum yield of formation of triplet anthracene in viscous paraffin solution. We obtained a value of 0·6, in reasonable agreement with the measurement of Wilkinson, and in good accord with the fluorescence yield of 0·33.

One of us (P. G. B.) wishes to thank the Science Research Council for a N.A.T.O. Fellowship, during the tenure of which this research was carried out.

REFERENCES

Amster, R. L. & Porter, G. 1966 *Proc. Roy. Soc.* A **296**, 38.

Bridge, N. K. & Porter, G. 1958 *Proc. Roy. Soc.* A **244**, 276.

Fernandez, J. & Becker, R. S. 1959 *J. Chem. Phys.* **31**, 467.

Godfrey, T. S. & Porter, G. 1966 *Trans. Faraday Soc.* **62**, 7.

Hatchard, C. G. & Parker, C. A. 1956 *Proc. Roy. Soc.* A **235**, 518.

Katz, J. J., Pennington, F. C., Thomas, M. R. & Strain, H. H. 1963 *J. Am. Chem. Soc.* **85**, 3801.

Kutyurin, V. M., Ulebekova, M. V. & Artamkina, I. Yu. 1962 *Fiziologiya Rast.* **9**, 115.

Lamola, A. A. & Hammond, G. S. 1965 *J. Chem. Phys.* **43**, 2129.

Linschitz, H. & Sarkanen, K. 1958 *J. Am. Chem. Soc.* **80**, 4826.

Livingston, R. 1960a *Quart. Rev. Chem. Soc.* **14**, 174.

Livingston, R. 1960b *Encyclopedia of plant physiology*, vol. i, p. 830. Berlin: Springer.

Livingston, R. & Fujimori, E. 1958 *J. Am. Chem. Soc.* **80**, 5610.

Livingston, R., Porter, G. & Windsor, M. W. 1954 *Nature, Lond.* **173**, 485.

Livingston, R., Watson, W. F. & McArdle, J. 1949 *J. Am. Chem. Soc.* **71**, 1542.

Medinger, T. & Wilkinson, F. 1965 *Trans. Faraday Soc.* **61**, 620.

Müller, A., Rumberg, B. & Witt, H. T. 1963 *Proc. Roy. Soc.* B **157**, 313.

Parker, C. A. & Joyce, T. A. 1966 *Chem. Comm.* 234.

Porter, G. & Strauss, G. 1966 *Proc. Roy. Soc.* A **295**, 1.

Trurnit, H. J. & Colmano, G. 1959 *Biochem. Biophys. Acta* **31**, 434.

Weber, G. & Teale, J. W. F. 1957 *Trans. Faraday Soc.* **53**, 646.

Zcheile, F. P. & Comar, C. L. 1941 *Bot. Gaz.* **102**, 463.

Reprinted from the Proceedings of the Royal Society, A, *volume* 299, pp. 348–353, 1967

Triplet state quantum yields for some aromatic hydrocarbons and xanthene dyes in dilute solution

By P. G. Bowers

Department of Chemistry, University of British Columbia

and G. Porter, F.R.S.

The Royal Institution, 21 Albemarle Street, London, W. 1

(*Received* 15 *December* 1966)

Quantum yields of triplet state formation and extinction coefficients of the triplet states have been determined by direct depletion methods for solutions of anthracene, phenanthrene, 1,2,5,6-dibenzanthracene, fluorescein, dibromofluorescein, eosin and erythrosin. The values obtained for the hydrocarbons are in reasonable agreement with those obtained by other workers using energy transfer and heavy atom perturbation techniques.

In all cases which we have studied, the sum of the quantum yields of fluorescence and triplet state formation is equal to unity within the limits of experimental error, showing that radiationless transfer from the excited singlet to the ground state is negligible.

INTRODUCTION

In a previous paper (Bowers & Porter 1966) we have shown how it is possible to measure quantum yields for triplet state formation (ϕ_T) using direct flash photolysis, by observing the transient absorption changes under conditions which enable the light absorption to be precisely calculated. In this way it was found that the sum of the fluorescence and triplet yields for the chlorophylls in polar solvents was near unity, but that only a small amount of intersystem crossing occurred in non-polar solvents.

This paper reports some further ϕ_T determinations in dilute solution. Polyacene hydrocarbons have been the subject of several other studies (Lamola & Hammond 1965; Medinger & Wilkinson 1965; Parker & Joyce 1966), and it seemed desirable to verify that our procedure gave similar results. The xanthene dyes were thought to be particularly suited to triplet yield determination by the direct method, because of their intense visible absorption bands. In addition, a knowledge of ϕ_T is the key to evaluation of many of the other rate constants for the complex photochemical processes of these substances.

EXPERIMENTAL

The main features of the method were similar to those which have been described previously for determinations with chlorophyll (Bowers & Porter 1966). A parallel, suitably filtered flash (70 to 200 J) was incident normally on the face of a 5 cm quartz cell of 1 cm square cross-section. Triplet absorption or ground state depletion was monitored perpendicular to the direction of excitation using a conventional photo-electric arrangement. The incident flash intensity was measured by ferrioxalate actinometry.

349   *Triplet state quantum yields*

The dyes were studied as their dianions in boric acid + borax pH 9 buffer solutions. Fluorescein and erythrosin had been purified chromatographically. Aromatic hydrocarbons were recrystallized and sublimed before use. 3-Me pentane and liquid paraffin was purified by the procedure of Godfrey & Porter (1966). All samples were subjected to thorough degassing before flashing. For liquid paraffin solutions, the degassing was facilitated by maintaining the liquid at 80 to 100 °C in a water bath.

The following filter combinations were used: anthracene, OX7 glass + saturated $CuSO_4$ (2 cm); phenanthrene, OX7 glass + Pyrex plate (2 mm) + 1·4 M $NiSO_4$ and 0·06 M $CuSO_4$ (2 cm); 1,2,5,6-dibenzanthracene (DBA), OX7 glass + saturated $Cu(NO_3)_2$ (2 cm) xanthene dyes, Ilford 303 glass + 2 cm water.

Extinction coefficients

The hydrocarbons were studied by excitation into the weak bands in the near ultra violet, and by monitoring the intensity of triplet-triplet absorption in the visible. Maximum triplet extinction coefficients in each case were found by subjecting a dilute sample ($ca.$ 10^{-7} M) to an unfiltered flash of 1000 J, in order to achieve complete conversion to the triplet state. Owing to some decomposition, the initial (i.e. zero-time) optical density change decreased by about 10 % with successive flashes, and was extrapolated to 'zero flash' for computation of the extinction coefficient. Several determinations using samples of different concentrations were carried out for each hydrocarbon.

Measurements on the dyes were made by observing depletion of their ground state absorption bands around 500 nm. The triplet extinction coefficients were estimated by taking the transient depletion spectrum (from a weak filtered flash) over a 30 nm range surrounding the maximum, and comparing this with the ground state absorption spectrum. Then, assuming ϵ_3 (triplet) to be approximately constant over the range, for two wavelengths λ_a and λ_b, giving transient optical density changes $\Delta\mu^a$ and $\Delta\mu^b$,

$$\epsilon_3 = \frac{\Delta\mu^b\epsilon_1^a - \Delta\mu^a\epsilon_1^b}{\epsilon_1^b - \epsilon_1^a}. \tag{1}$$

The relevant triplet extinction coefficients, and the ground state values on which they are based, are summarized in table 1.

TABLE 1. EXTINCTION COEFFICIENTS, $\epsilon \times 10^{-4}$ (M^{-1} cm^{-1}) and λ (nm).

compound	solvent	singlet		triplet	
anthracene	liquid paraffin	0·717	378	6·3 ± 0·5	424
phenanthrene	liquid paraffin	0·022	346	2·4 ± 0·2	480
1,2,5,6-dibenzanthracene	liquid paraffin	0·113	394	2·2 ± 0·3	535
fluorescein	aqueous, pH 9	8·8	489	1·5 ± 0·4	489
dibromofluorescein	aqueous, pH 9	9·5	506	1·8 ± 0·6	506
eosin	aqueous, pH 9	9·9	518	2·8 ± 0·4	518
erythrosin	aqueous, pH 9	*10·0	526	2·6 ± 0·7	526

* Assumed.

Sources of data on singlet states are: fluorescein (Lindquist 1960); dibromofluorescein (S. Emmons 1966, personal communication); eosin (Lindquist & Kasche 1965).

Calculation of quantum yields

The amount of triplet state formed per flash was found by extrapolating the triplet decay curve (hydrocarbons), or ground state recovery curve (dyes), back to zero time, and combining the initial optical density changes with the appropriate extinction coefficients.

The absorbed intensity I_a was calculated from the incident intensity I_0 by wavenumber integration

$$\frac{I_a}{I_0} = \frac{\int T_f(\nu)\,R(\nu)\,[1 - T_s(\nu)]\,\mathrm{d}\nu}{\int T_f(\nu)\,R(\nu)\,\mathrm{d}\nu};$$

T_f and T_s are the fractional transmissions of filter and sample respectively, and R is the relative output of the flash lamp (Shaw 1967). Typical curves are shown in figure 1. In the case of the filter used in irradiating the dyes, it was necessary to do

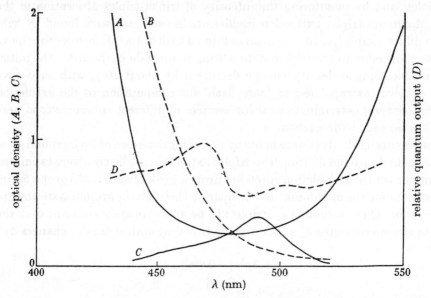

FIGURE 1. Absorption spectra: (*A*) Ilford 303 glass filter, (*B*) 0·15 M potassium ferrioxalate, 1 cm path; (*C*) fluorescein, 8×10^{-7} M, 5 cm path; (*D*) relative intensity of exciting flash.

a similar integration to calculate I_0 from the fraction of light absorbed by the actinometer solution. None of the compounds studied showed any appreciable permanent decomposition after exposure to about 20 flashes.

The hydrocarbon solutions were 10^{-4} to 10^{-5} M, and the dyes 10^{-6} to 10^{-7} M. These concentrations were chosen to give about 10 % absorption of the incident flash excitation. The value of the apparent triplet quantum yield, $\phi_{\mathrm{app.}}$, measured as the ratio of the amount of triplet formed to the calculated intensity of light absorbed, increased with decreasing intensity in the case of the dyes (figure 2). This was because at the highest intensities the calculated light absorbed corresponded to more than 20 % conversion of the ground state molecules initially present. The reduction in $\phi_{\mathrm{app.}}$ at such significant depletions is due to several factors such as re-circulation, and has been discussed previously (Bowers & Porter 1966). The true

351   *Triplet state quantum yields*

triplet yield was found by extrapolating $\phi_{app.}$ to zero intensity. As figure 2 shows, such a correction was not required for anthracene or the other two hydrocarbons, because the more concentrated solutions were only depleted by about 1 %.

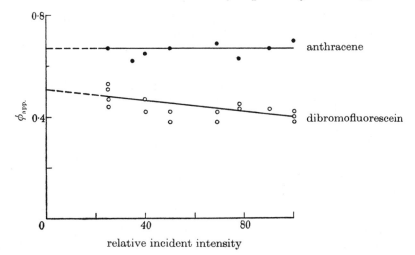

FIGURE 2. Typical plots of the apparent triplet quantum yield against incident intensity.

RESULTS AND DISCUSSION

The triplet extinction coefficients found for the hydrocarbons (table 1) may be compared with several literature values. Porter & Windsor (1958) obtained $7.15 \times 10^4 \mathrm{M}^{-1} \mathrm{cm}^{-1}$ for anthracene in liquid paraffin and more recently Wild & Gunthard (1965) have found $6.7 \times 10^4 \mathrm{M}^{-1} \mathrm{cm}^{-1}$ for glycerol solutions. A value of $2.7 \times 10^4 \mathrm{M}^{-1} \mathrm{cm}^{-1}$ for the maximum triplet extinction coefficient of phenanthrene in a rigid glass at low temperature has been measured by Keller & Hadley (1965). These are all in good accord with our determinations, although Porter & Windsor's value of $6.7 \times 10^4 \mathrm{M}^{-1} \mathrm{cm}^{-1}$ for 1,2,5,6-*DBA* in hexane is about three times as high as the value we obtained.

The method based on equation (1) for determining ϵ_3 for the dyes is not particularly precise, as the error limits show, but since the quantum yields are calculated from $\epsilon_1 - \epsilon_3$, this uncertainty is greatly reduced in the final results.

The triplet yields, generally the mean result of several experiments for each compound, are listed in table 2, together with the fluorescence yields under comparable conditions, where these have been measured. Notice that phenanthrene and 1,2,5,6-*DBA* were studied in 3-*Me* pentane, assuming the same triplet extinction coefficients at maximum absorption as in liquid paraffin.

The figure of 0.58 for anthracene compares with a value of 0.75 found by Medinger & Wilkinson (1965) using heavy atom quenching, and a fluorescence yield of 0.33. Lamola & Hammond (1965), using the method of sensitized *cis-trans* isomerization, obtained 0.76 and 0.89 for phenanthrene and 1,2,5,6-*DBA* respectively. The extent of agreement between these published figures and our values is some assurance that there are no large undetected sources of error in the method.

P. G. Bowers and G. Porter   352

TABLE 2. TRIPLET QUANTUM YIELDS AT 23 °C

compound	solvent	concentration (M)	Φ_T	Φ_F
anthracene	liquid paraffin	2×10^{-5}	0.58 ± 0.10	0.33
phenanthrene	3-Me pentane	5×10^{-4}	0.70 ± 0.12	0.14
1,2,5,6-dibenzanthracene	3-Me pentane	1×10^{-4}	1.03 ± 0.16	—
fluorescein (Fl)	aqueous, pH 9	8×10^{-7}	0.05 ± 0.02	0.92
$FlBr_2$	aqueous, pH 9	8×10^{-7}	0.49 ± 0.07	—
eosin ($FlBr_4$)	aqueous, pH 9	6×10^{-7}	0.71 ± 0.10	0.19
erythrosin (FlI_4)	aqueous, pH 9	9×10^{-7}	1.07 ± 0.13	0.02

There is a growing body of evidence that radiationless $S_1 \rightarrow S_0$ conversion plays at most only a minor part in the primary process of the polyacenes. However, until there is a great improvement in techniques it will remain very difficult to account with certainty for the 10 to 15 % of excited molecules covered by experimental error.

The value of ϕ_T for fluorescein is given within rather wide limits, because only a very small amount of depletion could be observed. The effect of halogenation is very marked in increasing the intersystem crossing efficiency (and also in decreasing the triplet lifetime, although detailed studies were not made). In all cases, the fraction of excited singlet molecules which does not undergo fluorescence or intersystem crossing must be quite small.

Previous estimates of ϕ_T for the xanthene dyes have been made from indirect measurements. From a study of their photoreduction by allyl thiourea, Adelman & Oster (1956) quote 0.02, 0.12, 0.09, and 0.05 for the triplet yields of the sodium salts of Fl, $FlBr_2$, $FlBr_4$, and FlI_4 respectively at pH 7. The differences between these figures and our values are much too large to be covered by experimental errors. In more recent work, Koizumi and his co-workers (Ohno, Usui & Koizumi 1965; Ohno, Kato & Koizumi 1966; Momose, Uchida & Koizumi 1965) have found that the ϕ_T values measured in photoreduction experiments appear to vary considerably with the nature of the reductant and with pH, and it seems that the overall photoreduction kinetics are more complicated than was originally thought. In view of this, the low ϕ_T values must be regarded with some suspicion.

In the solvents glycerol and ethanol, Parker & Hatchard (1961) have found ϕ_T for eosin, from delayed fluorescence measurements, to be 0.06 and 0.02 respectively. The corresponding fluorescence yields (0.72 and 0.45) are much higher than in aqueous solution.

REFERENCES

Adelman, A. H. & Oster, G. 1956 *J. Am. Chem. Soc.* **78**, 3977.
Bowers, P. G. & Porter, G. 1966 *Proc. Roy. Soc.* A **296**, 435.
Godfrey, T. S. & Porter, G. 1966 *Trans. Faraday Soc.* **62**, 7.
Keller, R. A. & Hadley, S. G. 1965 *J. Chem. Phys.* **42**, 2382.
Lamola, A. A. & Hammond, G. S. 1965 *J. Chem. Phys.* **43**, 2129.
Lindquist, L. 1960 *Ark. Kemi.* **16**, 79.
Lindquist, L. & Kasche, V. 1965 *Photochem. Photobiol.* **4**, 923.
Medinger, T. & Wilkinson, F. 1965 *Trans. Faraday Soc.* **61**, 620.
Momose, Y., Uchida, K. & Koizumi, M. 1965 *Bull. Chem. Soc. Japan* **38**, 1601.

Ohno, T., Kato, S. & Koizumi, M. 1966 *Bull. Chem. Soc.* Japan **39**, 232.
Ohno, T., Usui, Y. & Koizumi, M. 1965 *Bull. Chem. Soc. Japan* **38**, 1022.
Parker, C. A. & Hatchard, C. G. 1961 *Trans. Faraday Soc.* **57**, 1894.
Parker, C. A. & Joyce, T. A. 1966 *Chem. Commun.* 234.
Porter, G. & Windsor, M. W. 1958 *Proc. Roy. Soc.* A **245**, 238.
Shaw, G. 1967 Ph.D. Thesis, University of Sheffield.
Wild, U. & Gunthard, H. H. 1965 *Helv. Chem. Acta* **48**, 1843.

Reprinted from the *Journal of The Chemical Society, Faraday Transactions II*, 1973, vol. 69

Quenching of Aromatic Triplet States in Solution by Nitric Oxide and Other Free Radicals

By O. L. J. Gijzeman, F. Kaufman and George Porter *

Davy Faraday Research Laboratory of The Royal Institution,
21 Albemarle Street, London W1X 4BS

Received 8th November, 1972

The rate constants for quenching of aromatic hydrocarbon triplets by nitric oxide and by the free radical di-t-butyl nitroxide have been measured by the laser flash photolysis technique. For low energy triplets the efficiency of quenching is inversely proportional to triplet energy. At high triplet energy the rates are proportional to triplet energy and there is a large effect of solvent polarity.

These data are analysed in terms of an enhanced intersystem crossing process catalysed by the doublet state nitric oxide molecule. The data are shown to be consistent with important contributions from charge-transfer interactions for high triplet energy compounds in non-polar solvents.

The quenching of electronically excited states frequently involves energy or electron transfer to an acceptor molecule. However, an increasing number of examples in the literature indicates that efficient quenching can take place even when the quencher molecule is physically unchanged in the process.[1] One of the earliest examples of quenchers which act in this catalytic manner are species with unpaired electrons which can enhance intersystem crossing without energy transfer by allowing overall spin conservation.[2] Examples of this type of quencher include paramagnetic ions and paramagnetic molecules such as NO and O_2.

The interpretation of data from quenching by paramagnetic ions is complicated by solvation effects and electrostatic interactions due to ionic charges and quenching by oxygen can be accompanied by energy transfer. On the other hand, excited state quenching by nitric oxide, one of the simplest stable paramagnetic substances, does not suffer from these complications. Recently, measurements of the static NO quenching process in rigid solvent matrices have been reported.[3] These experiments gave ambiguous triplet quenching efficiencies due to the difficulty of separating triplet from singlet quenching. Our studies on the oxygen quenching of triplet states in fluid media led us to expect non-diffusion controlled quenching rates in the NO case.[4] To check these predictions we have studied the NO quenching of organic triplet states in solution by laser flash photolysis.

EXPERIMENTAL

The lifetimes of triplet states in the presence of free radical quenchers were determined by recording the decay of the triplet-triplet absorption following excitation with a frequency doubled ruby laser pulse. Details of the laser flash photolysis apparatus have been previously reported.[5] The concentration of quencher chosen ($10^{-2} - 10^{-3}$ M) resulted in triplet lifetimes that were easily measured with the instrument and which followed good (pseudo) first-order kinetics since $[Q] \gg [^3M^*]$. No corrections to the quenching rates due to other quenching processes or intrinsic decays had to be applied since, in the absence of added quencher, the triplet lifetimes were found to be much longer.

The aromatic hydrocarbons, used in concentrations of $\sim 10^{-4}$ M, were purchased from Aldrich or Koch Light Chemical Companies and were zone refined when necessary. Solvents of spectroscopic grade or previously distilled solvents were used in the quenching studies. Dimethoxymethane was treated with sodium to remove methanol impurity followed by distillation with collection of the 42°C fraction. Di-t-butyl nitroxide was obtained from Dr N. M. Atherton (Sheffield) and was found to require no further purification.

Nitric oxide (Matheson) was freshly purified for each quenching run by subjecting the required quantity of this quencher to repeated freeze-pump-thaw cycles in a vacuum line manifold fitted with greaseless high vacuum taps. The solidified NO, cooled to liquid argon temperature, was subsequently sublimed into a previously evacuated bulb at liquid nitrogen temperature. The solid colourless nitric oxide prepared in this way was immediately used in the quenching studies. Known pressures of the gas, measured with an Edwards barometrically compensated mechanical vacuum gauge, were added to rigorously degassed and frozen (liquid N_2) solutions of the hydrocarbon from a side arm bulb of known volume. After addition of the gas the frozen solutions were sealed off and kept frozen until the lifetime measurements were performed (<2 h). The solutions were brought to room temperature and shaken for 30 min to ensure complete equilibration of the gaseous quencher. No spectral changes before or after the lifetime determination were detected for samples prepared in this manner. The reproducibility of the measurements was ± 15 %.

The solubility of NO in the organic solvents used was determined by measuring the vapour pressure of a previously degassed solution to which a known quantity of the nitric oxide quencher had been added. An evacuable flask with 25 ml of solution (35 ml overhead volume) was fitted with a mercury manometer and a magnetic stirring bar. Upon addition of a known volume of gas at a pressure p the frozen solution was warmed and allowed to come to equilibrium (30 min) with vigorous stirring. Determination of the final pressure after correction for the measured vapour pressure of the solvent leads to the value of s(mol l.$^{-1}$ mm^{-1}) where the concentration of quencher $[Q] = sp$ (see table 1).

TABLE 1.—SOLUBILITY COEFFICIENTS (s) OF NITRIC OXIDE GAS IN VARIOUS ORGANIC SOLVENTS

solvent	$s \times 10^5$/mol l.$^{-1}$ mm^{-1}
hexane	4.5
dimethoxymethane	3.2
diethyl ether	6.3
tetrahydrofuran	3.9
acetone	5.3
acetonitrile	1.8

RESULTS

The physical quenching of electronically excited aromatic triplets by simple doublet state free radicals was studied as a function of triplet energy and solvent polarity. Tables 2 and 3 give the rate constants for the quenching in hexane by NO and by di-t-butyl nitroxide (DTBNO) respectively. It is seen that the triplet quench-

TABLE 2.—QUENCHING OF AROMATIC HYDROCARBON TRIPLET STATES BY NITRIC OXIDE IN HEXANE SOLUTION

compound	E_T/cm^{-1}	$k_q \times 10^{-7}$/l. mol^{-1} s^{-1}	$p \times 10^3$ [a]
triphenylene	23 300	72	72
phenanthrene	21 600	20	20
chrysene	20 000	10	10
coronene	19 100	4.6	4.6
1, 2 : 3, 4-dibenzanthracene	17 800	1.8	1.8
1, 2-benzanthracene	16 500	1.3	1.3
anthracene	14 700	0.84	0.84
3, 4 : 8, 9-dibenzpyrene	12 000	2.1	2.1
naphthacene	10 300	7.0	7.0

[a] Quenching probabilities calculated with $k_d = 3 \times 10^{10}$ l. mol^{-1} s^{-1}.

ing rates decrease by at least two orders of magnitude on going from high energy triplets ($E_T \approx 23\,000$ cm^{-1}) to intermediate energy triplet aromatics ($E_T \approx 14\,000$ cm^{-1}). In the lowest triplet energy region studied, however, quenching by NO and DTBNO leads to rates which increase with decreasing triplet energy. This behaviour contrasts strongly with that found [4] for quenching by triplet state molecular oxygen (see fig. 1).

TABLE 3.—QUENCHING OF AROMATIC HYDROCARBON TRIPLET STATES BY
di-t-BUTYL NITROXIDE IN HEXANE SOLUTION

compound	E_T/cm^{-1}	$k_q \times 10^{-7}$/l. mol^{-1} s^{-1}
triphenylene	23 000	217
phenanthrene	21 600	158
chrysene	20 000	96
coronene	19 100	33
1, 2 : 5, 6-dibenzanthracene	18 300	29
1, 2-benzanthracene	16 500	6.2
anthracene	14 700	1.2
3, 4 : 8, 9-dibenzpyrene	12 000	1.9
naphthacene	10 300	3.5

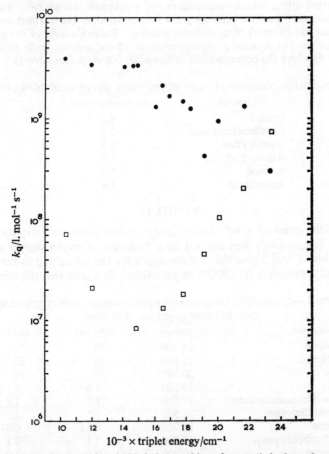

FIG. 1.—Rate constants for O$_2$ (○) and NO (□) quenching of aromatic hydrocarbon triplet states in hexane as a function of triplet energy. Oxygen quenching data from ref. (4).

QUENCHING OF TRIPLET STATES

The effect of changes in solvent polarity on the NO quenching reaction was studied by measuring the NO quenching rates for chrysene, 1, 2-benzanthracene and 3, 4 : 8, 9-dibenzpyrene triplets as a function of dielectric constant (table 4).   For

TABLE 4.—SOLVENT EFFECTS ON NITRIC OXIDE QUENCHING RATE CONSTANTS OF CHRYSENE (CHR), 1, 2-BENZANTHRACENE (BA) AND 3, 4 : 8, 9-DIBENZPYRENE (DBP)

solvent	η^a/cP	ε^a	$(\varepsilon-1)/(2\varepsilon+1)$	$k_q \times 10^{-7}$ /l. mol^{-1} s^{-1}		
				CHR	BA	DBP
hexane	0.300	1.88	0.185	10	1.3	2.1
dimethoxymethane	0.340	2.65	0.262	4.3	—	—
diethyl ether	0.242	4.34	0.345	1.3	0.83	—
tetrahydrofuran	0.460	7.58	0.406	1.6	0.58	—
acetone	0.304	20.7	0.463	1.7	0.31	2.1
acetonitrile	0.325	37.5	0.482	1.5	0.30	—

[a] J. A. Riddick and W. B. Bunger, *Organic Solvents—Physical Properties and Methods of Purification*, 3rd Edn., in *Techniques of Chemistry*, ed. H. Weissburger (Wiley Interscience, New York), vol. 2.

chrysene the quenching rate is seen to decrease by approximately a factor of six in going from hexane to acetone.   The decrease with increasing solvent polarity is less pronounced for 1, 2-benzanthracene and quenching of 3, 4 : 8, 9-dibenzpyrene appears to be independent of solvent polarity.   The qualitative influence of solvent on the quenching rates is different from the results obtained in the oxygen quenching case, where the quenching rates were found to increase with increasing solvent polarity.[4]

In view of the unexpected differences between our results for the quenching of triplets by O_2 and by NO, quenching constants of the excited singlet states by NO were also measured in collaboration with G. Beddard using the fluorescence lifetime method.   Fluorescence quenching constants by NO for a few aromatic hydrocarbons in hexane and in acetone were identical within experimental error with those for quenching by oxygen.

KINETIC SCHEME AND THEORY

Intermolecular enhancement of radiationless decay by species which contain unpaired electrons is thought to be due to electronic interactions between the excited state and the quencher in a collision complex.[2, 6]   For the quenching of excited triplet states by doublet molecules the following possible kinetic reaction scheme can be considered :

$$^3M^* + {}^2Q \underset{k_{-d}}{\overset{\frac{1}{4}k_d}{\rightleftharpoons}} {}^2(^3M^* \dots {}^2Q) \overset{k_{et}}{\to} {}^1M + {}^2Q^* \qquad (1)$$

$$^2(^3M^* \dots {}^2Q) \overset{k_{isc}}{\to} {}^1M + {}^2Q \qquad (2)$$

$$\underset{k_{-d}}{\overset{\frac{3}{4}k_d}{\rightleftharpoons}} {}^4(^3M^* \dots {}^2Q) \overset{k_{et}}{\to} {}^1M + {}^4Q^*. \qquad (3)$$

The quenching by doublet species is analogous to that of the oxygen (triplet) case except that, in the former, doublet and quartet state collision complexes are formed at the appropriate spin statistical factor times diffusion controlled rate.   The energy transfer paths (1) and (3) can be unambiguously ruled out on energetic grounds for NO, since the energy of the lowest lying available state ($^4\Pi$) has been estimated from experiment to be at 35 000 cm^{-1}.[7]   This leaves the enhanced intersystem crossing process (2) as the only available physical mechanism for quenching of excited triplet states by nitric oxide.

In this interpretation the " forbidden " intersystem crossing process in the excited state becomes more allowed due to collisions with a quencher which enable spin to be conserved in the process. The experimentally measured triplet lifetime τ for nitric oxide quenching of $^3M^*$ is given by:

$$1/\tau = k_0 + p[NO]k_d/3 \qquad (4)$$

where k_0 is the rate constant for intrinsic decay of the unperturbed triplet state and p is the probability of quenching per encounter. Since k_0 is negligible under the conditions of the experiments the rate constant for quenching is:

$$k_q = \tfrac{1}{3}k_d\left(\frac{k_{isc}}{k_{isc}+k_{-d}}\right). \qquad (5)$$

The large range of rate constants observed for this quenching process indicate substantial reaction inefficiencies which cannot be due to changes of k_d or k_{-d} with molecular parameters.[4] Therefore, the variation in k_q with triplet energy is due to non-diffusion controlled reactions. The reaction probability p can be evaluated from the assumed value of 3×10^{10} for k_d. The small p values obtained in this way (table 2) imply that $k_{isc} \ll k_{-d}$. This means that over most of the triplet energy range p is proportional to k_{isc}/k_{-d} with the result that the quenching should be independent of viscosity for moderately viscous solvents. This has been corroborated by experiment, where it was found that the rates were virtually the same in hexane and dodecane.

One theoretical treatment that can be used to try to understand the observed variation of k_{isc} in eqn (5) involves consideration of the transient bimolecular complex as a species which undergoes radiationless decay between the initial triplet and final ground state of the aromatic hydrocarbon. The rate [8] is given by:

$$k_{isc} = 2\pi/\hbar \; \beta^2 \rho F \qquad (6)$$

where β is the electronic matrix element for the mixing of initial and final states, ρ is the density of final states and F is the Franck–Condon factor. Inasmuch as there is no energy transferred in this quenching process the latter term is determined solely by the hydrocarbon and is given (for these large energy gaps) by the usual Siebrand expression [9]:

$$F(E) = F(E_0)10^{-((E-E_0)/\eta)10^{-4}}. \qquad (7)$$

In the present case of triplet state decay, E equals the energy of the triplet state and it is seen that eqn (6) and (7) predict increasing rates of quenching with decreasing triplet energies.

The β term in eqn (6) is determined by contributions from two possible mechanisms of quenching arising from exchange (β_{ex}^2) or charge-transfer (β_{ct}^2) interactions [6, 10]:

$$\beta_{ex}^2 = |\langle\psi_i|\mathscr{H}|\psi_f\rangle|^2 \qquad (8)$$

$$\beta_{ct}^2 = \left|\frac{\beta'_{ct}}{\Delta E}\right|^2 = \frac{|\langle\psi_i|\mathscr{H}|\psi_{ct}\rangle\langle\psi_{ct}|\mathscr{H}|\psi_f\rangle|^2}{(E_{ct}-E_T)^2} \qquad (9)$$

Qualitatively at least, the difference between these two mechanisms is indicated in fig. 2. The charge-transfer mechanism involves indirect mixing of the initial and final states of the complex via the charge transfer state $^2(M^+ NO^-)$, the extent of mixing being dependent on the energy difference $E_{ct} - E_T$. The importance of each quenching process will depend on the relative magnitudes of β_{ex}^2 and β_{ct}^2.

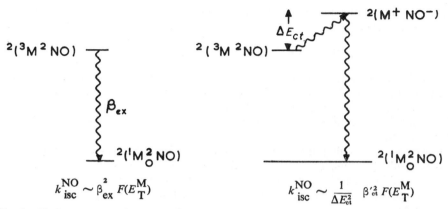

$$k_{isc}^{NO} \sim \beta_{ex}^2 \, F(E_T^M) \qquad\qquad k_{isc}^{NO} \sim \frac{1}{\Delta E_{ct}^2} \, \beta_{et}'^2 \, F(E_T^M)$$

Fig. 2.—Exchange versus charge-transfer quenching mechanism for NO quenching of organic triplet states.

LOW TRIPLET ENERGY BEHAVIOUR

At triplet energies below $15\,000$ cm^{-1} the observed rate constant k_q for NO quenching of triplets decreases with increasing triplet energy. The similar behaviour for oxygen quenching (fig. 1) at higher values of E_T has been attributed to an energy transfer reaction via electron exchange interaction.[4] It was shown that eqn (6), (7) and (8) could adequately fit the experimental data for the increase in oxygen quenching probability as a function of decreasing triplet energy. We may assign the same mechanism to the low energy quenching data for NO.

For quenching by the exchange interaction eqn (6) and (8) give

$$k_{isc} = 2\pi/\hbar |\beta_{ex}|^2 \rho F. \tag{10}$$

Values of $p(= k_{isc}/k_{-d})$ for NO in this triplet energy region are less than 0.01 (table 2). These values can be compared with the values of k_{et}/k_{-d} for oxygen in the energy range $18\,000$-$22\,000$ cm^{-1} to ensure equal energy gaps. In the oxygen case k_{et}/k_{-d} has been found to be 0.1-0.5.[4] Thus it follows that $\beta_{ex}(O_2) > \beta_{ex}(NO)$.

It should be remembered, however, that the value for oxygen refers to an energy transfer process whereas the NO value applies to enhanced intersystem crossing. Assuming the matrix element for enhanced intersystem crossing for oxygen to be of the same order of magnitude as that for NO it follows that this pathway of deactivation of the complex ($^3M^{*3}O_2$) can be neglected compared to the energy transfer process. This has been found experimentally.[4]

On the other hand, if exchange interactions are less important in NO charge-transfer interactions could be more important. In the case of oxygen quenching this effect appeared only in higher dielectric constant solvents. For NO even low dielectric constant solvents (such as hexane) could be " polar " enough to make charge-transfer interactions possible.

HIGH TRIPLET ENERGY BEHAVIOUR

The unexpected increase in nitric oxide and DTBNO quenching rates at higher triplet energies cannot be explained by variations in the Franck–Condon factor, since this gives the opposite behaviour, when exchange interactions are dominant (see above). On the other hand, the charge-transfer mechanism imposes an additional term on the rate expression that is an explicit function of the nergy of the triplet state.

The importance of this mechanism is dependent on the energy of the charge-transfer state which is given by:

$$E_{ct}(M^+Q^-) = I_p^M - E_a^Q - C - \frac{2\mu^2}{\sigma^3}\left\{\frac{\varepsilon-1}{2\varepsilon+1} - \frac{n^2-1}{4n^2+2}\right\} \qquad (11)$$

where I_p is the ionisation potential of the " donor " molecule, E_a the electron affinity of the acceptor and C the Coulomb stabilisation energy of the complex. The last term takes into account the effect of the solvent (dielectric constant ε and index of refraction n) which is supposed to provide a cavity of radius σ for the complex with dipole moment μ.

For oxygen quenching of triplet aromatics it was found that an increase in solvent polarity resulted in higher reaction probabilities.[4] This was interpreted as an effect due to quenching contributions from the charge-transfer mechanism since an increase in solvent polarity ε lowers the value of E_{ct} in (11), thus enhancing the size of β_{ct} by lowering ΔE.

Although $^3(M^+O_2^-)$ and $^2(M^+NO^-)$ are expected to have similar values of C and interaction distance σ, differences in electron affinity $[E_a(O_2) = 0.74$ eV, $E_a(NO) = 0.9$ eV ref. (1))] and the increased dipole moment μ of the complex would lead to a lower value of E_{ct} for the (M^+NO^-) complex. These factors could be large enough to make the charge-transfer contribution to the quenching process dominant, even in non polar solvents.

The rate expression for quenching by this charge-transfer mechanism is, from (6) and (9):

$$k_{isc} = \frac{2\pi}{\hbar}\rho\left|\frac{\beta'_{ct}}{\Delta E}\right|^2 F. \qquad (12)$$

It can be expected that any variation in ΔE with triplet energy would be important if the β_{ct}^2 term influences the observed quenching. Since only I_p and E_T should change significantly as a function of the aromatic hydrocarbon, a plot of $I_p - E_T$ against E_T may indicate the approximate triplet energy dependence of the charge-transfer contribution to k_{isc}. This plot, shown in fig. 3, reveals an inverse dependence of

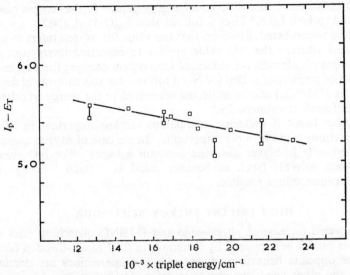

FIG. 3.—Ionisation potentials of aromatic hydrocarbons minus triplet energy (in eV) as a function of triplet energy. Ionisation potentials as determined from charge-transfer absorption (cf. ref. (1)).

energy gap $E_{ct} - E_T$ on triplet energy assuming small changes in C, μ and σ with hydrocarbon. The data indicate that the energy gap between the triplet and the charge-transfer state decreases with increasing triplet energy. In going from anthracene to triphenylene, fig. 3 shows that the charge-transfer state moves about 2500 cm^{-1} closer to the triplet state. Smaller energy gaps lead to increased mixing with the charge transfer state with correspondingly larger contributions from β_{ct}^2 to the observed quenching.

Therefore it seems reasonable to suggest that the experimentally observed increase in quenching rate with triplet energy can be understood in terms of a direct dependence of β_{ct}^2 on E_T. At sufficiently large triplet energies this may well overwhelm the underlying reverse dependence of F on E_T.

COMPARISON OF THEORY WITH EXPERIMENT

Eqn (5), (7), (11) and (12) can be combined to give an expression for k_q for high energy triplets quenched by NO via the charge-transfer mechanism:

$$k_q = \frac{k_d}{3k_{-d}} k_{isc} = \frac{k_d}{3k_{-d}} \frac{2\pi F(E_0)\rho}{\hbar} \left| \frac{\beta_{ct}'}{E_{ct} - E_T} \right|^2 10^{-((E-E_0)/\eta)10^{-4}}$$

$$= \frac{\gamma}{(I_P - E_T - C')^2} 10^{-((E-E_0)/\eta)10^{-4}}. \tag{13}$$

The constants γ and C' can be adjusted to give reasonable agreement with experiment (fig. 4). The constant C' is found to be 5 eV. This energy has to be equal to the Coulomb stabilisation energy of the complex plus the electron affinity of NO plus the solvent term. For the iodine–benzene complex this Coulomb term has been estimated to be 3-3.5 eV.[11] For the smaller NO molecule this term might increase to

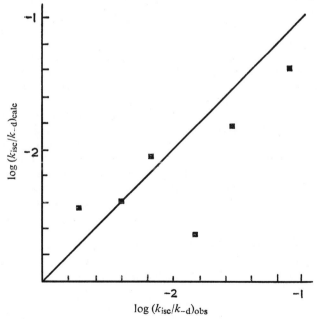

FIG. 4.—Comparison of calculated against observed values of k_{isc}/k_{-d} for NO quenching of organic triplets ($E_T \geqslant 15\,000$ cm^{-1}).

3.5-4 eV, which, when added to the—unknown—solvent term and $E_a(NO)$ gives approximately the observed value of C'.

From the constant γ a rough estimate of β'_{ct} can be obtained. This yields, when divided by the energy gap $E_{ct} - E_T$ from eqn (11), $\beta_{ct}(NO) \geqslant 5$ cm^{-1}, which seems a reasonable value for weakly interacting molecules.

The main difficulty and uncertainty in the present analysis is in the exact location of the charge-transfer state. Considering the poorly known values of I_p (error ~ 0.1 eV) and the questionable assumption of a constant C' for all complexes we feel that the analysis of our data supports our conclusion, that at least for triplet energies above 15 000 cm^{-1} charge-transfer interactions are dominant and can explain the experimental observations.

The values of β_{ex} for oxygen aromatic complexes can be obtained from the previously published quenching data.[4] One then obtains $\beta_{ex}(O_2) = 1 \pm 0.1$ cm^{-1} for solutions in hexane, cyclohexane and benzene. The constancy of these values supports our previous conclusion that, in these low dielectric constant solvents, exchange interactions are dominant. The values are also in agreement with the prediction that, for NO, charge-transfer interactions should be more important.

SOLVENT EFFECTS

The charge-transfer interaction mechanism at high triplet energies suggests that quenching by NO should be strongly dependent on the polarity of the solvent. The experimental data for chrysene triplet (table 4) indicate that in solvents of moderate to high dielectric constant the quenching rate decreases to a constant value which is almost an order of magnitude less than the rate in the low polarity case, hexane. This decrease in rate is not consistent with either direct formation of a charge-transfer complex or an increase in the stability of the complex, both of which would be favoured by polar solvents. The large observed decrease in going from hexane to diethyl ether is in contrast to about the 1.5 fold increase observed in the oxygen quenching case.[4] This corroborates the greater importance of β^2_{ct} in NO quenching for the reasons given above.

At lower triplet energies (1, 2-benzanthracene) the solvent effect is less pronounced whereas at very low triplet energies (3, 4 : 8, 9-dibenzpyrene) no solvent effect at all is found. This is in agreement with the idea that charge-transfer effects become more important at higher triplet energies.

The most puzzling fact, however, is the decrease in quenching rate with increasing solvent polarity. The data in table 4 indicate that this can not be due to specific NO solvent interactions, since in that case all aromatic molecules would be affected to the same extent. The possibility that the charge-transfer state actually shifts below the triplet level seems to be ruled out by the results on 1, 2-benzanthracene, where this would require a shift of more than 5000 cm^{-1}, which seems rather unlikely.

A possible explanation could be that there is a considerable solvent effect on μ, the dipole moment of the complex. This could be due to the fact that NO itself has a small, but finite, dipole moment. This may or may not increase the dipole moment of the complex depending on the orientation of the NO dipole in the complex. This explanation implies that the solvent effect is really more complicated than eqn (11) suggests and that no conclusions can be drawn with respect to solvent effects.

DISCUSSION

The quenching of excited triplet states by paramagnetic species involves a catalytic enhanced radiationless decay process when energy transfer reactions do not occur.

The results reported here indicate that under appropriate conditions of solvent and triplet energies, quenching of organic triplets by nitric oxide depends strongly on interactions with the charge-transfer state of the collision complex. This factor becomes important enough to yield rate constants for NO quenching of high energy triplets that exceed the values of energy transfer quenching rates for molecular oxygen (fig. 1). Previous comparisons of NO against O_2 static quenching of excited naphthalene singlet and triplet states indicated higher quenching probabilities for nitric oxide.[3] This agrees with our interpretation, since the conditions of the static experiments, with a high triplet energy molecule in a non polar solvent matrix, favour the charge-transfer mechanism.

The free radical DTBNO gives qualitatively the same quenching behaviour as NO, although the observed effect at low triplet energy is less striking for the nitroxide. This may mean that DTBNO also quenches by a catalytic mechanism. However, the increased molecular size, dipole moment and the potential availability of acceptor energy levels in the nitroxide [12] means that simple comparisons between these two species cannot be made at present.

Recently the dynamic quenching probabilities by NO and O_2 of several aromatic hydrocarbons adsorbed on a fluffy (large surface to weight ratio) polystyrene matrix were reported.[13] The O_2 (NO) quenching rates were found to increase (decrease) with decreasing triplet energy for high energy triplets. In addition the $k_q(O_2)/k_q(NO)$ ratios for high energy triplets were found to increase with decreasing triplet energy. These results qualitatively confirm our experimental findings in solution. Moreover, the quantitative agreement between these two types of experiments, for NO and O_2, was good. The relative quenching efficiencies from the solution and the polystyrene fluff experiments were found to agree within experimental error.

It has been found that there is no effect of an applied magnetic field on oxygen quenching rate.[14] This was interpreted by one of two mechanisms, involving either substantial charge-transfer interactions in the non-polar matrix used or large O_2 quenching inefficiencies. The latter interpretation agrees with our evidence for a polar solvent requirement in the charge-transfer assisted quenching by oxygen.[4] Furthermore, at the low oxygen pressures used (1-20 μmHg) the small observed oxygen quenching rate constants are most consistent with substantial collisional inefficiencies ($p = 10^{-2} - 10^{-3}$) which would lead to no magnetic field effects.[14]

Inasmuch as high triplet energies favour the charge-transfer mechanism with nitric oxide, one might expect the same effect at very high triplet energies for quenching by oxygen. The rate of O_2 quenching of benzene ($E_T = 29\,500$ cm^{-1}) has been estimated as 2×10^{10} l. mol^{-1} s^{-1}.[15] This is approximately 1/10 the value expected for a collision controlled reaction, and is substantially larger than the rates for oxygen quenching of lower energy triplets such as anthracene and naphthalene.[16] If oxygen quenching of benzene is a purely physical process, the increased quenching rate could reflect enhancement from charge transfer interactions. However, since literature values of I_p (benzene) yield relatively large $I_p - E_T$ factors, other terms in eqn (13), such as γ or C' may be contributing to the quenching if charge transfer effects are important.

We wish to thank the National Institutes of Health (U.S.A.) for the award of a Fellowship to F. K. and the Royal Society European Exchange Programme for the award of a Fellowship to O. L. J. G.

[1] J. B. Birks, *Photophysics of Aromatic Molecules* (Wiley Interscience, New York, 1970).
[2] G. Porter and M. R. Wright, *Disc. Faraday Soc.*, 1959, **27**, 18.

O. L. J. GIJZEMAN, F. KAUFMAN AND G. PORTER   737

[3] P. F. Jones and S. Siegel, *J. Chem. Phys.*, 1971, **54**, 3360.

[4] L. K. Patterson, G. Porter and M. R. Topp, *Chem. Phys. Letters*, 1970, **7**, 612; O. L. J. Gijzeman, F. Kaufman and G. Porter, *J.C.S. Faraday II*, 1973, **69**, 708.

[5] G. Porter and M. R. Topp, *Proc. Roy. Soc. A*, 1970, **315**, 163.

[6] G. J. Hoytink, *Acc. Chem. Res.*, 1969, **2**, 114.

[7] G. W. Robinson and R. P. Frosch, *J. Chem. Phys.*, 1964, **41**, 367.

[8] G. W. Robinson and R. P. Frosch, *J. Chem. Phys.*, 1963, **38**, 1187.

[9] W. Siebrand, *J. Chem. Phys.*, 1967, **47**, 2411.

[10] J. N. Murrell, *Mol. Phys.*, 1960, **3**, 314.

[11] R. S. Mulliken and W. B. Person, *Molecular Complexes* (Wiley Interscience, New York, 1969).

[12] J Murata and N. Mataga, *Bull. Chem. Soc. Japan*, 1971, **44**, 354.

[13] N. E. Geacintov, R. Benson and S. B. Pomeranz, *Chem. Phys. Letters*, 1972, in press.

[14] N. E. Geacintov and C. E. Swenberg, *J. Chem. Phys.*, 1972, **57**, 378.

[15] D. R. Snelling, *Chem. Phys. Letters*, 1968, **2**, 346; A. Morikawa and R. T. Cvetanovic, *J. Chem. Phys.*, 1970, **52**, 3237.

[16] G. Porter and P. West, *Proc. Roy. Soc. A*, 1964, **279**, 302.

Proc. R. Soc. Lond. A. **340**, 519–533 (1974)

Vibrational energy dependence of radiationless conversion in aromatic vapours

By G. S. Beddard,† G. R. Fleming, O. L. J. Gijzeman‡
and Sir George Porter, F.R.S.

*Davy Faraday Research Laboratory of The Royal Institution,
21 Albemarle Street, London W1X 4BS*

(*Received 29 March* 1974)

Under collision-free conditions the fluorescence lifetimes of naphthalene, 1–2 benzanthracene, chrysene and pyrene decrease with increasing excitation energy within the first excited singlet state. In anthracene the fluorescence lifetime is constant within the first excited singlet state. An explanation of these findings in terms of bond length changes between the excited singlet state and the first triplet state, and also the symmetry of the first singlet state is given. The effect of added foreign gas on fluorescence lifetimes and triplet yields is discussed in terms of a model which allows the non-radiative decay probability to vary with the degree of vibrational excitation. Excellent agreement is obtained with the experimental results on naphthalene fluorescence. The decrease of triplet yield as foreign gas pressure is reduced is also interpreted in terms of this model.

Introduction

A great deal of experimental and theoretical effort has recently been directed towards an understanding of the radiative and non-radiative properties of excited states, as a function of excess vibrational energy (Spears & Rice 1971; Heller, Freed & Gelbart 1972; Fleming, Gijzeman & Lin 1973). Both experiment and theory have focused attention on the low pressure (collision free) gas phase, and on the study of single vibronic levels. The effect of collisions on the properties of excited states has been studied experimentally by several authors (Neporent 1950; Stevens 1957; Ashpole, Formosinho & Porter 1971; Formosinho, Porter & West 1973; Beddard, Formosinho & Porter 1973b) and a detailed theoretical approach to this problem has recently been attempted (Lin 1972; Von Weyssenhoff & Schlag 1973; Freed & Heller 1974).

Most aromatic molecules in the gas phase show a rather broad absorption spectrum, and the excitation of discrete, single vibronic, levels is possible in only a few cases. It has been accomplished for benzene by Spears & Rice (1971) and for naphthalene by Hseih, Laor & Ludwig (1971) and by Knight, Selinger & Ross (1973).

† Present address: Physical Chemistry Department, South Parks Road, Oxford.
‡ Present address: Laboratory for Physical Chemistry, Nieuwe Prinsengracht 126, Amsterdam (c), The Netherlands.

G. S. Beddard and others

However, relatively broad band excitation, involving many levels, may still yield valuable information on the photophysical processes of excited aromatic molecules, especially when performed in conjunction with studies of the pressure dependence.

We report the results of our study of the decay rates in the excited state of several large aromatic molecules as a function of excess vibrational energy. The previous work of Beddard et al. (1973 b) and Ashpole et al. (1971) on the effects of pressure and wavelength upon naphthalene fluorescence and triplet yields, and later work on other molecules, is discussed with particular reference to the model of Freed & Heller (1974). This model seems suitable for the present purpose since it deals exclusively with the gross features of the relaxation process, which are expected to be revealed in these studies.

Experimental

The photon counting method was used for measuring the fluorescence decays. The fluorescence was detected at right angles by an E.M.I. 9813 kB photomultiplier. The band width of the exciting light was 0.5 nm or less. Measurements were made with the pressure of aromatic $\leqslant 27\,\mathrm{Pa}$ (0.2 Torr), the compounds being heated, where necessary, to provide this pressure. Where the decays were exponential, a weighted least-squares estimation of the lifetime was used. For non-exponential decays a combination of two exponential decays was used to fit the measured decay curve.

For naphthalene all the emission above 320 nm was monitored. For anthracene, chrysene, pyrene and 1–2 benzanthracene all the emission above 380 nm was monitored. In anthracene the fluorescence lifetime is short and convolution was performed to determine the true lifetime.

Results

The lifetimes under collision-free conditions of naphthalene, anthracene, 1–2 benzanthracene, chrysene and pyrene as a function of excitation wavelength are given in tables 1–5. A plot of inverse lifetime (decay rate) against excess energy is shown in figure 1 where benzene (Spears & Rice 1971) has been included for comparison purposes. The decays of naphthalene, anthracene, 1–2 benzanthracene and chrysene were all exponential, whereas the pyrene decays were all non-exponential except at the shortest wavelength used (265 nm). The pyrene decays were analysed as two exponentials, the long component being measured at 1 μs in each case. In table 5 both long and short components are given; in figure 1 only the long component is shown. The pyrene decay curves fall into two types, at 265 nm the decay is exponential for two decades decrease in intensity, after this a slight tailing is seen, possibly due to pyrene–pyrene collisions. In the region 330–370 nm the decays are non-exponential initially, becoming exponential at longer times. From 280 to 330 nm the total decays are a mixture of these two types.

For both chrysene and 1–2 benzanthracene the lifetimes measured at the longest wavelength are very similar to the solution values (Birks 1970), indicating that the vapour phase quantum yields are also similar to their solution values (0.14 and *ca.* 0.2 respectively).

The anthracene lifetimes are very similar to those recently published by Laor, Hseih & Ludwig (1973) and they show a remarkable constancy for excitation in S_1,

TABLE 1. FLUORESCENCE LIFETIMES OF NAPHTHALENE VAPOUR

(23 °C, 0.0091 kPa (0.07 Torr))

wavelength/nm	excess energy†/cm^{-1}	lifetime‡/ns
308	466	220
300	1313	150
290	2463	83.2
278	3951	63.1
268	5293	46.3
246	8630	26.2

† Origin taken as 312.5 nm.
‡ Precision ± 2 ns.

TABLE 2. FLUORESCENCE LIFETIMES OF ANTHRACENE VAPOUR

(105 °C, 0.0052 kPa (0.04 Torr))

wavelength/nm	excess energy/cm^{-1}	lifetime†/ns
370	0	5.03
362	597	5.05
347	1,791	5.05
330	3 276	5.07
300	6 306	5.05
240	14 640	3.9

† Precision ± 0.15 ns.

TABLE 3. FLUORESCENCE LIFETIMES OF 1–2 BENZANTHRACENE VAPOUR

(140 °C, 0.013 kPa (0.1 Torr))

wavelength/nm	excess energy/cm^{-1}	lifetime†/ns
380	0	45.6
370	711	45.0
360	1 462	44.9
350	2 255	42.2
340	3 096	41.5
330	3 981	40.8
320	4 934	39.3
310	5 942	38.7
300	7 017	38.2
270	10 721	37.3
260	12 145	31.6
250	13 684	30.5
240	15 351	29.2

† Precision ± 1 ns.

G. S. Beddard and others

and an abrupt change when excitation is into S_2. In naphthalene, 1–2 benz-anthracene, chrysene and pyrene, the decay rate changes smoothly as excitation changes from S_1 to S_2.

TABLE 4. FLUORESCENCE LIFETIMES OF CHRYSENE VAPOUR

(190 °C, 0.026 kPa (0.2 Torr.))

wavelength/nm	excess energy/cm^{-1}	lifetime†/ns
355	0	43.8
345	817	41.5
330	2 134	39.5
320	3 081	35.4
310	4 089	34.7
270	8 868	32.2
250	11 831	30.6
240	13 498	29.2

† Precision ± 1.5 ns.

TABLE 5. FLUORESCENCE LIFETIMES OF PYRENE VAPOUR

(120 °C, 0.013 kPa (0.1 Torr.))

wavelength/nm	excess energy/cm^{-1}	fast component lifetime/ns	slow component lifetime/ns
370	0	97	512 ± 10
360	751	90	440 ± 5
347	1 791	78	388
330	3 226	90	308
322	4 029	130	308
310	5 223	110	286
300	6 306	97	240
285	8 061	93	200
265	10 709	—	113

DISCUSSION

Vibrational energy dependence of fluorescence lifetimes

Except for anthracene, the rate of fluorescence decay increases with energy of excitation. This we attribute to an increased probability of radiationless conversion from higher vibrational levels.

The constancy of the fluorescence lifetime of anthracene over a range of pressures and excitation wavelengths, compared with the other molecules shown in figure 1, is very striking. In anthracene, apparently both the non-radiative and radiative rates are insensitive to the vibrational energy content, unless an increase in one exactly balances a decrease in the other, which seems very unlikely. The radiative process in anthracene is strongly allowed in contrast to the other molecules shown in figure 1. From our previous calculations (Fleming, Gijzeman & Lin 1973) we expect less energy dependence for an allowed transition than for the case where a promoting mode is involved.

The model of Heller *et al.* (1972) can be applied to the variation of the non-radiative rate with vibrational energy, provided rapid vibrational energy redistribution is not occurring. In this model, totally symmetric C—C modes are optically populated whilst totally symmetric C—H modes act as accepting modes to take up the electronic energy gap. The increasing radiationless rate with excess energy,

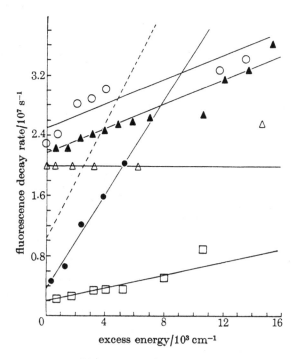

FIGURE 1. Plot of fluorescence decay rate against excess energy for: ---, benzene; ●, naphthalene; △, anthracene; ▲, 1–2 benzanthracene; ○, chrysene; □, pyrene. The decay rates for anthracene have been divided by 10 to bring them on scale.

within this model, is due to the increasing ability of the C—C modes to act as accepting modes as the number of quanta in the C—C modes in the initial state is increased. In other words, the Franck–Condon factor $\langle m_{\mathrm{opt}} | m + a_{\mathrm{opt}} \rangle$ increases with m_{opt}. For example, for an undistorted harmonic oscillator we have

$$\frac{|\langle m | m + 1 \rangle|^2}{|\langle m | m \rangle|^2} \simeq X(m+1), \tag{1}$$

where X is the dimensionless displacement parameter. The C—H Franck–Condon factors are of the form $\langle 0 | n \rangle$, and since the C—H displacements are very small in aromatic hydrocarbons (Byrne, McCoy & Ross 1965) they decrease very rapidly as n increases. Thus if, for higher initial optical levels, the C—H modes are in a position to accept less energy than they do for low initial levels, the radiationless decay will be accelerated for the higher levels. As equation (1) shows, the larger X becomes the

G. S. Beddard and others

more likely will this effect become, for a given m. The X used in equation (1) is related to bond length changes by the formulae (McCoy & Ross 1962)

$$X = M\omega(\Delta Q)^2/2\hbar,$$

and
$$\Delta Q = (\sum_j (\Delta r_j)^2)^{\frac{1}{2}},$$

where Δr_j is the change in bond length (in ångströms) of the jth C—C bond. Thus, the larger the differences in bond lengths between the two states involved in the radiationless transition, the faster the rate of increase will be as the vibrational energy in the initial state is increased.

Let us assume that the dominating process in the decay of the anthracene S_1 state is $S_1 \rightarrow T_1$ intersystem crossing. The S_1 state has L_a symmetry as does the T_1 state. (All T_1 states of the polyacenes are of L_a symmetry.) Hückel theory would predict these two states to have identical bond lengths, as no distinction is made between singlet and triplet states in this approximation (see, for example, Streitweiser 1961). Thus Hückel theory would predict no increase in S_1–T_1 intersystem crossing rate in anthracene, since $X \simeq 0$. In all other cases in figure 1 the symmetries of S_1 and T_1 are different. Thus, according to Hückel theory, a non-zero value of X will result, leading to an increasing radiationless rate with increasing vibrational energy. These conclusions are supported by the more sophisticated calculations of McCoy & Ross (1962), who used Pariser's (1956 a, b) wave functions to calculate bond orders in the excited states of benzene, naphthalene, anthracene and azulene. By using table 3 of McCoy & Ross (1962) it is possible to calculate ΔQ (R in their notation) for S_1–T_1 in anthracene. The result is $\Delta Q(S_1$–$T_1) = 0.0038$ Å.† The corresponding value for naphthalene is $\Delta Q(S_1$–$T_1) = 0.0765$ Å consistent with simple ideas from Hückel theory. So we find $\Delta Q^2_{\text{naphthalene}}/\Delta Q^2_{\text{anthracene}} \simeq 400$ and hence, on the basis of our Franck–Condon factor argument, we can infer that the rate of increase of S_1–T_1 intersystem crossing in anthracene would be negligible compared with naphthalene. On the basis of our previous calculations (Beddard, Fleming, Gijzeman & Porter 1973 a), if internal conversion (S_1–S_0) were significant in anthracene, we would expect this process to increase with increasing vibrational energy since $\Delta Q(S_1$–$S_0)$ is of reasonable magnitude (0.087 Å). Thus we conclude that S_1–T_1 intersystem crossing is the major decay route in anthracene. These arguments should be applicable to all aromatics where S_1 is of L_a symmetry. Perylene falls into this category and Ware & Cunningham (1966) have reported that the fluorescence lifetime is independent of excitation wavelength between 380 and 320 nm. We would also predict that naphthacene, for example, would show no variation of intersystem crossing rate with excess energy within the first absorption band.

We should point out that a small $\Delta Q(S_1$–$T_1)$ for C—C bonds does not imply that the absolute value of the rate S_1–T_1 is small. This is primarily determined by C—H bond length changes and the energy gap.

All the molecules shown in figure 1, with the exception of anthracene, have S_1

† 1 Å $= 10^{-10}$ m $= 0.1$ nm.

states of L_b symmetry and thus would be expected to show increases in their S_1–T_1 intersystem crossing rates with increasing excess energy. In addition, the radiative rate would be expected to be more sensitive to vibrational energy since there are now forbidden transitions (Fleming *et al.* 1973). However, we expect the variation of the fluorescence lifetime to be dominated by variations in the non-radiative decay rate. This is borne out for the two molecules for which quantum yields are known, naphthalene (Beddard *et al.* 1973 *b*; Uy & Lim 1972) and benzene (Spears & Rice 1971). We will assume it to be true for the other molecules. Setting aside pyrene for the moment (the only one of these molecules with a non-exponential decay curve) it is seen from figure 1 that the decrease in lifetime, with vibrational energy, becomes less rapid as molecular size is increased. This seems reasonable, because as molecular size is increased, the antibonding effect of excitation is spread out over more bonds. Thus in the series benzene, naphthalene, 1–2 benzanthracene S_1 and T_1 will become progressively closer in bonding to S_0 (and hence to each other) and so will become progressively less sensitive to vibrational energy content in the processes of $S_1 \rightarrow T_1$ and $S_1 \rightarrow S_0$. Also, increase of molecular size will increase the number of high frequency totally symmetric C—H accepting modes, making the C—C modes less effective in competing for the vibrational energy. This effect will also reduce the sensitivity of the radiationless decay on excess vibrational energy. Lowering the symmetry of the molecule, e.g. by substitution, would have the same effect. This argument seems to be borne out by the cases of 1–2 benzanthracene and chrysene which have the same number of C—C bonds and, within experimental error, exactly the same rate of change of fluorescence lifetime with excess energy.

Pyrene seems to provide a rather special case. The fluorescence decays are non-exponential at wavelengths 370–285 nm. This has been explained by Werkhoven *et al.* (1971) as being due to 'slow' ($k = 10^7 \mathrm{s}^{-1}$) vibrational redistribution. If vibrational redistribution is occurring in pyrene, then neither the theory of Heller *et al.* (1972) (no redistribution) or that of Fischer (1970) (fast redistribution) would apply. However, the same general considerations of C—C bond length changes given for the other molecules should hold. Now, since the 'average' population of totally symmetric C—C modes in S_1 would be less than predicted for a given excitation energy, the lifetime, as measured by the slow decay (decay of the redistributed levels (Werkhoven *et al.* 1971)) might be expected to change more slowly with excitation energy than for a molecule of equivalent size where vibrational redistribution is not occurring. As figure 1 shows, the lifetime of pyrene (19 C—C bonds) changes approximately half as rapidly ($4 \times 10^5 \mathrm{s}^{-1}$ per $10^3 \mathrm{cm}^{-1}$) as it does for 1–2 benzanthracene and chrysene (21 C—C bonds) ($8 \times 10^5 \mathrm{s}^{-1}$ per $10^3 \mathrm{cm}^{-1}$).

Pressure effects in naphthalene fluorescence

Recent measurements by Beddard *et al.* (1973 *b*) of the fluorescence lifetimes and quantum yields of naphthalene as a function of excitation energy and foreign gas pressure have shown that, for high excitation energy, the normal stabilization effect is observed but that, for low excitation energy, there is a decrease in the lifetime and quantum yield with increasing foreign gas pressure. This second effect, it was considered, could not entirely be attributed to collisional *activation* of the initial state–'going up to a Boltzmann distribution'. It was proposed that collisional vibrational redistribution to singlet vibrational levels with large intersystem crossing rates was responsible.

To test this hypothesis we attempted to fit the lifetime measurements with the step-ladder model of vibrational relaxation of Freed & Heller (1974). This model assumes that the molecule can be replaced by one N-fold degenerate oscillator. Collisions induce transitions to adjacent levels only, and the decay rate of each level is assumed to increase linearly with quantum number. Thus $k_m = d + ml$, where k_m is the decay rate of level m, d being the decay rate of the lowest level and l the incremental decay rate. Explicit expressions for the lifetime or quantum yield as a function of pressure are given in the appendix. The parameters required are:

(1) The average amount of energy removed per collision. This determines the spacing of the levels of the effective oscillator. To take into account the effect of temperature, a parameter $\theta = \hbar\omega_{\text{effective}}/kT$ is defined.

(2) The decay rate of the lowest level of the effective oscillator (d). This is determined from the zero pressure lifetime at the longest wavelength.

(3) The incremental decay rate (l) which is determined from the zero pressure lifetime at the shortest wavelength combined with the value of $\hbar\omega_{\text{effective}}$. To a reasonable approximation the variation of decay rate with excess energy in naphthalene can be equated with changes in the non-radiative decay rate (Uy & Lim 1972).

(4) The collision frequency (= rate of transition $v = 1 \rightarrow v = 0$). This is used as a variable. The collision frequency is directly proportional to foreign gas pressure since we are considering only excited aromatic–foreign gas collisions. The proportionality constant, i.e. the efficiency compared with gas kinetic collision frequency, is determined empirically by comparison of experimental and calculated curves.

(5) The degeneracy of the effective oscillator, obtained from the high pressure lifetime, l, and θ. A degenerate oscillator was used to mimic the density of states function of the molecule. We would expect that the degeneracy required to fit the experimental results would be less than the degeneracy required to give the true density of states at any energy. In naphthalene an oscillator of degeneracy ≈ 11 gives a reasonable approximation to the density of states as calculated by the method of steepest descents (Lau & Lin 1971). The degeneracy of the effective oscillator for our best fit is twofold.

Once a particular θ ($\hbar\omega_{\text{effective}}$) has been chosen, the rest of the parameters are determined via the experimental lifetimes. The best fit of the experimental results, with argon as foreign gas, is shown in figure 2, where the calculated rates have been determined from the tangent to the slope of the (non-exponential) decay curve at 400 ns. The same method was used by Beddard *et al.* (1973 b) to extract lifetimes from the experimental results. As can be seen from the figure, the model reproduces

Radiationless conversion in aromatic vapours **527**

both the increase in lifetimes (decrease in rate) at high excess energies and the decrease in lifetime (increase in rate) at low excess energy. (The numbers above the curves in figure 2 denote the initial quantum number in the effective oscillator, i.e. $E_{\text{excess}}/\hbar\omega_{\text{effective}}$.) Our conclusion is, then, that the experimental results are consistent with the idea that the Boltzmann population can be approached 'from below' as well as from above. The approach from below gives rise to the observed decrease in lifetimes as foreign gas pressure is increased.

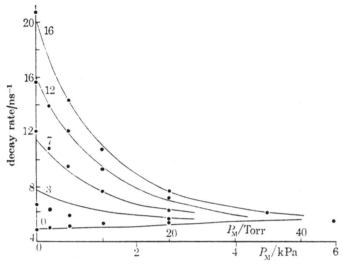

FIGURE 2. Plot of fluorescence decay rate of naphthalene vapour (k) against argon gas pressure (P_M). Solid lines represent calculated values, points experimental values. $D = 2$, $\theta = 1.5$, $l = 0.945$ ns^{-1}, $d = 4.88$ ns^{-1}.

A possible objection to this argument might be that, since the ground state is in Boltzmann equilibrium, excitation at the wavelength of the 0–0 transition will produce a Boltzmann distribution in the excited state, immediately on absorption. In other words, light of frequency equivalent to the 0–0 transition will produce not only the $v_{\text{opt}} = 0$, $\{v\} = 0$ level (v_{opt} denotes the optical mode and $\{v\}$ denotes the vibrational quantum number of all other modes), but also the $v_{\text{opt}} = 0$, $\{v\} = 1, 2, \ldots$ levels. Now, if decay rate increases, with increasing vibrational energy in the excited state, in the absence of collisions, it will tend towards a $v_{\text{opt}} = 0$, $\{v\} = 0$ population, as the level $v_{\text{opt}} = 0$, $\{v\} = 1$ will decay more rapidly than $v_{\text{opt}} = 0$, $\{v\} = 0$. Collisions will then be required to return the excited state to a Boltzmann distribution. This process will cause a decrease in the lifetime as molecules are constantly being pumped into shorter-lived levels.

It is generally considered that the amount of energy removed per collision for aromatic–inert gas collisions is of the order of a few hundred wavenumbers (see, for example, Stevens (1957)). In other words a strong collisional process, where the molecule is equilibrated in one collision, is not operating. Accordingly we attempted to fit the data with $\hbar\omega_{\text{effective}}$ varied between 100 and 600 cm^{-1}. The best fit (figure 2)

528   G. S. Beddard and others

was for $\hbar\omega_{\text{effective}} = 300\,\text{cm}^{-1}$. An effective frequency of 200 or $400\,\text{cm}^{-1}$ gave considerably poorer fits with the data. Once $\hbar\omega_{\text{effective}}$ is known, the efficiency of the collisions can be obtained. It was found that $k(v = 1 \to v = 0)/Z_{gk}$ was $ca.$ 2 %. Since the rate of going from level n to level $n-1$ is proportional to n, the efficiency will increase for depopulation of higher levels. An average efficiency taken over the first 10 levels would be $ca.$ 10 %. This is of the same magnitude as the average efficiency calculated for the relaxation of β-naphthylamine by argon, by Von Weyssenhoff & Schlag (1973), of 14–21 %.

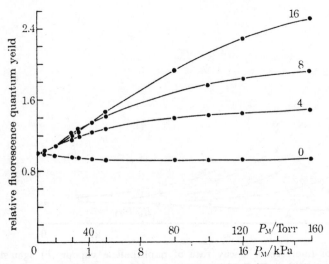

FIGURE 3. Calculated relative fluorescence quantum yield F_M/F_0 naphthalene vapour against argon gas pressure (P_M). The numbers above the curves are the initial quantum number in the effective oscillator. The parameters used for the calculation are the same as for figure 2.

Figure 3 shows the variation of quantum yield with gas pressure calculated with the same parameters as figure 2. This should be compared with figure 3 of Beddard $et\ al.$ (1973b) (here the results are with methane, but the qualitative similarity is very marked). The yields are plotted as ratios of the yield, F_M, at a given pressure divided by the zero pressure yield, F_0, of that particular level. Thus all the points on the high pressure side represent the same quantum yield. A comparison of figures 2 and 3 shows that a higher pressure of buffer gas is required to equilibrate the quantum yields than to equilibrate the lifetimes. This is to be expected since the quantum yield is the area under the fluorescence decay curve and will be primarily determined by the first part (short time) of the decay. The lifetimes (both experimental and calculated) were, however, determined when the fluorescence intensity had declined to $ca.$ 1 % of its initial value (400 ns).

The low pressure region of the curves for levels 8 and 16 in figure 3 show an interesting effect which may be amenable to experimental observation. The ratio F_M/F_0 for level 8 initially changes slightly more rapidly than for level 16. This can be rationalized since level 8 has a longer lifetime than level 16 and thus will be affected at lower collision frequency. However, since level 16 differs in quantum yield from the equilibrium value, by a larger amount, the relative effect will be greater, when the collision frequency becomes large enough to affect this level significantly. Inspection of the upper two curves of figure 2 of Beddard *et al.* (1973b) certainly does not rule out that this effect is occurring, but the data are not sufficiently accurate to confirm this prediction unambiguously.

Pressure effects in the triplet state

The first investigations of the effect of added foreign gas on the formation of triplet states in aromatic hydrocarbons were carried out by Porter & Wright (1955). Their findings were later confirmed quantitatively by Ashpole *et al.* (1971). All experiments show that the observed optical density of the triplet state increases to a maximum value with increasing pressure of added gas. Since the values of the absorbance were obtained at times, long enough for a complete thermal equilibration of the triplet state to occur, there appear to be two explanations for the observed behaviour. The first, originally proposed by Ashpole *et al.* (1971), attributes the fall off at low pressure to the occurrence of reversible intersystem crossing. This theory has been criticized by Soep, Michel, Tramer & Lindquist (1973), who argued that this model must lead to non-exponential fluorescence decay, which is not observed.

An alternative explanation again invokes enhanced intersystem crossing from higher vibrational levels, this time from the triplet state (Freed & Heller 1974). In this model, vibrationally excited triplet states are lost during vibrational relaxation via intersystem crossing to the ground state. If we assume that the intersystem crossing rate ($T_1 \to S_0$) increases with increasing vibrational energy, the amount of triplets lost will be increased at lower pressures. This in turn will lead to a lower optical density for the equilibrated or relaxed triplet. This explanation is supported by the experimental results of Ashpole *et al.* (1971), who found that the rate of decrease of the triplet absorbance with decreasing pressure depended on the nature of the added foreign gas and was greater for gases with a higher efficiency of vibrational energy removal.

Further support for this hypothesis can be obtained from the experiments where only the first excited singlet state of anthracene is irradiated, by using conventional flash photolysis with filtered flash sources (Ashpole *et al.* 1971) and from the laser flash photolysis experiments of Formosinho *et al.* (1973). In both cases we expect the triplet states to be formed with less excess vibrational energy than in the case of a wide band flash. This implies that less triplets are lost during vibrational relaxation, because of a lower probability of crossing to the ground state. Thus, the change in absorbance between high and low pressure regions should be less, as is observed experimentally.

As can be seen from the discussion given so far, the behaviour of the triplet state is intimately linked with the presence of an incremental decay rate. In view of our previous discussion of pressure effects in the excited singlet state, we expect this incremental decay rate to be determined by the displacement of the C—C modes between T_1 and S_0. By reasoning analogous to that given earlier for the excited singlet states, we then find that the incremental decay rate should be less for the larger molecules pyrene and perylene than for naphthalene and anthracene. Thus pressure effect will be less pronounced for the former molecules, again in agreement with the results of Ashpole *et al.* (1971).

So far our discussion has been concerned only with the long time, equilibrated, behaviour of the triplet state. Recently, experiments relating to the short time properties of the triplet yield have also been reported (Formosinho *et al.* 1973; Soep *et al.* 1973). In these experiments the triplet–triplet absorption is monitored immediately after intersystem crossing. Unfortunately, owing to experimental difficulties, only changes within one vibronic band could be determined as a function of pressure.

Immediately after intersystem crossing the triplet state will be formed with a large amount of excess vibrational energy. According to recent theories of radiation-less conversion (Heller *et al.* 1972), part of this energy will be contained in the optical (C—C stretching) modes, if these modes were originally excited in the singlet state. Thus the observed T—T absorption band will be a mixture of the $n \rightarrow n$, $n-1 \rightarrow n-1 \ldots 0 \rightarrow 0$ vibrational transitions, where n is determined by the singlet excitation process. If one assumes that the vibrational band shape is determined entirely by sequence congestion, and that this effect is identical for the $n \rightarrow n$ and $0 \rightarrow 0$ transitions, the height of the T—T absorption band in the initially formed triplet state will be determined by n, the number of C—C quanta excited. This argument has been used previously by Fleming *et al.* (1974) to explain the results of Formosinho *et al.* (1973).

However, if the sequence congestion effects are different for the $n \rightarrow n$ and $0 \rightarrow 0$ transitions, this argument will no longer be valid. The integrated absorption band will still provide a rough measure of the $n \rightarrow n$ and $0 \rightarrow 0$ Franck–Condon factors, provided the concentration of triplets are the same. In general we expect $F(n \rightarrow n) < F(0 \rightarrow 0)$ (Fleming *et al.* 1974).

In the experiments of Formosinho *et al.* (1973) the absorption was monitored at low pressure as a function of time (and wavelength). Under these conditions many triplet molecules will undergo radiationless transitions to the ground state during the (slow) vibrational relaxation process. This effect will offset the increased Franck–Condon factor for the $0 \rightarrow 0$ transition, and may lead to the similar inte-grated intensity ratio for the $n \rightarrow n$ and $0 \rightarrow 0$ transitions which was observed.

In the experiments of Soep *et al.* (1973), the triplet–triplet absorption was monitored at a fixed time (30 ns) while varying the pressure. In the high pressure region we expect the intersystem crossing to originate from a vibrationally relaxed singlet state (Beddard *et al.* 1973 b). This implies that the triplet state is formed with no C—C quanta excited. Furthermore, because of the high pressure, virtually no triplet molecules will decay during vibrational relaxation. The observed integrated intensity will thus correspond to the $0 \rightarrow 0$ absorption band only.

In the low pressure region there are two possible explanations. The first is that, since no vibrational relaxation occurs in the singlet manifold, intersystem crossing will occur via $S_2 \rightsquigarrow S_1^* \rightsquigarrow T_1^{**}$ transitions. We expect this process to be faster than intersystem crossing from a thermalized S_1 state, as occurs in the high pressure case (Heller *et al.* 1972; Beddard *et al.* 1973 *a*). Thus more triplet molecules will be formed at low pressure but the integrated intensity will be given by the product of the high triplet concentration and a small $(n \rightarrow n)$ Franck–Condon factor. Thus an explanation of the integrated intensities in the high and low pressure limits requires consideration of both Franck–Condon factors and concentration of triplet molecules, as in the experiments of Formosinho *et al.* (1973). A second explanation could be that the low pressure intersystem crossing process $S_2 \rightsquigarrow S_1^* \rightsquigarrow T_1^{**}$ actually prepares the triplet state with no C—C quanta excited. Since the incremental decay rate in the singlet state is primarily determined by the number of totally symmetric C—C quanta excited, this process will produce the same triplet concentration as in the high pressure case $(S_2 \rightsquigarrow S_1^* \rightsquigarrow S_1 \rightsquigarrow T_1^* \rightsquigarrow T_1)$. We would then expect that both the high and low pressure cases will give the same integrated T—T absorption intensity. Obviously, only a detailed time *and* pressure study can further clarify these results, and elucidate the possible involvement of internal conversion from high vibrational levels of S_1 (Beddard *et al.* 1973 *a*).

Conclusions

Changes in non-radiative decay of vibrationally excited states with excess energy are intimately linked with the displacements of the optical (C—C) modes. The magnitude of the incremental decay rate can be correlated with the symmetry of the S_1 and T_1 (or T_1 and S_0) states and with the size of the molecule. Pressure effects on fluorescence lifetimes (or on the observed absorbance of $T_1 \rightarrow T_n$ absorptions) can then be understood on the basis of the differences of relaxation rates into lower levels. This approach provides a coherent description of the pressure and energy dependence of electronic relaxation in aromatic vapours.

We wish to thank Professor Karl Freed for sending us details of his stochastic model of vibrational relaxation before publication, and for helpful discussions on this subject. We thank the S.R.C. for the award of studentships to G. S. B. and G. R. F. and the Royal Society European Exchange Programme for the award of a Fellowship to O. L. J. G.

Appendix

As has been shown by Freed & Heller (1974), the population of the various levels of the effective oscillator are obtained from the generating function $G_m(s,t)$ as:

$$P_n(m,t) = \frac{1}{n!} \left(\frac{\partial^n}{\partial s^n} G_m(s,t) \right)_{s=0}, \tag{A 1}$$

where $P_n(m, t)$ is the probability of finding the molecule in level n of the oscillator at time t, if the initial excitation is in level m. The generating function is given by:

$$G_m(s, t) = \frac{e^{-\alpha_D t}}{[\cosh{(bt)}]^D (1 + Zy)^D} \left[\tfrac{1}{2}\delta \left(\frac{\gamma}{\delta} - \frac{Z + Y}{1 + ZY}\right)\right]^m, \tag{A 2}$$

where
$$b = \tfrac{1}{2}\delta v, \quad \alpha_D = d + Dv(1 - \tfrac{1}{2}\gamma), \quad Y = \tanh{(bt)},$$

$$Z = \frac{\gamma - 2s}{\delta}, \quad \gamma = 1 + \rho + \frac{l}{v}, \quad \delta^2 = \gamma^2 - 4\rho. \tag{A 3}$$

Here, d is the decay rate of the lowest level of the oscillator, l is the incremental decay rate, v is the collision frequency for the transition $v = 0 \rightarrow v = 1$, D is the degeneracy of the oscillator and $\rho = \exp{(\hbar\omega_{\text{effective}}/kT)}$.

Neglecting changes in the radiative decay rate, the total unresolved emission intensity is given by:

$$I_m(t) = f \sum_{n=0}^{\infty} P_n(m, t) = f G_m(1, t)$$

$$= f \frac{e^{-\alpha_D t}}{[\cosh{(bt)}]^D (1 + XY)^D} \left[\tfrac{1}{2}\delta \left(\frac{\gamma}{\delta} - \frac{X + Y}{1 + XY}\right)\right]^m, \tag{A 4}$$

where $X = (\gamma - 2)/\delta$ and f is the radiative rate constant for all levels. The decay rate is then obtained as $k(t) = -\mathrm{d}/\mathrm{d}t \ln I_m(t)$

$$k(t) = -\frac{\mathrm{d}}{\mathrm{d}t} \ln I_m(t)$$

$$= \alpha_D + b \left[DY + \frac{1 - Y^2}{1 + XY} \left(DX + m\delta(1 - X^2) \frac{1}{\gamma(1 + XY) - \delta(X + Y)}\right)\right]. \tag{A 5}$$

In general the decay rate given by equation (A 5) will be time-dependent, leading to non-exponential decay. However, it can easily be shown that in the limit $v \rightarrow \infty$, $k(t) \rightarrow d + (Dl/\rho - 1)$, which is the decay rate for an ensemble of equilibrated degenerate harmonic oscillators. If $v \rightarrow 0$, $k(t) \rightarrow d + ml$ as expected.

References

Ashpole, C. W., Formosinho, S. J. & Porter, G. 1971 *Proc. R. Soc. Lond.* A **323**, 11.

Beddard, G. S., Fleming, G. R., Gijzeman, O. L. J. & Porter, G. 1973*a* *Chem. Phys. Lett.* **18**, 481.

Beddard, G. S., Formosinho, S. J. & Porter, G. 1973*b* *Chem. Phys. Lett.* **22**, 235.

Birks, J. B. 1970 *The photophysics of aromatic molecules*. New York: Wiley.

Byrne, J. P., McCoy, E. F. & Ross, I. G. 1965 *Australian J. Chem.* **18**, 1589.

Fischer, S. F. 1970 *J. chem. Phys.* **53**, 3195.

Fleming, G. R., Gijzeman, O. L. J. & Lin, S. H. 1973 *Chem. Phys. Lett.* **21**, 527.

Fleming, G. R., Gijzeman, O. L. J. & Lin, S. H. 1974 *J. C. S. Faraday II*, **70**, 1074.

Formosinho, S. J., Porter, G. & West, M. A. 1973 *Proc. R. Soc. Lond.* A **333**, 289.

Freed, K. F. & Heller, D. F. 1974 *J. chem. Phys.* (In the Press.)

Heller, D. F., Freed, K. F. & Gelbart, W. M. 1972 *J. chem. Phys.* **56**, 2309.

Hseih, J. C., Laor, U. & Ludwig, P. K. 1971 *Chem. Phys. Lett.* **10**, 412.

Knight, A. E. W., Selinger, B. K. & Ross, I. G. 1973 *Australian J. Chem.* **26**, 1159.

Laor, U., Hseih, J. C. & Ludwig, P. K. 1973 *Chem. Phys. Lett.* **22**, 151.

Lau, K. H. & Lin, S. H. 1971 *J. phys. Chem.* **75**, 2458.

Lin, S. H. 1972 *J. chem. Phys.* **56**, 4155.

McCoy, E. F. & Ross, I. G. 1962 *Australian J. Chem.* **15**, 573.

Neporent, B. S. 1950 *Zh. Fiz. Khim.* **24**, 1219.

Pariser, R. 1956a *J. chem. Phys.* **24**, 250.

Pariser, R. 1956b *J. chem. Phys.* **25**, 1112.

Porter, G. & Wright, F. J. 1955 *Trans. Faraday Soc.* **51**, 1205.

Soep, B., Michel, C., Tramer, A. & Lindquist, L. 1973 *Chem. Phys.* **2**, 293.

Spears, K. G. & Rice, S. A. 1971 *J. chem. Phys.* **55**, 5561.

Stevens, B. 1957 *Chem. Rev.* **57**, 439.

Streitweiser, A. 1961 *Molecular orbital theory for organic chemists.* New York: Wiley.

Uy, J. O. & Lim, E. C. 1972 *J. chem. Phys.* **56**, 3374.

Ware, W. R. & Cunningham, P. T. 1966 *J. chem. Phys.* **44**, 4364.

Werkhoven, C. J., Deinum, T., Langelaar, J., Rettschnick, R. P. H. & Van Voorst, J. A. D. 1971 *Chem. Phys. Lett.* **11**, 478.

von Weyssenhoff, H. & Schlag, E. W. 1973 *J. chem. Phys.* **59**, 729.

Further publications

THE TRIPLET STATE IN FLUID MEDIA.
G. Porter and M. W. Windsor, *Proc. Roy. Soc.*, 1958, **A245**, 238.
Comment on this paper. Citation Classic, G. Porter, *Current Contents*, 1984, **15**, 18.

SPECTROSCOPIC STUDIES OF THE PHOSPHORESCENT
STATES OF AROMATIC HYDROCARBONS.
G. Porter and M. W. Windsor, "Molecular Spectroscopy", Institute of Petroleum,
London, 1955, p. 6.

STUDIES OF THE TRIPLET STATE IN FLUID SOLVENTS.
G. Porter and M. W. Windsor, *Disc. Faraday Soc.*, 1954, **17**, 178.
(Part 1 of Primary Photochemical Processes in Aromatic Molecules)

PHOTOTROPY OF CHLOROPHYLL SOLUTIONS.
R. Livingston, G. Porter and M. Windsor, *Nature*, 1954, **173**, 485.

FREE RADICALS AND TRIPLET STATES IN AROMATIC VAPOURS.
G. Porter, *Chem. Soc. Special Pub.*, 1958, no. 9, 139.

VIBRATIONAL RELAXATION FOLLOWING INTER-SYSTEM CROSSING.
S. J. Formosinho, G. Porter and M. A. West, *Chem. Phys. Letters*, 1970, **6**, 7.

TRANSFERT D'ENERGIE DE L'ETAT DE TRIPLET EN SOLUTION.
G. Porter and M. R. Wright, *J. Chim. Phys.*, 1958, **55**, 705.

ENERGY TRANSFER FROM MOLECULES IN THE TRIPLET STATE.
G. Porter, *Pure App. Chem.*, 1962, **4**, 141.

EFFECT OF TEMPERATURE AND VISCOSITY ON THE TRUE
FIRST ORDER DECAY OF THE TRIPLET STATE OF AROMATIC
MOLECULES.
G. Porter and L. J. Stief, *Bull. Soc. Chim. Belg.*, 1962, **71**, 641.

RADIATIONLESS CONVERSION FROM THE TRIPLET STATE.
M. Z. Hoffman and G. Porter, *Proc. Roy. Soc.*, 1962, **A268**, 46.

VISCOSITY DEPENDENCE OF UNIMOLECULAR CONVERSION
FROM THE TRIPLET STATE.
G. Porter and L. J. Stief, *Nature*, 1962, **195**, 991.

γ-EXCITATION OF THE SINGLET AND TRIPLET STATES OF
NAPHTHALENE IN SOLUTION.
B. Brocklehurst, G. Porter and J. M. Yates, *J. Phys. Chem.*, 1964, **68**, 203.

PROTON TRANSFER DURING REACTIONS IN THE EXCITED STATE.
T. S. Godfrey, G. Porter and P. Suppan, *Disc. Faraday Soc.*, 1965, **39**, 194.

ABSORPTION SPECTRUM OF TRIPLET BENZENE.
T. S. Godfrey and G. Porter, *Trans. Faraday Soc.*, 1966, **62**, 7.

THE TRIPLET STATE OF CHLORANIL.
D. R. Kemp and G. Porter, *Chem. Comm.*, 1969, 1029.

TRIPLET FORMATION IN THE VAPOUR PHASE.
C. W. Ashpole, S. J. Formosinho and G. Porter, *Chem. Comm.*, 1969, 1305.

REACTIVITY, RADIATIONLESS CONVERSION AND
ELECTRON DISTRIBUTION IN THE EXCITED STATE.
G. Porter, Proceedings of the Thirteenth Conference on Chemistry at the University
of Brussels, 1965, Interscience, London, 1967, p. 79.

DETECTION AND LIFETIME OF THE TRIPLET STATE OF
ACETONE IN SOLUTION.
G. Porter, R. W. Yip, J. M. Dunston, A. J. Cessna and S. E. Sugamori,
Trans. Faraday Soc., 1971, **67**, 3149.

OXYGEN QUENCHING OF SINGLET AND TRIPLET STATES.
L. K. Patterson, G. Porter and M. R. Topp, *Chem. Phys. Letters*, 1970, **7**, 612.

PRESSURE DEPENDENCE OF INTERSYSTEM CROSSING IN
AROMATIC VAPOURS.
C. W. Ashpole, S. J. Formosinho and G. Porter, *Proc. Roy. Soc.*, 1971, **A323**, 11.

PRESSURE EFFECTS ON THE FLUORESCENCE FROM
NAPHTHALENE VAPOR.
G. S. Beddard, S. J. Formosinho and G. Porter, *Chem. Phys. Letters*, 1973, **22**, 235.

OXYGEN QUENCHING OF AROMATIC TRIPLET STATES IN SOLUTION.
PART 1.
O. L. J. Gijzeman, F. Kaufman and G. Porter, *J.C.S. Faraday II*, 1973, **69**, 708.

VIBRATIONAL RELAXATION IN THE TRIPLET STATE.
S. J. Formosinho, G. Porter and M. A. West, *Proc. Roy. Soc.*, 1973, **A333**, 289.

TRIPLET STATE OF ACETONE IN SOLUTION – DEACTIVATION
AND HYDROGEN ABSTRACTION.
G. Porter, S. K. Dogra, R. O. Loutfy, S. E. Sugamorie and R. W. Yip,
J.C.S. Faraday I, 1973, **69**, 1462.

TRIPLET STATE OF α-NITRONAPHTHALENE.
C. Capellos and G. Porter, *J.C.S. Faraday II*, 1974, **70**, 1159.

ROLE OF THE $^3(n\text{-}\pi^*)$ STATE IN THE PHOTOREDUCTION OF ACRIDINE.
E. Vander Donckt and G. Porter, *J. Chem. Phys.*, 1967, **46**, 1173.

CONCENTRATION QUENCHING AND EXCIMER FORMATION
BY PERYLENE IN RIGID SOLUTIONS.
J. A. Ferreira and G. Porter, *J.C.S. Faraday Trans. II*, 1977, **73**, 340.

INTERNAL CONVERSION FROM VIBRATIONALLY EXCITED LEVELS.
G. S. Beddard, G. R. Fleming, O. L. J. Gijzeman and G. Porter, *Chem. Phys. Letters*,
1973, **18**, 481.

THE EFFECT OF CONCENTRATION ON THE TRIPLET YIELD
OF PYRENE IN POLY (METHYLMETHACRYLATE).
E. Avis, P. Avis and G. Porter, *J.C.S. Faraday Trans. II*, 1976, **72**, 511.

ACIDITY CONSTANTS OF ANTHRACENE DERIVATIVES IN
SINGLET AND TRIPLET EXCITED STATES.
E. Vander Donckt and G. Porter, *Trans. Faraday Soc.*, 1968, **64**, 3218.

DECAY OF THE TRIPLET STATE. I. FIRST-ORDER PROCESSES
IN SOLUTION.
J. W. Hilpern, G. Porter and L. J. Stief, *Proc. Roy. Soc.*, 1964, **A277**, 437.

DECAY OF THE TRIPLET STATE.
II. RATE AND MECHANISM IN THE GAS PHASE.
G. Porter and P. West, *Proc. Roy. Soc.*, 1964, **A279**, 302.

Chapter 5

AROMATIC FREE RADICALS

The spectra of aromatic free radicals, like the spectra of aromatic molecules, are usually complex, many-line and many-band, structures and can rarely be identified on spectroscopic grounds alone. Many of them are labile in the excited state and therefore non-fluorescent so that flash photolysis and absorption spectroscopy provides almost the only method of preparation and detection.

The assignment of these spectra to their radical source is greatly facilitated, and indeed made possible, by comparative studies of the flash photolysis of a series of closely related molecules. The use of isotopes is well established in this connection but similar spectral perturbations can be made by substitution of hydrogen with an alkyl group for example, which also will have only a small effect on the energy of the electronic transition. Aromatic radicals such as triphenyl methyl are stable in equilibrium with their dimers at room temperature and have been known since the time of Gomberg (1900). Many of the host of aromatic radicals detected for the first time by flash photoysis in the 1950's are related to the Gomberg radical but are less stable. The prototypes of these are benzyl and the isoelectronic anilino and phenoxy which are, in turn, related to the Würster salts and the semiquinones respectively. The simplest aromatic radical, phenyl, is highly reactive and, unlike the benzyl type radicals, cannot be observed in solution but has been observed and studied in the gas phase.

The main use of these spectra, once identified unequivocably, is for the study of the photochemical processes by which they are formed, their physico-chemical properties and their chemical reactions.

Reprinted from the Transactions of the Faraday Society, No. 395, Vol. 51, Part 11,
November 1955

PRIMARY PHOTOCHEMICAL PROCESSES IN AROMATIC MOLECULES

† PART 3. ABSORPTION SPECTRA OF BENZYL, ANILINO, PHENOXY AND RELATED FREE RADICALS

BY GEORGE PORTER * AND FRANKLIN J. WRIGHT

Physical Chemistry Dept., University of Cambridge

Received 14th *March,* 1955

Photolysis, in the vapour phase, of toluene, ethyl benzene, benzyl chloride and other benzyl derivatives, results in the formation of a common transient species, with a characteristic narrow-banded electronic absorption spectrum, which must be identified with that of the free benzyl radical. Anilino, phenoxy, *p*-xylyl and similar free radicals have been detected in the same manner. Molecules from which such spectra have been observed are all characterized by the possibility of photolytic fission of a relatively weak bond in the side chain to yield a radical which is stabilized by resonance between benzenoid and quinonoid canonical forms. The lifetimes were always less than 10^{-4} sec which, at the concentrations employed, indicates a high collisional efficiency of recombination.

The resonance stabilization which is to a large extent responsible for the existence, at normal temperatures, of triphenylmethyl, and similar radicals of the type first described by Gomberg,[1] results in a reduction of the bond dissociation energy in many aromatic molecules,[2] although it may not be sufficient to give detectable equilibrium concentrations of free radicals. It is generally found that the probability of photolytic bond fission is greatest at the weakest bond in a molecule, and we might therefore expect that a common photochemical dissociation process in aromatic molecules will be that which results in the formation of two radicals possessing, together, a greater resonance energy than that of the parent molecule. The simplest examples of this behaviour would be provided by toluene,

* present address : Chemistry Dept., University of Sheffield.

† Parts 2 and 3 of this work were supported by a grant from the Department of Scientific and Industrial Research.

PRIMARY PHOTOCHEMICAL PROCESSES

aniline and phenol which might yield the benzyl, anilino and phenoxy radicals respectively. Each of these radicals has both benzenoid and quinonoid canonical forms, e.g.,

·CH₂   CH₂   CH₂

which result in a considerable resonance stabilization. The resonance energy of benzyl has been given as 24·5 kcal/mole which corresponds to the low C—H bond dissociation energy [3] in toluene of 77·5 kcal/mole.

There is considerable evidence for the existence of the benzyl radical, and for its low reactivity with other molecules. Thus it was one of the radicals detected by Paneth and Lautsch,[4] using the mirror technique, and its stability is the basis of the toluene carrier technique of Szwarc.[5] Nevertheless, it does not exist in significant equilibrium concentration at normal temperatures ; it has therefore never been detected by physical methods, and its spectrum is unknown. Indeed, very few polyatomic free radical spectra, other than those which can be obtained in equilibrium, have been recorded, and in the gas phase only four, all of them triatomic, can be considered established.[6] Quite recently Schuler and his colleagues [7] have detected a number of emission spectra during the passage of a mild electrical discharge through aromatic vapours, and some of these are very probably attributable to free radicals. In particular, the " V spectrum ", which has also been reported by Walker and Barrow,[8] has been assigned to $C_6H_5C\cdot$, $C_6H_5CH\cdot$, $C_6H_5CH_2$ " or perhaps a form containing less hydrogen and perhaps charged ".[8]

In the course of our study of the triplet states of aromatic molecules in the vapour phase (part 2) [9] we investigated the primary products of photolysis of monocyclic compounds. No triplet state spectra were found, and benzene itself showed no transient species and no dissociation. Toluene and a number of other compounds gave sharp band spectra which, as we hope to show, are to be attributed to benzyl and other free radicals of the type discussed. The experimental conditions were identical with those used during our investigations [9] of the triplet state. The path length was 1 m, and all experiments were conducted in the presence of inert gas (carbon dioxide or nitrogen) at a pressure at least one hundred times greater than that of the aromatic vapour.

RESULTS

Provided a sufficient excess of inert gas was present benzene itself showed no decomposition and no new bands were observed. Most other compounds, whether they gave new band spectra or not, were decomposed photochemically to compounds which gave rise to a continuous absorption lasting several minutes. This absorption was not observable until a few milliseconds after the end of photolysis ; it reached a maximum after about 10^{-1} sec and then decayed slowly over a period of several minutes. This behaviour has frequently been observed in other flash photolysis work and is characteristic of the condensation of a supersaturated vapour to small particles which scatter the light whilst they remain in the light path and which eventually condense on the wall. These continuous spectra are insufficiently specific for identification of the products and have not been studied further. Fortunately the time interval which elapses, before their appearance causes nearly complete extinction, is long enough to permit the detection and study of most free radicals and other intermediates of interest.

The phenyl derivatives bromobenzene, chlorobenzene, fluorobenzene, nitrobenzene, and benzonitrile, like benzene itself, gave no new spectra during or immediately after photolysis. Reaction occurred in most cases, evidenced by the later appearance of continuous absorption, and it is possible that the primary decomposition of most of these molecules resulted in the formation of the phenyl radical. If so, the phenyl radical must differ from those radicals whose spectra are described later, either in being more reactive or in having no transition of comparable intensity above 2700 Å. Although the spectral region investigated was 2200 to 5000 Å, spectra below 2700 Å could have been obscured by the absorption of the parent molecule.

THE BENZYL RADICAL

The flash photolysis of toluene vapour at 2 mm pressure, in the presence of 700 mm of carbon dioxide or nitrogen resulted in the appearance of a new banded spectrum which was observed only during irradiation by the photolysis flash. The spectrum consisted of one very sharp band at 3052 Å and several much weaker bands showing no obvious regularity. The 3053 Å band was 3 Å wide at half-maximum extinction, and had a rather sharp head degraded to longer wavelengths. Its appearance was therefore different from that of triplet spectra which all gave broader bands with an intensity distribution nearly symmetrical about the centre.

Investigation of related compounds immediately confirmed that this spectrum could not be identified with the triplet state of toluene. The same spectrum was observed with benzyl chloride, ethyl benzene, o-chlorotoluene, benzylamine, diphenyl methane, and benzyl alcohol. In the three latter molecules the spectrum was very weak and only the 3053 Å band was detectable. The high resolving power of the spectrograph and the sharpness of the band made it possible to compare the wavelengths of the spectra to better than 0·2 Å, using an optical comparator, and there can be no doubt that the same species is responsible for the absorption in each case. Spectra taken before and during photolysis are shown in fig. 1 and positions of the bands are given in table 1. Except with o-chlorotoluene, which will be discussed separately, no other spectra were observed during photolysis of these molecules.

Since no common spectrum can arise from each of these parent molecules the new bands must arise from a dissociation product. The intensity of the spectra did not increase with successive flashes and the absorption cannot therefore be attributed to secondary photolysis of a product. Except for o-chlorotoluene, all the molecules being considered have the generic formula $C_6H_5CH_2X$ and possible primary photochemical processes giving rise to a common species are as follows:

$$C_6H_5CH_2X \rightarrow C_6H_5CH_2 + X \qquad (1)$$

$$C_6H_5CH_2X \rightarrow C_6H_5CH + HX \qquad (2)$$

$$C_6H_5CH_2X \rightarrow C_6H_5 + CH_2X \qquad (3)$$

Ionic species, and processes involving fission of several bonds, are eliminated on energetic grounds.

There are strong arguments against supposing that the spectrum is that of the phenyl radical formed by reaction (3), since it has been shown that no spectrum is obtained from compounds such as chlorobenzene and bromobenzene. Dissociation occurs in these molecules, and the C—Cl and C—Br dissociation energies are less than that of C—CH_3 so that the phenyl radical should be observed during photolysis of the phenyl compounds at least as readily as from the benzyl derivatives. Secondly, if this spectrum were attributed to phenyl a second dissociation process would have to be invoked to account for the later results. Reaction (2) might be supported owing to the possibility of formation of a spin-paired molecule containing a divalent carbon atom. No analogous process could occur, however, in compounds such as phenol which, it will shortly be shown, behave in a similar manner. Definite evidence against type 2 decomposition is provided by recent work of Porter and Strachan who have shown that toluene, benzal chloride and benzotrichloride each give a different radical spectrum. We can therefore identify the dissociation process with reaction (1) and the observed spectrum with the benzyl radical.

The case of o-chlorotoluene requires special consideration. Two band systems were observed, one of which was identical with that assigned to benzyl and the second, of comparable intensity, consisted of four bands at longer wavelengths. At first sight the appearance of benzyl during photolysis of o-chlorotoluene is contrary to expectation, but a careful consideration of the primary products of photolysis provides the explanation. There are two bonds which, having low dissociation energy, might be expected to break on excitation of this molecule, firstly the C—Cl bond and secondly a C—H bond of the methyl group. The former will result in a tolyl radical of similar stability to phenyl, having little additional resonance energy. The o-tolyl radical is, however, isomeric with the more stable benzyl, and construction of a molecular model shows that migration of a hydrogen atom from the side chain to the ring may occur very easily. The converse

PRIMARY PHOTOCHEMICAL PROCESSES

transfer will be so endothermic that when equilibrium between the two forms is reached we shall expect the benzyl form to be greatly predominant. The second spectrum will then be assigned to the product of the second dissociation process, i.e. to o-chlorobenzyl. Its position, to long wavelengths of benzyl, is typical of the influence of chlorine substituted in the benzene ring.

TABLE 1.—SPECTRA OBSERVED DURING PHOTOLYSIS AND THEIR PROBABLE ASSIGNMENTS

molecule	λ(Å)	ν(cm^{-1})	I relative	radical
$C_6H_5CH_3$	3068	32,590	2	$C_6H_5CH_2$
	3053	32,760	10	
	2966	33,720	3	
$C_6H_5CH_2CH_3$	3068	32,590	2	$C_6H_5CH_2$
	3053	32,760	10	
	2966	33,720	3	
$C_6H_5CH_2Cl$	3100	32,260	1	$C_6H_5CH_2$
	3068	32,590	2	
	3053	32,760	10	
	2966	33,720	3	
	2936	34,060	2	
	2918	34,270	1	
$C_6H_5CH_2NH_2$	3053	32,760	—	$C_6H_5CH_2$
$C_6H_5CH_2OH$	3053	32,760	—	$C_6H_5CH_2$
$C_6H_5CH_2C_6H_5$	3053	32,760	—	$C_6H_5CH_2$
o-Cl . $C_6H_4CH_3$	3153	31,720	2	o-ClC$_6$H$_4$CH$_2$
	3157	31,680	4	
	3161	31,640	5	
	3165	31,600	6	
	3068	32,590	2	$C_6H_5CH_2$
	3053	32,760	10	
	2966	33,720	3	
	2936	34,060	1	
p-CH$_3$C$_6$H$_4$CH$_3$	3100	32,260	—	p-CH$_3$C$_6$H$_4$CH$_2$
$C_6H_5NH_2$	3008	33,250	—	$C_6H_5NH_2$
C_6H_5OH	3920	34,250 (wide)	—	C_6H_5O
$C_6H_5OCH_3$	2920	34,250 (wide)	—	C_6H_5O
C_6H_5SH	3100 to shorter wavelengths	32,260 limit	—	C_6H_5S

ANILINO AND PHENOXY

Photolysis of aniline vapour at 0·2 mm pressure in the presence of 600 mm of carbon dioxide resulted in the appearance of a strong band at 3008 Å which, like benzyl, was only present during irradiation by the photolysis flash (fig. 2). Although no reduction in the absorption of the aniline molecule was detectable the extinction at the maximum at 3008 Å was greater than at the maximum of the aniline spectrum near 2935 Å, showing that the extinction coefficient of the new species was many times greater than that of aniline. There were signs of structure at lower wavelengths which were too weak to measure reliably. The 3008 band was rather more diffuse than the strong band of benzyl and about double the width. Though similar in appearance to many of our triplet spectra, recent experiments in rigid glass by Norman and Porter [10] have shown that a similar spectrum is obtained whose lifetime is much too long for the triplet state. It must, there-fore, be a dissociation product and by analogy we shall assign this spectrum to the anilino radical. In this case no confirmation by photolysis of other molecules has yet been obtained. Dimethyl aniline gave negative results and negative results were also obtained

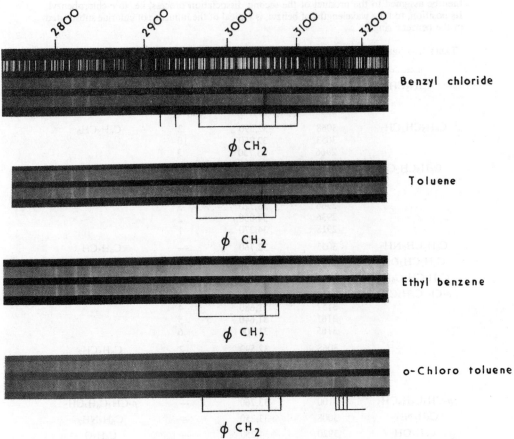

FIG. 1.—Absorption spectra of free benzyl. In each case the first spectrum is taken before, and the second during, photolysis.

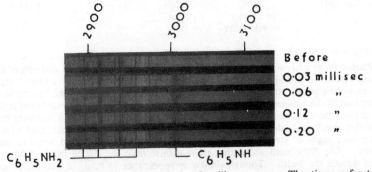

FIG. 2.—Spectra recorded during photolysis of aniline vapour. The times refer to the intervals between the peaks of the two flashes.

with phenylhydrazine (using nitrogen as inert gas) which might have been expected to yield the same product on photolysis. The non-appearance of a certain radical in cases like this cannot, however, give much indication of whether it is formed in the primary act. All the radicals with which we are concerned have lifetimes so short that they are only just detectable by the present techniques and a slight additional reactivity would be sufficient to reduce their lifetime below the experimental limit of detection. It should also be noted that whilst hydrogen abstraction by anilino from aniline produces no change, abstraction from phenylhydrazine produces a different species.

Photolysis of phenol resulted in a transient spectrum consisting of one diffuse band, about 100 Å wide, with a maximum at 2920 Å. In spite of its diffuseness the intensity distribution was characteristic, with a definite maximum, and it was possible to obtain clear evidence of the identity of this spectrum by photolysis of anisole. A similar transient absorption was obtained which density comparisons showed was identical with that from the photolysis of phenol, and which is therefore to be assigned to the phenoxy radical.

OTHER MOLECULES

Other molecules giving negative results, in addition to those already mentioned, were benzaldehyde, acetophenone, benzophenone, o-bromotoluene, o-toluidine, diphenyl, and dibenzyl. α-Methyl naphthalene gave a triplet spectrum (part 2) but no radical was observed. Two other substances gave transient spectra which we believe are to be assigned to radicals. Thiophenol showed a continuous absorption, which was present only during irradiation, beginning at 3400 Å and extending to shorter wavelengths with no definite maximum. p-Xylene gave a sharp band at 3100 Å, similar in appearance to that of benzyl. These are probably the spectra of thiophenoxy and p-xylyl respectively.

The positions of all spectra found in this investigation are recorded in table 1. The exact agreement between measurements from different molecules arises from the fact that it was possible to establish the identity of the spectra to a greater accuracy than the band centre position could be defined.

DISCUSSION

The lifetimes of all radical spectra observed were less than the time resolution of the method, i.e. 10^{-4} sec and therefore no kinetic investigations have been possible. It is worthy of note, however, that the observations indicate a high efficiency of radical recombination. All reactions of benzyl and similar free radicals with the parent molecule, except that which regenerates the same species, would be too endothermic to occur at the observed rate at room temperature and the radicals must disappear either by dimerization or by recombination with a hydrogen atom. Even at the highest concentration of radicals no decrease in absorption by the parent molecule was detectable which sets an upper limit to the radical concentration of about 5 % of that of the molecule, i.e. 0·1 mm for benzyl and 0·01 mm for aniline. This leads to a lower limit for the collision efficiency of recombination of radicals, or radical and atom, of the order of 10^{-2} and 10^{-1} for benzyl and anilino respectively. These radicals therefore show a reactivity with respect to recombination which is typical of free radicals such as methyl, though their reactivity with other molecules must be far less.

The appearance of the spectra is interesting. In each case a single band is predominant and the width of this band increases in the order benzyl, anilino, phenoxy, whilst the separation from the spectrum of the parent molecule decreases. The very sharp line-like structure of benzyl and p-xylyl is also found in triphenyl methyl [11] even in solution, and seems to be characteristic of this type of molecule. It is probably a reflection of the loose coupling between the odd electron and the rest of the molecule which results in an " atomic line " spectrum with little vibrational excitation. The increasing diffuseness in anilino and phenoxy implies a shorter lifetime of the upper state such as could result from an increasing probability of dissociation.

There is little doubt that these investigations could be extended to a number of similar molecules. Unfortunately the radical lifetimes are beyond the time resolution of the technique employed and, therefore, not only are kinetic studies impossible but it is probable that in many cases free radicals are formed which escape detection. As a result of this work, however, similar studies have been initiated in solution and in rigid media where these disadvantages are less evident. Studies of this kind, which will be reported later, fully confirm the assignments given here and extend the observations to a wide range of similar radicals.

NOTE ADDED IN PROOF

The ' V spectrum ' referred to in the introduction has recently been reinvestigated by Schüler and Michel [12] who have put forward new evidence supporting the assignment to benzyl. The intensity maximum of this system occurs at 4477 Å (22,330 cm⁻¹). Longuet-Higgins and Pople [13] have predicted that the two lowest transitions in benzyl should occur at 27,900 and 33,700 cm⁻¹ and that the latter should be much the stronger. Bingel [14] has reached similar conclusions but predicts the lower transition at 21,500 cm⁻¹.

As pointed out by Schüler and Michel, the emission spectra of polyatomic molecules almost invariably consists of the transition from the lowest excited state only. On the other hand, in absorption, we shall observe most readily the system with highest transition probability. The observation of two different systems from benzyl, at 22,300 and 32,600 cm⁻¹, by Schüler and ourselves respectively is therefore readily understood and is in fair quantitative agreement with the predictions of molecular orbital calculations.

[1] Gomberg, *Ber.*, 1900, **33**, 3150; *J. Amer. Chem. Soc.*, 1900, **22**, 757.
[2] Wheland, *The Theory of Resonance* (Wiley, 1944).
[3] Szwarc, *J. Chem. Physics*, 1948, **16**, 128.
[4] Paneth and Lautsch, *J. Chem. Soc.*, 1935, 380.
[5] Szwarc, *Chem. Rev.*, 1950, **47**, 75.   [6] Porter, *J. Phys. Radium*, 1954, **15**, 113.
[7] Schüler, Reinebeck and Köberle, *Z. Naturforschung*, 1952, **7a**, 421, 428.
[8] Walker and Barrow, *Trans. Faraday Soc.*, 1954, **50**, 541.
[9] Porter and Wright, *Trans. Faraday Soc.*, 1955, **51**, 1205.
[10] Norman and Porter, *Proc. Roy. Soc. A*, 1955, **230**, 399.
[11] Anderson, *J. Amer. Chem. Soc.*, 1935, **57**, 1673.
[12] Schüler and Michel, *Z. Naturforschung*, 1955, **10a**.
[13] Longuet-Higgins and Pople, *Proc. Physic. Soc.*, 1955, **68**, 591.
[14] Bingel, *Z. Naturforschung*, 1955, **10a**.

Spectrochimica Acta, 1958, Vol. 12, pp. 299 to 304. Pergamon Press Ltd.

The electronic spectra of benzyl

G. Porter
Department of Chemistry, University of Sheffield

and

Emma Strachan
British Rayon Research Association

(*Received* 7 *March* 1958)

Abstract—Studies of spectra of the photolysis products of toluene and related molecules in low temperature rigid glasses have confirmed the assignment of the 3053 Å system to benzyl. A new absorption system has been observed, with maximum at 4527 Å. The two absorption systems are in excellent agreement with recent molecular orbital calculations and with emission spectra previously assigned to this radical.

INFORMATION concerning the electronic spectra of benzyl is now available from three sources:

(1) Emission spectra obtained by SCHULER and collaborators [1] and by WALKER and BARROW [2] in mild electrical discharges through mixtures of aromatic vapours and inert gases.

(2) Absorption spectra obtained by PORTER and WRIGHT [3] following flash photolysis of aromatic vapours, by PORTER and WINDSOR [4] following flash photolysis of aromatic molecules in solution and by NORMAN and PORTER [5] by photolysis of rigid solutions at low temperatures.

(3) Theoretical calculations of DEWAR, LONGUET-HIGGINS, POPLE [6, 7] and of BINGLE [8].

The emission spectra lie in the region of 4477 Å whilst the absorption spectra have maxima at 3053 Å, 3178 Å and 3187 Å in the gas phase, in paraffin solution at room temperature and in rigid E.P.A. glass at −197°C respectively. These findings may be explained in three ways:

(1) The emission or the absorption or both spectra have been incorrectly assigned to benzyl.

(2) Owing to operation of the Franck-Condon principle the spectra, although of the same electronic transition, are widely separated in emission and absorption. This is very improbable in view of the small number of bands and the clear evidence of progression limits.

(3) Both spectra are due to benzyl but they correspond to separate electronic transitions.

SCHULER and MICHEL [9] have recently supported the third of these possibilities and here we present new data which confirm this interpretation.

Assignment of spectra to benzyl

We have repeated and extended the measurements of NORMAN and PORTER on the spectra of the products of photochemical dissociation of aromatic molecules in rigid glasses using, in the first place, identical techniques. Our results leave no doubt that the 3187 Å system is to be assigned to the benzyl radical. Some of our earlier results were used by Porter and Wright in making this assignment and further details of our measurements will be published shortly. For the present purpose it is sufficient to note two confirmations of this assignment which we have obtained in addition to that which is referred to by Porter and Wright.

(1) The series toluene, ethyl benzene, *iso*propyl benzene and *tert*-butyl benzene give radical spectra in E.P.A. which are almost identical in structure. Taking into account the energetic limitations the only possible assignments are as follows:

Molecule	λ_{max} of radical (Å)	Radical
ϕCH_3	3187	ϕCH_2
$\phi CH_2 CH_3$	3222	$\phi CHCH_3$
$\phi CH(CH_3)_2$	3242	$\phi(CH_3)_2$
$\phi C(CH_3)_3$	3242	$\phi(CH_3)_2$

(2) The series toluene, diphenyl methane (and benzhydrol) and triphenyl methane give radical spectra with similar structure with maxima as follows:

Molecule	λ_{max} of radical	Radical
ϕCH_3	3187	ϕCH_2
ϕCH_2 (and $\phi_2 CHOH$)	3355	$\phi_2 CH$
$\phi_3 CH$	3415	$\phi_3 C$

The assignments given are the only self-consistent set and here, in addition, one of the radical spectra, that of triphenyl methyl, is already well established, indeed it was the first free radical to be detected and the first recorded radical spectrum. The 3187 Å system must therefore be considered established as part of the benzyl spectrum and in view of the experimental conditions the transition must involve the ground state.

The assignment of the 4477 Å emission system to benzyl is less certain owing to the complexity of excitation conditions in the gaseous discharge. It has been assigned to other species by Walker and Barrow and also by Schüler and his collaborators, although Schüler and Michel have recently supported the assignment to benzyl. They pointed out that this assignment was not necessarily inconsistent with the fact that the absorption system appears at much shorter wavelengths. It is well known from fluorescence studies that the emission spectra of polyatomic molecules consist almost exclusively of combinations between the lowest excited state and the ground state. Furthermore, if the transition probability of this system is low, it may be much weaker in absorption than transitions to higher levels so that the lower transition would appear only in emission and the upper system only in absorption. Evidence that benzyl has two transitions in the region investigated, the lower of which has low transition probability, is derived from two sources; first from analogy with the well-known spectra of the related radical triphenyl methyl, which shows a weak visible system and a strong system in the ultraviolet, and secondly by theoretical calculations which will be referred to later.

Proof that the carrier of the 4477 Å system of SCHÜLER is benzyl would be given if the same system could be found from the benzyl radical in absorption. Although the 3187 Å system may never appear in emission the 4477 Å system must, if it is due to benzyl, appear in absorption provided a sufficient density of benzyl can be attained. Careful examination of all our plates of the benzyl spectra showed no visible absorption bands. If the ratio of extinction coefficients of the visible and ultraviolet systems was comparable with that of triphenyl methyl a tenfold increase in benzyl radical concentration should be sufficient to make detection of the visible system possible. Previous experience had shown that the optimum concentration of benzyl attainable by photochemical methods had already been attained and therefore the only means of obtaining increased absorption was by increasing the path length above that of the 1 cm cell used in all experiments so far described.

Long cell for irradiation of low temperature glasses

The cell used was 20 cm long and of the design shown in Fig. 1. It, and the surrounding Dewar vessel, were constructed entirely of quartz and the Dewar, which was unsilvered, had double quartz flats in the base. The reaction vessel was filled with solution to a point well above the middle window so that, when contraction took place on cooling, no miniscus was formed in the optical path. The whole length of the frozen glass was irradiated by two U-shaped low pressure mercury vapour lamps and the spectrum was examined by means of a hydrogen or tungsten source and a small Hilger prism spectrograph.

This arrangement has been used many times without damage to the cell, the frozen solvent being in all cases M.P. (*isopentane* three parts, methyl *cyclo*hexane two parts). It is anticipated from experience with other arrangements that more trouble might be encountered if E.P.A. were used as the solvent.

Absorption spectra of benzyl in the long path

Traces of the spectra obtained from photolysis of a 10^{-2} molar solution of toluene in M.P. at $-197°C$ and in the 20 cm cell are shown in Fig. 2. The 3178 Å system now appears with very high intensity and, in addition, a weak but very sharp series of bands is present in the visible region. The two systems appear together on irradiation and disappear on warming the glass. The positions of the main band maxima, are given in Table 1.

Both systems consist of a series of rather sharp bands with one strong band at long wavelengths. The strong band at 4527 Å must lie close to the origin of the visible system and is therefore to be compared with the maxima of the shortest wavelength band in the emission system, at 4477 Å. The shift of 50 Å to longer wavelengths in solutions is quite normal, the corresponding shift in the U.V.

G. Porter and Emma Strachan

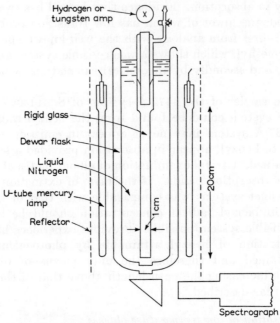

Fig. 1. Arrangement for irradiation and spectroscopy of low temperature solutions in
20 cm path.

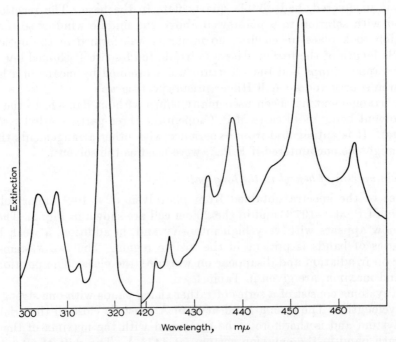

Fig. 2. Densitometer traces of spectra of benzyl obtained from 10^{-2} molar solutions of
toluene in M.P. at $-197°$C using a 20 cm path.

The electronic spectra of benzyl

Table 1. Absorption spectra of benzyl in M.P. glass

U.V. system		Visible system	
(Å)	(cm⁻¹)	(Å)	(cm⁻¹)
3178	31,470	4635	21,580
3125	32,000	4527	22,090
3078	32,480	4460	22,420
3039	32,900	4375	22,850
		4330	23,090
		4245	23,550
		4220	23,700

system being 134 Å. There is some uncertainty as to what should be taken as the origin of the emission bands and the true origin of the absorption system is probably the weaker 4635 Å band, in which case the shift from the strongest emission band becomes 158 Å. If we consider only the main progression, in both emission and absorption, we obtain the following scheme of vibrational levels.

$v'v''$	2,0		1,0		0,0		0,1		0,2
Absorption (glass)	23,550	(700)	22,850	(760)	22,090		—		—
Emission (gas)	—				22,330	(944)	21,386	(933)	20,453

Of course v' and v'' do not necessarily refer to the same vibrational mode in emission and absorption. In spite of uncertainties as to the exact origin of the band systems there can be little doubt that the emission and visible absorption spectra involve the same transition and that both are to be assigned to benzyl.

Comparison with molecular orbital calculations.

The appearance of a weak transition in the visible and a strong one in the near ultraviolet is fully in accordance with molecular orbital calculations. Benzyl is an alternant hydrocarbon radical with seven π electrons, the three lowest configurations of which will be:

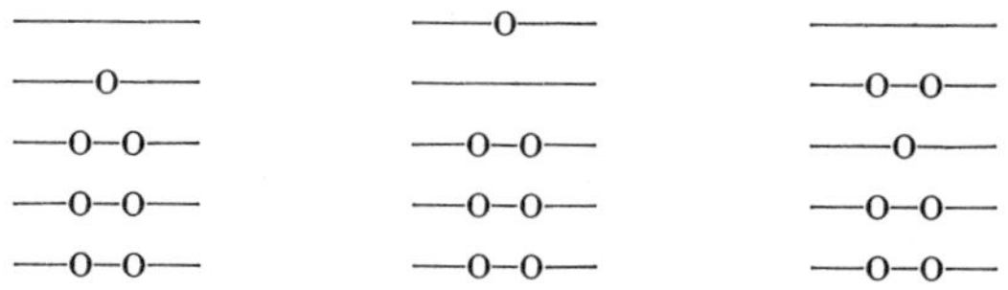

As pointed out by DEWAR and LONGUET-HIGGINS the configurations χ_1 and χ_2 are degenerate in the first approximation but when configuration interaction is introduced we obtain the wave functions $\Psi_1 = (1/\sqrt{2})\,(\chi_1 + \chi_2)$ and $\Psi_2 = (1/\sqrt{2})\,(\chi_1 - \chi_2)$.

G. PORTER and EMMA STRACHAN

The former has the lower energy and the transition from the ground state is weak. Calculations of the energies of these levels have been made by LONGUET–HIGGINS and POPLE and similar calculations have been given by BINGEL using both molecular orbital and free electron approaches. The results of these calculations and the experimental data from emission and absorption are compared in Fig. 3.

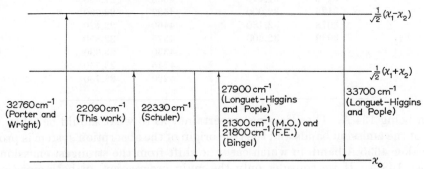

Fig. 3. Scheme of observed and calculated electronic transitions in benzyl.

The agreement is very satisfactory and lends support both to the assignments and to the methods of calculation.

Acknowledgement—The work described in this paper forms part of a programme of fundamental research being carried out by the British Rayon Research Association.

References

[1] SCHÜLER H., REINEBECK L. and KÖBERLE R., *Z. Naturf.* 1952 **7a** 421.
[2] WALKER S. and BARROW R. F., *Trans. Faraday Soc.* 1954 **50** 541.
[3] PORTER G. and WRIGHT F. J., *Trans. Faraday Soc.* 1955 **51** 1469.
[4] PORTER G. and WINDSOR M. W., *Nature, Lond.* 1957 **180** 187.
[5] PORTER G. and NORMAN I., *Proc. Roy. Soc.* A 1955 **230** 399.
[6] DEWAR M. J. S. and LONGUET–HIGGINS H. C., *Proc. Phys. Soc. Lond.* A 1954 **67** 795.
[7] LONGUET–HIGGINS H. C. and POPLE J. A., *Proc. Phys. Soc. Lond.* A 1955 **68** 591.
[8] BINGEL W., *Z. Naturf.* 1955 **10a** 462.
[9] SCHÜLER H. and MICHEL P., *Z. Naturf.* 1955 **10a** 459.

Extrait du *Journal de Chimie Physique*, 1964, p. 1517.

THE HIGH RESOLUTION ABSORPTION SPECTROSCOPY OF AROMATIC FREE RADICALS,

by G. PORTER and B. WARD.

[Department of Chemistry, The University, Sheffield 10.]

SUMMARY

The flash photolysis of aromatic molecules at high resolution, using long path lengths, has enabled us to observe a wealth of new spectra. In particular the weak, long wavelength transition of benzyl and substituted benzyl radicals has been observed. Their assignment to benzyl, rather than, for example, to an isomeric tropyl radical is confirmed and the structure of the spectrum is discussed. Brief reference is made to other transient spectra which we have recently recorded, including the spectrum of the phenyl radical.

Introduction.

The first recorded spectrum of an aromatic free radical was that of triphenyl methyl by GOMBERG ([1]) in 1900. In this classical communication GOMBERG stated « I wish to reserve the field for myself ». After the respectable period of more than half a century, we may perhaps be permitted to look further into the problem.

Apart from the work of GOMBERG and a few similar spectra of rather stable free radicals in solution, polyatomic free radical spectroscopy is a very recent affair made possible mainly by two experimental techniques: flash photolysis ([2]) and matrix stabilization ([3]). Matrix stabilization was introduced by LEWIS and LIPKIN in the early forties but was applied to radicals which were fairly stable even in ordinary solutions. It was shown to be generally applicable to most radicals, with a suitable choice of conditions, by NORMAN and PORTER ([4]) in 1954. The method of flash photolysis was introduced in 1949 and has since been used for the study of all kinds of transient species in all types of medium.

Both of these methods have been useful in the study of aromatic free radicals and about 140 of these have now been characterised. The radicals are of interest for several reasons; they have fairly sharp spectra in easily accessible regions of the spectrum; they are reasonably stable and above all the possibi-lity of studying a large range of related compounds, which is unique to aromatic molecules, makes identification and interpretation relatively easy.

The first class of aromatic free radicals to be extensively studied by spectroscopy comprised those radicals of which benzyl is the prototype. This class includes most of the stable free aromatic radicals such as triphenyl methyl and Würster's salts and those radicals isoelectronic with ben-zyl, namely phenoxyl and anilino. PORTER and WRIGHT ([5]) showed that flash photolysis of toluene, phenol and aniline vapours and many of their derivatives produced transients having characteristic absorption spectra around 3 000 Å, which were assigned to the benzyl, phenoxyl and anilino radicals respectively. These radicals are formed by the loss of an atom or group of atoms β to the ring following predissociation of the parent molecule in the excited state. Further studies of these radicals in rigid matrices ([6]) and in solution have been carried out. CHILTON and PORTER ([7]) showed that ionizing radiations were a powerful alternative method of preparation. Flash photolysis has recently been employed to study the spectra of both acidic and basic forms of phenoxyl and anilino radicals in solution ([8]), and to determine acidity constants and rates of reaction. Similar studies on such related radicals as the semiquinones ([9]) and ketyls ([10]) have also been carried out.

High Resolution Studies.

Experimental.

Photochemical studies of aromatic molecules in the gas phase have been hindered by the low vapour pressures of many compounds, by the low quantum yields of the photolytic processes and by the short life-times and weak absorptions of many of the resulting

G. PORTER AND B. WARD

THE BENZYL RADICAL

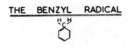

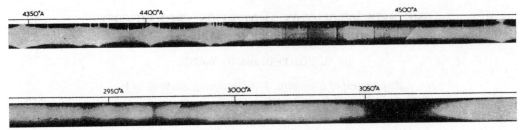

Fig. 1. — The visible and UV absorption spectra of the benzyl radical.

transients. These difficulties have been overcome to a certain extent by using a more sensitive flash photolysis apparatus whose main points are :

1) a reaction vessel containing a multiple reflection mirror system ([11]) giving path lengths of up to 10 metres;

2) fairly high energies (4 000 J. in 50 μsec.) and short delays, and

3) the spectra are photographed on a 21 ft. grating spectrograph so that weak banded absorptions are well resolved.

Many new spectra have been observed at high resolution and much new information obtained about the primary photochemical processes in aromatic molecules. Here we describe some studies of benzyl and related radicals, and mention briefly some new spectra which have been obtained recently.

The Benzyl Radical ΦCH₂.

Benzyl has been the subject of numerous experimental and theoretical investigations and the 3 000 Å transition is well characterized. A photograph at high resolution (fig. 1) shows that the spectrum is predissociated yet contains some potentially analysable structure. In overall appearance it resembles that of a monosubstituted benzene.

Some halogenated toluenes were photolysed and the short wavelength transitions of the resulting benzyl radicals observed. These spectra resembled those of the unsubstituted radical but were shifted to the red (see table I).

Intensity measurements on the spectra showed that the half lives of the radicals were about 100 μsec. It may be noted that the three different species obtained with o, m and p substituents confirm the benzyl structure and eliminate the possibility that isomerisation to tropyl has occurred. Ortho and

meta bromo toluenes behaved in an anomalous manner since we failed to detect any benzyl radicals on photolysis of these compounds. Benzyl bromide is also known to give a negative result ([5]) and it is possible that the increased spin-orbit interaction arising from the presence of the heavy bromine atom results in rapid deactivation of the excited molecules by intersystem crossing.

TABLE I

Molecule	Radical	λ_{max} (Å)
$C_6H_5CH_3$	$C_6H_5CH_2$	3 053
$C_6H_5CHCl_2$	C_6H_5CHCl	3 098
$C_6H_5CCl_3$	$C_6H_5CCl_2$	3 106
$o\text{-}ClC_6H_4CH_3$	$o\text{-}ClC_6H_4CH_2$	3 153
$m\text{-}ClC_6H_4CH_3$	$m\text{-}ClC_6H_4CH_2$	3 130
$p\text{-}ClC_6H_4CH_3$	$p\text{-}ClC_6H_4CH_2$	3 073,5

An emission system from toluene and many of its derivatives in electric discharges was first reported by SCHULER ([12]), who tentatively assigned it to the benzyl radical, although it occurred at 4 500 Å. This apparent discrepancy was resolved when it was realized, as a result of calculations by LONGUET-HIGGINS and POPLE ([13]), that the expected spectrum of benzyl should consist of an allowed transition at short wavelengths and a forbidden one at longer wavelengths. Since emission occurs from the lowest electronically excited state, the high energy transition at 3 000 Å will not appear in emission but the emission system at 4 500 Å should appear in absorption. Eventually PORTER and STRACHEN ([14]), using the matrix stabilization technique with long pathlengths, observed the long wavelength transition in absorption.

Confirmation of the assignment of the emission and absorption systems around 4 500 Å to benzyl

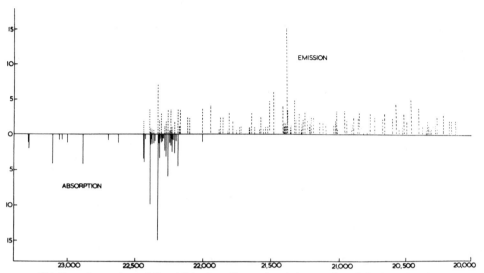

Fig. 2. — Comparison of the visible absorption and emission spectra of the benzyl radical.

has now been obtained following a study of toluene and some of its derivatives. The flash photolysis with an 8 m. pathlength of 14 mm. toluene vapour resulted in the appearance of a very sharp banded spectrum, with a maximum at 4 477 Å, which is shown in figure 1. Identical but weaker spectra were given by benzyl chloride, ethyl benzene and benzyl methyl ether. The wavelengths and wavenumbers of the bands are given in table III, which includes visual estimates of the band intensities. Part of the spectrum was found to coincide with bands in the emission system previously attributed to benzyl (see fig. 2) and therefore it is evident that the absorption and emission systems are due to the same species. Definite evidence that this species is benzyl was given by a study of some halogenated toluenes. Photolysis of these compounds gave spectra similar to that from toluene and the results are summarised in table II.

TABLE II

Parent	Radical	λ_{max} (Å)
$C_6H_5CH_3$	$C_6H_5CH_2$	4 477,2
$C_6H_5CH_2Cl$	$C_6H_5CH_2$	4 477,2
$C_6H_5CHCl_2$	C_6H_5CHCl	4 493,4
$C_6H_5CCl_3$	$C_6H_5CCl_2$	4 517,1
$p\text{-}ClC_6H_4CH_3$	$p\text{-}ClC_6H_4CH_3$	4 600,0

The only photochemical process which is consistent with these results is a β bond fission of the parent molecule :

$$C_6H_5CH_2X \quad \rightarrow \quad C_6H_5CH_2 + X$$

to give the benzyl radicals shown in the table. Since benzal chloride and p-chloro toluene give different radical spectra, the possibility of an isomerization to tropyl is again ruled out - in agreement with the conclusions drawn from our observations on the short wavelength transition.

Careful examination of the plates showed that in molecules containing more than one kind of β bond, the fission was confined solely to one particular bond. For example, the spectrum from benzyl chloride contained no bands due to α-chloro benzyl.

No other transient spectra were given by toluene between 2 800 and 6 800 Å, although several attempts were made to detect a forbidden doublet-quartet transition of benzyl, which is of some theoretical interest.

Attempted Analysis of the Spectrum.

The benzyl radical has been successfully treated as a 7 π electron system having C_{2v} symmetry by several workers. Following Mulliken's recommendations, molecular orbital theory predicts that the spectrum under discussion is due to a $^2A_2 — ^2B_1$ transition, whereas valence bond theory suggests that it is $^2B_1 — ^2B_1$. Since the vibronic interactions will be different in the two cases, a correct analysis of the spectrum would, in principle, allow the correct symmetry to be assigned to the lowest excited state.

The transition is symmetry allowed on group theoretical grounds but it is predicted to have a very small transition moment. One might expect, then, that a weak 0-0 band should appear in a region which is common to both emission and absorption. Analyses of the emission spectrum have been attempted by several workers who chose the following 0-0 bands :

SCHULER ([12]) . . . .   21,805 cm^{-1}
LEACH ([15]) . . . .   22,002 cm^{-1}
WALKER ([16, 17]) . . . .   22,324 cm^{-1}

If a mirror image symmetry is postulated for the absorption and emission spectra, then the 0-0 band should lie in the region 22,200-21,700 cm^{-1}

TABLE III

Absorption Spectrum of the Benzyl Radical.

Wavelength (Å) in Air	Vacuum Wavenumber (cm^{-1})	Intensity	Difference (cm^{-1})
4 543,83	22 001,7	1	0
4 506,90	22 182,0	4	180
4 503,90	22 196,7	0	195
4 502,16	22 205,3	3	204
4 499,54	22 218,2	0	217
4 498,15	22 225,1	2	223
4 497,28	22 229,4	1	228
4 496,34	22 234,1	2	232
4 494,94	22 241,0	1	239
4 492,33	22 253,9	3	252
4 492,01	22 255,5	6	254
4 489,36	22 268,6	3	267
4 487,29	22 278,9	2	277
4 485,45	22 288,0	0	286
4 483,80	22 296,3	1	295
4 482,51	22 302,7	0	301
4 482,31	22 303,7	2	302
4 480,62	22 315,1	0	313
4 479,35	22 318,4	3	317
4 478,52	22 322,5	1	321
4 477,25	22 328,9	10	327
4 472,67	22 351,8	0	350
4 470,09	22 364,6	1	363
4 468,45	22 372,8	1	371
4 467,22	22 379,0	0	377
4 466,30	22 383,6	1	382
4 465,06	22 389,9	6	388
4 457,98	22 425,4	0	424
4 455,96	22 435,6	4	434
4 455,04	22 440,2	4	439
4 419,56	22 620,4	1	619
4 405,13	22 694,4	0	693
4 368,73	22 883,5	4	882
4 346,90	22 998,5	1	997
4 339,78	23 036,2	0	1 035
4 335,95	23 056,5	0	1 055
4 326,47	23 107,0	4	1 105
4 292,99	23 287,2	1	1 286
4 291,97	23 292,8	2	1 291

Intensities estimated visually on a 1-10 scale.

(see fig. 2). This would seem to rule out 22,324 cm^{-1}, which is the strongest band in absorption. LEACH and his co-workers have been able to analyse almost completely the emission spectrum with 22,002 as their 0-0 band, whereas SCHULER was able to make only a partial analysis with 21,805 cm^{-1}. The fact that 22,002 cm^{-1} is the band of longest wavelength in the absorption spectrum supports the assignment of LEACH. Frequency differences are given in table III.

The spectrum is composed of very sharp line like bands and broader diffuse ones, but the physical significance of this is not clear. In the region of the strongest band, the structure is quite complicated and here some of the features may be rotational heads and not separate vibrational structure. The extent of the spectrum is rather small, which probably means that there is little change of shape on excitation. This was confirmed by a simple M.O. calculation based on bond orders which predicted only a slight lengthening of the ring along the C_2 axis.

One might expect the spectrum to analyse in terms of frequencies similar to those active in toluene, but this is not the case. There is very little regularity in the spectrum and no obvious progressions are apparent. A frequency of 180 cm^{-1} is the most common feature but it is not known whether this is a fundamental or not, since experiments with and without a diluent gas proved inconclusive. One of the most puzzling aspects is the absence of the very strong 327 cm^{-1} frequency in combination. The strong band at 882 cm^{-1} is probably the totally symmetric ring breathing frequency. In common with many other substituted benzenes, benzyl exhibits a difference frequency of 61 cm^{-1} which is usually attributed to a 1-1 transition of low frequency.

A consistent and complete analysis of the spectrum has not proved possible. One explanation of the complexity might be that the symmetries of the radical are different in the two states, in which case many more vibrations will become allowed. The spectrum may also be complicated by the presence of olefinic type vibrations of the CH_2 group which is partially conjugated to the ring.

Phenoxyl and Anilino Radicals.

The transitions of these radicals resemble those of benzyl since they have the same number of π electrons. Perturbation theory has some success in predicting the relative intensities of and energy differences between the long and short wavelength transitions of the members of the isoelectronic series.

The Phenoxyl Radical Φ 0.

The long wavelength transition of phenoxyl, previously observed by LAND and PORTER [18] in the gas phase, was photographed at high resolution in the hope that some fine structure might be visible, similar to that in benzyl. This was not realized and the spectrum consisted of two broad diffuse bands having maxima at 3 800 and 3 950 Å. It seems that this type of spectrum is characteristic of phenoxyl type radicals. Similar spectra were given by methyl substituted phenols, showing that phenoxyl and not benzyl type radicals were formed. Band maxima of some phenoxyl radicals in the gas phase are given in the following table IV.

TABLE IV

Parent	Radical	λ_{max} (Å)
C_6H_5OH	C_6H_5O	3 800, 3 950, 6 000
$C_6H_5OCH_3$	C_6H_5O	3 800, 3 950, 6 000
$C_6H_5OC_2H_5$	C_6H_5O	3 800, 3 950
$o\text{-}MeC_6H_4OH$	$o\text{-}MeC_6H_4O$	3 900
$p\text{-}MeC_6H_4OH$	$p\text{-}MeC_6H_4O$	4 000
2,6 $diMeC_6H_3OH$	2,6 $diMeC_6H_3O$	3 890

A transient diffuse spectrum in the region 5 300-6 100 Å was detected after the flash photolysis of phenol and anisole. It is not due to the phenyl or cyclopentadienyl radicals (see later) and it is best assigned to an n-π^* transition of phenoxyl. A similar transition of 2,4,6 tritertiary butyl phenoxyl is known in solution [18].

The Anilino Radical ΦNH.

Experiments with aniline showed that the long wavelength transition of anilino in the region of 4 000 Å was even more diffuse than that of phenoxyl and devoid of any spectroscopic interest. The lack of structure in the spectra of phenoxyl and anilino compared to benzyl is rather surprising. Anilines do, however, give sharp spectra which cannot be assigned to anilino and these will be discussed briefly later.

The Phenyl Radical Φ.

Many unsuccessful attempts have been made to detect the phenyl radical spectroscopically, although it is well established as an intermediate in organic chemistry. We have recently obtained a common spectrum from benzene and halogenated benzenes in the region 4 300-5 300 Å, which we attribute to the phenyl radical. The spectrum has been almost completely analysed in terms of frequencies which are very similar to the main frequencies in the benzene spectrum. The transition occurs from a $\pi^6 n$ ground state configuration to a $\pi^5 n^2$ exited state configuration.

The Cyclopentadienyl Radical C_5H_5.

A common spectrum between 2 900 and 3 500 Å is given on photolysis by phenol, phenolic ethers, aniline, nitrobenzene and cyclopentadiene. Similar spectra are obtained from halogen substituted derivatives of phenol, aniline and nitrobenzene. On the basis of the substituent effects and particularly that the same spectrum is obtained irrespective of the position of the halogen substituted, these spectra are assigned to the cyclopentadienyl radical and its halogen substituted derivatives. The spectra are very extensive and well resolved, and the two strongest bands are identical with those reported by THRUSH [19] upon photolysis of cyclopentadiene and ferrocene.

Details of the spectra of phenyl, cyclopentadienyl and its derivatives, and of other transient species derived from aromatic molecules, will be published shortly.

BIBLIOGRAPHY

(1) M. GOMBERG. — J.A.C.S., 1900, 12, 757.
(2) G. PORTER. — Proc. Roy. Soc., 1950, A 200, 284.
(3) G. N. LEWIS and D. LIPKIN. — J.A.C.S., 1942, 64, 2801.
(4) I. NORMAN and G. PORTER. — Proc. Roy. Soc., 1955, A 230, 399.
(5) G. PORTER and F. J. WRIGHT. — Trans. Faraday Soc., 1955, 51, 395.
(6) G. PORTER and E. E. STRACHEN. — Trans. Faraday Soc., 1958, 54, 431.
(7) H. T. J. CHILTON and G. PORTER. — J. phys. Chem., 1959, 63, 904.
(8) E. J. LAND and G. PORTER. — Trans. Faraday Soc., 1963, 59, 2016, 2027.
(9) N. K. BRIDGE and G. PORTER. — Proc. Roy. Soc., 1958, A 244, 259, 276.
(10) A. BECKETT and G. PORTER. — Trans. Faraday Soc., 1963, 59, 2 038, 2051.
(11) J. U. WHITE. — J. Opt. Soc. Amer., 1942, 32, 285.
(12) H. SCHULER. — Z. Naturf., 1952, 79, 421.
    H. SCHULER and J. KUSJAKOW. — Spectrochim. Acta, 1961, 17, 356.
(13) H. C. LONGUET-HIGGINS and J. A. POPLE. — Proc. Phys. Soc., 1955, 68, 591.
(14) G. PORTER and E. E. STRACHEN. — Spectrochim. Acta, 1958, 12, 299.
(15) S. LEACH, L. GRAJCAR and J. ROBERT. — To be published.
(16) T. F. BINDLEY and S. WALKER. — Trans. Faraday Soc., 1962, 58, 217.
(17) A. T. WATTS and S. WALKER. — J. chem. Soc., 1962, p. 4323.
(18) E. J. LAND, G. PORTER and E. E. STRACHEN. — Trans. Faraday Soc., 1963, 57, 1885.
(19) B. A. THRUSH. — Nature, 1955, 178, 155.

DISCUSSION

G. Giacometti. — Pr. PORTER's evidence of the existence of benzyne in his flash photolized systems is very beautiful indeed and his tentative suggestion of a para-structure as

gains support the organic work by VAN TAMELEN (JACS, 1963) who prepared Dewar's benzene (bi-cyclo-butene). A not planar structure for para-benzyne seems to be at least much reasonable.

G. Porter. — The « Dewar » form of benzene is, of course, a quite different substance from para-phenylene or para-benzyne which contains only four hydrogen atoms. The stability of Dewar benzene is not surprising since the molecule is surely non-planar and is not a resonance structure of benzene, indeed the name « Dewar benzene » is not correct for the stable molecule which has recently been prepared. Para-benzyne, if it exists, is likely to be planar or very nearly so.

C. A. McDowell. — Since VAN TAMELEN and his colleagues have shown by N.M.R. and other physical evidence that they have undoubtly synthesised the Dewar form of benzene this surely removed any conceptional difficulty about accepting a structure such as is proposed for p-phenylene, i.e. :

In fact on theoretical grounds one could argue that there may be better reasons for the existence of p-phenylene than the Dewar form of the benzene molecule.

S. Leach. — Phenyl must have C_{2v} symmetry, formally at least, and so I would expect a 0,0 band to appear, perhaps very weakly, this band being strictly forbidden for the 2 600 Å benzene transition. Furthermore, the e_g^+ vibrations in benzene should have their degenerescence lifted in C_{2v} symmetry and should each give rise to two vibrations of a_1 and b_1 symmetries for C_{2v}. The doublet structure which you mentioned as observing in phenyl might be due to this splitting of the degeneracy. However one should expect this behaviour only for bands derived from e_g^+ type vibrations but not totally symmetrical progressions.

G. Porter. — I think it likely that the doublet structure which is present in the bands of the phenyl spectrum is rotational fine structure since it is similar in all the bands including that at the origin.

S. Leach. — I would like to ask Pr. PORTER whether he observed any phenyl transitions to shorter wavelengths, in particular, in the benzene region itself?

In photolysis experiments on benzene in an argon matrix at 20 °K, carried out at Berkeley in 1957, I observed the formation of two unstable species. One gives a series of broad bands in the 3 300-2 700 Å region, which I believe may be due to the hexatrienyl diradical. The other gives a series of well resolved bands in the 2 600-2 200 Å region, distinct from the benzene bands in the same region, which might possibly be due to the phenyl radical. Both spectra disappear on warming to about 70 °K.

G. Porter. — We observed no higher energy transitions of phenyl but the 2 600-2 200 Å region is difficult to study. We have never observed the 5 000 Å system of phenyl in our low temperature stabilisation work.

Reprinted from the Proceedings of the Royal Society, A, *volume* 287, pp. 457–470, 1965

The electronic spectra of phenyl radicals

By G. Porter, F.R.S. and B. Ward

Department of Chemistry, University of Sheffield

(*Received* 22 *January* 1965)

[Plate 4]

An electronic absorption spectrum, attributed to phenyl, has been observed in the visible region with origin at 18 908 cm^{-1} after flash photolysis of benzene and halogenobenzenes. Similar spectra of fluoro, chloro and bromo phenyl are observed after flash photolysis of disubstituted benzenes. The vibrational structure of the phenyl spectrum has been analysed in terms of two fundamental frequencies at 571 and 896 cm^{-1} which correspond to the e_{2g} and a_{1g} frequencies of the B_{2u} state of benzene.

The ground state of phenyl has a $\pi^6 n$ electronic configuration and the observed transition is interpreted as $^2A_1 \rightarrow {}^2B_1$ resulting from a $\pi \rightarrow n$ excitation.

Phenyl, the simplest of all aromatic radicals, has hitherto evaded spectroscopic detection although it has been well characterized by its chemical reactions. Evidence for its existence in the gas phase was obtained by Dull & Simons (1933) who pyrolysed lead tetraphenyl in the presence of mercury vapour and found diphenyl and diphenyl mercury among the products. Glazebrook & Pearson (1939) studied the radicals produced on photolysis of acetophenone and benzophenone in the gas phase by the metal mirror technique and from an analysis of the products showed that free phenyl was probably formed. Later work (Duncan & Trotman-Dickenson 1962) on the photolysis of acetophenone established that the main products of reaction were carbon monoxide, benzene, diphenyl, toluene, methane and ethane. Kinetic studies of the same system indicated that the reactivities of phenyl radicals with different substrates were parallel to those of methyl radicals. Phenyl is formed on pyrolysis of iodobenzene, bromobenzene and diphenyl mercury at 600 °C (Jaquis & Szwarc 1952) and also in the mercury photosensitized decomposition of benzaldehyde (Harrison & Lossing 1959).

There is a large amount of evidence for the participation of phenyl radicals in solution reactions which is summarized by Waters (1946) and described in some detail in a recent monograph by Williams (1960). In aromatic solvents, arylation of the substrate is the main reaction but in compounds such as toluene and in aliphatic solvents, hydrogen abstraction takes place. Dimerization occurs only to a small extent. Generally, the decomposition of a suitable source of phenyl radicals in an aromatic solvent results in the formation of a mixture of *ortho*, *meta* and *para* substituted diphenyls, irrespective of any substituent group in the substrate. This is indicative of substitution by an electrically neutral radical and not by charged species. Substituted phenyl radicals can be formed by the decomposition of appropriately substituted parent molecules. In these cases, the distribution of isomers among the reaction products is determined by the relative charge distributions in both radical and substrate.

G. Porter and B. Ward

Phenyl radicals are produced in solution on photolysis of such organo-metallic compounds as triphenyl bismuth, diphenyl mercury and tetraphenyl lead. McDonald-Blair & Bryce-Smith (1960) have shown that liquid iodobenzene, when photolysed in the presence of silver powder, forms phenyl radicals—as evidenced by the appearance of biphenyl, benzene and the three isomeric iododiphenyls among the reaction products.

In 1949, Schuler & Reinbeck decomposed several benzene derivatives by electron impact and observed an emission spectrum which they originally attributed to phenyl. Later work, and particularly our recent observation of this spectrum at high resolution in absorption (Porter & Ward 1964a) has shown that the carrier is the benzyl radical. It is interesting to note, especially in the light of the results to be described shortly, that Schuler did not observe emission from any decomposition product of the halogenated benzenes.

Most of the free radical spectra which are known have been obtained by the method of flash photolysis (Porter 1950) and we have recorded the absorption spectra of about one hundred aromatic free radicals in this way. The simplest of all aromatic radicals, phenyl, has, however, been conspicuously absent from this list. Most of our work has been carried out in solution, where phenyl radicals would be too reactive for detection after several microseconds, or in the gas phase at low resolution. The importance of high resolution as a means of increasing sensitivity has been demonstrated by the success of Herzberg and his collaborators in the application of flash photolysis to the detection of smaller radicals in the vapour phase. There was good reason to suppose, from what was already known of the spectra of aromatic radicals that, unlike most large aliphatic radicals, many of them would have systems of narrow bands and fine structure which would only be resolvable at the highest dispersion.

Using path lengths of 8 m and a 21 ft. grating spectrograph we have studied the transient products of flash photolysis of sixty-eight aromatic molecules in the vapour phase. Many new spectra, not observable at lower resolution, have been identified (Porter & Ward 1964a, b) and, of these, the spectra of phenyl and its derivatives are, perhaps, the most important.

Experimental

The Sheffield spectrograph has an Eagle mounting and was constructed in our workshops to designs of the Ottawa instrument, kindly lent to us by Dr A. Douglas. The plateholder is 4 ft. in length and can be raised and lowered remotely. Owing to local vibration difficulties a special mounting was devised which consists of a 20-ton block of concrete hanging on four springs. This technique, which is in some cases less expensive than further excavation and sealing of a building, has been entirely satisfactory and effective in damping out the low frequency disturbances which are predominant in this location. The instrument is fitted with a Bausch and Lomb concave grating, of 21 ft. radius of curvature having 105 000 lines and blazed for 6000 Å in the first order. Part of the grating was masked to eliminate satellite lines. The whole spectrograph is housed in a thermally insulated room.

Porter & Ward   *Proc. Roy. Soc. A, volume 287, plate 4*

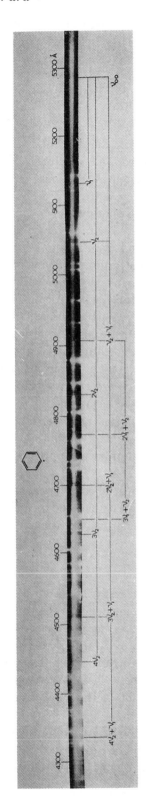

Figure 1. Absorption spectrum of the phenyl radical and the assignment of its principal bands. $\nu_1 = 571\ \mathrm{cm}^{-1}$; $\nu_2 = 896\ \mathrm{cm}^{-1}$.

Wavelength calibration was carried out by superposition of light from a neon-filled, hollow cathode iron arc. Spectra were enlarged photographically and wavelengths determined by reference to the M.I.T. wavelength tables. Final measurements were made by means of a photoelectric measuring comparator. For the region 2800 to 4600 Å Kodak I.O. plates were used and, for the region 4000 to 6500 Å, either Kodak I.F. or Ilford H.P. 3 plates.

The flash photolysis apparatus employed two flash lamps, 50 cm in length, through which 4000 J was discharged giving a half peak duration of 25 μs. The quartz reaction vessel was 5 cm in diameter and incorporated a multiple reflexion system of a design similar to that described by Bernstein & Herzberg (1948). Initially, front surface mirrors were used, because of their better optics, but owing to the problem of polymer deposition from products of photolysis it became necessary to change to quartz meniscus mirrors, back surfaced with silver or aluminium. The surfaces were easily cleaned with a hydrocarbon solvent.

A source flash of 150 J and 10 to 15 μs duration, argon filled, recorded the absorption spectrum of transient products after a delay which was usually adjusted to be between 0 and 200 μs. Sixteen traversals of the tube and a slit width of 100 μm were used in most experiments which required 80 to 100 separate flashes to give adequate plate density. The reaction vessel was filled with fresh gas mixture between each flash and, when necessary, the mirrors were cleaned during the experiment.

To avoid heating effects after the flash the aromatic vapour was mixed with an excess of argon at 400 mmHg pressure. The usual procedure was to flow the argon over the solid or liquid aromatic compound at a rate slow enough to ensure saturation. The reaction vessel was evacuated after each flash and fresh mixture admitted. In order to prevent condensation on the mirrors and to obtain a higher vapour pressure of the less volatile compounds, the laboratory was operated at a temperature of 30 to 35 °C.

Aromatic materials were obtained from a variety of commercial sources. Most of them were recrystallized or redistilled and their purity checked by melting or boiling point and, in important cases, by vapour phase chromatography.

Results

The flash photolysis of benzene, chlorobenzene, bromobenzene and iodobenzene, in the presence of excess argon, gave rise to an extensive transient banded spectrum in the region 4400 to 5300 Å. There was exact coincidence in the wavelengths of all the bands from the four molecules and no other spectra were observed except in the case of iodobenzene which gave the spectrum of I_2 and, when oxygen was not thoroughly excluded, showed the transient spectrum of IO. The spectrum from benzene was the weakest and the spectra from the halobenzenes were of similar intensity, the greater proportional decomposition of the iodobenzene being compensated for by its lower vapour pressure.

This spectrum, which we shall later attribute to the phenyl radical, is shown in figure 1, plate 4 and measurements of all the bands are given in table 2. The lifetime of the radical was about 50 μs. The spectrum was not observed from any of the other substances investigated.

460   G. Porter and B. Ward

The products from flash photolysis of benzene and the three halogenobenzenes were collected in a cold trap and analysed by Dr Tsang in this laboratory. The principal product detectable by vapour phase chromatography was, in each case, diphenyl.

Flash photolysis of a number of dihalogenated benzenes gave similar, but not identical, spectra in the region 5000 to 6000 Å. The results are summarized in table 1 where the positions of the origins of the band systems, identified as described later, are given. Measurements of all these spectra are given in tables 3 to 7.

TABLE 1. SUMMARY OF RESULTS ON PHENYL RADICALS

parent	radical	0–0 band (cm^{-1})
benzene	phenyl	19 908
chlorobenzene	phenyl	19 908
bromobenzene	phenyl	19 908
iodobenzene	phenyl	19 908
o. fluorobromobenzene	o. fluorophenyl	16 060
m. fluorobromobenzene	m. fluorophenyl	18 375
p. fluorobromobenzene	p. fluorophenyl	17 436
p. fluorochlorobenzene	p. fluorophenyl	17 436
p. bromochlorobenzene	p. chlorophenyl	18 425
p. dichlorobenzene	p. chlorophenyl	18 425
m. bromoiodobenzene	m. bromophenyl	18 534
p. dibromobenzene	p. bromophenyl	18 551

A careful search for phenyl-type radical spectra was made, with negative results, from the following compounds: cyclopentadiene, phenol and *para*-chlorophenol, anisole, toluene and chlorinated toluenes, *ortho*- and *meta*-dichlorobenzenes, *para*-bromoiodobenzene and *para*-diiodobenzene.

DISCUSSION

Assignment of the spectra

The common spectrum from benzene and its halogenated derivatives indicates that the carrier must contain no more than six carbon and five hydrogen atoms and no halogen. Cyclopentadienyl can be ruled out because of the negative result from cyclopentadiene which gave the spectrum of cyclopentadienyl strongly at lower wavelengths. In view of our negative results from many other aromatic molecules and the analytical evidence the only possible assignment of these spectra seems to be to the phenyl radical and its halogenated derivatives, formed by photolytic dissociation of a carbon–halogen bond or carbon–hydrogen bond in the case of benzene:

$$C_6H_5X \xrightarrow{h\nu} C_6H_5 + X,$$

$$(X = Cl, Br, I, H).$$

Since diphenyl is the major product, the radicals must disappear mainly by dimerization and not by arylation of the parent molecules.

The assignment of the substituted radical spectra requires a little more discussion since there are no *a priori* grounds, except bond energy considerations, for predicting

TABLE 2. ABSORPTION SPECTRUM OF PHENYL

λ_{air} (Å)	vacuum wavenumber (cm^{-1})	intensity	difference (cm^{-1})	assignment
5291·01	18894·7	4	13·3	—
5290·18	18897·7	6	10·3	—
5287·30	18908·0	6	0	0–0
5190·05	19262·3	1	354·3	—
5168·30	19343·4	1	435·4	—
5154·03	19396·9	1	488·9	—
5134·74	19469·8	4	561·8	$\nu_1 - 10$
5132·21	19479·4	4	571·4	ν_1
5092·90	19629·7	2·5	721·7	ν_3
5079·81	19680·3	1·5	772·3	—
5069·03	19722·1	1	814·1	—
5051·30	19791·4	10	883·4	$\nu_2 - 13$
5048·00	19804·3	10	896·3	ν_2
4910·35	20359·5	6	1451·5	$2\nu_3$
4909·19	20364·3	10	1456·3	$\nu_1 + \nu_2 - 9$
4906·95	20373·6	10	1465·6	$\nu_1 + \nu_2$
4871·16	20523·3	1·5	1615·3	$\nu_2 + \nu_3$
4859·13	20574·1	2	1666·1	$772 + \nu_2$
4856·98	20583·2	1	1675·2	—
4853·53	20597·8	1	1689·8	—
4834·67	20678·2	1	1770·2	—
4832·58	20687·1	10	1779·1	$2\nu_2 - 10$
4830·93	20694·2	5	1786·2	—
4830·16	20697·5	10	1789·5	$2\nu_2$
4809·46	20786·6	1	1878·6	—
4777·37	20926·2	1	2018·2	$\nu_1 + 2\nu_3$
4773·93	20941·3	6	2033·3	—
4773·38	20943·7	6	2035·7	$2\nu_1 + \nu_2$
4713·21	21211·1	1	2303·1	—
4704·03	21252·4	1·5	2344·4	$\nu_2 + 2\nu_3$
4703·72	21253·8	8	2345·8	$\nu_1 + 2\nu_2 - 12$
4701·14	21265·5	10	2357·5	$\nu_1 + 2\nu_2$
4681·37	21355·3	3	2447·3	—
4670·18	21406·5	1	2498·5	—
4647·15	21512·8	3	2604·8	$3\nu_1 + \nu_2$
4636·09	21563·9	1	2655·9	—
4631·07	21587·2	9	2679·2	$3\nu_2$
4638·58	21552·3	1	2644·3	—
4614·52	21664·7	2·5	2756·7	$2\nu_1 + \nu_2 + \nu_3$
4560·68	21920·4	1	3012·4	—
4514·54	22144·4	7	3236·4	$\nu_1 + 3\nu_2 - 9$
4512·67	22153·6	7	3245·6	$\nu_1 + 3\nu_2$
4496·70	22232·3	1	3324·3	$3\nu_1 + \nu_2 + \nu_3$
4494·27	22244·3	1	3336·3	—
4450·65	22462·3	5	3554·3	$4\nu_2 - 11$
4448·51	22473·1	5	3565·1	$4\nu_2$
4339·24	23038·9	1	4130·9	—
4339·15	23039·4	1	4131·4	$\nu_1 + 4\nu_2$
4323·58	23122·4	2	4214·4	$3\nu_1 + 2\nu_2 + \nu_3$

TABLE 3. ABSORPTION SPECTRUM OF O. FLUOROPHENYL

band	λ_{air} (Å)	vacuum wavenumber (cm^{-1})	intensity	assignment
1	6224·90	16060·1	1	0–0
2	6214·18	16087·8	1	—
3	6197·39	16131·4	6	—
4	6195·04	16137·5	3	—
5	6187·54	16157·1	4	—
6	6167·86	16208·6	9	—
7	6147·32	16262·7	1	—
8	6038·08	16557·0	3	—
9	6036·11	16562·4	3	1. +502
10	6028·50	16583·3	4	2. +497
11	6010·15	16633·9	9	3. +503
12	6000·65	16660·2	7	5. +503
13	5951·62	16797·5	1	—
14	5886·47	16983·4	1	—
15	5885·23	16987·0	1	—
16	5877·12	17010·4	1	—
17	5875·42	17015·3	1	—
18	5869·21	17033·4	1	—
19	5860·07	17059·9	9	9. +498
20	5850·82	17086·9	9	10. +504
21	5834·42	17134·9	10	11. +501
22	5817·46	17184·9	2	—
23	5815·60	17190·4	2	—
24	5717·57	17485·1	5	14. +502
25	5709·08	17511·1	6	16. +499
26	5692·93	17560·8	6	19. +499
27	5683·91	17588·6	4	20. +502
28	5603·62	17840·7	1	—
29	5581·08	17912·7	2	—
30	5558·11	17986·7	7	24. +502
31	5550·22	18012·3	5	25. +501
32	5534·49	18063·5	5	26. +503
33	5473·15	18267·2	1	—
34	5465·26	18292·3	1	—
35	5451·77	18337·6	1	28. +497
36	5443·78	18364·5	1	—
37	5429·04	18414·4	2	29. +502
38	5422·41	18436·9	2	—
39	5408·29	18485·0	3	30. +498
40	5400·89	18510·3	1	36. +498
41	5392·80	18538·1	2	—

TABLE 4. ABSORPTION SPECTRUM OF M. FLUOROPHENYL

λ_{air}(Å)	vacuum wavenumber (cm^{-1})	intensity	difference (cm^{-1})
5440·58	18375·3	5	0
5219·49	19153·6	10	778·3
5220·07	19151·5	8	776·2
5186·51	19275·4	3	900·1
5183·38	19287·1	3	912·0
4996·55	20013·8	1	1648·5
4980·65	20072·1	1	1696·8
4979·97	20075·6	1	1700·3
4827·46	20709·0	3	2333·7
4797·73	20837·3	1	2462·0
4700·89	21266·6	2	2891·3

TABLE 5. ABSORPTION SPECTRUM OF *P.* FLUOROPHENYL

λ_{air} (Å)	vacuum wavenumber (cm^{-1})	intensity	difference (cm^{-1})	assignment
5749·00	17389·5	2	−46·4	—
5746·94	17395·8	2	−40·1	—
5735·11	17431·6	4	−4·3	—
5733·70	17435·9	4	0	0–0
5572·62	17939·9	3	504	ν_1
5538·37	18050·8	1	615	ν_6
5515·34	18126·2	2	690	ν_3-46
5500·70	18174·5	6	739	ν_3
5488·77	18213·6	2	778	ν_5
5486·36	18222·0	2	786	ν_8
5484·12	18229·4	2	794	ν_4-46
5469·60	18277·8	2	842⎫	
5469·05	18279·6	2	844⎭	ν_4
5453·81	18330·7	7	895⎫	
5450·77	18341·0	7	905⎭	ν_2
5309·09	18830·4	3	1395	$\nu_5+\nu_6$
5307·62	18835·6	4	1400	$\nu_1+\nu_2$
5276·30	18947·4	1	1512	$\nu_2+\nu_6$
5272·76	18960·1	1	1524	$\nu_3+\nu_8$
5259·65	19007·4	1	1572	$2\nu_8$
5242·80	19068·5	6	1633	$\nu_2+\nu_3$
5232·47	19106·1	5	1670	$\nu_2+\nu_5$
5229·29	19117·7	7	1682⎫	$2\nu_4$ and
5226·57	19127·7	7	1693⎭	$\nu_2+\nu_8$
5215·48	19168·4	1	1733	$\nu_2+\nu_4$
5199·66	19226·7	10	1791⎫	$2\nu_2$
5197·22	19235·7	10	1800⎭	
5173·16	19325·2	2	1889	—
5132·34	19478·9	0	2043	—
5126·00	19503·1	0	2067	—
5111·06	19559·9	3	2124	—
5070·61	19716·0	2	2280	—
5067·00	19730·2	4	2294	$\nu_1+2\nu_2$
5035·42	19853·8	5	2418	$2\nu_2+\nu_6$
5031·48	19869·3	2	2433	—
5028·33	19881·8	3	2446	—
5008·07	19962·2	5	2526	$2\nu_2+\nu_3$
4996·40	20008·8	5	2573	$2\nu_2+\nu_5$
4993·98	20018·5	6	2583	$\nu_2+2\nu_4$
4969·75	20116·1	6	2680⎫	$3\nu_2$
4966·95	20127·5	6	2692⎭	
4894·90	20423·7	1	2988	—
4888·64	20449·9	1	3014	—
4819·21	20744·5	1·5	3309	$3\nu_2+\nu_6$
4793·69	20854·9	2	3419	$3\nu_2+\nu_3$
4784·49	20895·0	1	3459	$3\nu_2+\nu_5$
4756·43	21018·3	3	3582	$4\nu_2$
4746·37	21062·8	3	3627	—
4410·33	22674·0	1	5238	—
4363·70	22916·3	1	5480	—

G. Porter and B. Ward

TABLE 6. ABSORPTION SPECTRUM OF P. CHLOROPHENYL

λ_{air} (Å)	vacuum wavenumber (cm^{-1})	intensity	difference (cm^{-1})	assignment
5426·51	18423·0	7	0	0–0
5425·56	18426·2	7	3·2	—
5375·37	18598·2	8	175·2⎫	ν_1
5374·76	18600·3	10	177·3⎭	
5244·05	19063·9	1	640·9	—
5232·13	19107·3	2	684·3	—
5175·97	19314·7	5	891·7⎫	ν_2
5172·21	19328·7	7	905·7⎭	
5138·20	19456·7	1	1033·7	—
5128·03	19495·2	3	1072·2	—
5124·53	19508·5	3	1085·5	$\nu_1 + \nu_2$
4943·37	20223·5	3	1800·5	$2\nu_2$
4903·4	20388·3	2	1965·3	$\nu_1 + 2\nu_2$
4788·6	20877·1	1	2454·1	—

TABLE 7

ABSORPTION SPECTRUM OF M. BROMOPHENYL

λ_{air} (Å)	vacuum wavenumber (cm^{-1})	intensity
5395·50	18534·0	1
5049·56	19797·9	6
5047·02	19807·9	6

ABSORTION SPECTRUM OF P. BROMOPHENYL

λ_{air} (Å)	vacuum wavenumber (cm^{-1})
5389·01	18551·1

(Only one very weak band was visible in this spectrum.)

which carbon–halogen bond will cleave on excitation of a molecule such as *para*-bromochlorobenzene. For energetic reasons one might expect the weakest bond to break and this is what happens, very specifically in practice. (The dissociation energies of (Trotman-Dickenson 1955) benzene and its derivatives are (kcal/mole): C—H: 102, C—Cl: 86, C—Br: 71 and C—I: 57.) For example, *para*-fluorochloro and *para*-fluorobromobenzenes give the same spectrum and so the fluorine atom must be retained and the chlorine and bromine atoms eliminated on photolysis. A similar argument applies to the common spectra from *para*-bromochlorobenzene and *para*-dichlorobenzene. Once the principle that the weakest bond breaks has been established by experiment, the other spectra can be assigned to the appropriate radical. The bond breakage is confined to the weakest bond as evidenced by the fact that the spectra from *para*-dichlorobenzene and *para*-fluorochlorobenzene contain no common bands.

It is noteworthy that the *ortho, meta* and *para* disubstituted isomers yield characteristically different mono substituted phenyl radicals having no bands in common, showing that isomerization by hydrogen atom migration does not occur in phenyl and its derivatives, not even in the hot radical which must be present immediately after photodissociation. This result is in agreement with all previous work on phenyl radicals in solution including experiments with ^{14}C labelled phenyl radicals (Razuvajev & Zatajev 1963, 1964) which were designed specifically to detect any isomerization.

Analyses of the band structure
Phenyl

The spectrum of phenyl consists of a system of sharp, double-headed bands with a characteristic splitting of about $10 \, \text{cm}^{-1}$. The principal features of the spectrum are readily interpreted in terms of two vibration frequencies $571 \, \text{cm}^{-1}$ (ν_1) and $896 \, \text{cm}^{-1}$ (ν_2) which, because of the structure of the spectrum, probably correspond to the e_{2g} and a_{1g} frequencies of the B_{2u} state of benzene with frequencies 521 and $923 \, \text{cm}^{-1}$, respectively. The general vibrational structure of the spectrum resembles that of a mono substituted benzene and consequently it is possible to fix the 0–0 band unambiguously at $19\,908 \, \text{cm}^{-1}$. This strong band serves as the origin of a four-membered progression in the $923 \, \text{cm}^{-1}$ vibration which is probably the totally symmetric ring breathing frequency. The band at $19\,479 \, \text{cm}^{-1}$, distant $571 \, \text{cm}^{-1}$ from the origin, does not appear alone in a progression but serves as the origin of another progression in the 923 frequency with members $\nu_1 + n\nu_2$ ($n = 0$ to 4). In this respect, the 571 fundamental resembles the e_{2g} vibration in benzene. Another progression involving ν_1 and ν_2 is $\nu_2 + n\nu_1$ ($n = 0$ to 3) and the same vibrations also appear in combination with a frequency of $722 \, \text{cm}^{-1}$ (ν_3). The main progressions in the spectrum are indicated in figure 1, plate 4, and full details of the analysis are given in table 2.

Substituted phenyl spectra

Para-fluorophenyl has the same symmetry (C_{2v}) as the unsubstituted radical and its spectrum, some of whose bands exhibit the $10 \, \text{cm}^{-1}$ splitting, can be interpreted in a similar manner with combinations of frequencies 504, 895 and 739 forming the main progressions. The 0–0 band can be fixed unequivocably at $17\,436 \, \text{cm}^{-1}$.

Ortho- and *meta*-fluorophenyl have lower symmetry (C_s) than phenyl and hence have more totally symmetric vibrations. As a result their spectra are quite complex. The spectrum of *meta*-fluorophenyl does not show any regularities and no progressions are obvious. The position of the 0–0 band is uncertain but is assumed to be the strong band of longest wavelength at $18\,375 \, \text{cm}^{-1}$. The position is a little clearer in the *ortho*-fluorophenyl spectrum which consists of six groups of bands at intervals of $501 \, \text{cm}^{-1}$. The 0–0 band is almost certainly a member of the first group of bands and so is assigned to the band of longest wavelength in this group at 16 060.

The outstanding features of the spectrum of *para*-chlorophenyl are two short progressions in a $900 \, \text{cm}^{-1}$ frequency which have their origins $170 \, \text{cm}^{-1}$ apart. In this case the band of longest wavelength at $18\,423 \, \text{cm}^{-1}$ has been chosen as the origin although the significance of the $170 \, \text{cm}^{-1}$ difference remains uncertain.

G. Porter and B. Ward

The bromophenyl spectra have very weak intensities because of the low vapour pressures of the parent molecules and few bands are visible. Consequently, assignments of the 0–0 bands are doubtful.

The nature of the electronic transition

To a first approximation, the π orbitals of phenyl are almost identical with those of benzene. Depending upon the relative energies of the π and the non-bonding sp^2 orbitals, the ground state electronic configuration of the radical is either $\pi^6 n$ or $\pi^5 n^2$. These two possibilities can be represented as:

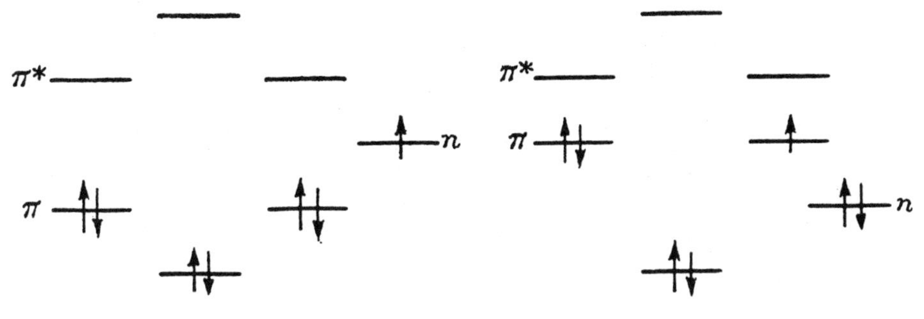

(a) (b)

The excitation of an electron in case (a) could give rise to the following types of transition:

$$\pi \to \pi^*,$$
$$\pi \to n,$$

and

$$n \to \pi^*,$$

while from case (b), the transitions

$$\pi \to \pi^*,$$
$$\pi \to \pi,$$
$$n \to \pi,$$

and

$$n \to \pi^*$$

could arise. Transitions involving bonding σ electrons may be ignored since they will be of too high energy. From the nature and magnitude of the shifts of the origin of the phenyl spectrum caused by halogen substitution, it should be possible to decide what transition is responsible for the radical absorption; table 8 presents the available data on the probable 0–0 bands and frequency shifts of phenyl radical spectra and it is clear that the magnitudes of the shifts vary in the order: F > Cl > Br. This is the opposite to that found in the $\pi \to \pi^*$ transitions of the halogenated benzenes and also bears no resemblance to the pattern of substituent effects in the cyclopentadienyl radical spectra which, theory (Longuet-Higgins & McEwen 1957) indicates, are caused by a $\pi - \pi$ transition. For these reasons, the transition in phenyl is, in all probability, not due to a $\pi \to \pi^*$ or a $\pi \to \pi$ combination.

The 0–0 bands of the $n \to \pi^*$ spectra of some halogenated pyridines have been estimated from diagrams reproduced in an article by Stephenson (1954) and are given in table 9. In the 2-substituted compounds, the $n \to \pi^*$ band is shifted so much as to be hidden under the $\pi - \pi^*$ spectrum and so the long wavelength tail of the $\pi - \pi^*$ band has been given as a lower limit for the origin of the $n \to \pi^*$ transition. 4-substituted pyridines are unstable and their spectra are unknown.

TABLE 8. PHENYL RADICAL SPECTRA

radical	0–0 band (cm.$^{-1}$)	shift (cm^{-1}) to red
phenyl	19 908	0
o. fluorophenyl	16 060	3848
m. fluorophenyl	18 375	1533
p. fluorophenyl	17 436	2472
p. chlorophenyl	18 423	1483
m. bromophenyl	18 534	1374
p. bromophenyl	18 551	1357

TABLE 9. $n - \pi^*$ SPECTRA OF SOME HALOGENATED PYRIDINES

compound	0–0 band (cm^{-1})	shift (cm^{-1}) to blue
pyridine	34 769	0
2. fluoropyridine	> 36 500	> 1730
3. fluoropyridine	35 100	330
2. chloropyridine	> 35 200	> 430
3. chloropyridine	34 800	30
2. bromopyridine	> 35 000	> 230
3. bromopyridine	34 600	− 170

Table 9 shows that the $n \to \pi^*$ spectrum of pyridine is shifted to the blue by halogen substitution and since the phenyl spectra exhibit a red shift, one can rule out $n \to \pi^*$ and, most probably, $n \to \pi$ transitions for phenyl. That the phenyl and pyridine spectra involve similar types of electronic states is indicated by the fact that the general nature of the shifts caused by halogen substitution is rather similar in the two cases but of opposite sign. The assignment of the phenyl radical absorption to a $\pi \to n$ transition is, therefore, in accordance with these shifts and supported by a consideration of the electronic properties of halogen atoms.

The effects of halogen substitution on the spectra of aromatic molecules show that, although they attract σ electrons, halogen atoms repel π electrons, by the inductive effect, in the order

$$F > I > Cl \sim Br.$$

The mesomeric effect of halogens, which is essentially the ability to donate π electrons into the aromatic nucleus, is inversely proportional to the ionization potential of the substituent and varies through the series:

$$I > Br > Cl > F.$$

The delocalization of electrons by a positive mesomeric effect causes a change in the electron distribution similar to that caused by a $+I\pi$ effect. However, since the shifts of the phenyl spectra decrease through the series $F > Cl > Br$, the main perturbations of the electronic states of phenyl by halogen atoms would seem to arise from the inductive effect. This is in agreement with Murrell (1963) who has suggested that when there is a change in electron density during a transition, the inductive effect is more important than the mesomeric.

Through the inductive effect, halogens deplete the position of substitution of π electrons and increase the π electron density at other carbon atoms in the ring, especially at positions *ortho* and *para* to the halogen. Thus a transition which involves moving a π electron from the ring into a non-bonding orbital will require less energy when the ring is substituted by a halogen atom. Therefore, the effect of halogen substitution on a $\pi \to n$ transition in an aromatic molecule is to cause a red shift whose magnitude depends upon the inductive interaction of the halogen with the π electrons of the ring, i.e. it varies in the order

$$F > Cl > Br.$$

The inductive effect leads to a build up of charge preferentially at positions *ortho* and *para* to the halogen and so one might expect the shifts caused by substitution at these positions, relative to the carbon atom with the non-bonding sp^2 electron, to be greater than the shift caused by *meta* substitution.

It is encouraging, in view of the conclusions reached previously, to find that the observed effects of different halogens and *ortho*, *meta* and *para* substitution are almost all in agreement with the assignment of the phenyl spectrum to a $\pi \to n$ transition arising from a $\pi^6 n$ radical ground state. However, one or two points do need a little elaboration.

The shift caused by *ortho* substitution in fluorine is much greater than that caused by *para* substitution and so, as well as the inductive interactions on the π electrons, some other effect must be operative. This may be due to a stabilization of the non-bonding orbital resulting from the inductive attraction of σ electrons by the fluorine atom. The effect will be transmitted through the bonding σ electrons and will consequently be greatest at a position *ortho* to the non-bonding orbital.

The relative positions of the *meta* and *para* bromophenyl spectra are not in accord with the assignment outlined above but this is not serious since the spectra were very weak and their origins are not known with any great degree of certainty.

Having assigned the spectrum to a $\pi \to n$ transition, one can enquire from which π orbital is the electron excited. The filled orbitals of highest energy have the following forms and symmetries:

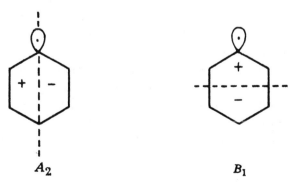

A_2 B_1

Although they are degenerate in benzene, the B_1 oribital in phenyl probably has lower energy than the A_2. The C—H bond in benzene is thought to be slightly positively polarized towards the hydrogen atom and so its removal will leave a small positive charge on the carbon atom, resulting in the stabilization of the B_1 orbital relative to the A_2.

Since the ground state of phenyl is $\pi^6 n$ it will be of $^2A_1$ symmetry and depending on whether the electron is removed from the A_2 or B_1 orbital, the excited state will have $^2A_2$ or $^2B_1$ symmetry. The analysis of the spectrum showed quite definitely that a 0–0 band was present and, therefore, the transition is symmetry allowed. The excited state must then be of $^2B_1$ symmetry, the transition having a moment perpendicular to the plane of the ring. The symmetry forbidden transition to the $^2A_2$ state must be at higher energies than the observed spectrum because the attraction between the hole in the π system and the electron in the non-bonding orbital is less in this state than in the $^2B_1$ owing to the node through the carbon atoms. However, no other spectra which could be attributed to phenyl were observed in the region 2900 to 7000 Å.

Photochemical aspects

On the basis of scanty experimental evidence, the formation of phenyl radicals from halogenated benzenes has long been postulated as a highly probable photochemical process. It was not surprising, then, to observe phenyl radicals, following photolysis of these compounds but the negative results from *ortho* and *meta* dichlorobenzenes are rather strange, in view of the positive result from the *para* isomer.

The appearance of phenyl absorption and the product diphenyl from benzene is interesting in view of the conflicting evidence concerning the photolysis of this compound. Krassina (1939) investigated the photolysis of benzene vapour and showed that only radiation of wavelength less than 2000 Å was effective in causing photochemical change. A deposit on the walls of the reaction vessel was thought to be diphenyl and so it was suggested that the main reactions were

$$C_6H_6 \xrightarrow{h\nu} C_5H_5 + H,$$

followed by
$$2C_5H_5 \rightarrow (C_5H_5)_2.$$

Similar conclusions were reached by Forbes & Cline (1941) who also showed that benzene is decomposed by mercury atoms excited by 1849 Å radiation but not by 2537 Å. Bates & Taylor (1927) investigated the decomposition of benzene in the presence and absence of mercury vapour and reported that hydrogen, methane and a tarry residue were formed especially in the photosensitized reaction. Unfortunately, no mention was made of the wavelengths of light used in their experiments. Both Lane & Noyes (1932) and West (1935) mention that photolysis of benzene with 2537 Å radiation produces negligible change. Wilson & Noyes (1941) were unable to detect diphenyl when benzene vapour was exposed to light of wavelength less than 2000 Å although the products did include acetylene, hydrogen and cuprene. In view of these results, they suggested, tentatively, that the primary process was

$$C_6H_6 \xrightarrow{h\nu} 3C_2H_2.$$

470   G. Porter and B. Ward

In considering the positive result from benzene in the present work it must
be noted that the photolysis was carried out with unfiltered radiation which
undoubtedly includes wavelengths down to 2000 Å. Equally significant, however,
is the high intensity of flash experiments which would have the effect of increasing
the probability of dimerization of phenyl over reactions with other molecules which
may result in the formation of polymer and other complex products.

The negative results for phenyl from phenol, anisole and *para* chlorophenol and
for cyclopentadienyl from the mono halogenated benzenes suggest that the two
photochemical processes are mutually exclusive. Yet the dihalogenated benzenes
seem to be exceptions to this rule and v.p.c. analysis showed that the products from
the flash photolysis of phenol contained a little diphenyl. Toluenes form only
benzyl radicals.

This work raises interesting questions about photochemical processes and
mechanisms and their relative probabilities of occurrence in aromatic molecules.
Even when several modes of decomposition are possible, each molecule seems to
exhibit one only and in any particular photochemical dissociation process, the
cleavage is confined specifically to the weakest bond. Further investigation of these
reactions should provide a better general understanding of unimolecular processes
in electronically excited states.

We are grateful to Dr R. N. Dixon for help with spectroscopic techniques and
to Dr J. N. Murrell for many helpful discussions.

REFERENCES

Bates, J. R. & Taylor, H. S. 1927 *J. Amer. Chem. Soc.* **49**, 2438.
Bernstein, H. J. & Herzberg, G. 1948 *J. Chem. Phys.* **16**, 30.
Dull, M. F. & Simons, J. H. 1933 *J. Amer. Chem. Soc.* **55**, 3898.
Duncan, F. J. & Trotman-Dickenson, A. F. 1962 *J. Chem. Soc.* 4672.
Forbes, G. S. & Cline, J. E. 1941 *J. Amer. Chem. Soc.* **63**, 1713.
Glazebrook, H. H. & Pearson, T. G. 1939 *J. Chem. Soc.* 589.
Harrison, A. G. & Lossing, F. P. 1959 *Can. J. Chem.* **37**, 1696.
Jaquis, M. T. & Szwarc, M. 1952 *Nature, Lond.*, **170**, 312.
Krassina, G. I. 1939 *Acta Physiochim. U.S.S.R.* **10**, 189.
Lane, C. E. & Noyes, W. A. Jr 1932 *J. Amer. Chem. Soc.* **54**, 161.
Longuet-Higgins, H. C. & McEwen, K. C. 1957 *J. Chem. Phys.* **26**, 719.
McDonald-Blair, J. & Bryce-Smith, D. 1960 *J. Chem. Soc.* 1788.
Murrell, J. 1963 *The theory of the electronic spectra of organic molecules.* London: Methuen.
Porter, G. 1950 *Proc. Roy. Soc.* A, **200**, 284.
Porter, G. & Ward, B. 1964a *J. Chim. Phys.* **61**, 1517.
Porter, G. & Ward, B. 1964b *Proc. Chem. Soc.* 288.
Razuvajev, G. A. & Zatajev, B. G. 1963 *Pap. Acad. Sci. U.S.S.R.* **148**, 863.
Razuvajev, G. A. & Zatajev, B. G. 1964 *Pap. Acad. Sci. U.S.S.R.* **154**, 164.
Schuler, H. & Reinbeck, Z. 1949 *Naturforschg.* **4**a, 560.
Stephenson, H. P. 1954 *J. Chem. Phys.* **22**, 1077.
Trotman-Dickenson, A. F. 1955 *Gas kinetics.* Butterworth.
Waters, W. A. 1946 *The chemistry of free radicals.* Oxford: Clarendon Press
West, W. 1935 *J. Amer. Chem. Soc.* **57**, 1931.
Williams, G. H. 1960 *Homolytic aromatic substitution.* London: Pergamon Press.
Wilson, J. E. & Noyes, W. A. Jr 1941 *J. Amer. Chem. Soc.* **63**, 3025.

Reprinted from Proceedings of the Chemical Society, September, 1964, page 288

Photochemical Formation of Cyclopentadienyls from Benzene Derivatives

By G. PORTER and B. WARD*

A NEW type of reaction has been observed in the photolysis of anilines, phenols, thiophenols, nitrobenzenes, and their halogenated derivatives in which a carbon atom is eliminated from the ring, along with the nitrogen-, oxygen-, or sulphur-containing side-group, to form cyclopentadienyl or halogenated cyclopentadienyls. This reaction occurs in addition to the formation of phenoxyl and anilino-radicals previously reported[1] and is in marked contrast to the behaviour of benzene and halogenated benzenes which form phenyls[2] and of alkylbenzenes which form benzyl-type radicals.[1]

All the molecules mentioned above give, on flash photolysis in the vapour phase, well-resolved transient band spectra in the region 3000—3900 Å. The evidence for the assignment of these spectra to cyclopentadienyl radicals and their halogenated derivatives is as follows: (i) A common spectrum with strongest band at 3343 Å, attributed to cyclopentadienyl is observed from aniline, phenol, nitrobenzene, thiophenol, anisole, and phenetole. (ii) A spectrum attributed to fluorocyclopentadienyl, with strongest band at 3250 Å, is observed from p-fluoronitrobenzene. (iii) A common spectrum attributed to chlorocyclopentadienyl, with strongest band at 3424 Å, is observed from m- and p-chlorophenol, m- and p-chloroaniline, o- and p-chloronitrobenzene, and o-chloroanisole. (iv) A common spectrum attributed to bromocyclopentadienyl, with strongest band at 3550 Å, is observed from o- and p-bromophenol, o-, m-, and p-bromonitrobenzene, and p-bromoaniline.

The spectrum derived from the unhalogenated compounds mentioned in (i) above is also obtained, much more strongly, from cyclopentadiene. Product analysis after flash photolysis of phenol shows carbon monoxide and hydrogen in approximately equimolar amounts, and further products not yet identified, but no biphenylene and only a trace of biphenyl.

The fact that all three halogenated isomers give identical spectra shows either that rapid isomerisation occurs or that all positions of substitution are equivalent. Since the spectra consist of many sharp bands which appear in the same intensity ratio from the different isomers, we believe that the latter interpretation is the more probable, and this leads almost unequivocally to the assignment of our spectra to cyclopentadienyl and its derivatives. The two strongest bands of our spectrum from cyclopentadiene are identical with two bands reported by Thrush[3] to be obtained on flash photolysis of this compound, and our observations support his assignment of these bands.

Although certainly unexpected, the ring fission of phenols, anilines, thiophenols, and nitrobenzenes is energetically feasible. Heats of formation of cyclopentadienyl and its derivatives are, of course, not yet known, but there is little doubt that the energy of formation of carbon monoxide, carbon monosulphide, or hydrocyanic acid is more than adequate to compensate for the energy required to break the aromatic ring.

We are indebted to Dr. S. M. Tsang for carrying out the product analysis of phenol.

(*Received, July 29th, 1964.*)

* Department of Chemistry, The University, Sheffield 10.
[1] Porter and Wright, *Trans. Faraday Soc.*, 1955, **51**, 395.
[2] Porter and Ward, *Proc. Chem. Soc.*, 1964, 288.
[3] Thrush, *Nature*, 1955, **178**, 155.

Reprinted from Journal of the Chemical Society, Section A, 1968

Rate of Dimerisation of Gaseous Benzyne

By **George Porter,** Davy Faraday Research Laboratory, The Royal Institution, 21 Albemarle Street, London W.1
Jeffrey I. Steinfeld, Department of Chemistry, Massachusetts Institute of Technology, Cambridge, Mass. 02139, U.S.A.

The flash photolysis of phthalic anhydride vapour results in the formation of benzyne and subsequently of biphenylene. The rate constant of dimerisation of benzyne has been determined directly, from the rate of appearance of biphenylene, as $(4 \cdot 6 \pm 1 \cdot 2) \times 10^9$ l. mole^{-1} sec.$^{-1}$. This is in accordance with calculations based on a transition state having free internal rotation about the bond connecting the two benzyne moieties.

A RECENT observation [1] of products in the pyrolysis of phthalic anhydride which indicate benzyne as an intermediate suggests that photolysis of phthalic anhydride vapour should be a good source of gaseous benzyne. The only other method known for producing this species in the gas phase is the flash vaporisation of certain solid aromatic materials,[2,3] in which it is difficult to observe the transient spectra or to carry out reliable kinetic studies, and by pyrolytic procedures.[4] The object of investigating this system was to search for the expected long-wavelength spectrum of the benzyne biradical; in the analogous long-wavelength transitions in the phenyl radical,[5] a sharp banded spectrum is observed making it possible to carry out a vibrational analysis. In the course of this investigation, the rate of dimerisation was determined, and is reported here.

EXPERIMENTAL

Several milligrams of vacuum-sublimed phthalic anhydride were introduced into a quartz photolysis cell of length 1 m., and from 30 to 300 torr of buffer gas (Ne or Ar) was added before sealing off. The cell was surrounded by an oven [6] which could be heated to a temperature sufficient to vaporise the organic materials, and maintained between 90° and 200°. The photolysis flash was produced by discharging 19 μF at 16 kv through two 1 m. quartz lamps, connected in series, and filled with 60 torr of krypton. Transient spectra were recorded by photographing a synchronised spectroflash source on a small Hilger quartz spectrograph or a Hilger–Littrow spectrograph. The kinetics of formation of biphenylene were determined by monitoring the long-wavelength biphenylene absorption bands, using an 'iodine–quartz' lamp as a continuous source, dispersed with a 250 mm. Bausch and Lomb monochromator, detected with an RCA 931A photomultiplier, and displayed on a Tektronix 545A oscilloscope.

RESULTS AND DISCUSSION

As the phthalic anhydride vapour was flashed, a new absorption built up in the 3500 Å region; the bands were easily identified as those of biphenylene by reference to published spectra.[7] The amount of biphenylene formed per flash could be estimated from the incremental absorption, by use of the vapour extinction coefficient determined for biphenylene in this region. This determination was carried out by Mr. C. W. Ashpole,

[1] E. K. Fields and S. Myerson, *Chem. Comm.*, 1965, 474.
[2] R. S. Berry, G. N. Spokes, and M. Stiles, *J. Amer. Chem. Soc.*, 1962, **84**, 3570.
[3] R. S. Berry, G. N. Spokes, and M. Stiles, *J. Amer. Chem. Soc.*, 1964, **86**, 2738.

[4] H. F. Ebel and R. W. Hoffman, *Annalen*, 1964, **673**, 1.
[5] G. Porter and B. Ward, *Proc. Roy. Soc.*, 1965, *A*, **287**, 457.
[6] S. K. Hussain, Ph.D. Thesis, Sheffield University, 1965.
[7] R. Hochstrasser, *Canad. J. Chem.*, 1961, **39**, 768.

3 L

J. Chem. Soc. (A), 1968

at the Davy Faraday Research Laboratory, using a sample of biphenylene kindly supplied by Dr. J. F. W. McOmie. The extinction coefficient at the wavelength of maximum absorption (3520 Å) was 8200 l. mole^{-1} cm.$^{-1}$ and at the second maximum (3360 Å) was 4600 l. mole^{-1} cm.$^{-1}$, the determinations being made at 150° and an argon pressure of 10 cm. Hg. The equivalent incremental extinction coefficient, in the region covered by the monochromator during the experiments, was 4100 l. mole^{-1} cm.$^{-1}$.

The yield of biphenylene per flash, based on the initial amount of phthalic anhydride, was of the order of a few per cent. (1—4 × 10^{-7} mole/l.). The absorption showed a second-order increase in the period 0·5 to 5·0 msec. following the flash, and bimolecular rate coefficients were determined over a range of pressure and temperature.

There was no significant variation in rate coefficient with temperature over the range 90—200° or with inert gas pressure over the range 0—300 torr. Variation of monochromator slit width did not produce any measurable change in the values. When biphenylene alone was flashed under the same conditions, there was no transient effect over the time range investigated in this work.

The mean of 22 determinations led to a value for the dimerisation rate coefficient of benzyne of (4·6 ± 1·2) × 10^9 l. mole^{-1} sec.$^{-1}$. This compares with the value of (8 ± 1) × 10^8 l. mole^{-1} sec.$^{-1}$ determined by Schafer and Berry [8] using flash vaporisation. The discrepancy may arise from the method used by Schafer and Berry for

estimating concentrations and difficulties arising from scattering by condensed material.

Our rate coefficient for benzyne dimerisation leads to an entropy of activation $\Delta S_c^{\ddagger} = 17\cdot75$ e.u. for a standard state of one mole/cm.^{3}. Since there are four equivalent ways of forming a bond between two biradicals, the quantity R log$_e$4 should be subtracted from this value, leaving −10·5 e.u. to be associated with the formation of the transition state. Convertion to an ideal-gas standard state of 1 atm. leads to $\Delta S_p^{\ddagger} = -31\cdot1$ e.u. A reasonable model for the transition state would be a biphenyl molecule with radical sites replacing one of the hydrogens *ortho* to the C–C single bond on each of the rings; there would presumably be completely free rotation about this bond. From the tabulated ideal-gas standard entropies of benzene [9] and biphenyl [10] one can estimate $\Delta S_p^{\ddagger}$ for this model of −34·9 e.u.; this is close enough to the observed value to indicate that such a model for the transition state, but with a slightly looser coupling than that in the biphenyl molecule, is probably correct.

The only transient spectrum observed following the photolysis flash was the long-wavelength end of the previously observed [2] diffuse 2600 Å band. Absorption to a lower-lying state in the visible part of the spectrum has not yet been seen, and the search for this spectrum is being continued with a multiple-traversal photolysis cell.

One of us (J. I. S.) would like to thank the National Science Foundation for a Postdoctoral Fellowship during the year 1965–66.

[7/1152   *Received, August 31st, 1967*]

[8] M. Schafer and R. S. Berry, *J. Amer. Chem. Soc.*, 1965, **87**, 4497.
[9] F. Rossini and his co-workers, 'Selected Values of Thermodynamic Properties of Hydrocarbons', American Petroleum Institute, Carnegie Press, 1953.

[10] (*a*) J. E. Katon and E. R. Lippincott, *Spectrochim. Acta*, 1959, **15**, 650; (*b*) M. Rolla, *Bol. sci. Fac. Chim. ind. Bologna*, 1941, **2**, 181.

Further publications

ABSORPTION SPECTRA AND ACIDITY CONSTANTS OF
PHENOXYL RADICALS.
E. J. Land, G. Porter and E. Strachan, *Trans. Faraday Soc.*, 1961, **57**, 1885.

SPECTRA AND KINETICS OF SOME PHENOXYL DERIVATIVES.
E. J. Land and G. Porter, *Trans. Faraday Soc.*, 1963, **59**, 2016.

ABSORPTION SPECTRA AND ACIDITY CONSTANTS OF
ANILINO RADICALS.
E. J. Land and G. Porter, *Trans. Faraday Soc.*, 1963, **59**, 2027.

THE ELECTRONIC SPECTRA OF BENZYL – A NEW TRANSITION.
G. Porter and M. I. Savadatti, *Spectrochim. Acta*, 1966, **22**, 803.

THE SPECTRUM OF PHENYL.
G. Porter and B. Ward, *Proc. Chem. Soc.*, 1964, **28**, 288.

PRIMARY PHOTOCHEMICAL PROCESSES IN AROMATIC MOLECULES.
PART 13. BI-PHOTONIC PROCESSES.
B. Brocklehurst, W. A. Gibbons, F. T. Lang, G. Porter and M. I. Savadatti,
Trans. Faraday Soc., 1966, **62**, 1793.

PRODUCTION OF BENZYL RADICALS BY IONIZING AND
ULTRA-VIOLET RADIATION.
B. Brocklehurst, G. Porter and M. I. Savadatti, *Trans. Faraday Soc.*, 1964, **60**, 2017.

THE PHOTOLYTIC PREPARATION OF CYCLOPENTADIENYL
AND PHENYL NITRENE FROM BENZENE DERIVATIVES.
G. Porter and B. Ward, *Proc. Roy. Soc.*, 1968, **A303**, 139.
NB. For a later interpretation of the spectrum attributed to Phenyl
Nitrene see Terry A. Miller *et al.*, *J. Phys. Chem.* 1990, **94**, 3387.

RELATIVE ELECTRON SPIN DISTRIBUTION IN ANILINO
AND PHENOXYL RADICALS.
N. M. Atherton, E. J. Land and G. Porter, *Trans. Faraday Soc.*, 1963, **59**, 818.

PROBLEMES D'ACTUALITE DANS LA SPECTROSCOPIE DES
RADICAUX LIBRES.
G. Porter, *J. Phys. Radium*, 1954, **15**, 497.

Chapter 6

TRAPPED RADICALS

A second method for the detection of unstable or highly reactive species is to prevent their unimolecular decay by holding them at low temperatures whilst bimolecular reactions are prevented by keeping the molecules separate from each other in a rigid cage. The idea occured to me "through a stained glass window" as I thought how some of the images formed by atoms and their complexes trapped in inorganic glasses were centuries old and had lost none of their clarity by diffusion.

In the 1950s various methods were developed for this purpose and one of the first was the method described here. This was merely the photolysis, at liquid nitrogen temperature, of the parent molecules trapped in an inert glass matrix composed of a mixture of organic solvents, which had been previously described by G. N. Lewis and his colleagues. Most organic radicals were readily stabilised in this way and later the work was extended to inert gas matrices and liquid helium temperatures for the more reactive species.

Reprinted from the Proceedings of the Royal Society, A, volume 230, pp. 399–414, 1955

Trapped atoms and radicals in rigid solvents

By I. Norman* and G. Porter†

Department of Physical Chemistry, University of Cambridge

(*Communicated by R. G. W. Norrish, F.R.S.—Received* 16 *February* 1955)

(Plates 11 and 12)

A number of very reactive radicals and atoms prepared by photolysis in rigid hydrocarbon glasses at the temperature of liquid nitrogen have been detected by their absorption spectra. Thus ethyl iodide dissociates to iodine atoms which only recombine to give I_2 when the glass is softened. Similarly CS_2 and ClO_2 give CS and ClO which can be detected spectroscopically in the frozen glass. Toluene and other benzyl derivatives yield the benzyl radical, and the spectra of a number of similar unstable aromatic radicals have been recorded for the first time. Most of these radicals have lifetimes of less than 1 ms at comparable concentrations in the gas phase or in ordinary solutions, but have been observed for many hours in the rigid glass. The method should be of general applicability for the study of the primary products of photochemical or radiation chemical processes.

Introduction

The majority of odd-electron molecules either cannot exist under conditions of thermodynamic equilibrium except at concentrations too small to be detectable experimentally, or can exist at measurable concentrations in equilibrium only under conditions of temperature or pressure which are not readily investigated. Furthermore, under normal conditions, the rate of establishment of equilibrium is so rapid that the average lifetime of these species, at concentrations high enough to be observed directly, rarely exceeds a few milliseconds and may be much less. Direct observation of these intermediates during the course of a chemical reaction, for the purposes of studying the kinetics of the reaction or the physical properties of the intermediate itself, has therefore usually had to depend on the use of special rapid recording techniques. The necessity for rapid recording reduces the sensitivity of measurement, and the number of physical parameters which can be recorded in this manner is limited. The difficulties of applying infra-red spectroscopy and similar techniques to free radicals have recently been reviewed (Porter 1954). It would clearly be advantageous to be able to prepare free radicals and similar reactive molecules under conditions such that their lifetime was extended to a period long enough to enable any physical property to be measured by conventional methods without recourse to rapid recording, i.e. to a period of minutes or even hours.

The short lifetime of a free radical is generally a consequence of one or other of two classes of rapid reaction. First, the free radical may recombine with a second radical, usually with nearly unit collision efficiency, and secondly, most free radicals react rapidly with other molecules, especially when these molecules are in

---

* Present address: Courtaulds Alabama Inc., Mobile, Alabama.

† Present address: British Rayon Research Association, Heald Green Laboratories Manchester.

I. Norman and G. Porter

high concentration such as when they form the solvent. The latter type of reaction is usually accompanied by a considerable activation energy, and, in any case, a solvent could probably always be found for which the activation energy of reaction with a particular radical was of the order of a few kilocalories or greater. In principle, therefore, the reactions of the second kind could always be reduced to negligible rates by working at a sufficiently low temperature. On the other hand, the recombination of two atoms or radicals may have a zero or even a negative temperature coefficient, and recombination would therefore occur with high collision efficiency at all temperatures. Clearly, this type of reaction can only be prevented by inhibiting collisions between the radicals. This would be the case if each radical were separated from its neighbours by a viscous medium through which diffusion was very slow, such as a hard glass. It is a common experience that molecules embedded in such a glass retain their spatial position for very long periods.

In order to produce a solution of radicals trapped in a rigid medium two approaches are possible. First, the radicals may be trapped by rapid freezing after preparation. This method has been applied to isolate rather unstable species, such as peroxides, from the products of a reaction, but considerable experimental difficulties arise when the lifetime of the intermediate is very short, especially when it is necessary to prepare a transparent homogeneous solution. A second method is to produce the radicals *in situ* from a parent substance already dissolved in the rigid medium at low temperatures. This obviously cannot be done by thermal methods, and recourse must be had to radiation of some kind. Photolysis methods should usually be applicable if dissociation occurs as readily in rigid media as in fluid solvents or gases.

A considerable amount of attention has recently been devoted to the optical behaviour of aromatic molecules dissolved in rigid glasses at low temperatures. This work has been directed mainly to the study of fluorescence and phosphorescence but one notable exception is the investigation of Lewis & Lipkin (1942) on the primary photochemical processes of some aromatic molecules and dyestuffs in rigid media. These authors found evidence for photo-ionization as well as dissociation into radicals, and were able to observe the absorption spectrum of the labile substances so formed for long periods after irradiation. Their observations were limited, however, to relatively stable radicals such as triphenyl methyl, and the above discussion suggests that the phenomenon should not be limited only to compounds of this type. There seems to have been practically no previous work on the photochemistry, in rigid media, of simpler molecules, whose photochemistry in other phases is better understood. The present investigation was undertaken, first, to discover how far the primary photochemical processes and efficiencies of the gas and liquid phases are applicable also to rigid media, and, secondly, to test the feasibility of, and the experimental conditions necessary for, the preparation of solutions of trapped radicals.

With regard to the question of quantum efficiencies of dissociation in rigid media our reasoning was as follows (Norman & Porter 1954). If radiation of a frequency greater than that required for dissociation is used, the excess energy of the quantum absorbed must be dissipated as heat. This excess energy, in a typical case such as the photolysis of ethyl iodide using radiation of 2537 Å, will amount to over 50 kcal/mole, and, since a temperature rise of only 10° C is sufficient to produce noticeable softening of the glasses to be described, each quantum absorbed by a molecule of ethyl iodide should result in a viscosity-lowering of several hundred molecules of the solvent sufficient to enable the radicals to diffuse apart. Heat will rapidly be conducted away and the glass will again become rigid with consequent trapping of the primary products. An apparent objection to this argument is to be found in the primary recombination effect of Franck & Rabinowitch (1934) which might be expected to operate very efficiently in glassy solvents. We shall find no serious limitations due to this effect.

Experimental

To be suitable for our purpose a solvent at low temperature must form a glass which remains rigid during irradiation and which is transparent throughout the visible and quartz ultra-violet regions. A number of pure hydrocarbons form glasses at 77° K, but they tend to crack or crystallize more readily than mixtures. We have used pure *iso*pentane and pure 3-methyl pentane; the former will still flow slowly at 77° K under manual pressures, and most attempts to use it for radical trapping were unsuccessful. The latter was quite satisfactory except under intense illumination when it softened. This trouble was also experienced, though to a lesser extent, with many of the hydrocarbon mixtures. When substances other than hydrocarbons can be used, mixtures of *iso*pentane, diethyl ether and ethyl alcohol, which have been previously described many times, were found to be most satisfactory. The mixture volume ratios of solvents used in all the experiments to be described are given below along with the abbreviations used:

(1) $3MeP.P$ 3-methyl pentane (3 parts), *iso*pentane (2 parts).
(2) $P.MeH$ *iso*pentane (3 parts), methyl *cyclo*hexane (2 parts).
(3) $E.P.A$ ether (5 parts), *iso*pentane (5 parts), ethyl alcohol (2 parts)
　　　　　　　(Lewis, Magel & Lipkin 1940).

All solvents must be spectroscopically pure and must be very dry, since small quantities of water seem to initiate crystallization. Methods of purification have been described by Potts (1953), and essentially the same procedure was used in this work. Ethyl alcohol was purified by the method of Lund & Bjerrum (1931).

The experimental arrangement which was eventually used for recording absorption spectra of the reaction mixture after intervals of irradiation is shown in figure 1.* A quartz cell, 17 mm in length and 15 mm in diameter, fitted closely between the two inner windows of a Dewar flask which was constructed entirely of quartz, and was silvered except for the four windows. The cell was clamped rigidly to the flask and the whole apparatus was firmly attached to a turn-table on an optical bench which had two fixed positions for irradiation of the sample and spectroscopic recording, respectively. For the latter purpose a hydrogen lamp was used; the photolysis source was either a low-pressure mercury lamp filtered

* We are indebted to Mr F. Webber for construction of the quartz Dewar flask.

402   I. Norman and G. Porter

to give radiation of 2537 Å or a 1 kW high-pressure mercury point source. The light was focused on to the reaction vessel by means of a quartz lens of 10 cm diameter and passed through a water filter and a copper sulphate solution to reduce heating of the reaction cell. The spectrograph was a small Hilger quartz instrument and Selochrome plates were used.

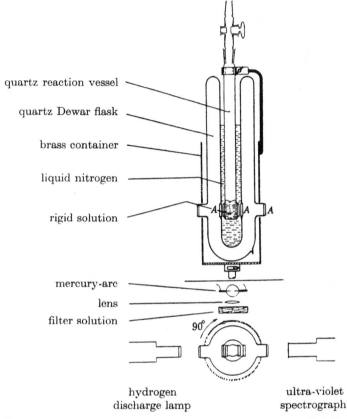

quartz reaction vessel

quartz Dewar flask

brass container

liquid nitrogen

rigid solution

mercury-arc

lens

filter solution

hydrogen
discharge lamp

ultra-violet
spectrograph

FIGURE 1. Arrangement for irradiation and spectroscopy of rigid solutions.

Although the apparatus is very simple in design and operation, there are several points about it which might usefully be mentioned. Any of the mixtures described above will form a perfectly satisfactory glass if a test-tube containing the mixture is plunged into liquid air. When it is required to form such a glass in a cell designed for spectroscopy, however, a few simple precautions must be taken. First, if cooling is rapid a very deep meniscus is formed which extends below the centre of the cell and makes accurate measurements impossible. This difficulty was mentioned by Potts, who preferred to use the less rigid glasses to reduce the trouble. This is an unsatisfactory compromise, and the difficulty can be completely avoided by cooling slowly from the bottom of the cell first. To make this possible it is desirable that the cell be removable from the Dewar flask without emptying the liquid nitrogen, and this has not usually been a feature of previous arrangements. A removable cell has several other advantages; rapid transfer to a bath at another temperature is possible, and if, as happens not infrequently, the cell is broken by expansion of the solvent on warming, the more valuable Dewar flask remains. The 25 % expansion of the glass which accompanies warming to room temperature must be allowed to occur freely, and it is advisable to keep the neck of the reaction vessel as nearly as possible equal to the size of the cell itself; in our apparatus the inside diameter of this side tube was 12 mm.

Gas bubbles from the liquid nitrogen were kept out of the light path by designing the cell to fit closely between the inner windows of the Dewar flask. Condensation of water vapour on the cell windows was eliminated, except for very long runs, by means of a loose plug in the neck of the Dewar flask. No condensation ever occurred on the outer windows. In all the experiments to be described the solution was freed from dissolved air and the cell was sealed off before irradiation, although separate experiments indicated that dissolved air had no effect either on the properties of the glasses formed or on the spectrum observed.

RESULTS

Ethyl iodide

Ethyl iodide is known (Norton 1934) to have a fairly high quantum of yield dissociation in solution ($\phi_{2537} = 0.41$), and the energy per einstein, at its absorption maximum near 2537 Å, is greatly in excess of that required for dissociation to an ethyl radical and an iodine atom. The fact that the iodine molecule absorbs in the visible region at wave-lengths to which both the iodine atom and ethyl iodide are transparent makes ethyl iodide a convenient substance for investigation. We shall first inquire whether photochemical dissociation of ethyl iodide occurs in the rigid glass and, if so, how the quantum yield of dissociation compares with that in ordinary solvents.

A solution of ethyl iodide was made up in one of the solvents described and was irradiated for a given time at 20° C under standard conditions of light intensity and optical arrangement. The iodine liberated was determined spectrophotometrically after transferring the solution to a standard cell. A second portion of the solution was now introduced to the reaction cell, care being taken that the geometrical arrangement was unchanged, and frozen down in liquid nitrogen to a clear glass. After irradiation for a known period of time the solution was brought to a temperature of 20° C and the liberated iodine again estimated. The absorption of ethyl iodide at 2537 Å was unchanged on cooling to 77° K, apart from the increase in concentration caused by volume contraction, for which a small correction was applied.

The relative quantum yields determined under two sets of experimental conditions were as follows:

solvent	[EtI] (10^{-4}mole/l.)	light source	decomposition range (%)	$\dfrac{\phi_{77°\,\mathrm{K}}}{\phi_{293°\,\mathrm{K}}}$
3-methyl pentane	9·1	low pressure	8 to 9	0·55
$E.P.A$	8·4	high-pressure Hg	2 to 2·5	0·51

The quantum yield of iodine formation is less in the viscous glass in both experiments, but the decrease is not great enough to introduce any serious practical limitation.

404   I. Norman and G. Porter

Observations on the reaction in rigid media

(1) 3-*Methyl pentane*-iso*pentane glass*. Photolysis of a 10^{-3} M-ethyl iodide solution in this hydrocarbon mixture at room temperature, using radiation of 2537 Å, showed visible formation of iodine after a few minutes. Two hours' irradiation of the same solution at 77° K, using the same light intensity, produced no visible effect, and the solution remained transparent in the quartz ultra-violet region down to the beginning of absorption by ethyl iodide. The transparent glass was now raised above the surface of the liquid nitrogen and allowed to warm slowly. Immediately, while the solution was still quite viscous, the strong coloration of the iodine molecule appeared. Recooling to 77° K had no further effect.

No spectrum other than that attributed to the iodine molecule was detected, and this was confirmed by comparison with solutions of iodine in the same solvent. It was noticed, both with the iodine prepared by photolysis of ethyl iodide and with the iodine solutions prepared separately, that a colour change occurred on cooling, the solution changing from violet at room temperature to yellow as it approached the temperature of liquid air. This is characteristic of iodine complex formation, and spectral evidence of such complexes, even with saturated hydrocarbons, has recently been reported by other workers (Hastings, Franklin, Schiller & Matsen 1953). A second explanation of the colour change, that the iodine was being precipitated from solution at low temperatures, was also considered, since extrapolations of high-temperature solubility data for iodine in saturated hydrocarbons would indicate that at 77° K the solubility was much less than the concentration in our solutions. The extrapolation is a long one, however, and the explanation is improbable, since the same colour change was observed over a wide range of iodine concentrations, and when precipitation of iodine did occur at higher concentrations it resulted in a typical cloudy suspension.

Irrespective of the nature of the spectral shift our experiments show clearly that the iodine molecule, complexed or otherwise, is absent after photolysis in the rigid glass and appears instantly the glass is softened. This is in accordance with our expectations if the iodine atom produced by dissociation in the glass is trapped therein. No trace of recombination could be detected after several hours at 77° K. The absence of any new spectrum which could be attributed to ethyl indicates either that this radical has no strong absorption in the region investigated or that, unlike the iodine atom, it is removed even at 77° K by reaction with the solvent.

(2) 3-*Methyl pentane glass*. Providing the intensity was not high enough to produce softening of the glass, the observations on ethyl iodide solutions in pure 3-methyl pentane were identical with those just described.

(3) Iso*pentane glass*. It has already been mentioned that the viscosity of iso-pentane at 77° K is less than that of its mixtures with 3-methyl pentane. It was found that the photolysis of ethyl iodide solutions in iso-pentane resulted in iodine molecule formation during photolysis at 77° K. Though diffusion in the glass was slow, as evidenced by the restriction of the iodine molecules to the path of the light beam, it is sufficiently rapid in this case to allow recombination of iodine atoms during irradiation.

(4) E.P.A *glass*. It has long been known that solution of iodine in alcohol not only results in a shift of the visible absorption maximum to 445 mμ, but also that two strong ultra-violet regions of absorption appear with maxima at 290 and 360 mμ. Batley (1928) showed that the ultra-violet absorption should be attributed to hydrogen tri-iodide, which was formed rapidly both by thermal and photo-chemical reactions.

A rigid solution of ethyl iodide in *E.P.A*, after photolysis for 1h, remained transparent throughout the visible and ultra-violet regions down to the beginning of absorption by ethyl iodide. On warming to room temperature, weak visible colour appeared as well as strong absorption bands with maxima at 290 and 360 mμ, these effects being shown in figure 2, plate 11. The absorption of the final products was identical with that of a solution of iodine in *E.P.A* at room temperature. The development of visible colour and ultra-violet absorption was not instantaneous on warming, however, but occurred over a period of about 1h at 20° C. It was possible to transfer the solution to the cell of a spectrophotometer and to follow the appearance of I_3^- by its absorption at 360 mμ. Its formation was apparently first order with a rate constant at 3° C of $9.4 \times 10^{-4} s^{-1}$. Photolysis of ethyl iodide in *E.P.A* at 195° K showed visible iodine formation after 2 min, but when a solution which had been irradiated for 30 min at 77° K was brought to 195° K no iodine formation was observed after 1h at the higher temperature.

It is clear from these observations that neither I_2 nor HI_3 are formed in the rigid glass, but their non-appearance immediately on softening shows that some other species than iodine atoms is then formed, since the recombination of iodine atoms in fluid solvents occurs practically at every bimolecular collision (Davidson, Marshall, Larsh & Carrington 1951). This species may be HI, which absorbs in the same region as ethyl iodide, its slow reoxidation at room temperature resulting eventually in the appearance of iodine. These reactions require further investigation but are rather irrelevant to the present discussion.

Iodine

The photolysis of iodine itself was investigated in rigid media, since dissociation to atoms should result in a disappearance of the molecular spectrum. The 1 kW high-pressure mercury lamp was used, and the intensity necessary to produce decomposition caused softening of most glasses, even while immersed in liquid nitrogen. *E.P.A* was found to be most suitable, and in this case complete trans-parency of a 10^{-4}M solution of iodine was readily attained and recorded photo-graphically, after irradiation for 1h. The iodine colour reappeared on warming, the behaviour being in every way analogous to that of iodine atoms prepared by photolysis of ethyl iodide.

Carbon disulphide

In the gas phase, carbon disulphide dissociates photochemically to carbon monosulphide, and a sulphur atom and the spectra of CS and of S_2 have been detected after photolysis (Porter 1950). The carbon monosulphide eventually disappears by a wall reaction with a half-life which may be as long as several minutes.

406   I. Norman and G. Porter

In solution the radical has never been detected, although we have found that photolysis of hydrocarbon solutions of carbon disulphide at 20° C results in decomposition and the separation of a yellowish solid product. We were unable to observe any radical spectra in absorption several minutes after photolysis, and recent investigations, by Porter & Windsor, of carbon disulphide dissolved in spectroscopically pure hexane showed no absorption by CS at times as short as $20\,\mu s$ after photolysis. Therefore, although carbon disulphide, in solution, undergoes photochemical decomposition, the lifetime of any CS formed is too short for detection by the fastest available techniques.

The course of photolysis of a 10^{-2}M solution of carbon disulphide in a rigid glass (3 $MeP.P$) is shown in figure 3, plate 11. After a few minutes' irradiation at 77°K a very strong band system appeared whose absorption maxima, compared with the heads of the bands of CS vapour, were as follows:

band	λ gas Å	λ glass Å	λ glass $-\lambda$ gas (Å)
0, 0	2575·6	2621	45
1, 0	2507·3	2553	46
2, 0	2444·8	2493	48

Since the intensity distribution is different in the two media no exact comparison of wave-length is possible, but the shift to longer wave-lengths in denser media is a general phenomenon. There can be no doubt as to identity of the bands observed in the glass with the spectrum of the CS molecule.

On warming the glass, the spectrum of CS completely disappeared, and a weak general absorption, probably caused by the solid products, was observed. Absorption spectra of the solution at 195° K, after photolysis at 77° K, showed no trace of absorption by CS. At no time was the spectrum of S_2 observed, which is in accordance with expectations, since S atoms will not be able to recombine in the glass, and as soon as the glass becomes fluid the polymerization of S_2 molecules can occur very rapidly.

Chlorine dioxide

In the gas phase chlorine dioxide dissociates photochemically into a ClO radical and an oxygen atom (Lipscomb, Norrish & Porter 1954). The ClO radical has a banded spectrum, with its intensity maximum at 2700 Å, and it disappears by a rapid bimolecular reaction to form molecular oxygen and chlorine.

The spectrum of a solution of chlorine dioxide in a hydrocarbon solvent is shown in figure 4, plate 11. The banded structure is retained but becomes almost entirely diffuse when the solution is cooled to the temperature of liquid nitrogen. Photolysis of this glassy solution results in the rapid destruction of chlorine dioxide, its spectrum being replaced by a diffuse spectrum, of similar intensity but at lower wave-lengths. If photolysis is continued the second spectrum also disappears, indicating that the product itself is photochemically dissociated, but no further absorption appears. On melting the glass, the new absorption spectrum is completely removed and most of the original absorption by chlorine dioxide reappears.

The wave-lengths of the intensity maximum of the new spectrum, as well as the limits of absorption of the first peak (2550 and 3000 Å), agree exactly with those

326

Norman & Porter   *Proc. Roy. Soc. A, volume* 230, *plate* 11

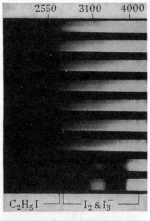

2550   3100   4000   Å

liquid before irradiation

solid before irradiation

15 min irradiation

30 min irradiation

45 min irradiation

60 min irradiation

80 min irradiation

liquid after melting

solid after refreezing

C_2H_5I — I_2 & I_3^-

FIGURE 2. Photolysis of ethyl
iodide in *E.P.A.*

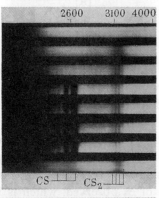

2600   3100   4000   Å

cell empty

liquid before irradiation

solid before irradiation

30 min irradiation

60 min irradiation

120 min irradiation

+ 60 min unexposed

liquid after melting

CS — CS_2

FIGURE 3. Photolysis of carbon
disulphide in *3MeP.P.*

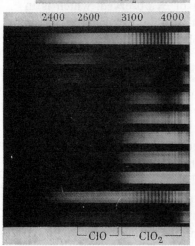

2400   2600   3100   4000   Å

liquid before irradiation

solid before irradiation

1 min irradiation

3 min irradiation

5 min irradiation

10 min irradiation

15 min irradiation

25 min irradiation

liquid after melting

solid after refreezing

ClO — ClO_2

FIGURE 4. Photolysis of chlorine
dioxide in *P.MeH.*

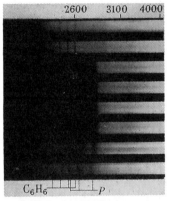

Å

liquid before irradiation

solid before irradiation

30 min irradiation

60 min irradiation

90 min irradiation

120 min irradiation

solid after refreezing

liquid after melting

FIGURE 5. Photolysis of benzene in *E.P.A.*

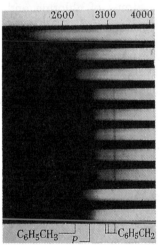

Å

solvent

liquid before irradiation

solid before irradiation

15 min irradiation

30 min irradiation

45 min irradiation

60 min irradiation

+ 60 min unexposed

liquid after melting

solid after refreezing

FIGURE 6. Photolysis of toluene in *E.P.A.*

of the spectrum of ClO within the accuracy of measurement of such quantities. The fact that no banded structure is observed may be a result of solvent interaction, especially since this region of ClO absorption shows strong predissociation in the gas phase. The reappearance of chlorine dioxide on warming the glass is then to be attributed to recombination of oxygen atoms and ClO radicals, since no ClO_2 is formed by reaction between two ClO radicals (Porter & Wright 1953).

Though this explanation is satisfactory, an alternative one must be considered. The observed spectrum also agrees closely with that of ClO_3, especially in that there is a second region of absorption beginning at 2500 Å and extending to shorter wave-lengths. Although ClO_3 can hardly have been formed in the rigid glass from the products of dissociation of ClO_2, it is possible that chlorine dioxide is present, at very low temperatures, mainly in its dimeric form, and this is in accordance with the spectral change which occurs on cooling the solution. We then have the possibility of the following primary process:

$$Cl_2O_4 + h\nu = ClO + ClO_3.$$

In this event the observed spectrum would be composed of the spectra of both radicals, and on warming, the reverse reaction could be responsible for the reappearance of ClO_2. Whether the product is entirely ClO, or is partly ClO_3, prolonged photolysis would result in the formation of atoms of chlorine and oxygen which are transparent throughout the region investigated. The main objection to the suggested formation of ClO_3 is that this radical dimerizes in the liquid phase, even at room temperature, to form Cl_2O_6, which has a strong absorption in the visible region. This spectrum was never observed even when the products in the glass were brought to 195° K, at which temperature there can have been no possibility of thermal dissociation of Cl_2O_6.

Whichever of the above two primary processes is predominant, photolysis of ClO_2 in the rigid glass results in the formation of unstable free radicals which remain trapped as long as the solution is held rigid.

Photolysis of simple aromatic molecules

Porter & Wright (1955) have recently recorded the spectra of a number of unstable aromatic radicals by the flash photolysis of aromatic vapours. Most of these spectra had a sharp banded structure and they were all assigned to resonance-stabilized radicals of the benzyl type. They could never be detected for longer than a fraction of a millisecond, however, and were, in this respect, more analogous to free radicals such as methyl than to stable radicals of the triphenyl methyl type. We were therefore led to investigate whether these radicals could be prepared and trapped by photolysis in an organic glass.

Benzene

Irradiation of benzene, in the vapour or liquid phases, with light absorbed in the quartz ultra-violet region, does not result in any observable reaction. This fact has been frequently noted in the literature and is also true of transient effects. Flash photolysis, both of benzene vapour and of its solutions in hydrocarbons, shows no spectral change even at the shortest times which can be studied. After a few minutes' irradiation in a rigid medium at 77° K, however, a product is formed which has a very intense banded spectrum and which is stable at room temperature for many hours after melting the glass. The effect is illustrated in figure 5, plate 12, which shows the spectra obtained on irradiation of a solution of benzene in $E.P.A$ at 77° K. Although there is a decrease in the amount of absorption after warming, much of the spectrum P remains and must therefore be identified with some permanent product of photolysis. Gibson, Blake & Kalm (1953) have recently reported this effect and our measurements in solution at room temperature agree closely with theirs. The intensity maxima measured in the rigid glasses were at the following wave-lengths (Å):

*iso*pentane glass	2781	2672	2559
$E.P.A$ glass	2766	2659	2552

The Berkeley workers identified the bands with those of hexatriene, formed by ring-opening of the benzene followed by hydrogen abstraction from the solvent. Here is an example of abstraction from the solvent even in the glass at 77° K, and we have found that similar spectra are obtained from many substituted benzenes. Melting the glass and refreezing always resulted in a decrease in the absorption by the hexatriene or its derivative, sometimes almost to zero, though further melting or holding at room temperature for several hours had no effect. Hexatriene polymerizes readily, and its rapid removal immediately after the first melting may be due to the presence of trapped radicals or other reactive species which initiate the polymerization reaction.

In addition to the intense spectrum of hexatriene, two weak sharp bands were observed after long photolysis of benzene at 77° K. They were no longer visible after warming, but a decrease proportional to that of the hexatriene spectrum would have reduced them below the limit of detectability. The wave-lengths of these bands were 3101 and 2926Å, the former being the stronger; they were observed only in $E.P.A$ glass and no assignment is possible at present.

Irradiation of solutions of naphthalene and anthracene in rigid glasses for comparable times with those used in the benzene experiments produced no observable spectral change, and substituted benzenes, such as halogenobenzenes, benzonitrile and nitrobenzene, showed only diffuse absorption which was also present after melting the glass. The photolysis of benzyl derivatives, on the other hand, resulted in the appearance of sharp-banded spectra which were only present in the rigid solvent.

The benzyl radical

A typical series of spectra obtained on photolysis of toluene in $E.P.A$ at 77° K is shown in figure 6, plate 12. After 15 min. irradiation two types of spectra are present: first, there is a diffuse spectrum beginning at 2839Å and extending to shorter wave-lengths until it overlaps the spectrum of toluene, and secondly, there is a sharp-banded spectrum which consists of a strong line-like band at 3187Å and two weaker bands at 3082 and 3039Å. The former, marked P, which is still visible, though much reduced in intensity, after the glass has been melted and refrozen, is probably the spectrum of a methyl-substituted hexatriene by analogy with the similar behaviour of benzene. The latter, which disappears completely when the glass is softened, is the spectrum of the benzyl radical, as will shortly be made clear. Its lifetime in the glass is many hours, as is shown by the unchanged intensity of its absorption after remaining unirradiated for 1 h.

A spectrum identical in all respects with the sharp spectrum described was also obtained, both in $E.P.A$ and hydrocarbon glasses, on photolysis of three other substances—benzyl chloride, benzyl alcohol and benzylamine. With the alcohol the spectrum was rather weak, but the other two substances gave a similar radical concentration to toluene. The wave-lengths of the bands, measured by comparison with an iron arc, are given in table 1. By direct comparison of the plates, using a comparator, it was shown that the position of the bands was identical in the four substances to within 1 Å. There can therefore be no doubt that the bands are identical and, by arguments similar to those used in connexion with the gas-phase spectra, we conclude that the absorption is that of the free benzyl radical. The sharp-banded structure is very similar to that of benzyl in the gas phase apart from a shift to longer wave-lengths of 137 Å, such shifts being characteristic of spectra in rigid media.

Other hydrocarbon radicals

Photolysis of a number of other aromatic hydrocarbons gave spectra which were very similar in appearance to that of benzyl but which were at a different wave-length. The spectra were all characterized by a narrow line-like structure in which one line was very much stronger than the rest of the spectrum, and in this respect resemble the well-known spectrum of the triphenyl methyl radical. The spectra showed no decrease in intensity over a period of 1 h in the glass, but were removed completely by melting and refreezing. They were readily distinguished from the spectra of permanent products by the fact that the latter were quite diffuse and could be observed after the glass was melted. The positions of these spectra are given in table 1 along with an identification of the radical to which the spectrum should probably be assigned. These assignments are less certain than in the case of benzyl, since each of the other spectra in table 1 was obtained from one substance only. They are based on a consideration of spectral position, band width and resonance structure of the radicals, and the internal consistency of the results lends support to the assignments. Thus the absorption maxima of the radicals of the benzyl series lie in order of increasing resonance stabilization:

$\phi CH_2\cdot$	$pCH_3C_6H_4\dot{C}H_2$	$\phi_2\dot{C}H$	$\phi\dot{C}HCH_2\phi$	$\phi_3C\cdot$
3187	3225	3351	3625	5180

No trace of the benzyl spectrum was observed in the photolysis products of diphenyl methane or dibenzyl. It appears that the separation of a hydrogen atom occurs more readily, in the rigid glass, than that of more bulky radicals such as benzyl. The photolysis of ethyl benzene gave a band which, although very close to the position of benzyl, was not identical with it. The two sharp bands obtained from this molecule appear to belong to different species, since their relative intensity was markedly affected by the solvent, but the assignment is doubtful.

The behaviour of fluorene is interesting, since the sharp spectrum attributed to the free radical was not obtained either in the gas phase or in solution but instead, in the fluid media, an intense absorption identified with the triplet state was observed. The times between irradiation and investigation are of course too long in the present experiments for the detection of absorption by the triplet state.

Photolysis of aniline

In addition to benzyl and its derivatives Porter & Wright detected the anilino and phenoxy radicals during the gas-phase photolysis of aniline and of phenol and anisole respectively. Photolysis of aniline in $E.P.A$ glass resulted in the appearance of two new regions of absorption in different parts of the spectrum. The first of these was a strong band at 3088 Å, rather broader than the bands so far described, which is to be compared with the band at 3008 Å, also broader than the benzyl bands, found in the photolysis of aniline vapour, and attributed to anilino. It was apparent that the whole of the region of absorption by aniline from 3000 Å to below 2500 Å showed a greatly increased intensity after irradiation which disappeared completely on warming the glass. As a result of this absorption no further increase in the intensity of any of the new spectra occurred after the first 15 min irradiation, and the strong fluorescence of aniline at 3100 Å, which could be photographed both before irradiation and after the solution had been melted and refrozen, was absent in the frozen glass after 15 min irradiation.

We must now discuss the second region of absorption which occurred in the visible region. As a result of this the irradiated glass was coloured yellow and remained so until melted. The spectrum consisted of a single sharp band at 4288 Å, comparable in intensity with the band attributed to anilino, and a weaker continuous absorption extending from 4050 Å to shorter wave-lengths. This can hardly belong to the same species as the ultra-violet spectrum, since it was always absent in the gas phase even when the ultra-violet spectrum was very intense. We believe that this is the spectrum of the aniline radical cation for the following reasons. First, although it is a transient species the position of its absorption is anomalous when compared with the spectra of neutral-free radicals such as benzyl and phenoxy. Secondly, Lewis & Lipkin demonstrated the occurrence of photo-oxidation processes in $E.P.A$ glass by loss of an electron from triphenylamine and other substances containing an electronegative atom. Thirdly, the non-appearance of the spectrum in the gas phase is explained, since only when solvation of the ion and electron are possible will the energy requirements for photo-ionization (photo-oxidation) be satisfied.

Details of spectra obtained with all other molecules investigated are given in table 1. The assignment of the two bands obtained from o-toluidine is based on comparison of spectral position and band width. The diffuse spectrum given by phenol and anisole is assigned to phenoxy on the basis of comparisons with the gas phase spectrum.

TABLE 1. DESCRIPTION OF SPECTRA OBSERVED AFTER PHOTOLYSIS
OF AROMATIC SUBSTANCES IN RIGID GLASSES

molecule	solvent	radical spectra band centre (Å)	probable radical	diffuse spectra of permanent products (Å)
C6H5–CH3	E.P.A	3187 (5) 3082 (1) 3039 (1)	C6H5–CH2·	2830 limit (strong)
C6H5–CH2Cl	E.P.A	3182 (5) 3082 (1) 3047 (1)	C6H5–CH2·	2830 limit (weak)
C6H5–CH2NH2	E.P.A	3186 (6) 3078 (1) 3043 (1)	C6H5–CH2·	none
C6H5–CH2OH	E.P.A	3184 (2)	C6H5–CH2·	none
C6H5–CH2CH3	P.MeH and E.P.A	3228 (2) E.P.A 3210 (1) P.MeH 2902 (2) E.P.A 2896 (4) P.MeH	? C6H5–CḢ·CH3 ?	2850 limit E.P.A 2900 (1) 2786 (4) } P.MeH
CH3–C6H4–CH3	P.MeH	3225 (8) 3167 (1) 3097 (3) 3044 (1)	CH3–C6H4–CH2·	3004 (10) 2853 (7)
C6H5–CH2–C6H5	P.MeH	3351 (5)	C6H5–CḢ–C6H5	2874 (1) 2801 (4)
C6H5–CH2CH2–C6H5	E.P.A	3625 (3)	C6H5–CH2ĊH–C6H5	2800 limit (weak)
fluorene (CH2)	P.MeH	3450 (4) 3380 (1)	fluorenyl (ĊH)	3520 (0) 3300 (1)
C6H5–NH2	E.P.A	3088 (8) w 4288 (4) w	C6H5–NH· C6H5–NH2+	none
CH3–C6H4–NH2	E.P.A	3217 (1) 3119 (4) w	ĊH2–C6H4–NH2 CH3–C6H4–NH·	2800 limit (weak)
C6H5–OH	E.P.A	2870 limit	C6H5–O·	2600 limit
C6H5–OCH3	E.P.A	2860 limit	C6H5–O·	2600 limit

Notes. (1) The following molecules showed no change: chlorobenzene, bromobenzene, benzophenone, naphthalene, anthracene.

(2) The following molecules gave spectra of stable products only: benzene, nitrobenzene, benzonitrile, o-chlorotoluene.

(3) All radical spectra recorded were line-like bands, similar to those in figure 6, unless marked otherwise. Bands marked w were about twice this width. Wave-lengths marked 'limit' refer to long wave-length limit of a continuous absorption.

(4) Numbers in brackets are intensities, on scale 0 to 10, of spectra obtained after 30 min irradiation.

Discussion

The original purpose of this work was to investigate the possibility of trapping atoms and free radicals in rigid media, and our results have shown that the method is applicable to a very wide range of such species. In the course of establishing this fact results have been obtained which are interesting in themselves and which clearly deserve further study, particularly in the field of aromatic free radical spectra. Though further work may modify some of the details of our conclusions in particular cases, the experimental proof which we have given of the general possibilities of the method is unlikely to be affected.

Although Lewis and his co-workers confined themselves to the study of relatively complex molecules and relatively stable radicals and ions, the general conclusion of our work was already implicit in their publications. That the implication was not generally appreciated is evident from the numerous unsuccessful attempts which have been made recently to trap radicals by more elaborate methods. It now seems likely that any free radical which can be prepared by photolysis can be stabilized in this manner, and there would not appear to be any insuperable difficulties to the use of the method for the investigation of infra-red spectra and other physical properties. It is therefore important to consider what are the limitations of the method and how general can be its application.

There are three conditions which must be satisfied before a rigid solution of free radicals can be obtained:

(1) Photochemical dissociation must occur in the rigid medium.

(2) Recombination of the radicals by diffusion must be slow.

(3) Reaction with the solvent must be slow.

Although exceptions may yet be found, every substance which we have investigated which dissociates photochemically in the gas phase dissociated also in rigid media. The times of irradiation were similar to those required for comparable decomposition in the gas phase, and in one case it was shown that the quantum yield was also comparable. It is probably necessary to use light quanta of energy well in excess of the dissociation energy, but since it is always possible in principle, and usually possible in practice, to use such an absorption region it is unlikely that any limitation of the method will arise from this cause.

The requirement of rigidity of the glass, sufficient to prevent any significant diffusion during a period of hours, apparently presents no difficulty. Although in some instances, such as with pure *iso*pentane, recombination did occur at 77° K, the glasses in which this occurred had a relatively low rigidity. We intend to investigate the rate of recombination as a function of viscosity, but it can be stated at this stage that the glasses used for the major part of this work should be sufficiently rigid to prevent diffusion of any free radical, since they have been shown to be suitable for single atoms.

The third criterion is the most important one, since it is already apparent that reaction with the solvent can occur at 77° K in the rigid glass, one example being the formation of hexatriene during benzene photolysis. Since contact between the radical and the solvent is inherent in the method, the prevention of reaction must depend on energetic considerations, and conditions must be chosen such that the half-time of reaction is of the order of 1 h or greater. The first point to be considered is the possibility of hot radical reactions, that is, reaction between the highly energetic radical immediately after photolysis, and a solvent molecule. There is no reason why this should be of any greater importance in rigid solvents than in gases, since in both cases it is a question of the relative rates, on collision, of chemical reaction and of energy transfer without reaction. Since thermal energy equilibration is almost complete after a few collisions, and since hydrogen abstraction reactions, by free radicals, are usually accompanied by steric factors of 10^{-3} or less, reactions of hot radicals have not generally been found to contribute significantly to the mechanism of photochemical reactions in fluid media and can probably be ignored in rigid solvents also.

The frequency factors applicable to reactions in solution are not well understood, and nothing is known about such factors in rigid glasses. It is probable, however, that the frequency factor will be comparable with or less than a vibration frequency, and for the purpose of discussion we shall consider it to be 10^4. (We may consider this as applying to dissociation of the activated radical-solvent complex.) Then, for a half-time of reaction of 1 h at $77°$ K we find that the activation energy must be 6·3 kcal/gmole. If the activation energy is reduced to 5 kcal the half-life becomes about 1 s. Now the activation energies of hydrogen abstraction by alkyl radicals from hydrocarbons, ethers and alcohols usually lie between 6 and 10 kcal/mole, so that, in view of the uncertainty in the appropriate frequency factor, these form a borderline case. All the radicals which we have detected would, however, be expected to have an activation energy for hydrogen abstraction from a saturated hydrocarbon considerably in excess of the above figure. Thus the bond dissociation energy of toluene into benzyl and a hydrogen atom has been given as 77·5 kcal/mole (Swarc 1950) and the C—H bond dissociation energy in the paraffins is usually 90 kcal or greater. The activation energy for hydrogen abstraction by benzyl is therefore at least 13 kcal, and the rate will be negligible at $77°$K. In $E.P.A$ the activation energy may be less but is still high enough to prevent reaction. On the other hand, it is quite probable that much lower activation energies apply to the reaction of phenyl and similar less stable radicals with the solvent, and this may explain our failure to detect any spectra which could be assigned to such radicals. Again, we were unsuccessful in attempts to observe the spectra of HS and OH after photolysis of H_2S and H_2O_2 in rigid glasses, although flash photolysis in the vapour phase results in the appearance of these spectra. Though it is possible that the fine-line spectra of these radicals in the vapour become a diffuse absorption over a wide region which is consequently difficult to observe, it is also probable that these radicals react with the solvent too rapidly for detection under our conditions. The failure to detect the ethyl radical after photolysis of ethyl iodide may be explained in a similar way, though again it is by no means certain that ethyl has a transition in the region investigated.

Though reaction with the solvent clearly introduces a limitation to the trapped radical technique as practised in the present work, the limitation can readily be removed either by using lower temperatures or different solvents. Thus in liquid helium the activation energy for a 1h half-life, calculated as above, becomes 0·25 kcal/mole. Mixtures of fluorocarbons form good glasses at the temperature of liquid nitrogen and should be suitable for trapping the alkyl and most other radicals of interest at present. Glasses formed of substances with even stronger bonds, the oxides for example, may also prove useful. It is probable that with a suitable choice of experimental conditions the primary products of any photo-chemical or radiation chemical process could be trapped in glassy media for further investigation.

One of us (I.N.) is grateful to the Fulbright Commission for a fellowship, during the tenure of which this work was carried out.

REFERENCES

Batley, A. 1928 *Trans. Faraday Soc.* **24**, 438.
Davidson, N., Marshall, R., Larsh, A. E. & Carrington, T. 1951 *J. Chem. Phys.* **19**, 1311.
Franck, J. & Rabinowitch, E. 1934 *Trans. Faraday Soc.* **30**, 120.
Gibson, G. E., Blake, N. & Kalm, M. 1953 *J. Chem. Phys.* **21**, 1000.
Hastings, S. H., Franklin, J. L., Schiller, J. C. & Matsen, F. A. 1953 *J. Amer. Chem. Soc.* **75**, 2900.
Lewis, G. N. & Lipkin, D. 1942 *J. Amer. Chem. Soc.* **64**, 2801.
Lewis, G. N., Magel, T. T. & Lipkin, D. 1940 *J. Amer. Chem. Soc.* **62**, 2975.
Lipscomb, F. J., Norrish, R. G. W. & Porter, G. 1954 *Nature, Lond.*, **174**, 785.
Lund, H. & Bjerrum, J. 1931 *Ber. dtsch chem. Ges.* **64**B, 210.
Norman, I. & Porter, G. 1954 *Nature, Lond.*, **174**, 508.
Norton, B. M. 1934 *J. Amer. Chem. Soc.* **56**, 2294.
Porter, G. 1950 *Proc. Roy. Soc.* A, **200**, 284.
Porter, G. 1954 *J. Phys. Radium*, **15**, 113.
Porter, G. & Wright, F. J. 1953 *Disc. Faraday Soc.* **14**, 23.
Porter, G. & Wright, F. J. 1955 *Trans. Faraday Soc.*
Potts, W. J. 1953 *J. Chem. Phys.* **21**, 191.
Swarc, M. 1950 *Chem. Rev.* **47**, 75.

Reprinted from the *Transactions of the Faraday Society*, No. 431, Vol. 54, Part 11, November, 1958

PRIMARY PHOTOCHEMICAL PROCESSES IN AROMATIC MOLECULES

PART 4.—SIDE-CHAIN PHOTOLYSIS IN RIGID MEDIA

By G. Porter and E. Strachan

Chemistry Dept., The University of Sheffield
British Rayon Research Association

Received 14th April, 1958

Two processes of side-chain fission have been identified in the photolysis of substituted aromatic molecules in rigid solvents at low temperatures. One involves fission of a β bond to yield two radicals and the other results directly in the formation of two molecules, one of which is styrene. Quantum yields at 2537 Å in one example of each process, were $1 \cdot 1 \times 10^{-2}$ and $3 \cdot 6 \times 10^{-2}$ respectively. The radical products have been identified spectroscopically and conditions necessary for their stabilization have been investigated. Generalizations concerning the relative probability of dissociation of different bonds in equivalent positions are applicable to the forty molecules which have been studied.

The primary photolytic bond dissociation processes in aromatic compounds have received relatively little attention owing to the low quantum yields of dissociation and the difficulties of analysis of the complex products which are often formed. Previous studies have been based mainly on analysis of the gaseous products of reaction and the work most relevant to the present discussion is that of Hentz, Sworski and Burton [1, 2] who found evidence for fission in the side chain of toluene with a quantum yield of gaseous products of about 1 %.

The method which we have used here is one which eliminates the complexities of secondary reactions. It is based on the observations of Norman and Porter [3] who found that toluene and related molecules dissociate photochemically in rigid media and that aromatic free radicals remain trapped in the matrix at low temperatures, and can be observed spectroscopically. The spectra were first observed following flash photolysis of aromatic vapours by Porter and Wright (part 3) [4] and were attributed to benzyl and its derivatives. It has recently been shown that they are also formed during photolysis of ordinary solutions at normal temperatures.[5] Whilst further studies in gases and liquids will be necessary for the interpretation of the kinetics of the radical reactions, the matrix isolation technique is in many ways more suitable for the identification of the radical spectra and of primary photochemical processes owing to the elimination of most secondary reactions.

EXPERIMENTAL

The apparatus and procedure used in the main part of this investigation were identical with those described by Norman and Porter,[3] the cell length being 1·5 cm. In later experiments a different arrangement incorporating a cell 20 cm in length was used to detect weaker transitions.[6] Spectra were recorded by means of the Hilger medium and small quartz instruments and scanned by microdensitometer. All solvents were spectroscopically pure and were carefully dried in order to prevent crystallization of the glass. The aromatic compounds available as commercial products were purified by fractional distillation, recrystallization or vacuum sublimation and boiling points or melting points in every case were within one degree of those reported in the literature. Some of the particular methods used were : toluene, ethyl benzene, *iso*propyl benzene, benzyl chloride, benzyl cyanide, anisole, phenetole purified by the method of Vogel,[7] benzyl alcohol

PRIMARY PHOTOCHEMICAL PROCESSES

according to Mathews,[8] aniline according to Knowles,[9] o-xylene purified by fractional distillation, m- and p-xylenes by fractional crystallization. Phenyl ethyl chloride was prepared from the corresponding alcohol by the method of Norris and Taylor.[10]

A high-intensity light source was used in most experiments involving the 1·5 cm cell. This was a 1 kw high-pressure mercury arc, type ME/D (combined with a 2-cm water filter), which emits most lines of mercury, except for the 2537 line which is reversed, and also a continuous spectrum throughout the visible and ultra-violet region. For quantum-yield determinations a low pressure mercury-vapour lamp, combined with a 1-cm filter of 4 N acetic acid was used from which the only significant radiation in the region of aromatic absorption was at 2537 Å. The 20-cm cell was irradiated from the side by two U-tube low pressure mercury-vapour lamps; the length of each limb was 20 cm and the whole apparatus was surrounded by a reflector.

The solvents generally used to form glasses were E.P.A. (ether, isopentane and ethanol in proportions 5 : 5 : 2) and M.P. (methyl cyclohexane and isopentane in proportions 2 : 3). Solutions were always outgassed and sealed off before irradiation and all experiments were performed at − 197°C.

RESULTS

Four types of primary dissociation process have been identified in this work. These are as follows.

(i) RING FISSION

This has previously been suggested to explain the formation of hexatriene during photolysis of benzene in rigid glasses.[11,3] Spectra very similar to that of hexatriene are also observed in the permanent products of photolysis of toluene and related molecules in rigid glasses and these are probably substituted hexatrienes formed by ring fission followed by abstraction of two atoms from the solvent. The quantum yield of this process in benzene is approximately 0·01 at 2537 Å (Anderton and Porter, unpublished work).

(ii) ELECTRON EJECTION

This was first clearly established by Lewis and his collaborators in a number of aromatic molecules.[12] We have found spectra of radical cations which must result from this process, after photolysis of many amines. Unlike processes (i), (iii) and (iv), electron ejection (photo-oxidation in the terminology of Lewis) does not readily occur in non-polar glasses (e.g. M.P.), and it is normally encountered only in basic molecules.

(iii) SIDE-CHAIN DISSOCIATION AT THE β BOND

This is exemplified by the formation of the benzyl radical from toluene. Benzyl and its derivatives have strong banded spectra in the near u.-v. region, which are formed on irradiation of rigid solutions, remain as long as the glass is kept rigid and disappear completely and irreversibly on warming.[3]

(iv) SIDE-CHAIN DISSOCIATION INTO TWO MOLECULES, ONE OF WKICH IS STYRENE

This occurs rather generally in compounds of the type $C_6H_5CHXCH_2Y$. Styrene is observed in the rigid glass before warming and is therefore formed in the primary act.

A fifth process, dissociation into two ions, was discussed by Lewis but we have found no evidence for this reaction in any of the systems considered here.

Processes (i) and (ii) will be discussed further in later communications and the present work is concerned mainly with processes (iii) and (iv).

ASSIGNMENTS OF RADICAL SPECTRA

In part 3 arguments were given for assigning the common transient spectrum, observed on photolysis of toluene and other benzyl compounds in the gas phase, to benzyl and it was mentioned that more definite evidence was available from our work in rigid solvents. This will now be given. Common transient spectra, satisfying all energetic requirements, could arise from the primary photochemical dissociation of molecules of formula $C_6H_5CH_2X$ in only three ways :

$$C_6H_5CH_2X \longrightarrow C_6H_5CH_2 + X, \tag{1}$$

$$C_6H_5CH_2X \longrightarrow C_6H_5CH + HX, \tag{2}$$

$$C_6H_5CH_2X \longrightarrow C_6H_5 + CH_2X. \tag{3}$$

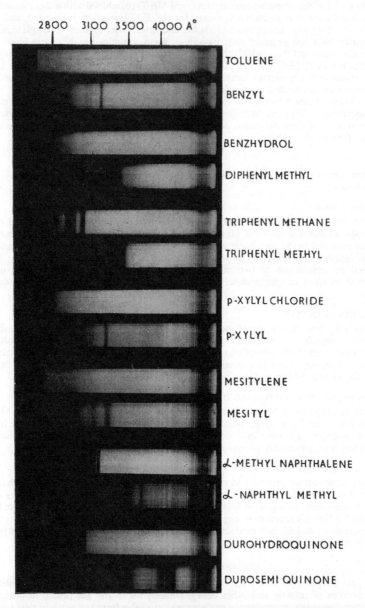

FIG. 1.—Radical spectra resulting from the photolysis of aromatic molecules at − 197°C. In each case the first spectrum is recorded before and the second after irradiation of the rigid solution for 15 min.

If we consider the series $C_6H_5CH_3$, $C_6H_5CH_2X$. $C_6H_5CHX_2$ and $C_6H_5CX_3$ process (iii) would result in one common spectrum, process (ii) could result in no more than two radical spectra and process (i) in no more than three radical spectra from the four molecules. We have investigated three such series in which X was a chlorine atom, a methyl radical and a phenyl radical respectively. The wavelength maxima of the radical absorption spectra in E.P.A. were as follows.

molecule	max. (Å)	radical
$C_6H_5CH_3$	3187	$C_6H_5CH_2$
$C_6H_5CH_2Cl$	3187	$C_6H_5CH_2$
$C_6H_5CHCl_2$	3231	C_6H_5CHCl
$C_6H_5CCl_3$	3238	$C_6H_5CCl_2$
$C_6H_5CH_2CH_3$	3222	$C_6H_5CHCH_3$
$C_6H_5CH(CH_3)_2$	3242	$C_6H_5C(CH_3)_2$
$C_6H_5C(CH_3)_3$	3242	$C_6H_5C(CH_3)_2$
$C_6H_5CH_2C_6H_5$	3355	$C_6H_5CHC_6H_5$
$C_6H_5CH(C_6H_5)_2$	3415	$C_6H_5C(C_6H_5)_2$

In each series three distinct spectra were observed, providing excellent confirmation that the dissociation process involved is β bond fission. The spectra can be assigned with some confidence to the above radicals with the exception of the 3231 Å band observed from benzal chloride. This spectrum was rather weak and, although definitely different from any other band of the series, its position and appearance were rather sensitive to concentration and other factors and its assignment should be regarded as uncertain pending further examination.

Further confirmation of the assignments is afforded by the fact that one spectrum in the series, that of triphenyl methyl, was previously well known and established being, in fact, the first free radical spectrum to be observed by Gomberg in 1900.[13] The colour of this radical is caused by a weaker absorption band at 5110 Å and we have detected this band in the present work by use of the 20 cm cell. Finally we have considered the relationship of the absorption bands of benzyl and its derivatives to the emission spectra of Schuler and his collaborators [14] and the molecular orbital calculations of Dewar, Longuet-Higgins and Pople [15, 16] and of Bingel.[17] This led us to search for the predicted weak long-wavelength system of benzyl and we have recently recorded this spectrum by use of the 20 cm cell.[6] Comparison of the data from all these sources gives a consistent account of the spectrum of benzyl and final confirmation of our assignments.[6]

Typical free-radical spectra, obtained after 15 min irradiation are shown in fig. 1. The solvent was E.P.A. in each case except α-methyl naphthalene for which the solvent was M.P. The principal absorption maxima of all spectra which we have assigned to neutral free radicals, and the molecules from which they have been observed after photolysis at − 197°C are given in table 1. The intensities are visual estimates referred to the strongest band as 10. The wavelength measurements refer to radicals in E.P.A. in all cases except the two naphthyl methyl radicals which were in M.P.

In most cases the bands were sharp and maxima could be estimated to ± 5 Å whilst comparison could be made to about ± 1 Å. The basis of the assignments has been described and although it has not always been possible to obtain a radical from more than one molecule, the close resemblances of the radical spectra in position, band width and structure are strong evidence that radicals of the benzyl type are involved and this is usually sufficient to define the carrier. The spectra all fall in a quite narrow spectral region and some chance coincidences may be expected. The only clear case of coincidence of different radical spectra is found in the three xylyl radicals but, even here, there are distinct differences in band structure ; the 3230 Å band in *m*-xylene, for example, is a doublet.

Only spectra attributed to neutral radicals are recorded in table 1 but in all the aromatic amines other transient spectra which we have assigned to the radical cations appeared on photolysis. Our reasons for not attributing these spectra to neutral radicals must be mentioned briefly. The neutral radicals formed from aniline and its N-methyl derivatives were identified in the same manner as benzyl. Thus the three aniline derivatives

TABLE 1.—FREE RADICAL SPECTRA

radical	parent molecule	$\lambda A°$	$v\,cm^{-1}$	relative intensitie
$C_6H_5CH_2$	$C_6H_5CH_3$, $C_6H_5CH_2Cl$,	3187	31380	10
	$C_6H_5CH_2OH$, $C_6H_5CH_2NH_2$,	3082	32450	3
	$C_6H_5CH_2CN$, $C_6H_5CH_2COOH$,	3047	32810	3
	$C_6H_5CH_2CH_2NH_2$, $C_6H_5CH_2CH_2OH$,			
	$C_6H_5CH_2CH_2CH_2OH$.			
$C_6H_5CH \cdot CH_3$	$C_6H_5CH_2CH_3$, $C_6H_5CHOHCH_3$,	3222	31040	10
	$C_6H_5CHNH_2CH_3$, $C_6H_5CH_2CH_2OH$.	3167	31580	5
		3129	31960	6
		3083	32440	6
$C_6H_5C(CH_3)_2$	$C_6H_5CH(CH_3)_2$, $C_6H_5C(CH_3)_3$	3242	30840	
$(C_6H_5)_2CH$	$(C_6H_5)_2CH_2$, $(C_6H_5)_2CHOH$	3355	29810	10
		3305	30260	8
		3240	30860	5
		3180	31450	3
		3122	32030	4
$(C_6H_5)_3C$	$(C_6H_5)_3CH$	3415	29290	10
		3358	29780	8
		3303	30270	2
$C_6H_5CH_2CHC_6H_5$	$(C_6H_5CH_2)_2$	3625	27580	
o—$CH_3C_6H_4CH_2$	o—$CH_3C_6H_4CH_3$	3230	30960	10
		3170	31550	1
		3100	32260	1
m—$CH_3C_6H_4CH_2$	m—$CH_3C_6H_4CH_3$	3230	30960	10
		3100	32260	6
p-$CH_3C_6H_4CH_2$	p-$CH_3C_6H_4CH_3$	3230	30960	10
	p-$CH_3C_6H_4CH_2Cl$	3170	31550	1
		3100	32260	3
		3047	32810	2
$1, 3(CH_3)_2 5CH_2, C_6H_3$	$1, 3, 5(CH_3)_3C_6H_3$	3249	30770	10
		3109	32170	6
		2964	33730	2
C_6H_5CHCl	$C_6H_5CHCl_2$	3231	30950	10
		3190	31350	5
		3100	32260	4
$C_6H_5CCl_2$	$C_6H_5CCl_3$	3238	30880	10
		3108	32180	8

		3700	27030	7
		3554	28140	5
		3500	28570	7
		3424	29210	10
		3840	26040	7
		3647	27420	5
		3687	27120	4
		3500	28570	8
		3424	29210	10
C_6H_5NH	$C_6H_5NH_2$	3109	32060	

TABLE 1.—(*Cont.*)

radical	parent molecule	λ Å	ν cm^{-1}	relative intensities
$C_6H_5NCH_3$	$C_6H_5NHCH_3$ $C_6H_5N(CH_3)_2$	3166	31590	
$C_6H_5CH_2NC_6H_5$ or $C_6H_5CHNHC_6H_5$	$C_6H_5CH_2NHC_6H_5$	3745 3635	26700 27510	10 8
C_6H_5O	C_6H_5OH, $C_6H_5OCH_3$ $C_6H_5OCH_2CH_3$	2870 limit	34840	
p-HOC_6H_4O	*p*-HOC_6H_4OH	4140	24150	10
	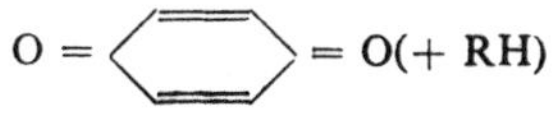	3550	28170	6
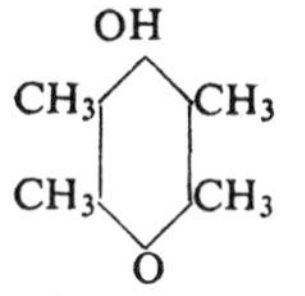	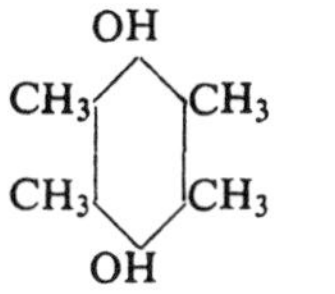	4220 4041 3300	23700 24740 30300	10 4 8
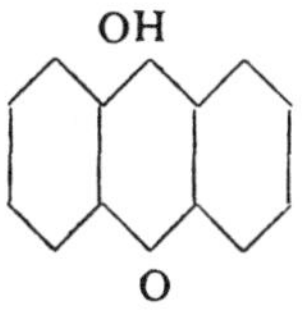	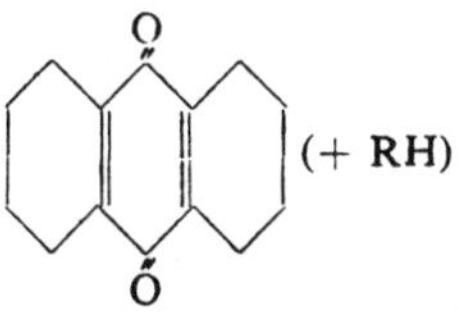	5100 3830 3531	19610 26110 28320	10 8 5

can give only two aniline type radicals and the near u.-v. spectra were therefore assigned as follows.

molecule	max. (Å)	radical
$C_6H_5NH_2$	3109	C_6H_5NH
$C_6H_5NHCH_3$	3166	$C_6H_5NCH_3$
$C_6H_5N(CH_3)_2$	3166	$C_6H_5NCH_3$.

These spectra were observed in both the polar E.P.A. glass and the non-polar M.P. glass. Other bands were observed in the visible region which were different for each of the three molecules and could not therefore be spectra of the two anilino radicals. Furthermore the visible bands were not observed in gas-phase flash photolysis and were absent or very weak in M.P. glass. Irradiation of N : N′-dimethyl *p*-phenylene diamine in E.P.A. yielded the well-known spectrum of the radical cation [18] at very high intensity whilst no such spectra were observed after irradiation in the M.P. glass. Other spectra showing these characteristics are therefore also attributed to radical cations and will be discussed in a later communication.

SIDE-CHAIN DISSOCIATION INTO TWO MOLECULES

A number of substances, after photolysis in E.P.A., gave a common banded spectrum which was still present after warming the glass. The spectrum was similar to that of styrene reported in the literature, and its identity was confirmed by comparison with a solution of pure styrene in our solvents both at room temperature and in the glass at − 197°C. The substances from which styrene was formed on irradiation in the glass were ϕCH_2CH_3, $\phi CH(OH)CH_3$, $\phi CH(NH_2)CH_3$, ϕCH_2CH_2OH, $\phi CH_2CH_2NH_2$, ϕCH_2CH_2Cl and ϕCH_2CH_2Br. Some of these also gave radical products which have been discussed; others such as ϕCH_2CH_2Cl and ϕCH_2CH_2Br gave no radical products and styrene was the only spectrum observed. Permanent products were also observed in the spectra of other compounds, some of which, although they have not been identified, were probably not of the hexatriene type, but no compounds, other than those mentioned above, showed the styrene spectrum. The only substances investigated which gave neither transient spectra not styrene were benzyl bromide and duroquinone.

DETERMINATION OF QUANTUM YIELDS

Quantum yields of processes (iii) and (iv) were determined in a representative example of each type of decomposition. For process (iii) we had only one possibility—the photolysis of triphenyl methane to give triphenyl methyl—since extinction coefficients of none of the other radicals are yet known. For process (iv) phenyl ethyl bromide was chosen because styrene is the only product detected in the photolysis of this molecule and extinctions can therefore be determined without interference from other products.

The source of radiation in both cases was the filtered low-pressure mercury arc which gave essentially monochromatic light at 2537 Å. The incident intensity was determined by conventional uranyl oxalate actinometry, and the concentrations of products were found by photometry of the photographic absorption spectra.

β BOND DISSOCIATION IN TRIPHENYL METHANE

A solution of triphenyl methane (2·44 g/l.) in E.P.A. was irradiated for 30 min and the concentration of triphenyl methyl determined from its extinction at 3450 Å using the extinction coefficients given by Chu and Weissman.[19] (These authors used a toluene + triethylamine solvent and the extinction coefficients may therefore differ slightly from those in E.P.A.) Three determinations of quantum yield gave the values $1·10 \times 10^{-2}$, $1·13 \times 10^{-2}$ and $1·08 \times 10^{-2}$ and a mean value,

$$\phi = 1·11 \times 10^{-2}.$$

STYRENE FORMATION FROM PHENYL ETHYL BROMIDE

A 10^{-3} M solution of phenyl ethyl bromide in E.P.A. was irradiated at the temperature of liquid nitrogen for 2 h and the styrene formed was determined spectrophotometrically after warming the glass to room temperature. The whole spectrum was measured to confirm its identity and quantitative estimation was based on the first absorption maximum at 2905 Å. The parent substance does not absorb at this wavelength. The concentration of styrene was determined by direct comparison with extinction measurements made on solutions of pure styrene in the same solvent. Three determinations of quantum yield gave the values $3·6 \times 10^{-2}$, $4·2 \times 10^{-2}$ and $3·0 \times 10^{-2}$ and a mean value,

$$\phi = 3·6 \times 10^{-2}.$$

The similarity of absorption intensities of the benzyl type radicals obtained from different molecules at comparable optical densities and times of irradiation and also of the amounts of styrene formed from the molecules listed in the last section suggest that the quantum yields determined probably represent the order of magnitude of quantum yields of processes (iii) and (iv) in most other molecules investigated.

MATRIX REQUIREMENTS FOR STABILIZATION

All experiments on the stabilization of aromatic radicals have been carried out in glasses at − 197°C. The low temperature has the twofold effect of ensuring high rigidity of the glass and so lowering the rate of diffusion and also of reducing the rate of chemical reactions which have a finite activation energy. It is interesting to enquire to what extent a low temperature is necessary for stabilization of benzyl type radicals in matrices which remain rigid at higher temperatures. Two systems were investigated in order to throw light on this question.

(i) Toluene was irradiated in liquid paraffin (nujol) at − 78°C at which temperature the glass is quite rigid. No benzyl radical nor any other transient species were observed.

(ii) α-Methyl naphthalene was irradiated in a matrix of polymethyl methacrylate. Solutions of the α-methyl naphthalene in polymethyl methacrylate were cast into films from chloroform and, after outgassing for several days, the film was cut into discs 1 mm thick and 1 cm in diameter. Five of these discs were placed together to form a cylinder 0·5 cm long, and this was irradiated in liquid nitrogen in the same manner as our solutions in E.P.A. or M.P. After irradiation in liquid nitrogen, spectra appeared which were identical with those found in the similar experiment in E.P.A. and described in table 1. On raising the temperature to − 78°C the spectra disappeared, although the matrix is

343

of course still rigid in a macroscopic sense, at much higher temperatures. These experiments show that high viscosity of the matrix is not a sufficient condition for stabilization of benzyl type radicals, even at − 78°C and that low temperatures are also necessary.

Photolysis of toluene in polymethyl methacrylate in a similar manner was unsuccessful because of the poor transparency of the polymer in the region of toluene absorption. No benzyl radical bands were observed but a region of absorption with maximum at 350 mμ, and half-width 35 mμ, appeared after 5 min irradiation and disappeared completely when the polymer was allowed to come to room temperature. An identical spectrum was obtained from the photolysis, at liquid nitrogen temperature, of polymethyl methacrylate alone. That the transient spectrum was not a product of photochemical dissociation of chloroform, from which the film had been cast, was established by the fact that the identical spectrum was obtained when the polymethyl methacrylate film was cast from ethyl acetate solution and also the photolysis of chloroform in E.P.A. at − 197°C gave no transient spectrum. The 350 mμ spectrum is therefore to be attributed to a primary product of photodecomposition of polymethyl methacrylate which is unstable at room temperature. Its assignment must await further investigations on related molecules.

DISCUSSION

Of the forty molecules investigated all except benzyl bromide undergo side-chain fission by process (iii) or (iv) or both, on irradiation in the near ultra-violet region. The fission processes which have been identified are summarized in table 2. Of the two primary products of dissociation, only one has been observed in each case and the other is inferred on the grounds that the process written is probably the only one which is energetically possible.

In both types of dissociation, energy is transferred intramolecularly from an excited π electron to the β bond. Light of wavelength 2537 Å is quite close to the origin of the first singlet-singlet transition in the benzene derivatives and there is therefore little excess energy of vibration in the excited molecule. Side-chain fission must therefore occur by a predissociation mechanism, i.e. a crossing to a second electronic state which is repulsive in the bond parameter concerned. For β bond fission the state concerned is probably the triplet formed by combination of the ground doublet states of the two radicals. The potential energy curves for this process, in particular for toluene, have recently been discussed by one of us.[20]

Comparison of β bond fission processes in the series of related molecules given in table 2 allows some interesting conclusions to be reached concerning the relative probabilities of dissociation of different bonds at equivalent positions. In order to make such comparisons we shall assume that the extinction coefficients of the benzyl-type radicals are not greatly different so that the optical densities observed are a fair measure of radical concentrations. The spectra in the cases to be considered are so similar as to be nearly indistinguishable so that this assumption is reasonable. Now if we study the radicals which are found after dissociation of those molecules which possess more than one kind of β bond we find, in nearly all cases, that only one of these bonds is dissociated. Furthermore all the data of table 2, including dissociation of the amines as well as the hydrocarbons, are in accordance with the statement that the probability of separation of a radical at the β bond lies in the order :

$$\begin{bmatrix} OH, NH_2, Cl. CN, \\ COOH, CH_2NH_2, CH_2OH, \end{bmatrix} > H > \begin{bmatrix} CH_3, C_2H_5, C_6H_5, C_6H_5CH_2 \end{bmatrix}.$$

This result is somewhat unexpected. It would not have been surprising if no regularities had been found since small differences in potential energy curves could greatly affect crossing probabilities. But, if, as we find, the bonds can be arranged in an order of dissociation probability which is applicable to a large number of different molecules it might be expected that this order would be the order of bond dissociation energies. This is not the case. For example, the bond energies of $C_6H_5CH_2$—H, $C_6H_5CH_2$—CH_3 and $C_6H_5CH_2$—$CH_2C_6H_5$ in the gas phase are 77·5, 63 and 47 kcal/mole respectively [21] and there is little doubt

TABLE 2.—SIDE-CHAIN FISSION PROCESSES

$C_6H_5CH_3 \rightarrow C_6H_5CH_2 + H$

$C_6H_5CH_2Cl \rightarrow C_6H_5CH_2 + Cl$

$C_6H_5CH_2OH \rightarrow C_6H_5CH_2 + OH$

$C_6H_5CH_2CN \rightarrow C_6H_5CH_2 + CN$

$C_6H_5CH_2NH_2 \rightarrow C_6N_5CH_2 + NH_2$

$C_6H_5CH_2COOH \rightarrow C_6H_5CH_2 + COOH$

$C_6H_5CH_2CH_2NH_2 \rightarrow C_6H_5CH_2 + CH_2NH_2$

$\rightarrow C_6H_5CH{=}CH_2 + NH_3$

$C_6H_5CH_2CH_3 \rightarrow C_6H_5CHCH_3 + H$

$\rightarrow C_6H_5CH{=}CH_2 + H_2$

$C_6H_5CHOHCH_3 \rightarrow C_6H_5CHCH_3 + OH$

$\rightarrow C_6H_5CH{=}CH_2 + H_2O$

$C_6H_5CHNH_2CH_3 \rightarrow C_6H_5CH \cdot CH_3 + NH_2$

$\rightarrow C_6H_5CH{=}CH_2 + NH_3$

$C_6H_5CH_2CH_2OH \rightarrow C_6H_5CH_2 + CH_2OH$

$\rightarrow C_6H_5CH \cdot CH_3 + OH$

$\rightarrow C_6H_5CH{=}CH_2 + H_2O$

$C_6H_5CH_2CH_2CH_2OH \rightarrow C_6H_5CH_2 + CH_2CH_2OH$

$C_6H_5CH_2CH_2Cl \rightarrow C_6H_5CH{=}CH_2 + HCl$

$C_6H_5CH_2CH_2Br \rightarrow C_6H_5CH{=}CH_2 + HBr$

$C_6H_5CH(CH_3)_2 \rightarrow C_6H_5C(CH_3)_2 + H$

$C_6H_5C(CH_3)_3 \rightarrow C_6H_5C(CH_3)_2 + CH_3$

$C_6H_5CHCl_2 \rightarrow C_6H_5CHCl + Cl$

$C_6H_5CCl_3 \rightarrow C_6H_5CCl_2 + Cl$

$C_6H_5CH_2CH_2C_6H_5 \rightarrow C_6H_5CH_2CHC_6H_5 + H$

$C_6H_5CH_2C_6H_5 \rightarrow (C_6H_5)_2CH + H$

$(C_6H_5)_3CH \rightarrow (C_6H_5)_3C + H$

$o\text{-}, m\text{- and } p\text{-}CH_3C_6H_4CH_3 \rightarrow o\text{-}, m\text{- and } p\text{-}CH_3C_6H_4CH_2 + H$

$p\text{-}CH_3C_6H_4CH_2Cl \rightarrow p\text{-}CH_3C_6H_4CH_2 + Cl$

$1:3:5(CH_3)_3C_6H_3 \rightarrow 1:3:(CH_3)_25, CH_2 \cdot C_6H_3$

$C_6H_5NH_2 \rightarrow C_6H_5NH + H$

$C_6H_5NHCH_3 \rightarrow C_6H_5NCH_3 + H$

$C_6H_5N(CH_3)_2 \rightarrow C_6H_5NCH_3 + CH_3$

$C_6H_5CH_2NHC_6H_5 \rightarrow C_6H_5CH_2NC_6H_5 + H$

or

$C_6H_5CHNHC_6H_5 + H$

$C_6H_5OH \rightarrow C_6H_5O + H$

$C_6H_5OCH_3 \rightarrow C_6H_5O + CH_3$

$C_6H_5OC_2H_5 \rightarrow C_6H_5O + C_2H_5$

$HOC_6H_4OH \rightarrow HOC_6H_4O \cdot + H$

$2:3:5:6\text{-}(CH_3)_41:4(OH)_2C_6 \rightarrow 2:3:5:6(CH_3)_41, (OH)4(O), C_6 + H$

$\alpha \text{ and } \beta \ CH_3C_{10}H_9 \rightarrow \alpha \text{ and } \beta \ CH_2C_{10}H_9$

that the dissociation energy of a β C—H bond in the gas phase is greater, in all the molecules of table 2 than that of a β C—C bond. The probability of dissociation found in this work is, however, consistently greater for the β C—H bond.

There is evidence, very limited at present, that the same will be found to hold for the photolysis of these compounds in solution but not in the gas phase. Thus Porter and Windsor [5] found that the flash photolysis of solutions of diphenyl methane in paraffin at normal temperatures gave the diphenyl methyl radical but Porter and Wright [4] found that flash photolysis of vapours of both ethyl benzene and diphenyl methane gave only the benzyl radical. It seems that photolysis in the gas phase may, for equivalent bonds, occur in accordance with bond-energy (gas-phase) considerations but that in solution, and in rigid solvents at low temperatures, other factors are important. There are two obvious factors which should be considered. First, bond-dissociation energies are almost unknown in solution and may be very different from those in vapours owing to differences in solvation energies of both the parent molecules and the radical products. Secondly, the activation energy of dissociation in solution may exceed the bond energy owing to a cage effect and the additional energy required to separate the dissociation products in the presence of the solvent molecules. This would be greater for the larger radicals and might account for the greater probability of separation of the smaller hydrogen atom.

Of the radical dissociation processes listed in table 2 there is one which cannot result directly from β bond fission. This is the formation of the $C_6H_5CHCH_3$ radical from $C_6H_5CH_2CH_2OH$. The most probable explanation of this observation is that, in addition to fission of the β bond to give benzyl and a type (iv) dissociation to give styrene, this molecule undergoes a γ bond fission by ejection of an OH radical and that the $C_6H_5CH_2CH_2$ radical so formed rapidly isomerizes to the more stable radical $C_6H_5CHCH_3$. Perhaps the most exceptional result in all the molecules investigated is the absence of any radicals from the photolysis of benzyl bromide. This confirms earlier results of Porter and Wright in the gas phase and is now well established since we made many attempts to detect radicals from this substance. The C—Br bond is one of the weakest investigated and therefore our failure to detect benzyl radicals from this substance is remarkable. It may be that dissociation of the β bond does occur but that, because the C—Br bond is exceptionally weak, sufficient energy remains in the benzyl radical after dissociation for this " hot " radical to react with the solvent before thermal equilibration can occur. Alternatively it may be that the increased spin-orbit interaction resulting from the presence of the heavy bromine atom results in rapid deactivation by inter-system crossing.

All molecules which give styrene as a primary product in the rigid glass are characterized by the fact that they can form a 4-centre transition state, dissociation of which leads directly to styrene and a second molecule :

$$C_6H_5CH \!=\!=\!=\! CH_2 \rightarrow C_6H_5CH = CH_2$$
$$\begin{array}{ccc} | & & | \\ X\; \text{-}\;\text{-}\;\text{-}\;\text{-Y} & + & X\text{------}Y \end{array}$$

Although intermolecular processes are excluded in the rigid glass the reaction could take place by fission of a single—probably a β—bond followed by reaction of radical X within the solvent cage to give styrene and XY. There is no means of distinguishing between this cage reaction and a true intramolecular mechanism in the present experiments but a distinction would be possible in the vapour phase where only the true intramolecular mechanism could be operative. Two molecules of structure $C_6H_5CHXCH_2Y$ did not show styrene in the products of dissociation. In both cases styrene formation would have necessitated the separation of a somewhat more complex molecule, viz.. C_6H_6 and CH_3OH.

Hydroquinones dissociate by a β bond fission in the same way as phenols. The semiquinone radical is also formed from quinones but this occurs by hydrogen abstraction from the solvent and is therefore quite separate from the other processes discussed in this paper. It has recently been the subject of a detailed investigation.[22]

This work forms part of the programme of fundamental research undertaken by the British Rayon Research Association.

1604   PRIMARY PHOTOCHEMICAL PROCESSES

1 Hentz and Burton, *J. Amer. Chem. Soc.*, 1951, **73**, 532.
2 Hentz, Sworski and Burton, *J. Amer. Chem. Soc.*, 1951, **73**, 578.
3 Norman and Porter, *Proc. Roy. Soc. A*, 1955, **230**, 399.
4 Porter and Wright, *Trans. Faraday Soc.*, 1955, **51**, 1469.
5 Porter and Windsor, *Nature*, 1957, **180**, 187.
6 Porter and Strachan, *Spectrochim. Acta*, 1958.
7 Vogel, *J. Chem. Soc.*, 1948, 607, 616, 674.
8 Mathews, *J. Amer. Chem. Soc.*, 1926, **48**, 562.
9 Knowles, *Ind. Eng. Chem.*, 1920, **12**, 881.
10 Norris and Taylor, *J. Amer. Chem. Soc.*, 1924, **46**, 753.
11 Gibson, Blake and Kalm, *J. Chem. Physics*, 1953, **21**, 1000.
12 Lewis and Lipkin, *J. Amer. Chem. Soc.*, 1942, **64**, 2801.   Lewis and Bigeleisen, *J. Amer. Chem. Soc.*, 1943, **65**, 2424.
13 Gomberg, *Ber.*, 1900, **33**, 3150.
14 Schuler and Michel, *Z. Naturforsch.*, 1955, **10a**, 459.
15 Dewar and Longuet-Higgins, *Proc. Physic. Soc. A*, 1954, **67**, 795.
16 Longuet-Higgins and Pople, *Proc. Physic. Soc. A*, 1955, **68**, 591.
17 Bingel, *Z. Naturforsch.*, 1955, **10a**, 462.
18 Michaelis, Schubert and Granick, *J. Amer. Chem. Soc.*, 1939, **61**, 1981.
19 Chu and Weissman, *J. Chem. Physics*, 1954, **22**, 21.
20 Porter, *Chem. Soc.*, *Special Publ.*, 1958.
21 Swarc, *Chem. Physics*, 1948, **16**, 128 ; 1949, **17**, 431.
22 Bridge and Porter, *Proc. Roy. Soc. A*, 1958, **244**, 259.

Further publications

TRAPPED ATOMS AND RADICALS IN A GLASS 'CAGE'.
I. Norman and G. Porter, *Nature*, 1954, **174**, 508.

FREE RADICAL STABILIZATION.
G. Porter, *Nature*, 1958, **182**, 1496.

RADIATION CHEMICAL PROCESSES IN RIGID SOLUTIONS.
H. T. J. Chilton and G. Porter, *J. Phys. Chem.*, 1959, **63**, 904.

STUDIES OF SOME NEW MATRICES FOR RADICAL STABILIZATION.
H. T. J. Chilton and G. Porter, Fourth International Symposium on
Free Radical Stabilization, Washington, D.C., 1959, p. 1.

STABILIZED FREE RADICALS IN SALT MATRICES.
H. T. J. Chilton and G. Porter, *Spectrochim. Acta*, 1960, **16**, 390.

PHOTOLYSIS OF BENZENE IN VISCOUS SOLVENTS.
E. J. Anderton, H. T. J. Chilton and G. Porter, *Proc. Chem. Soc.*, 1960, 352.

PHOTO-IONIZATION OF AROMATIC COMPOUNDS IN
HYDROCARBON GLASS AT 77°K.
W. A. Gibbons, G. Porter and M. I. Savadatti, *Nature*, 1965, **206**, 1355.

Chapter 7

MOLECULAR DYNAMICS IN SOLUTION

The following papers describe studies of essentially physical processes of molecules in solution which are closely involved with, and often rate determining, in simultaneous chemical processes. They are mainly diffusional processes of translational and rotational motion of molecules, most of them very fast in fluid solvents.

Reprinted from the Proceedings of the Royal Society, A, *volume* 284, pp. 9–16, 1965

Diffusion studies in viscous media

By A. D. Osborne and G. Porter, F.R.S.

Department of Chemistry, The University, Sheffield 10

(*Received* 27 *July* 1964)

The rate constants of quenching of the triplet state of naphthalene by α-iodonaphthalene, *t*-butyl hydroperoxide and oxygen have been studied in several solvents as a function of viscosity and temperature. The diffusion controlled rate constants for α-iodonaphthalene in hydroxylic media agree with those predicted by the equation $k_{calc.} = 8RT/2000\eta$ (η being viscosity). This is interpreted as evidence for the validity of the Stokes–Einstein expression, Stoke's law with 'slip' (i.e. coefficient of sliding friction zero) being used for diffusing species comparable in size to the molecules comprising the medium. In liquid paraffin/*n*-hexane mixtures, the ratio $k_{obs.} : k_{calc.}$ became progressively larger as the proportion of liquid paraffin:*n*-hexane was increased and in 100% liquid paraffin the ratio was 4·5. This is interpreted as being due to breakdown of the Stokes–Einstein expression in cases where the diffusing species is small compared with the molecules of the solvent. The observed rate constant for quenching by oxygen is also anomalously high, being over 100 times greater than predicted under conditions of high oxygen concentration and high solvent viscosity.

Introduction

A more quantitative understanding of diffusion controlled reactions would be of value for two reasons. First, such reactions provide a possible method for the measurement of diffusion coefficients under conditions where other methods are tedious or inapplicable and secondly, a comparison of measured rates with rate constants calculated from diffusion theory provides one of the few generally applicable methods for the measurement of concentrations and extinction coefficients of transient species.

While there are a number of ways of measuring diffusion coefficients in media of low viscosity (Tyrrell 1961) most methods would require prohibitively long times at high viscosities (*ca.* 2000 cP). Methods based on diffusion-controlled quenching of fluorescence have been used (Williamson & La Mer 1948; Ware 1962). This method also ceases to be useful at high viscosities because the lifetime of fluorescence (*ca.* 10^{-8} s) becomes appreciably shorter than the average time required for encounter with a molecule of quencher. This difficulty could be overcome by using longer lived excited species, and triplet states which have lifetimes in the range 10^{-3} to 10 s immediately suggest themselves. The observation of the rate of quenching of triplet states by their phosphorescence is not generally feasible since under fluid conditions most of the molecules decay by radiationless processes (Hilpern, Porter & Stief 1964) and the concentration of triplets is best monitored by observation of their absorption spectrum involving transitions to higher triplet levels.

Experimental

(a) *Apparatus and materials*

The general method and technique of the use of flash photolysis to measure rate constants by photoelectric monitoring of the transient absorption has already been described (Hoffman & Porter 1962).

Naphthalene was B.D.H. microanalytical grade and was not further purified. α-iodonapththalene was B.D.H. laboratory grade and was distilled in small quantities before use and stored over calcium. For experiments at 25 °C and for hydroxylic solvents, mixtures of isopropanol with 1,2-propanediol ($\eta = 2$ to 40 cP) and of 1,2-propanediol with glycerol ($\eta = 40$ to 512 cP) were used. For hydrocarbon solvents the viscosity was varied from 7 to 160 cP. by using mixtures of n-hexane and liquid paraffin. In each case the viscosity was measured by timing the flow of the solvent, thermostated at 25 °C, through a calibrated capillary. Values for the viscosity of 1,2-propanediol from 49 to -80 °C were those measured by Stief (1963).

Experiments at other temperatures employed the modified apparatus described by Hilpern *et al.* (1964). All solutions were rigorously outgassed.

(b) *Choice of quenchers*

Molecules having triplet levels of lower energy than triplet naphthalene act as efficient quenchers by energy transfer (Porter & Wilkinson 1961). α-iodonaphthalene is a particularly useful quencher because its triplet state, as well as being of lower energy, is very short-lived and the system is not, therefore, complicated by absorption of triplet α-iodonaphthalene formed in the energy transfer reaction

$$^3C_{10}H_8 + {}^1C_{10}H_7I \rightarrow {}^1C_{10}H_8 + {}^3C_{10}H_7I.$$

A disadvantage of the use of α-iodonaphthalene is that its absorption spectrum is similar to that of naphthalene. This places an upper limit on the concentration of α-iodonaphthalene which may be added to the system without appreciable direct absorption of light by the quencher. In practice it was found that α-iodonaphthalene could be used up to concentrations of 10^{-4} M which are useful for quenching studies at viscosities up to 63 000 cP. Quenchers which do not absorb in the visible and near ultra-violet do not suffer from this drawback. This precludes aromatic molecules and most aliphatic substances do not cause quenching of the triplet state. Since peroxides have a bond energy (35 kcal/mole) less than the energy of triplet naphthalene (55 kcal/mole), they might be expected to quench triplet naphthalene by energy transfer to the repulsive triplet of the peroxide. This was found to be the case (Lodhi, Osborne & Porter, unpublished work). At room temperature photosensitized decomposition of the peroxide occurs but at lower temperatures (< -30 °C, higher viscosities) no decomposition occurs presumably because of recombination of alkoxyl radicals trapped in the solvent cage.

Oxygen was chosen for study because of its obvious importance in the understanding of the decay of the triplet state.

(c) *Measurements of the solubility of oxygen in 1,2-propanediol*

For experiments using oxygen as quencher the concentration was varied by shaking the naphthalene solution, after outgassing, with a known pressure of oxygen. Measurement of the concentration of oxygen in the solution requires a knowledge of the solubility of oxygen as a function of pressure and, since these data were not available, the concentration of dissolved oxygen in equilibrium with various pressures of gaseous oxygen above the solution was determined with the apparatus shown in figure 1.

The design was such that a large surface area of solvent was exposed to the oxygen, and the solvent was stirred magnetically. Even so, attainment of equilibrium took up to 3 h as the curves in figure 2 show.

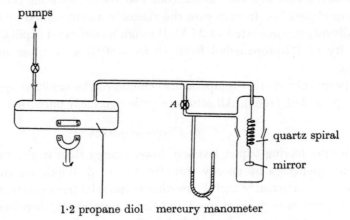

FIGURE 1. Apparatus for measurement of solubility of oxygen.

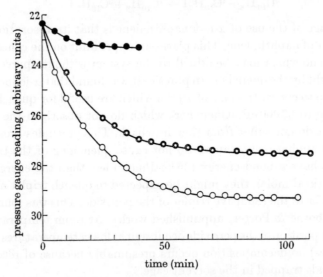

FIGURE 2. Oxygen uptake curves for various pressures (mmHg):
●, 138; ◎, 276; ○, 387.

The volume above the solution was kept small to maximize the fall in pressure on absorption of oxygen. The initial pressure was measured with a mercury manometer attached to the envelope of a Quartz spiral gauge. The fall in pressure following absorption of oxygen by the solvent was then followed on the spiral gauge. The procedure was to outgass the 1,2-propanediol (50 ml.) by pumping and stirring for 3 h. The required pressure of oxygen was admitted above the solution with the stopcock A open. This operation caused no deflexion of the gauge. The stopcock

12   A. D. Osborne and G. Porter

A was then closed and the deflexion followed with time. The resulting curves (figure 2) show the approach to and attainment of equilibrium. A plot of pressure of oxygen against concentration was linear (figure 3) and the solubility was found to be $2 \cdot 25 \times 10^{-6}$ mole/mmHg of oxygen. All solubility measurements were carried out at 25 °C. For quenching experiments at low temperature it was assumed that the oxygen concentration was the same as that in equilibrium at 25 °C with the pressure of oxygen above the solution. In view of the fact that, even at 25 °C in an apparatus designed for efficient uptake of oxygen, 3 h were required, it is reasonable to assume that at higher viscosities in cells having a small surface:volume ratio no significant further uptake of oxygen occurs.

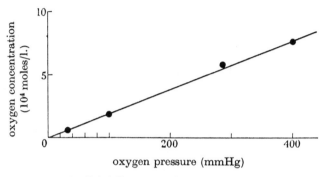

FIGURE 3. Solubility curve for oxygen in 1,2-propanediol.

RESULTS

(a) α-iodonaphthalene and t-butyl hydroperoxide

The observed pseudo first-order rate constant $k_{obs.}$ for the decay of the triplet state T in the presence of quencher Q is given (Porter & Wilkinson 1961) by

$$k_{obs.} = k_1 + k_Q(Q),$$

where k_1 is the rate constant in the absence of added quencher.

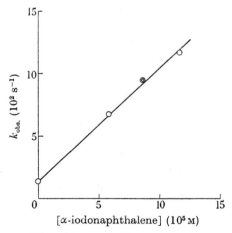

FIGURE 4. Plot of $k_{obs.}$ at 25 °C against [α-iodonaphthalene] in 6:1 glycerol/1,2-propanediol.

Values of the diffusion controlled second-order rate constant k_Q were obtained at 25°C with α-iodonaphthalene as quencher from the gradients of plots of $k_{\text{obs.}}$ against (Q) (figure 4) and these are compared in table 1 with $k_{\text{calc.}}$ derived from the expression $k_{\text{calc.}} = 8RT/2000\eta$ (see later discussion). In 1,2-propanediol, glycerol and 50/20 liquid paraffin/n-hexane the ratio $k_Q/k_{\text{calc.}}$ is close to unity. The low value for isopropanol is accounted for by the fact that the reaction is not strictly diffusion controlled at this low viscosity (2 cP). In liquid paraffin/n-hexane mixtures the ratio increased steadily with the proportion of liquid paraffin reaching 4·6 in 100 % liquid paraffin.

TABLE 1. SECOND-ORDER RATE CONSTANTS FOR THE QUENCHING OF
TRIPLET NAPHTHALENE BY α-IODONAPHTHALENE AT 25 °C

solvent	viscosity (cP)	k_1 (s^{-1})	k_Q	$k_{\text{calc.}}$	$k_Q/k_{\text{calc.}}$
			\$(10^6$ mole^{-1} s$^{-1})$		
hydroxylic					
isopropanol	2	2500	1700	4950	0·34*
1,2-propanediol	40	4500	280	248	1·13
6:1 glycerol/1,2-propanediol	512	140	21	19·5	1·08
hydrocarbon					
liquid paraffin/n-hexane					
50/20	7	4700	1600	1480	1·08
60/10	21	650	1200	495	2·43
65/5	55	550	540	189	2·86
70/0	160	350	280	61·5	4·56

* k_Q not fully diffusion-controlled.

The results of experiments where viscosity variation was effected in 1,2-propanediol by cooling are best displayed by an Arrhenius plot (figure 5). Such plots showed the now familiar low and high viscosity regions (Hilpern *et al.* 1964).

If $\quad k_Q = 8RT/2000\eta \quad$ then $\quad \log_{10} k_Q = \log_{10}(8R/2000) + \log_{10} T - \log_{10}\eta \quad$ and, neglecting the relatively small variation of $\log_{10} T$,

$$\log_{10} k_Q = \text{constant} - \log_{10}\eta$$
$$= \text{constant} - \Delta H/2\cdot303RT,$$

where ΔH is the activation energy for viscous flow. ΔH varies from 9 kcal/mole in the range 22 to 49 °C to 17 kcal/mole between -35 and -78 °C (Stief 1963) which explains the curvature of the rate constant plots above 0 °C. Above -40 °C quenching by peroxide was no longer diffusion controlled and at 25 °C it has a rate constant a factor of 10 less than the encounter rate. This is exhibited in the diagram by the fall off at temperatures above -40 °C.

The results show that for α-iodonaphthalene and t-butyl hydroperoxide the temperature coefficient of the rate constants for quenching is similar to that for viscous flow over the range 25 to -80 °C (40 to 31 000 000 cP) in 1,2-propanediol.

14   A. D. Osborne and G. Porter

(b) Oxygen

The results for quenching by molecular oxygen at pressures above the solution of 1·5, 32 mm and under atmospheric pressure of air are shown in figure 6. Comparisons of $k_{obs.}$ and $k_{calc.}$ are shown in table 2. At low concentrations of oxygen and

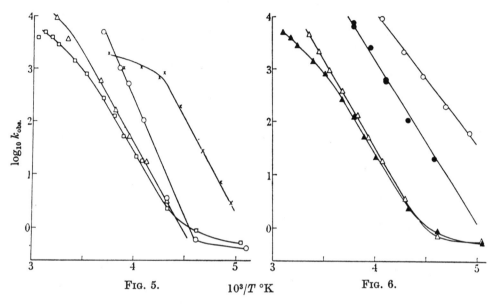

FIG. 5. $10^3/T$ °K   FIG. 6.

FIGURE 5. Arrhenius plot, $\log_{10} k_{obs.}$ against $1/T$. □, no added quencher; △, $7\cdot5 \times 10^{-6}$M α-iodonaphthalene; ○, $1\cdot65 \times 10^{-4}$M α-iodonaphthalene; ×, $5\cdot2 \times 10^{-3}$M t-butyl hydroperoxide.

FIGURE 6. Arrhenius plot for quenching by oxygen. ▲, no added quencher; △, 1·5 mm. oxygen ($3\cdot38 \times 10^{-6}$M); ●, 32 mm oxygen ($7\cdot2 \times 10^{-5}$M); ○, atmospheric air ($3\cdot6 \times 10^{-4}$M).

at low viscosities the ratio $k_Q : k_{calc.}$ was 6 to 8 but under atmospheric pressure of air the ratio became progressively greater with increasing viscosity rising to a value of 130.

TABLE 2

(O_2)	k_Q (10^6 mole^{-1} s^{-1})	η (cP)	$k_{calc.}$ (10^6 mole^{-1} s^{-1})	$k_Q/k_{calc.}$
$7\cdot2 \times 10^{-5}$M	90	596	14·7	6·1
	33·6	1995	4·2	8·0
	8·35	7079	1·14	7·3
	1·47	35480	0·22	6·7
	0·25	354800	0·021	12·2
	0·0052	21000000	0·00028	18·7
$3\cdot6 \times 10^{-4}$M	25	4300	1·89	13·2
	5·66	24000	0·323	17·5
	1·96	115000	0·0645	30
	0·55	770000	0·0092	54
	0·17	5200000	0·0013	131

Discussion

The simple expression $k = 8RT/3000\,\eta$ (Williamson & La Mer 1948; Debye 1942) results from the combination of the Smoluchowski equation (1917)

$$k = 10^{-34}\pi\sigma_{AB}N(D_A + D_B)$$

for the diffusion controlled reaction $A + B \to C$ (σ_{AB} = encounter cross-section; $D_A + D_B$ = diffusion coefficients of A and B; N = Avagadro's Number) with the Stokes–Einstein equation (Stokes 1850; Einstein 1905, 1906, 1908)

$$D_A = RT/6\pi r_A\,\eta$$

(r_A being the radius of A, assumed spherical). The assumption is made that

$$\sigma_{AB} = r_A + r_B = 2r_A.$$

The Smoluchowski expression is based on the validity of Fick's laws of diffusion and on the assumption that diffusion of A into B is spherically symmetric. These are generally good assumptions and the Smoluchowski equation is probably obeyed quite well. The Stokes–Einstein expression is more open to objection for species whose size is not appreciably bigger than that of the molecules of the surrounding medium.

The Stokes–Einstein expression is derived from the Stokes law ζ (the frictional coefficient) $= 6\pi\eta r$, which is a limiting ($\beta \to \infty$) form of the more rigorous expression (Tyrrell 1961, p. 127)

$$\zeta = 6\pi\eta r \,\frac{1 + 2\eta/\beta r}{1 + 3\eta/\beta r},$$

where β is the coefficient of sliding friction. If $\beta \to 0$, i.e. the species can 'slip' in contact with the medium

$$\zeta = 4\pi\eta r.$$

The modified Debye expression would then be $k_{\text{calc.}} = 8RT/2000\eta$ and the ratio of the expression with and without 'slip' $= 1.5$. Our finding that the experimental ratio $k_{\text{obs.}}/k_{\text{Debye}} \approx 1$ for $\beta = 0$, may be taken to mean that the molecules concerned can 'slip' in contact with the medium. A more satisfying approach is to measure D directly and this has been done at 35 °C for naphthalene and α-iodonaphthalene (Osborne, Tyrrell & Zaman 1964). This leads to value $\sigma = 4.2$ Å. This compares with a value of *ca.* 7 Å calculated from molar volumes assuming molecules to be spherical and 3·5 Å which is the distance between graphite plane. Although the Stokes–Einstein expression is probably only good to an order of magnitude the similarity of the temperature coefficient of k_Q to that for viscous flow shows that $D\eta\sigma_{AB}/T = $ constant (Walden's rule) over quite a large range of η and T, i.e. over the whole range covered (25 to -80 °C $\eta = 40$ to $30\,000\,000$ cP).

The much poorer agreement of the Debye equation with experiment in solutions containing large proportions of liquid paraffin may be attributed to the fact that liquid paraffin is composed of long chain molecules which offer high resistance to the passage of macroscopic bodies (e.g. the steel balls used in viscosity measurement) but considerably less resistance to the passage of molecules which are small

16   A. D. Osborne and G. Porter

compared to the size of chain. An extreme example of this is gelatin which has a high macroscopic viscosity but in which the rate constant of triplet decay is the same as water, the excited species and quencher molecules being able to diffuse through the water and around the large protein molecules. The same arguments apply to the diffusion of oxygen in 1,2-propanediol. Similar results have been obtained by Williamson & La Mer (1948) and more recently by Ware (1962) for low viscosity solvents from fluorescence quenching measurements.

The very much higher values of $k_Q/k_{calc.}$ for oxygen in high concentrations at high viscosities (table 2) may be attributed to static quenching in which molecules cause quenching over distances appreciably greater than 4 or 5 Å. For each separation of triplet from quencher there is a quenching probability. This falls off rapidly with distance and at low viscosities does not cause appreciable quenching at distances above 4 or 5 Å. At high viscosities where the times concerned are much greater, low quenching probabilities per unit time may contribute appreciably to the rate or, in other words, the effective σ in the Smoluchowski equation is increased.

REFERENCES

Debye, P. 1942 *Trans. Electrochem. Soc.* **82**, 265.
Einstein, A. 1905 *Ann. Phys., Lpz.,* **17**, 549.
Einstein, A. 1906 *Ann. Phys., Lpz.,* **19**, 371.
Einstein, A. 1908 *Z. Elektochem.* **14**, 235.
Hilpern, J., Porter, G. & Stief, L. J. 1964 *Proc. Roy. Soc.* A, **277**, 437.
Hoffman, M. Z. & Porter, G. 1962 *Proc. Roy. Soc.* A, **268**, 46.
Osborne, A. D., Tyrrell, H. J. V. & Zaman, M. 1964 *Trans. Faraday. Soc.* **60**, 395.
Porter, G. & Wilkinson, F. 1961 *Proc. Roy. Soc.* A, **264**, 1.
Smoluchowski, M. 1917 *Z. Phys. Chem.* **92**, 129.
Stief, L. J. 1963 To be published.
Stokes, G. G. 1850 *Mathematical and physical papers*, Cambridge University Press. London (1903), vol. III, pp. 1, 55.
Tyrrell, H. J. V 1961 *Diffusion and heat flow in liquids.* London: Butterworths.
Ware, W. R. 1962 *J. Phys. Chem.* **66**, 455.
Williamson, B. & La Mer V. K. 1948 *J. Amer. Chem. Soc.* **70**, 717.

PICOSECOND ROTATIONAL DIFFUSION
IN KINETIC AND STEADY STATE FLUORESCENCE SPECTROSCOPY

G. PORTER, P.J. SADKOWSKI and C.J. TREDWELL

Davy Faraday Research Laboratory of the Royal Institution, London W1X 4BS, UK

Received 4 May 1977

The rotational diffusion time constants of tetrachlorotetraiodofluorescein in a series of low viscosity, hydrogen bonding solvents have been determined by picosecond time-resolved fluorescence depolarisation spectroscopy. Steady state measurements of this dye and rhodamine 6G have been used to obtain results of comparable accuracy to the kinetic technique, provided that the fluorescence lifetimes are known. Data from both techniques indicate that Stokes—Einstein behaviour holds at low viscosities, but that in the higher viscosity solvents rotation times are longer than predicted.

1. Introduction

The excitation of an isotropic molecular system with a linearly polarised light beam produces a non-random distribution of excited states owing to the preferential absorption of light by those molecules with their absorption dipole lying in the plane of polarisation [1,2]. This anisotropic distribution may randomise in time via a number of processes, particularly molecular rotational diffusion, intermolecular energy transfer, and intramolecular relaxation. Provided that the absorber concentration is kept low ($<10^{-4}$ M) and excitation is to the first excited state, the latter two processes may be neglected. The rotational motion of excited state molecules may therefore be studied by probing the anisotropy of the excited state distribution.

Eisenthal et al. [3,4], and Lessing et al. [5—7] have investigated molecular rotation in a variety of low viscosity alcohols by picosecond time-resolved absorption spectroscopy. This technique utilises the anisotropy generated in the ground state population which is detected as a difference in the transmission (by the sample) of orthogonally polarised probe pulses. The rotation times reported for rhodamine 6G range from 100 ps in methanol to more than 1 ns in the higher alcohols [3—6]. Solvents with a molecular size smaller

than that of the solute give results compatible with the Stokes—Einstein relationship, in that the rotation time is linearly dependent upon the solvent viscosity [3,4]. However, the technique does require a large excited state population which complicates the kinetic analysis of the data and introduces the problem of stimulated emission [6].

The kinetic fluorescence depolarisation technique described by Fleming et al. [8] has the advantage that the excited state population can be kept low, thus minimizing the anisotropy of the ground state distribution and precluding stimulated emission. Their study of the dianionic fluorescein dyes, eosin and rose bengal, indicated that the Stokes—Einstein relationship was not an adequate description for the rotation of these molecules. They also suggested that solvent attachment via hydrogen bonding might play an important role in determining the rotation kinetics of molecules in solution, as have a number of other authors [3,4,7].

In this paper we report the use of the latter technique in the investigation of the validity of the Stokes—Einstein relationship for the dianion of 2,4,5,7-tetrachloro-2',3',4',5'-tetraiodofluorescein (TCTIF). We also examine steady state fluorescence depolarisation as a means of obtaining similar kinetic information from TCTIF and rhodamine 6G.

2. Experimental

The purity of 2,4,5,7-tetrachloro-2',3',4',5'-tetra-iodofluorescein (sodium salt of TCTIF) was confirmed by elemental analysis, thin layer chromatography, NMR and ^{13}C NMR; laser grade rhodamine 6G was used in the steady state measurements of this dye. 5×10^{-5} M solutions of the dyes were prepared in methanol, ethanol, n-propanol, n-butanol, n-pentanol, n-hexanol, n-heptanol, glycerol, and distilled water (pH 6), of these solutions only the first four alcohols and water were employed in the kinetic fluorescence depolarisation studies of TCTIF. Buffering the TCTIF in alkaline solution was found to have no significant effect upon the fluorescence lifetime. Only slight changes in the spectral characteristics of the dyes were observed in the various alcohols, but the aqueous solution of TCTIF showed a marked blue shift in both the emission and absorption spectra, and also a shortening of the fluorescence lifetime.

A complete description of the laser system has been given elsewhere [9], it consists of a mode-locked neodymium/glass laser oscillator which generates a train of over 100 pulses, each separated by 6.9 ns and approximately 6 ps (fwhm) in duration at the centre of the train. A temperature-tuned CDA crystal generates the 530 nm second harmonic prior to single pulse selection by a Pockels cell electro-optic shutter. The experimental arrangement for kinetic fluorescence depolarisation measurements is shown in fig. 1. Polariser Pl (HN22, Polaroid) is polarised in the same plane as the laser pulse to exclude any orthogonal component caused by scattering at the optical surfaces. Par-allel (0°) and perpendicular (90°) components of the fluorescence are resolved by polariser P2 (HN22) which may be rotated through 90°. The transmitted component of the TCTIF fluorescence (>570 nm — OG 570, Schott) is time resolved by an IMACON 600 ps streak camera (John Hadland (P.I.) Ltd.) with an S20 photocathode. A vidicon optical multichannel analyser (OMA 1205 A and B, Princeton Applied Research) detects the streak image and stores the intensity profile in a 500 channel digital memory. The linearity of the OMA/streak camera combination is better than ±3% within the range of 20 to 3000 counts in any channel of the memory. Digital data from the memory are subsequently transferred to punch tape for analysis, and are displayed on an oscilloscope so that the quality of the traces may be assessed. Fluorescence components polarised parallel ($I_{\parallel}$) and perpendicular ($I_{\perp}$) to the excitation pulse are recorded separately and normalised during the processing of the data.

Kinetic measurements of this type are normally performed with a 90° monitoring arrangement [8] as illustrated by the polarisation diagram in fig. 1. The excitation beam is polarised in the AB plane and directed along the B axis; fluorescence emitted along the C axis is monitored in this configuration. By monitoring the emission along the B axis (i.e. a 180° configuration) the streak camera optics are able to collect light more efficiently. Provided that care is taken to prevent stimulated emission and self-absorption, Cehelnik et al. [10] have shown that both configurations supply the same information. The use of low dye concentrations and low excitation intensities ($<10^{15}$ photons cm^{-2}) preclude these two possible sources of error.

Steady state depolarisation ratios were recorded on an MPF4 spectrofluorimeter (Perkin Elmer) using a 90° monitoring arrangement and HN22 polarisers. To correct for the inherent polarisation of the fluorimeter [10], the ratio $I_{\parallel}/I_{\perp}$ was determined for both 0° and 90° polarised excitation; these values were then multiplied together to give the corrected ratio. In all cases a 5 nm bandwidth 530 nm excitation beam was used in order to simulate the laser pulse. Fluorescence polarisation ratios were recorded at the maximum of the emission (λ_{max}); it should be noted that the fluorimeter polarisation is wavelength dependent and each measurement should be corrected independently.

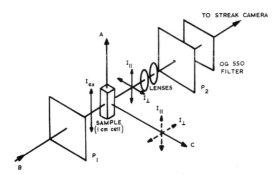

Fig. 1. The experimental arrangement for kinetic fluorescence depolarisation measurements.

3. Results and discussion

Theoretical treatments of fluorescence depolarisation by molecular rotation have been presented by a number of authors [11–13,8]; kinetic measurements are expressed in terms of the time dependent polarisation anisotropy, $r(t)$, where

$$r(t) = [I_\parallel(t) - I_\perp(t)]/[I_\parallel(t) + 2I_\perp(t)] \ ,$$

and $I_\parallel(t)$ and $I_\perp(t)$ are the respective values of $I_\parallel$ and $I_\perp$ at time t. It has been shown [11–13,8] that $r(t)$ is a single exponential function for spherical rotators and for symmetric oblate (disc shaped) and prolate (rod shaped) rotators where the transition moment is parallel to the symmetry axis. Fleming et al. [8] have demonstrated numerically that an oblate rotator with its transition moment perpendicular to the symmetry axis will have an $r(t)$ function that approximates to a single exponential for any value of the axial ratio. The only instance where $r(t)$ will be an obvious sum of exponentials is for a prolate rotator with a transition moment perpendicular to the symmetry axis. For a single exponential $r(t)$ function, τ_{rot} is defined as the $1/e$ lifetime of the decay curve.

The effect of molecular rotation on the decay kinetics of the two polarised fluorescence components is illustrated by fig. 2 for an aqueous solution of TCTIF (structure shown inset). Rotation out of the plane of polarisation is indicated by the rapid initial decay of

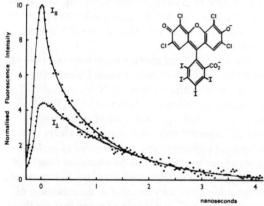

Fig. 2. The corrected fluorescence decay curves of TCTIF polarised parallel ($I_\parallel$) and orthogonal ($I_\perp$) to the excitation pulse. The molecular structure of TCTIF is shown inset.

the $I_\parallel(t)$ component, whereas the $I_\perp(t)$ component often shows a risetime due to rotation into the plane. Since the two components are recorded on separate laser shots, they are not directly comparable and must be normalised with respect to the excitation intensity. In a low viscosity solvent the rotation time is generally shorter than the fluorescence lifetime, consequently the system can randomise before the emission intensity becomes undetectable. For a normalised set of curves the ratio $I_\parallel(t)/I_\perp(t) = p(t)$ tends to unity as the system randomises; in practice, $p(t)$ tends to a constant value which may be used to normalise the data, provided that the time origin of both curves can be matched to within ±2 OMA channels. The curves shown in fig. 2 have been corrected in this manner, and the function $r(t)$ was obtained from the normalised values of $p(t)$ using the expression:

$$r(t) = [p(t) - 1]/[p(t) + 2] \ .$$

At time $t = 0$, the theoretical maxima of $r(t)$ and $p(t)$ are $r(0) = 0.4$ and $p(0) = 3.0$ [11–13,8]. These provide an additional check on the accuracy of the normalisation procedure; all of the results reported here gave values of 0.38 ± 0.02 for $r(0)$ on the leading edge of the curves. In all cases the function $r(t)$ was found to follow a single exponential decay law. A summary of the rotation times (τ_{rot}), spectral data, and fluorescence lifetimes of TCTIF in solutions of varying viscosity is given in table 1; a plot of τ_{rot} as a function of solvent viscosity is shown by the open circles in fig. 3a.

One of the main difficulties in the determination of molecular rotation times from steady state data has been that of obtaining an accurate value for the fluorescence lifetime (τ_{fl}) in any given solvent. The steady state polarisation ratio (P) is dependent on both of these relaxation times and is given by:

$$P = (I_\parallel - I_\perp)/(I_\parallel + I_\perp)$$
$$= P_0[1 + (1 - P_0/3)(\tau_{fl}/\tau_{rot}^*)]^{-1} \ ,$$

where P_0 is the fluorescence polarisation ratio in the absence of rotational diffusion, i.e. in glycerol and τ_{rot}^* is the rotation time determined by the steady state method. Theoretically P_0 is equal to 0.5, but owing to light scattering and birefringence in the cell, this value is difficult to attain; the value of 0.462 for P_0 recorded in these experiments corresponds to

Table 1

Solvent	η (cP) $20°C$	Abs. (nm)	Em. (nm)	τ_{rot} (ps) ±10%	τ_{rot}^* (ps) ±10%	τ_{rot} [b] (ps) (calc)	τ_{fl} (ps) ±5%	P
methanol	0.60	553	569	230	240	213 [a]	3400	0.035
water (pH 6)	1.00	538	562	340	250	355 [a]	1100	0.097
ethanol	1.20	552	568	480	450	426 [a]	3400	0.062
n-propanol	2.26	553	570	900	830	803	3400	0.103
n-butanol	2.95	554	576	1030	1030	1048	3400	0.122
glycerol	1000	552	568	–	–	–	–	$P_0 = 0.462$

[a] Ref. [8]   [b] Calculated for a sphere of radius 7 A.

$r(0) = 0.36$, in reasonable agreement with that obtained from the kinetic results. However, the value of $P_0 = 0.422$ for rhodamine 6G is probably a more typical result from this technique. Since the fluorescence lifetimes can be accurately measured with the streak camera [8,9], the rotation times (τ_{rot}^*) can be calculated from the above equation. Values of τ_{rot}^* and P for TCTIF in the lower viscosity alcohols are summarised in table 1; these rotation times are plotted as crosses in fig. 3a in order to demonstrate the agreement between steady state and kinetic results; fig. 3b shows a plot of all the steady state results for both dye molecules.

Applying the Stokes–Einstein relationship to the problem of a spherical rotating molecule [1,2], τ_{rot} (determined by kinetic or steady state methods) may

be expressed in terms of the diffusion coefficient (D), or the solvent viscosity (η) and the molecular radius (a):

$$\tau_{rot} = (6D)^{-1} = 4\pi\eta \, a^3/3kT \ ,$$

where k is the Boltzmann constant and T is the absolute temperature of the solvent. The rotation times should therefore be linearly dependent upon the viscosity of the solvent. Within the limits of experimental error the kinetic and steady state data, shown in fig. 3a, both show this behaviour. From the slope of the least squares fit to the kinetic data and the Stokes–Einstein equation, the apparent molecular radius is found to be 7.03 ± 0.03 Å. This is in excellent agreement with one of the calculated semi-axes [8] of rose bengal (oblate rotator 7 Å × 2 Å), which is the same

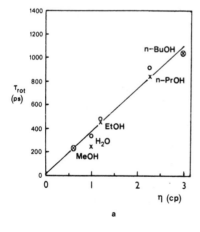

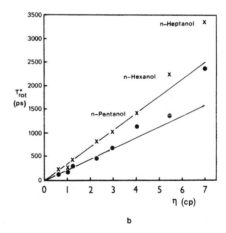

Fig. 3. (a) The rotation times of TCTIF as a function of viscosity (cP) kinetic method, τ_{rot} (○). Steady state method, τ_{rot}^* (×). (b) τ_{rot}^* as a function of viscosity (cP), for TCTIF (×) and rhodamine 6G (○).

Volume 49, number 3 CHEMICAL PHYSICS LETTERS 1 August 1977

as TCTIF except that the halogen substituents are exchanged. However, the rotation times are too long for an oblate rotator of these dimensions [8,12] and it must be concluded that solvent attachment via hydrogen bonding has increased the hydrodynamic volume of the molecule. A similar conclusion was reached by Lessing et al. [7] for fluorescein 27; the dye in methanolic solution was reported to have a rotation time of 200 ± 50 ps in agreement with our results and those of Fleming et al. [8]; the rotation time in n-decanol ($\eta = 13.8$ cP) was found to be 8.5 ± 1.0 ns, in comparison with 4.6 ns calculated for a Stokes–Einstein sphere of 7 Å radius.

Rhodamine 6G has a skeletal structure similar to these dyes and would therefore be expected to give the same experimental value for the apparent molecular radius. In fact at low viscosity the rotation time is twice as fast, although still a linear function of viscosity, as has previously been reported by Eisenthal et al. [3,4]. The apparent molecular radius calculated from these results is 6.0 ± 0.03 Å, which is close to the apparent molecular radius of 5.6 Å obtained from the theoretical rotation times for an oblate rotator with semi-axes of 7 Å and 2 Å [8]. Hydrogen bonding with the solvent must be far weaker in this particular instance to produce such a result. Although the differences in the functional groups of the rhodamines and the fluoresceins could cause some changes in the degree of hydrogen bonding, the major factor is probably the change from a cationic to a dianionic species. From the data available it is not possible to tell whether TCTIF behaves as a sphere of 7 Å radius, or whether solvent attachment merely enlarges the oblate rotator.

At higher viscosities, where the size of the solvent molecules exceeds that of the solute, the correlation between rotation time and viscosity breaks down; rhodamine 6G has been reported to rotate more rapidly than predicted [3,4], whereas Heiss et al. [14] find that it rotates more slowly, as does fluorescein 27 [7]. Our steady state measurements indicate that the latter is true for both TCTIF and rhodamine 6G; the former result may simply be due to the difficulty of measuring long rotation times with a solid state pulsed laser system.

4. Conclusions

We have shown that over a limited range of viscosities the Stokes–Einstein relationship is valid for TCTIF and Rh 6G, although the rotator may not be a true sphere. Steady state depolarisation measurements have been used to obtain useful rotational data in a region where the analysis of kinetic data would be most difficult. Provided that the $r(t)$ function is a single exponential and that the fluorescence lifetime is known, the accuracy is comparable with that of the kinetic technique.

Acknowledgement

We wish to thank the Science Research Council for overall support of this work and for the award of a Studentship to P.J.S., the Ministry of Defence for an award of a Fellowship to C.J.T., G.R. Fleming for his helpful comments and A.W. Ellis for the analysis of NMR spectra.

References

[1] J.R. Lombardi, J.W. Raymonda and A.C. Albrecht, J. Chem. Phys. 40 (1964) 1148.
[2] A.C. Albrecht, Progr. Reaction Kinetics 5 (1970) 301.
[3] T.J. Chuang and K. Eisenthal, Chem. Phys. Letters 11 (1971) 368.
[4] K. Eisenthal, Accounts Chem. Res. 8 (1975) 118.
[5] H.E. Lessing, A. von Jena and M. Reichert, Chem. Phys. Letters 36 (1975) 517.
[6] H.E. Lessing, Opt. Quantum Electron. 8 (1976) 309.
[7] H.E. Lessing and A. von Jena, Chem. Phys. Letters 42 (1976) 213.
[8] G.R. Fleming, J.M. Morris and G.W. Robinson, Chem. Phys. 17 (1976) 91.
[9] M.D. Archer, M.I.C. Ferreira, G. Porter and C.J. Tredwell, Nouv. J. Chim. 1 (1977) 9.
[10] E.D. Cehelnik, K.D. Mielenz and R.A. Velapoldi, J. Res. Natl. Bur. Std. US 79A (1975) 1.
[11] T.J. Chuang and K. Eisenthal, J. Chem. Phys. 57 (1972) 5094.
[12] T. Tao, Biopolymers 8 (1969) 609.
[13] F. Perrin, J. Phys. Radium 5 (1934) 497.
[14] A. Heiss, F. Dörr and I. Küln, Ber. Bunsenges. Physik. Chem. 79 (1975) 294.

Chemical Physics 61 (1981) 17–23
North-Holland Publishing Company

PICOSECOND FLUORESCENCE DEPOLARISATION MEASURED BY FREQUENCY CONVERSION

Godfrey S. BEDDARD, Tom DOUST and George PORTER

Davy Faraday Research Laboratory, The Royal Institution, London W1X 4BS, UK

Received 21 April 1981

A method of using sum frequency generation in a LiIO$_3$ crystal has been used with picosecond pulses from a dye laser to measure fluorescence decays in solution. Fluorescence can be detected at wavelengths greater than 1 μ. Some examples of the method to measure the rotational relaxation of dyes in solution are presented. In the dye cresyl violet two sites for solvent–solute interaction are proposed; one affects mainly the electronic properties of the molecule while the other hinders molecular rotation.

1. Introduction

The non-linear phenomenon of sum frequency generation in anisotropic crystals [1] has been used with a mode-locked synchronously pumped dye laser as an optical gate with picosecond time resolution [2–4]. When used to measure fluorescence decays it offers several advantages over time correlated single photon counting [5]. As well as greatly improved time resolution it is possible to detect fluorescence in the far red and infrared without using special red-sensitive photomultipliers; for example, fluorescence at 900 nm can be mixed with laser light at 600 nm to generate a sum frequency at 360 nm. Furthermore, the polarisation of the fluorescence can be analysed and hence the decay of the fluorescence anisotropy can be measured directly. This offers an alternative method to anisotropic absorption [6] and transient dichroism for the measurement of rotational diffusion on a picosecond time scale [7]. Thus, frequency conversion has almost as much importance in areas covered by existing techniques as it does in extending detection ranges in both the wavelength and time domains.

Two methods of collecting the fluorescence are described and compared. The rotational diffusion times of the dyes cresyl violet* and 3,3,3′,3′-tetramethyl-1,1′-dimethylindotricarbocyanine perchlorate (HITC) have been measured in various solvents. Both behave in a fashion consistent with solvent attachment and have slower rotation times than predicted by "stick" behaviour. Comparison of the rotational and spectral properties of cresyl violet in various solvents with those of the dye oxazine-1 indicates the existence of two possible sites for solvent attachment. These sites are identified as the N–O ring and the 2,2′-amino substituents.

2. Theory and description of the technique

The two optical configurations used are shown in figs. 1a and 1b. In both cases a synchronously pumped mode-locked rhodamine 6G dye laser producing pulses of about 4 ps (fwhm) and 80 mW average power was used. The pulse train is split on a 50% beam splitter and half of it passed down a variable optical delay line, the length of which is controlled by a translation stage driven by a stepping motor. The stepping

---

* Since this was completed similar results were reported by von Jena and Lessing [8] using a transient dichroism method.

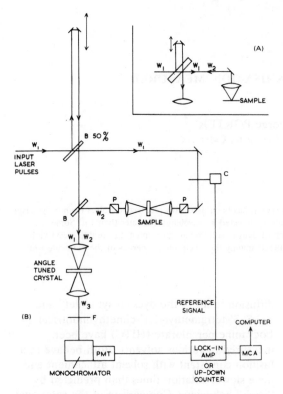

Fig. 1. Experimental arrangements used for time resolved fluorescence upconversion. F are filters, P are polarisers and C a sectored disc chopper connected to the lock-in amp. or photon counter. ω_1 is the laser beam and ω_2 the fluorescence. The insert shows optical arrangement 1, the main figure arrangement 2.

glass cell mounted perpendicularly to the exciting beam and spun about an axis parallel to the beam; the fluorescence is collected off the front face of the cell along the axis of the exciting beam. In method 2 (fig. 1b) the sample is flowed through a cell or pumped through a nozzle to form a jet. The fluorescence is collected at 180° to the exciting beam and the transmitted excitation light removed by filters.

The instantaneous power of the sum frequency in a crystal of length L is given approximately by [9]

$$P_\omega(L) \approx kL^2 P_1 P_2\, \omega_3/\omega_1, \qquad (1)$$

where the subscripts on the frequency ω and the power/unit area P refer to the gating pulse, the fluorescence and the sum frequency respectively. The frequencies are related by $\omega_3 = \omega_2 \pm \omega_1$. It can be seen that by appropriate phase matching the difference frequency can also be generated. This can be used to measure fluorescence in the UV [10] where the sum frequency would be at very short wavelengths where detection becomes difficult or where crystals transparent at such short wavelengths are not available. This may be of use in measuring fluorescence excited by the second harmonic of the laser or by multi-photon absorption.

Since the power of the laser pulses is constant to within a few percent, the sum frequency signal is proportional to the fluorescence intensity at a particular time delay. The time resolution in a thin crystal is only limited by the width of the laser pulse as the optical delay can be adjusted by increments corresponding to far smaller times than the duration of the laser pulse. By angle tuning the crystal the phase velocities of the laser pulse and the fluorescence can be matched for different fluorescence wavelengths. This allows decays to be measured at different wavelengths in the emission spectrum. In these experiments, type 1 phase matching was used with the two input beams on the ordinary ray and the sum frequency on the extraordinary ray. This phase matching condition also acts as a frequency selecting element; in the crystal used a bandwidth of about 3 Å was upconverted at a particular crystal orienta-

motor controller also indexes the memory of a multichannel analyser so that the signal at each delay is collected in consecutive channels of the analyser. The complete decay can thus be collected in one scan of the delay line. If the fluorescence is weak several scans can be averaged.

The undelayed portion of the pulse is used to excite the sample. The fluorescence is collected and focused into a $LiIO_3$ crystal (1 mm path length) along with the delayed laser pulse. This short path length, although it allows less frequency conversion is necessary to avoid group dispersion effects. In method 1 (fig. 1a), which is similar to that used by Hirsch et al. [4], the sample is contained in a 1 mm path length tion. Time gated spectra can be measured at a given time after excitation either by tuning the laser frequency [11] or at a fixed laser frequency by turning the crystal to select the fluorescence wavelength.

In our experiments the fundamental and second harmonic of the gating pulse are removed by filters and a monochromator is set to select the sum frequency. The beam exciting the fluorescence is modulated at about 700 Hz by a sectored disc chopper and the signal from the photomultiplier recovered by a lock-in amplifier. Alternatively the photomultiplier can be made to operate in a photon counting mode and the signal collected in an up–down counter triggered by the chopper. In both cases the signal can be normalised to the dye laser intensity which can be measured separately (not shown in fig. 1). To measure the autocorrelation function of the laser pulse the lens and sample cell can be replaced by a mirror in method 1 or the sample removed in method 2.

The fluorescence intensity I_s generated by the exciting pulse has a time profile given by

$$I_s(t) = \int_{-\infty}^{t} I_f(t') \, L(t-t') \, \mathrm{d}t', \qquad (2)$$

where t' is the time delay, $L(t)$ is the laser pulse shape and $I_f(t)$ the molecular fluorescence intensity at time t. The signal after the crystal $S(t)$ is given by

$$S(t) = \int_{-\infty}^{\infty} I_s(t') \, L(t-t') \, \mathrm{d}t', \qquad (3)$$

which combined with eq. (2) gives

$$S(t) = \int_{-\infty}^{t} I_f(t') \, A(t-t') \, \mathrm{d}t', \qquad (4)$$

where $A(t)$ is the measured autocorrelation function of the laser pulse. As the autocorrelation function is readily measured the signal can be fitted to the desired decay function at the shortest times with no a priori knowledge of the laser pulse shape. This offers a significant advantage over transient absorption methods [6, 7] where it is necessary to make assumptions about the laser pulse shape to facilitate curve fitting.

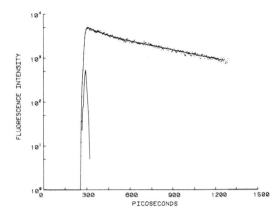

Fig. 2. Fluorescence from HITC in methanol detected at 770 nm and excited at 590 nm. The laser autocorrelation function has been displaced slightly for clarity. The solid line through the fluorescence is the convolution of the laser autocorrelation function with the sum of two exponential terms.

3. Results

Fig. 2 shows the fluorescence decay of HITC in methanol at 770 nm together with the laser autocorrelation function and the computed fit (sum of two exponentials) to the data. By angle tuning the crystal the fluorescence decay could be observed as far to the red as 980 nm, close to the limit of HITC emission. Fig. 3 shows fluorescence collected with polarisation parallel

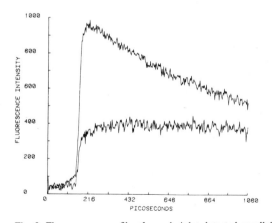

Fig. 3. Fluorescence profile of cresyl violet detected parallel top curve and perpendicularly to the excitation polarisation. The detection wavelength was 770 nm.

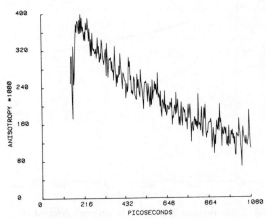

Fig. 4. Fluorescence anisotropy calculated from the data in fig. 3.

and perpendicular to the polarisation of the exciting light for cresyl violet in iso-propanol measured using method 2. The fluorescence anisotropy $r(t)$, fig. 4, was calculated from the equation [12, 13]

$$r(t) = [I_\parallel(t)G - I_\perp(t)]/[I_\parallel(t)G + 2I_\perp(t)], \qquad (5)$$

where

$$I_\parallel = [1 + 2r(t)]\, e^{-t/\tau_f}, \qquad (6a)$$

$$I_\perp = [1 - r(t)]\, e^{-t/\tau_f}. \qquad (6b)$$

$I_\parallel$ and $I_\perp$ are the fluorescence intensities polarised parallel and perpendicular to the excitation and τ_f is the fluorescence lifetime. The normalisation factor G was determined in two ways. Firstly, in very mobile liquids the signal at long times is free of the effects of rotational motion and the $I_\parallel$ and $I_\perp$ signals should have the same magnitudes so G can be determined by tail matching at long times. Secondly, if the laser pulse is short compared with the molecular events convolution does not significantly distort the signals at short times, thus the condition $I_\parallel = 3I_\perp$ at $t = 0$ can be used to calibrate the curves. The anisotropy $r(t)$ can then be related to the rotational relaxation lifetime depending upon its shape [12, 13]. The rotational relaxation lifetime can also be calculated from either the $I_\parallel$ or $I_\perp$ signals and if, as is the case, the fluorescence decay is single exponential the $I_\parallel$ signal is described by eq. (6a) and the $I_\perp$ signal by eq.

Table 1

Experimental excited state properties of cresyl violet, HITC and oxazine (all lifetimes given in picoseconds)

Solvent	Viscosity (cP)	Cresyl violet		HITC		Oxazine[a]	
		τ_R	τ_f	τ_R	τ_f	τ_R	τ_f
acetone	0.32	78 ± 4	3670			57	1122
methanol	0.55	134 ± 4	3650	226 ± 4	1080	84	813
water	1.03	130 ± 5	2390			141	552
ethanol	1.2	350 ± 14	3510	377 ± 11	1300	136	1024
propanol	2.2	696 ± 15	3870	436 ± 7	1500	237	1122

calculated stick lines for: cresyl violet (vol = 228 Å^3)   84 ps/cP
                       HITC (vol = 473 Å^3)   91 ps/cP
                       oxazine (vol = 317 Å^3)   120 ps/cP

measured r_0 values cresyl violet = 0.355 ± 0.02
                          HITC = 0.38 ± 0.02

[a] Taken from ref. [14].

(6b). In all cases so far reported [7, 13] in isotropic liquids $r(t)$ has been described by an exponential function i.e. $r(t) = r_0 \exp(-t/\tau_R)$. Hence, knowing τ_f either from up-conversion measurements with the polariser at 54.7° or from single photon counting allows τ_R and r_0 to be calculated. These values are very similar to those calculated by generating $r(t)$ directly from eq. (5). Table 1 shows the measured fluorescence decay times and rotational reorientation times for HITC and cresyl violet in different solvents. Fig. 5 shows a plot of τ_R versus solvent viscosity for these dyes together with the lines calculated on the basis of the Debye–Einstein theory at the "stick" boundary condition [13], for an oblate ellipsoid in the case of cresyl violet and a prolate ellipsoid for HITC. In both cases the deviation from the stick boundary condition is clear, the experimentally determined rotational relaxation times being slower than those predicted by theory. This contrasts sharply with the behaviour of oxazine 1 where the measured slope 87 ps/cP [14] is closer to

the calculated slip boundary condition where resistance to motion arises from solvent displacement in non spherical molecules. The calculated values for oxazine 1 are 60 ps/cP for slip and 110–134 ps/cP for stick depending on the exact molecular shape. Much stronger solvent–solute interactions are therefore present in cresyl violet than oxazine 1.

4. Discussion

The changes in emission maxima in both oxazine and cresyl violet indicate an interaction at a different site on the solute molecule than that suggested by the rotational relaxation measurements. For example, in both compounds the spectra in 1,4-dioxan (hydrogen bonding and polar) and acetonitrile (non hydrogen bonding but polar) showed shifts in the former but not in the latter. This suggests that hydrogen bonding provides a greater contribution to the spectral shifts than is caused by solvent polarity.

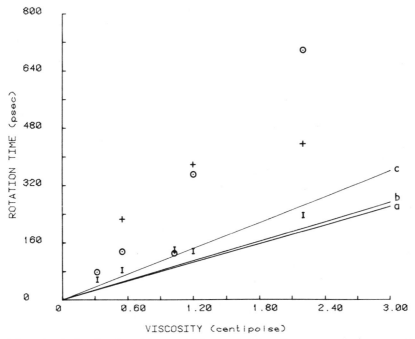

Fig. 5. Experimental rotational relaxation times (picoseconds) for cresyl-violet (○), HITC (+) and oxazine (I) [14] plotted against solvent viscosity (cP). Also shown are the calculated stick lines for the same solutes (a), (b) and (c), respectively.

Cresyl violet exhibits anomalous behaviour in water; a blue shift in fluorescence maximum compared to propanol is observed. Similar effects occur in merocyanine and have been attributed to changing and opposite contributions of the dipole moment and polarisability of the solute with the strength of the reaction field of the solvent [16].

The excited states of cresyl violet and oxazine also show similar behaviour both being quenched in water relative to the alcohols, table 1. It is probable that either intersystem crossing is more rapid in water than alcohols as occurs with xanthene dyes [17] or that internal conversion is more rapid in aqueous solution. A deuterium isotope effect τ_D/τ_H of 1.26 for oxazine in D_2O ($\tau_D = 695$ ps) and H_2O ($\tau_H = 552$ ps) supports the suggestion of Forster [18] that direct solute–solvent interaction enhances internal conversion.

These similarities in the ground and excited state properties of cresyl violet and oxazine are in contrast to their rotational diffusion behaviour. One factor causing this difference could be the change in shape between the two molecules, but HITC is more prolate than oxazine yet does not show similar behaviour but acts like other cyanines such as DODCI following stick behaviour. Specific hydrogen bonding to the amino group on cresyl violet although not to the diethyl amino group on oxazine will explain the difference in behaviour. A strong hydrogen bond will have the effect of slowing molecular rotation and will be particularly effective if bonding occurs at the end of the long axis of the ellipsoid of revolution. Solvent attachment on the central O, N containing ring will be near to the symmetry axis of the molecule and will impede molecular motion less. Rotation times slower than expected from stick behaviour have been observed in other compounds such as fluorescein derivatives [13, 20].

These differences and similarities between oxazine and cresyl violet can thus be explained qualitatively by two sites of solvent–solute interaction. The spectral and excited state properties are sensitive to interaction on the O and N con-

taining ring. These interactions, although large enough to change electronic properties are nevertheless sufficiently small that they do not affect rotational relaxation. The other interaction occurs on the amino substituent groups and while strong enough to slow rotation does not appear to interact appreciably with the pi-electron system. An exception to this behaviour appears to be cresyl violet in water where a rotation time faster than in alcohols of similar viscosity and which is near to the calculated stick line is observed. Since water is a good hydrogen bonding solvent the strong water–water hydrogen bonds may limit the hydrogen bond interaction with the solute. However the general reaction field of the solvent will still affect the electronic properties of the cresyl violet and oxazine molecules. This is supported by the similar rotation times of oxazine and cresyl violet in water.

Acknowledgement

We acknowledge the SRC for financial support for the work; G.S.B. for an Advanced Fellowship and T.D. for a Studentship.

References

[1] A Laubereau, L. Gleiter and W. Kaiser, Appl. Phys. Letters 25 (1974) 87.

[2] M. Dugay and J. Hansen, Appl. Phys. Letters 13 (1968) 178.

[3] H. Mahr and M.D. Hirsch, Opt. Commun. 13 (1975) 96.

[4] M.D. Hirsch and H. Mahr, Chem. Phys. Letters 60 (1979) 299;
G.S. Beddard, T. Doust and M.W. Windsor, in: Picosecond phenomena, Vol. 2, eds. R.M. Hochstrasser, W. Kaiser, C.V. Shank (Springer, Berlin, 1980) p. 167.

[5] G.S. Beddard, G.R. Fleming, G. Porter and R.J. Robbins, Phil. Trans. Roy. Soc. London A 298 (1980) 324.

[6] G.S. Beddard and M.J. Westby, Chem. Phys. 57 (1981) 121.

[7] D.P. Millar, R. Shah and A.Z. Zewail, Chem. Phys. Letters 66 (1979) 435;
A. von Jena and H.E. Lessing, Chem. Phys. 17 (1979) 91;

A. von Jena and H.E. Lessing, Laser handbook, Vol. 3, ed. M.L. Stitch (North-Holland, Amsterdam, 1979), and references therein.

[8] A. von Jena and H.E. Lessing, Chem. Phys. Letters 78 (1981) 187.

[9] A. Yariv, Quantum electronics (Wiley, New York, 1975) p. 454.

[10] L. Halliday and M.R. Topp, Chem. Phys. Letters 46 (1977) 8.

[11] H. Mahr, T. Daly and N.J. Frigo, Picosecond phenomena, Vol. 1, eds. C.V. Shank, E.P. Ippen, S.L. Shapiro (Springer, Berlin, 1978) p. 230.

[12] T. Tao, Biopolymers 8 (1969) 609.

[13] G.R. Fleming, J.M. Morris and G.W. Robinson, Chem. Phys. 17 (1976) 91.

[14] G.R. Fleming, D. Waldeck and G.S. Beddard, Il. Nuovo Cim. 63B (1981) 151.

[15] C. Hu and R. Zwanzig, J. Chem. Phys. 60 (1974) 4354.

[16] E. McRae, Spectrochem. Acta 12 (1958) 192.

[17] L.E. Cramer and K.G. Spears, J. Am. Chem. Soc. 100 (1978) 221.

[18] T. Forster, Chem. Phys. Letters 17 (1972) 309.

[19] G.R. Fleming, A.E.W. Knight, J.M. Morris, R.J. Robbins and G.W. Robinson, Chem. Phys. Letters 51 (1977) 399.

[20] G. Porter, P.J. Sadkowski and C.J. Tredwell, Chem. Phys. Letters 49 (1977) 416.

Further publications

AN INVESTIGATION OF THE VALIDITY OF THE DEBYE EQUATION
FOR ENCOUNTER CONTROLLED PROCESSES IN CONDENSED MEDIA.
A. D. Osborne and G. Porter, Sixth International Symposium on Free Radicals,
Cambridge, 1963, p. L1.97.

VISCOSITY-DEPENDENT INTERNAL ROTATION IN POLYMETHINE
DYES MEASURED BY PICOSECOND FLUORESCENCE SPECTROSCOPY.
A. C. Winkworth, A. D. Osborne and G. Porter in *Picosecond Phenomena III*.
Eds. K. B. Eisenthal, R. M. Hochstrasser, W. Kaiser and A. Laubereau,
Springer-Verlag, Berlin, 1982, p. 228.

SPECTRUM OF DELAYED FLUORESCENCE IN
PHENANTHRENE VAPOUR: A CRITERION OF PURITY.
B. Stevens, E. Hutton and G. Porter, *Nature*, 1960, **185**, 917.

FLUORESCENCE PROBE MOLECULE 1, 8-ANILINONAPHTHALENE
SULPHONATE (ANS)
G. R. Fleming, G. Porter, R. J. Robbins and J. A. Synowiec,
Chem. Phys. Letters, 1977, **52**, 228.

CONFORMATIONAL EFFECTS IN FLUORESCENT EXCITED
CHARGE-TRANSFER COMPLEX FORMATION.
X-J. Luo, G. S. Beddard, G. Porter, R. S. Davidson and T. D. Whelan,
J.C.S. Faraday Trans. I, 1982, **78**, 3467.

PHOTOPHYSICS OF 1, 1'-BINAPHTHYL AND ITS FORMATION
OF A COMPLEX WITH N-(1-PROPYL)-2, 5-DIMETHYLPYRROLE.
X-J. Luo, G. S. Beddard, G. Porter and R. S. Davidson,
J.C.S. Faraday Trans. I, 1982, **78**, 3477.

Chapter 8

PHOTOCHEMISTRY OF QUINONES AND KETONES

The carbonyl group is a chromophore which usually gives to a molecule the property of absorbing light in the conveniently near ultra-violet region of the spectrum. It is found in a vast range of organic compounds, particularly those of biological importance and it results in a wide variety of photochemical change. Its compounds have been more extensively studied than any other group and, among these, the ketones and quinones have received the most attention.

During the 1960s it became realised more and more clearly that an electronically excited molelcule is a new species with its own structure, physicochemical properties like dipole moment and pK and its own chemical reactivity, all different from those of the ground state. Furthermore, apart from the ground state, molecules could have several types of excited state each with its own properties. The first example of this was the distinction between the singlet and triplet states, common to all paired electron molecules. In addition, in carbonyl compounds, there would be π-π* and n-π* states and, more recently, charge-transfer states were recognised as a third, important class in substituted carbonyls.

It became clear that radiationless conversion between the different electronic states was very fast with the result that relaxation occured to the lowest level of a given multiplicity and subsequent reactions occured from this level. Flash photolysis made it possible to observe directly these excited states and the work below describes a comparative investigation of the different excited states that contribute to the rich photochemistry of carbonyl compounds. Some of the results seem contrary to chemical intuition but are now readily understood in terms of the type of electronic transition which lies lowest in energy.

The first paper which follows, by Bridge and Porter, broke new ground by describing the observation of both the triplet state absorption and the absorption of the radicals formed in its decay by reaction with the solvent by hydrogen abstraction. Furthermore the loss of a proton by the neutral semiquinone first formed to yield the semiquinone radical anion is observed via the spectra of both radicals.

The work began as a study of the practical problems of the dye-sensitised phototendering of fabrics, and why some dyes exhibited this effect strongly whilst others, often very similar in structure, did not. The extension of these studies to ketones and many of their substituents, and to the shorter times made available by nanosecond laser flash photolysis, led directly to a general understanding of the energy levels of the n-π*, π-π* and C-T states of carbonyl compounds and their differing chemical reactivities.

Reprinted from the Proceedings of the Royal Society, A, *volume* 244, pp. 259–275, 1958

Primary photoprocesses in quinones and dyes
I. Spectroscopic detection of intermediates

By N. K. Bridge and G. Porter*

*The British Rayon Research Association, Heald Green Laboratories,
Wythenshawe, Manchester*

(*Communicated by R. G. W. Norrish, F.R.S.—Received* 10 *October* 1957)

[Plates 4 and 5]

The flash photolysis of many quinones in solution produces a number of transient intermediates, each with its characteristic absorption spectrum. A preliminary study of these showed that the appearance of intermediates is clearly related to the efficiency of the quinone as a photosensitizer.

A detailed study of the transient spectra of duroquinone as a function of solvent and of pH enabled an unequivocal assignment to be made to all the principal absorptions. The main intermediates are

(1) The semiquinone radical with first absorption maximum at 410 mμ which appears in all solvents containing abstractable hydrogen at pH 7 and below.

(2) The semiquinone radical ion with first absorption maximum at 430 mμ which appears only at pH 7 and above.

(3) The triplet state which is only observed strongly in the viscous solvent liquid paraffin and has a relatively short life.

Analogous spectra of similar intermediates are reported for benzoquinone, toluquinone, xyloquinone, naphthaquinone and anthraquinones.

Introduction

In recent years strenuous efforts have been made to discover the primary processes following light absorption by dyes, principally because of their importance in dye fading and photodegradation of the dyed fibre (phototendering) (Symposium 1949). Dyes can act as photosensitizers of widely differing efficiency and the acceptable 'fast' dyes are those which not only do not fade but also cause little phototendering. The problem is so complicated that significant advances can only be expected by simplification of the experimental conditions, and Bolland & Cooper (1954) working in these laboratories studied the anthraquinone-sensitized photochemical oxidation of ethanol—a relatively simple system in which the sensitizer is representative of one of the most important classes of dyes. Their work led to a clearer understanding of the processes and accounted for many of the observations of previous workers (for example, Berthoud & Porret 1934; Bäckstrom 1944).

The mechanisms which they proposed to account for their observations in aqueous, non-alkaline solutions involved an extremely efficient cyclic process in which the excited sensitizer abstracted hydrogen from the ethanol to give a semiquinone radical, which was rapidly reconverted to the original anthraquinone by reaction with oxygen. However, by the very nature of their experiments, they

---

* Present address: Chemistry Department, The University, Sheffield, 10.

were unable to decide on the true identity of the excited sensitizer, nor could they give definite evidence for the presence of a semiquinone radical.

Michaelis and other workers (Michaelis 1938; Burstein & Davidson 1941; Venkataraman & Fraenkel 1955) have shown quite conclusively that one semiquinone—the semiquinone radical ion ($Q^{\cdot-}$)—can exist in equilibrium in solutions of some quinones of appropriate pH. The other semiquinone radical ($QH\cdot$) which has not previously been observed in solution, presumably exists in ionic equilibrium with $Q^{\cdot-}$ thus

$$Q^{\cdot-} + H^+ \rightleftharpoons QH\cdot, \quad \text{i.e.}$$

The reason for the stability of the $Q^{\cdot-}$ (see Michaelis above) is to be found in the additional resonance possibilities resulting from the alternative positions of the odd electron on the two oxygen atoms. This process is not possible in the case of $QH\cdot$ because of the greater mass of the hydrogen atom.

As regards the identity of the excited sensitizer, Bowen (1949) and others have indicated that this could be the molecule in either its excited singlet or triplet state. Lewis et al. (1941–5) first demonstrated the real existence of triplet energy levels by several methods and, in fact, a large amount of work since carried out shows that it is the exceptional molecule which is not capable of passing over into a triplet state on light absorption. In view of this, it seemed most likely that the primary act of photosensitization by dyes would be a transition—not necessarily directly—to the biradical or triplet state which, although short lived in solution, nevertheless, has a considerably longer life than the upper singlet state and consequently a higher probability of undergoing reaction.

From the above remarks it is clear that there may exist in these irradiated solutions several unstable, transient species, e.g. $Q^{\cdot-}$, $QH\cdot$ and the triplet. Application of the flash photolysis technique, devised by one of us (Porter 1950) for the study of such short-lived products, should enable us to detect them and so help greatly our complete understanding of the mechanism.

Porter & Windsor (1954) were the first to apply this technique to compounds in solution. They found, on flashing a large number of organic substances, that transient absorption spectra, of half lives between 10^{-2} and 10^{-5} s, appeared which were derived from molecules in their triplet states. In the present investigation the following two main problems were anticipated. First, is the triplet state formed initially and if so what governs its formation and decay? And secondly, is the product of the reaction of this, or any other excited state with the substrate, the radical ion formed by electron transfer, or the radical formed by hydrogen abstraction?

No previous observations have been made on the triplet states of quinone-type molecules in solution, apart from the work of Porter & Windsor (1957) who observed a transient when p-benzoquinone was flashed in liquid paraffin. The evidence for semiquinone radicals and radical ions, is more definite. In particular, Baxendale &

Hardy (1953) in a spectroscopic study of the equilibria of duroquinone (tetra-methyl *p*-benzoquinone) and durohydroquinone, showed that at a pH of approximately 10 to 11 the $Q^{\overline{\cdot}}$—for which they gave the absorption spectrum—was present in aqueous solutions, and they calculated that the equilibrium constant for the reaction $Q^{\overline{\cdot}} + H^{+} \rightleftharpoons QH\cdot$ was $> 10^{-11}$. Evidence for the presence of the radical, $QH\cdot$, is available but is not quite so conclusive. Lewis & Bigeleisen (1943) found that when hydroquinone was irradiated in a rigid ether-*iso*pentane-alcohol (e.p.a.) glass, an absorption band appeared at 411 mμ which they attributed to the semi-quinone radical. Michaelis & Woolman (1950) irradiated durohydroquinone and hydroquinone in rigid glass and found that several new absorption bands appeared in the blue and near ultra-violet region of the spectrum which lasted as long as the glass was kept rigid. They also attributed these bands to $QH\cdot$. Finally, Linschitz, Berry & Schweitzer (1954) observed similar bands for hydroquinone and *p*-benzo-quinone in rigid glasses containing triethylamine as well as in the conventional e.p.a. glass. They favoured the view that the radical ion was the absorbing species, but at the same time they admitted the possibility that it could be the radical. A 'delayed luminescence' which occurred on warming up the glass, was attributed by them to triplet-singlet emission (phosphorescence), caused by the population of the triplet state on recombination of the radical ion and trapped electron.

The experimental problem resolves itself into two parts, although each is dependent on the other. First, any transient spectra which are observed must be identified and assigned to the intermediates from which they arise. Secondly, each of the absorption spectra must be studied kinetically under a variety of conditions and so the whole series of changes elucidated. In principle this is always possible, since all the intermediates have absorption spectra, but in practice the spectra may not lie in an accessible region or may have lifetimes too short for detection. In the present study, experimental difficulties of this kind do arise, but nevertheless we shall see that conditions are quite favourable, and all expected species can be observed spectroscopically.

The present paper is concerned mainly with the unequivocal assignment of the spectra of all transients which are present. In part II (Bridge & Porter 1958) a preliminary kinetic investigation is described which shows the correctness of our assignment and gives an overall account of the reaction. Later work will deal with the kinetics of individual reactions in more detail.

EXPERIMENTAL

The photographic method

The general principle of the method has been outlined elsewhere (Porter & Windsor 1954). In the present case, the photolysis flash used either a 10 or a 2 μF condenser charged up to voltages between 4 and 10 kV; the spectroscopic flash was always used at 4 μF and 6 kV. The electronic circuits were also similar except for one or two items such as a built-in crystal-controlled time base and a brightening pulse for the c.r.o. The spectra were usually recorded on Selochrome photographic plates, with Kodak P 1500 or P 1600 for longer wavelengths. Oscillo-graphic traces of the flashes were always recorded, but the accuracy of timing was determined by the flash duration rather than fluctuations of the electronic appara-tus; consequently, as the half-height widths of the photolysis and spectro-flashes were approximately 20 μs, it was difficult to realize a resolution better than 5 to 10 μs. The reaction vessels were as described by Porter & Windsor, and in all cases the solutions were degassed by vigorous agitation, distillation and freezing under mercury diffusion pump pressure and then sealed off. The technique of evaporating excess solvent down to a mark on the side arm was used; with liquid paraffin the solid dye was placed in another smaller side-arm and the liquid paraffin heated (to degas it) in the main arm. When addition of acid or alkali was required it was put into the small side-arm and only allowed to contact the main solution a few minutes before an experiment began.

A Hilger intermediate quartz spectrograph was used. For accurate photometry, the plates were calibrated individually using a Hilger 7-step neutral filter of rhodium on quartz, the cell being filled with pure solvent for the purpose. Density measurements were made with a Hilger non-recording microdensitometer. For qualitative comparison of spectra only, the more easily calculated values of $\Delta \log_{10} R$ were used rather than the optical density (d). ($\Delta \log_{10} R = \log_{10} R_t - \log_{10} R_0$, where R_t and R_0 are microdensitometer readings at delay t and at the beginning of an experiment.)

The xenon lines, of known wavelength, obtained from the spectro-flash, provided an internal calibration of wavelength.

Materials

The solvents used were—water distilled in an all glass apparatus; n-hexane (B.D.H. special for spectroscopy); ethanol prepared from 95 % grade and stored under nitrogen, Pestemer (1951); liquid paraffin, medicinal B.P. grade purified by passing through a silica gel column—also another pure sample kindly supplied by Dr Morantz; Analar carbon tetrachloride; chloroform, M. and B. reagent grade purified, Vogel (1954); Analar dioxane purified (Vogel 1954). Several samples of duroquinone were used; we are grateful to Dr J. H. Baxendale for the first few batches. Later it was prepared in these laboratories from durene by the method of Smith (1930). Durohydroquinone was prepared from duroquinone by reduction with zinc in glacial acetic acid, washed and dried. When dry it was stable in air, m.p. 225° C. Other quinones were purified as follows: p-benzoquinone; steam distillation, water-washed and dried, sealed tube m.p. 115° C. Toluquinone, B.D.H. laboratory reagent sublimed to correct m.p. of 67·2° C. 2–6 meta-xyloquinone, Lights recrystallized from petroleum ether, dried over P_2O_5, sealed tube m.p. 55° C. 1-5 para-Xyloquinone, Baxendale, m.p. 125° C. Trimethyl-p-benzoquinone, pre-pared by the method of Smith et al. (1939) from Lights iso-pseudocumenol. 1-4 Naphthaquinone, purified by method of Fieser (1946) m.p. 124° C. The purified anthraquinones were kindly supplied by Dr H. R. Cooper of these laboratories. Unless otherwise stated, all other materials were of analytical grade and used without further purification.

Results

Preliminary observations

Cooper (1957, unpublished) studied many anthraquinones in his experiments on the photosensitized oxidation of alcohols and found that some were good sensitizers and others not. As a starting point in the present work, it was natural to look at these compounds to see if there was any clear distinction between sensitizers and non-sensitizers. It was found immediately that the former when flashed in ethanol or other solvents containing abstractable hydrogen (such as toluene or benzene), showed very strong transient absorptions (see, for example, figure 9, plate 4, anthraquinone 2-SO₃Na). On the other hand, the latter showed little, if any such absorption (see, for example, figure 10, plate 4, anthraquinone 2-6 diOH). Further study of several anthraquinones, the results of which are given in table 1, revealed many such species depending on solvent and pH, and it became increasingly clear that, even under these relatively simple conditions, the problem was very complex and required detailed study. It was, therefore, decided to search for a quinone which showed good transients and which was also stable under various solvent conditions and over a wide range of pH.

Table 1. Behaviour of anthraquinone derivatives when flashed

derivative	behaviour (Cooper 1957)	solvent	transients
2-6-diSO₃Na		ethanol/H₂O	
2-methyl		ethanol	
2-SO₃Na	sensitizer	ethanol	strong
1-chloro		hexane and toluene	
anthraquinone		ethanol	
1-5-diSO₃Na		H₂O + toluene	none
2-6-diOH	non-sensitizer	ethanol	none
2-amino		benzene	very weak
anthracene		ethanol	triplets only

Duroquinone

This substance was chosen mainly because of its known stability in alkaline solutions, and also because it had been the subject of previous work by Baxendale and others (see previous references). The spectra of the various species, both stable and unstable, that are known are given in figure 1. It was studied in the following solvents.

(a) *Liquid paraffin*

This was used because viscous inert solvents are known to prolong the life of the triplet state (Porter & Windsor 1954). Figure 11, plate 5, shows the result obtained. It was found from this, and similar plates, that there are two main transient absorptions.

(1) A very broad band T, with maxima at approximately 490 and 460 mμ which decreases rapidly in intensity with half-life of approximately 100 μs.

N. K. Bridge and G. Porter

(2) A narrower band R, with maximum at 409 mμ which decays more slowly with a half-life of about 2 ms.

Further bands at lower wavelengths, and one very weak one at *ca.* 435 mμ were obtained which will be discussed later so as to simplify the account. Figure 2 (*a*) shows the spectrum obtained at various delays, from which it is evident that R and T belong to two different species—further evidence for this will also be presented later.

Now, the possible species are: triplet, semiquinone radical and radical ion; and of these the triplet might be expected to have much the shorter life. Evidently this could be T rather than R. The radical-ion spectrum is known in aqueous solution

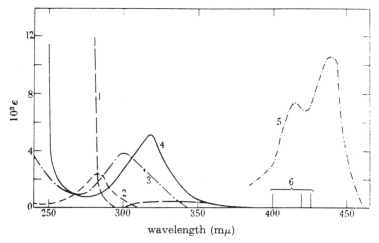

FIGURE 1. Duroquinone. durohydroquinone and duroquinone semiquinone radical ion spectra. 1, quinone in ethanol; 2, unionized hydroquinone (QH_2); 3, singly ionized hydroquinone (QH^-); 4, doubly ionized hydroquinone (QH^{2-}); 5, $Q^{\cdot}$ in aqueous solution (Baxendale); 6, $QH\cdot$ in e.p.a. (Michaelis).

(Baxendale), but it does not agree with either T or R: however, the solvent is very different so this is not conclusive. If R is the radical, hydrogen abstraction is occurring from the solvent. consequently the next step was to use solvents in which this is not possible, or less likely. (It was noted after several flashes in liquid paraffin that, at minimum delay, R is stronger and T weaker; this is in accordance with expectation if R is the radical and T the triplet.)

(b) Hexane. 3-methyl pentane and cyclohexane

In hexane, a very sharp band appeared (see figure 3 (*a*)) with half-life of approximately 100 μs. At first it was thought that this might be due to an impurity in the solvent; however, it was found by Dr Morantz (in this laboratory) in another pure hydrocarbon, 3-methyl pentane. The simplest explanation is that the R band is very narrow in such solvents and is broadened by perturbations of internal fields in viscous solvents. The total intensity of R is much less in hexane, possibly because the hydrogen atom is less easily abstracted. The absence of T (the triplet?) is in accordance with the known viscosity dependence of the non-radiative deactivation of the triplet state, and in fact T is only strongly observed in liquid paraffin (and in rigid glasses, q.v.). Experiments with mixtures of liquid paraffin and hexane showed a gradual transition from the narrow to the wider (R) band, although not quite as expected in that both seemed to be present. This sharp line in hexane and similar solvents is peculiar to duroquinone and is not yet entirely understood. However, the main point at present is the low intensity of all transients with the

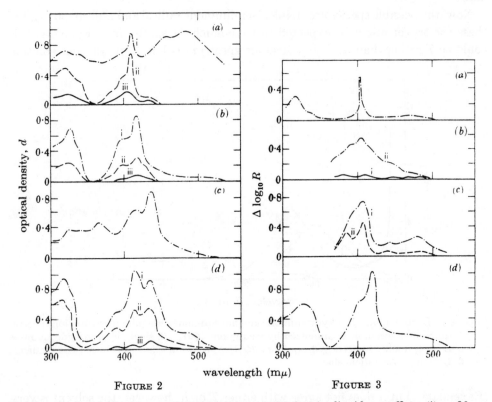

FIGURE 2   FIGURE 3

FIGURE 2. Duroquinone transient spectra. (a) 10^{-3} M in liquid paraffin. (i) < 30 μs; (ii) 200 μs; (iii) 3 ms. (b) $2 \cdot 5 \times 10^{-4}$ M in 50 % ethanol + 0·1 N-H_2SO_4. (i) < 30 μs; (ii) 2 ms; (iii) 19 ms. (c) $2 \cdot 5 \times 10^{-4}$ M in 50 % ethanol + 0·02 N-NaOH. 1 s delay. (d) 3×10^{-4} M in 96 % ethanol. (i) < 30 μs; (ii) 1·2 ms; (iii) 50 ms. Cell length 22 cm.

FIGURE 3. Duroquinone semiquinone radical spectra. (a) 5×10^{-4} M quinone in hexane; (b) 10^{-3} M quinone in (i) CCl_4 and (ii) $CCl_4 + 1$ % $CHCl_3$. (c) 10^{-3} M quinone in (i) dioxane, (ii) ether. (d) 5×10 M durohydroquinone after 30 min irradiation in e.p.a. (a, b and c all at < 30 μs delay).

appearance of a weak one in the region of absorption of R. *Cyclo*hexane showed a similar effect to the above, being very like hexane with about 1 % liquid paraffin in it.

(c) *Chloroform and carbon tetrachloride*

It was anticipated that the use of these would provide a clear-cut test of whether hydrogen abstraction was occurring from the solvent. It was found that if certain batches of Analar CCl_4 were used which contained < 0.01 % $CHCl_3$ (as determined

N. K. Bridge and G. Porter

by careful infra-red checks at the CH stretching frequency of 3010 cm^{-1}), then R was almost entirely absent at the earliest flashes. Addition of a trace (1 %) of CHCl$_3$ gave a very intense band with maximum at 409 mμ and similar in contour to the band in liquid paraffin (see figure 3 (b)). If the flashing of the CCl$_4$ solutions was continued some of the R band did ultimately appear, due presumably to hydrogen abstraction from duroquinone itself. However, the initial concentration of R was at least forty times greater in CHCl$_3$ than in CCl$_4$ (the optical density of absorption, d, at the maximum of R in CCl$_4$ was < 0.05, but in CHCl$_3$ there was complete absorption, i.e. $d > 2$). There was a very weak band in CCl$_4$ in the region of T.

(d) Ether and dioxane

In the former, the R band is similar in structure to that in CHCl$_3$, that is, somewhat more diffuse than in liquid paraffin. However, the band is quite characteristic with a maximum at 410 mμ and a second at $ca.$ 388 mμ (see figure 3 (c)). In dioxane the T band is reasonably strong with maximum at 485 mμ.

(e) Ethanol

The results obtained are more complex, the spectrum being very sensitive to pH. This may be so in some of the other solvents, although it is difficult to check except for ether where there were indications of a similar effect. Four cases were studied and are described in increasing order of complexity.

(i) Acid, 0.1 N-H$_2$SO$_4$, 50 % ethanol/H$_2$O. A band at 415 mμ only (with a subsidiary peak at 395 mμ). Half-life approximately 0.8 ms; the intensity increased on flashing. (Figure 2 (b).)

(ii) Alkali, 0.02 N-NaOH, 50 % ethanol/H$_2$O. A band with a higher wavelength peak than the above, at 438 mμ as well as the one at 415 mμ (and below). The intensity increased only with flashing and was independent of delay, lasting for at least 15 min. (Figure 2 (c).)

(iii) 50 % ethanol/H$_2$O. Peaks at 439 and 418 mμ (and below). These disappeared together with a half-life of approximately 1 ms.

(iv) 96 % Ethanol. See figure 12, plate 5. Peaks at 438 and 418 mμ (and below), whose relative intensity changed with delay quite clearly at less than about 20 % water. The low-wavelength peak (418 mμ) was initially greatest in intensity, but during the first 500 μs or so, the high-wavelength peak built up so that ultimately it was greater in intensity than the other. After this, they both decayed with a half-life of approximately 0.9 ms. As there was some overlapping of absorption this was not too easy to follow, but nevertheless it was quite definite. (Figure 2 (d).) (It was found that if alkali dissolved in about 4 % ethanol was added to ether then a high-wavelength band at $ca.$ 435 mμ appeared in addition to one at 410 mμ, just as with ethanol.)

Absorption by permanent products

To check that none of the observed transients could be due to decomposition of the quinone—in particular, ring scission— the spectrum of an ethanol solution before flashing twenty times, and after bubbling oxygen through it after flashing were measured and compared. (As hydroquinone is almost certainly the permanent product of flashing the oxygen reconverts it to duroquinone.) Figure 4 (a) gives the result, from which it can be seen that no detectable loss of duroquinone has occurred.

As the absorption of duroquinone and durohydroquinone overlap, it is difficult to compare quantitatively the amount of duroquinone disappearing and the amount of hydroquinone appearing after flashing. Figure 4(b) shows how, in

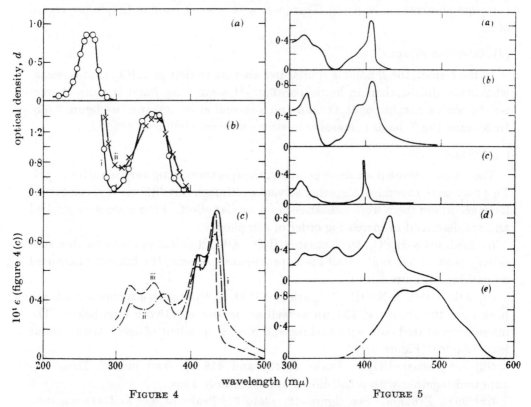

FIGURE 4   FIGURE 5

FIGURE 4. Duroquinone and semiquinone spectra. (a) 5×10^{-4} M quinone in ethanol, 1 mm cell. Full line, before; circles, after twenty flashes. (b) 3×10^{-4} M quinone in 50 % ethanol, 22 cm cell (i) before; (ii) after nine flashes. (c) (i) Durosemiquinone in aqueous solution (ϵ, Baxendale); (ii) 5×10^{-4} M quinone in 50 % ethanol + 0.02 N-NaOH, flashed at 510 μs delay in 22 cm cell; (iii) 10^{-4} M durohydroquinone in 50 % ethanol + 0.02 N-NaOH, air oxidation in 1 cm cell.

FIGURE 5. Duroquinone radical and triplet spectra. (a) $QH\cdot$ in liquid paraffin; (b) $QH\cdot$ in 50 % ethanol; (c) $QH\cdot$ in hexane; (d) $Q^{\overline{\cdot}}$ in 50 % ethanol; (e) triplet in liquid paraffin. Ordinates are the optical density, d.

a typical experiment, the density changes after nine flashes. The 340 mμ duroquinone peak is seen to drop due to disappearance of the quinone, and the 300 mμ minimum rises due to durohydroquinone formation (see figure 1). Assuming that the durohydroquinone does not absorb significantly at 340 mμ, it can be shown that the decrease in duroquinone concentration (2×10^{-5} M/flash) was accompanied by a comparable increase in absorption at 300 mμ by hydroquinone.

N. K. Bridge and G. Porter

The final product of radical attack on ethanol—acetaldehyde—absorbs very weakly ($\epsilon = 3\cdot4$) at 280 mμ. Even if every quinone molecule in these experiments produced one acetaldehyde molecule then the optical density change at the peak would still only be $\sim 0\cdot06$. (Actually acetaldehyde, over a period of approximately 90 min, converts into hemiacetal with even less absorption in this region.) Consequently, there can be no interference from these products. Before proceeding further the above results will be analyzed in terms of the expected unstable species.

Analysis of results

(a) The semiquinone radical ion ($Q^{\cdot-}$)

In view of the radical/radical-ion equilibrium $QH\cdot \rightleftharpoons Q^{\cdot-} + H^+$, the spectrum remaining after flashing in alkali, shown before in figure 2(c) and again in figure 4(c) is undoubtedly $Q^{\cdot-}$. It agrees almost exactly with that given by Baxendale & Hardy (1953) which is included in the above figure, as well as with a species produced by air oxidation of durohydroquinone. (See also figure 4(c).)

It is an unfortunate coincidence that the $Q^{\cdot-}$ band at 415 mμ falls so close to the R band at 418 mμ, but since the relative intensity at 438 and 415 mμ is independent of delay, no radical can be present in alkali. (This also is in accordance with the results of Baxendale & Hardy.)

(b) The semiquinone radical ($QH\cdot$)

In acid/ethanol and in all other neutral solvents except ethanol and possibly water, the 438 mμ band is absent (or very weak) so it must be concluded that the radical ion is absent. This is, of course, to be expected in acid and in non-polar solvents such as liquid paraffin. The band with maximum near 410 mμ which appears in all these solvents is very probably the same species ($QH\cdot$) in all cases. The band shape is very similar in $CHCl_3$, ether and dioxane, slightly sharper in liquid paraffin and much sharper in hexane and cyclohexane. This identification is supported by the observations of Michaelis & Woolman previously referred to, who found bands at 420, 415 and 398 mμ in the rigid glass solvent e.p.a. when durohydroquinone was irradiated. They attributed these bands to the uncharged radical, $QH\cdot$. Dr Strachan has repeated the experiments in these laboratories and a trace of the spectrum obtained is shown in figure 3(d), from which the great similarity to the radical, apart from the expected long wavelength shift in rigid media, is apparent.

The sharpness in hexane has already been discussed; the only other significant change was a shift of 16 mμ to longer wavelengths in ethanol. To investigate whether this was in fact a solvent shift, the spectrum was recorded (at minimum delay) in various mixtures of hexane/ethanol with the results shown in table 2. The smooth change observed clearly indicates that the same species is involved. The increase in intensity at higher-ethanol concentrations is accompanied by a broadening of the band to longer wavelengths, which is probably due to increasing proportions of the radical ion.

Since the band position will shift with solvent and pH, and since the ratio of radical ion changes with time in ethanol (this will be discussed in greater detail in part II) it is difficult to make exact comparisons between all the curves quoted. However, all the results in this region can be explained in terms of

(a) the semiquinone radical, favoured at low pH, with main maximum at 410 mμ in most solvents, with shift to 418 mμ in ethanol; and

(b) the semiquinone radical ion favoured at high pH, with main maximum at 438 and subsidiary maximum at 415 mμ in ethanol (and possibly water), somewhat lower in ether, and not observed in other solvents.

TABLE 2. SOLVENT EFFECT ON SEMIQUINONE RADICAL SPECTRUM

% ethanol in hexane	0·2	0·8	2·0	20	60	100
radical absorption at 30 μs delay (arbitrary units)	4	7	18	41	41	38
radical max. (mμ)	405	409	410	415	415	418

(c) *The triplet*

The very strong absorption, T, with maxima at 460 to 490 mμ observed in liquid paraffin cannot be due to either of the above species, since the absorption at 410, 418 and 438 mμ is less intense in liquid paraffin than at 460 to 490 mμ. It is difficult to reconcile it with any species other than a metastable excited state of duroquinone, in view of its immediate appearance and the fact that, unlike R, it is formed directly from the quinone more readily than by photolysis of the products. Further confirmation of this view of a metastable state—and in particular the triplet state—is offered by comparison of its occurrence with that of other triplets. The fact that it appears strongly only in viscous solvents is exactly similar to Porter & Windsor's observations on many other undoubted triplets.

The strongest evidence for the correctness of this assignment has recently been obtained by Dr D. J. Morantz in these Laboratories (unpublished work), using a flash photolysis apparatus designed for studies rigid in media at low temperature. T only was observed in liquid paraffin cooled to $-78°$ C. and its decay time was prolonged. This is entirely in accordance with our assignments and provides excellent confirmation of them, since alternative explanations of these facts would be very difficult. The absence of radicals agrees with additional observations by Strachan, and implies a low probability of hydrogen abstraction at low temperatures and presumably a significant activation energy for this process. The decay of the 490 mμ band is typical of triplet state behaviour and of nothing else. This longer life is characteristic of viscosity effects on triplet states, and the fact that there is a decay—in rigid media at low temperatures—confirms the assignment, since, in the many cases of radical behaviour studied in rigid media, not one has been found where a radical is observed to decay once it has been formed in the matrix.

The absence, or much lower concentration, of triplets in other media must be accounted for by radiationless deactivation (in solvents such as hexane) or to reaction with the solvent as well, in ethanol and most other media. In several cases where radical absorption is strong there is also a weaker absorption around 460 to 490 mμ, but it is difficult to decide whether this is due to a small amount of triplet formation or to a weaker band system of the radical.

(d) Low-wavelength and weak transient bands

It has been remarked that, in all cases, the prominent visible bands are accompanied by strong bands in the ultra-violet. Figure 2 shows this, the main group being at *ca.* 320 mμ. The peak at 320 mμ with a shoulder at 330 or 340 mμ, depending on solvent, would seem to be characteristic of the radical as it appears strongly under all conditions except in alkaline ethanol, where a weaker and more continuous absorption band is apparent. However, even though the general appearance of the bands does agree with the picture of the species involved, too much weight must not be attached to them as they are in a region of the spectrum difficult to study, mainly because of the changing absorption caused by the conversion of duroquinone to durohydroquinone. The very weak band observed around 435 mμ in liquid paraffin must, in view of the results just discussed, be identified with a small amount of radical-ion formation even in this non-ionic solvent.

(e) Summary of the radicals identified

The spectra of the three transient species of duroquinone in various solvents are given individually in figure 5 and a list of the maxima and photographically observed lifetimes of all of them is given in table 3 (*a*) and (*b*). The dashes indicate that nothing was observed.

TABLE 3*a*. DUROQUINONE TRANSIENT MAXIMA AND LIFETIMES
(NON-ALCOHOLIC SOLVENTS)

transient	hexane	liquid paraffin	ether	dioxane	CHCl$_3$
radical max. (mμ)	402	409, ~400	410, ~388	410, ~400	405, ~395
radical half-life (μs)	100	2000	300	350	300
radical-ion max. (mμ)	—	~435	—	~445	—
triplet max. (mμ)	—	460, 490	~485	485	—
triplet half-life (μs)	—	100	~30	~30	—

TABLE 3*b*. DUROQUINONE TRANSIENT MAXIMA AND LIFETIMES
(ALCOHOLIC SOLVENTS)

transient	96 % ethanol	50 % ethanol/H$_2$O pure	+ 0·1 N-H$_2$SO$_4$	+ 0·02 N-NaOH
radical max. (mμ)	418, ~395	418, ~395	415, ~395	—
radical half-life (μs)	900	750	600	—
radical-ion max. (mμ)	438, 418	440, 415	—	438, 415
radical-ion max. half-life (ms)	900	700	—	15 min
triplet max. (mμ)	~485	~485	~490	—
triplet max. half-life (μs)	~30	~30	30	—

OTHER METHYL QUINONES

The work just reported has established the species which are present, their spectra and the conditions of their appearance. In this section we shall demonstrate how the use of these conditions leads to the detection of similar species in other methyl quinones on the basis that similar qualitative behaviour is to be expected for related molecules. The various methyl quinones studied were *p*-benzo; tolu-; *m*- and *p*-xylo; trimethyl- and, of course, duroquinone, in the key media, acid/ethanol; alkali/ethanol and liquid paraffin.

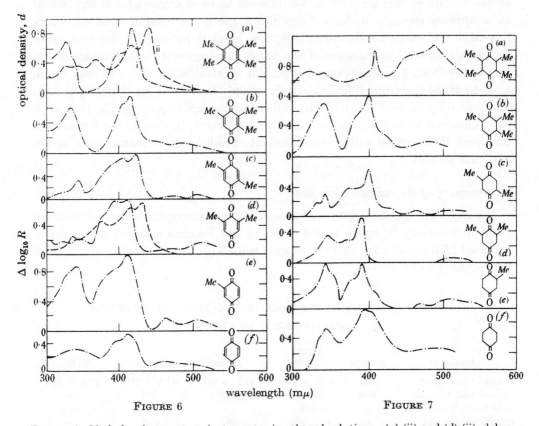

FIGURE 6

FIGURE 7

FIGURE 6. Methyl quinones, transient spectra in ethanol solutions. (*a*) (ii) and (*d*) (ii), delay 10 s, solvent 50 % ethanol + 0·02 N-NaOH. All others, 50 % ethanol + 0·1 N-H$_2$SO$_4$, delay < 30μs.

FIGURE 7. Methyl quinones transient spectra in liquid paraffin. All at < 30μs delay. The ordinates in figures 6 *a* and 7 *a* are optical density and in the other figures the ordinates are $\Delta \log_{10} R$.

Acid/ethanol behaviour

The results were entirely as expected. A single species was observed with a spectrum almost identical with that found for duroquinone under the same conditions, and of comparable lifetime; consequently it can, with some certainty, be identified as the semiquinone radical, QH·. The spectra of the six radicals are given in figure 6. The relative concentrations and extinction coefficients are unknown, but are probably very similar, since the intensities under similar conditions are nearly the same. The half-lives of the radicals are included in table 4.

Alkali/ethanol

As alkali rapidly attacks all but the most substituted methyl quinones—leading to hydroxylation and subsequent ring breakage—it was found impossible to detect the radical ion of any other methyl quinone apart from *m*-xyloquinone which was stable enough to allow time for about two flashes. Figure 6 includes the 'stable' species so obtained, and demonstrates the great similarity to the duroquinone Q^{-} so obtained as well as the difference from the spectrum in acid/ethanol. From this we conclude that it is the radical ion of *m*-xyloquinone. The other methyl quinones, when flashed in neutral ethanol, showed a similar behaviour to duroquinone in that a second peak at higher wavelengths than $QH\cdot$ became more pronounced at later times. This, most probably, was the radical ion of the other quinones.

TABLE 4. METHYL QUINONE RADICAL HALF-LIVES AND MAIN MAXIMA

quinone	liquid paraffin		ethanol + acid	
	(ms)	(mμ)	(ms)	(mμ)
duro-	2	409	0·8	413
3-methyl-	overlap prevents measurement	410	0·7	405
2-methyl-(p)	2·5	410	1·3	410
2-methyl-(m)	3·4	405	1·4	400
1-methyl-	overlap prevents measurement	405	0·8	403
benzo-	2	400	0·5	415

Liquid paraffin behaviour

All the quinones in liquid paraffin showed the radical spectrum observed in acid/ethanol, but shifted to slightly longer wavelengths and much sharper—exactly as occurs in duroquinone. There were, however, two important differences. First, no strong, high-wavelength transient such as was identified as the triplet of duroquinone was observed, although a very weak absorption was detected in this region at the shortest delays. It must be concluded that the triplet does not absorb in this region appreciably (although, by analogy, this is unlikely), that it is very rapidly deactivated, or that it is less readily formed.

Secondly, there was a definite broadening of the band to longer wavelengths as the reaction progressed with time. The sharp spectrum observed first must, on the basis of all our previous conclusions and its great similarity in structure with the radical of duroquinone, be identified as the neutral radical. The triplet can hardly increase in concentration after the flash, so, unless some other species is present—and it is difficult to postulate such a one under these conditions—this slight broadening effect and shift to longer wavelengths with time must be due to a tendency for the neutral radical to ionize and form the radical ion even in liquid paraffin solution, just as did duroquinone, but to a greater extent. There is another possibility to be considered in these cases, namely, the build-up of a quinhydrone formed by the combination of two semiquinone radicals. This species is known to be much more stable when formed from methyl quinones other than duroquinone, and, although it is present only in extremely small amounts when in solution (Michaelis & Granick 1944) yet it absorbs light in the region considered. For example, p-benzoquinhydrone has a very smooth and broad band stretching from 370 to 620 mμ with a maximum at 430 mμ. However, because of this very diffuseness of the spectrum it is not possible to decide with certainty between the two possibilities. The traces of the above radical spectra at minimum delay are given in figure 7 and the half-lives are included in table 4.

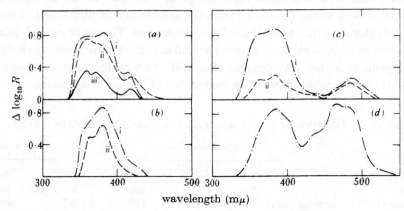

FIGURE 8. 2-Methyl anthraquinone transient spectra. (*a*) Liquid paraffin (i) < 30 μs; (ii) 140 μs; (iii) 1 ms. (*b*) 50 % ethanol + 0·1 N-H$_2$SO$_4$. (i) < 30 μs; (ii) 310 μs. (*c*) 96 % ethanol. (i) < 30 μs; (ii) 750 μs. (*d*) 50 % ethanol + 0·1 N-NaOH; 10 s delay.

Naphthoquinone

Observations were complicated in this case by interference of the products— mainly the hydroquinone—absorbing strongly in the region of the transients. The differences in spectra obtained in the three key solvents are not striking, though there is evidence for a weak triplet at high wavelengths in liquid paraffin. Addition of alkali to ethanol solutions was not successful because of hydroxylation of the quinone almost immediately.

Anthraquinone and derivatives

It has already been noted that these compounds were the first to be investigated by us partly because of the previous work of Cooper who classified them as either sensitizers or non-sensitizers of the photochemical oxidation of alcohol. The subsequent work described above has clarified the position regarding the complex transients we then obtained and which originally led to a search for simpler systems. Consequently, it is now possible to interpret some of the results obtained, although much work remains to be done on these more complicated molecules.

2-Methyl anthraquinone

This is a sensitizer. The various species obtained in liquid paraffin, ethanol, ethanol/acid and ethanol/alkali are given in figure 8. The main difference between them is the very intense band at 480 mμ in alkali/ethanol and other solutions which is evidently the radical ion. It is rather difficult to pick out any sign of a triplet, although there are indications of a band at 380 to 390 mμ in liquid paraffin which seems to decay faster than the main radical peaks at 360 and 380 mμ. Permanent change partly obscures the radical bands, especially in ethanol solutions.

Anthraquinone

Again large permanent change—to the hydroquinone—obscures the transients. It is not possible to pick out any one band as the triplet, although the distinction between the radical and radical ion is very clear. There is a strong Q^- band at 480 mμ and a group of bands at *ca.* 360 to 390 mμ due to the $QH\cdot$.

CONCLUSION

It is clear that, in spite of quantitative differences, the photolysis of a wide variety of quinones shows a behaviour which is general to this type of compound. There is a marked tendency to form the semiquinone radical by hydrogen abstraction and for this radical to exist in its relatively stable radical ion form at high pH and in its unchanged form at low pH. The spectrum of the ion is to be found always at longer wavelengths than that of the radical, an effect which is in accordance with the general conception of the free electron model and the increased resonance of the radical ion.

The triplet state is never observed for very long and is only clearly evident in viscous solvents in duroquinone. It is unlikely that the absorption spectra of the triplet states of these molecules are very different either in position or in extinction coefficient, and therefore the low concentration of triplets in other quinones is probably due to a higher reactivity or rate of deactivation, or to a smaller probability of non-radiative S-T conversion.

The general correlation between appearance of radicals and photosensitizing power is striking and of great interest. Its interpretation cannot, however, be attempted until kinetic studies have been carried out, and these are described in part II.

This work forms part of the programme of fundamental research undertaken by the British Rayon Research Association.

REFERENCES

Backstrom, H. L. J. 1944 *The Svedberg memorial volume*, p. 45. Uppsala: Almquist och Wiksells Boktryckeri.
Baxendale, J. H. & Hardy, R. 1953 *Trans. Faraday Soc.* **49**, 1433.
Berthoud, A. & Porret, D. 1934 *Helv. Chim. Acta*, **17**, 694.
Bolland, J. L. & Cooper, H. R. 1954 *Proc. Roy. Soc.* A, **225**, 405.
Bowen, E. J. 1949 *J. Soc. Dy Col.* **65**, 613.
Bridge, N. K. & Porter, G. 1958 *Proc. Roy. Soc.* A, **244**, 276 (part II).
Burstein, E. & Davidson, A. W. 1941 *Trans. Electrochem. Soc.* **80**, 175.
Fieser, L. F. 1946 *Organic syntheses*, collected Vol. I, p. 383. London: Chapman and Hall.
Lewis, G. N., Lipkin, D. & Magel, T. 1941 *J. Amer. Chem. Soc.* **63**, 3005.
Lewis, G. N. & Lipkin, D. 1942 *J. Amer. Chem. Soc.* **64**, 2801.

388

Bridge & Porter   *Proc. Roy. Soc. A, volume 244, plate 4*

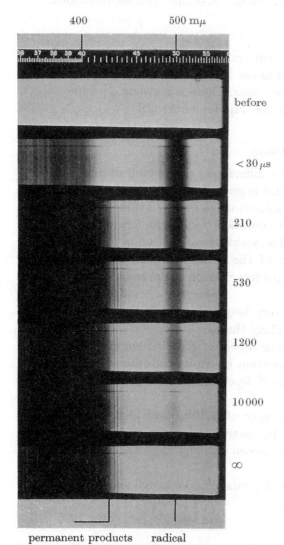

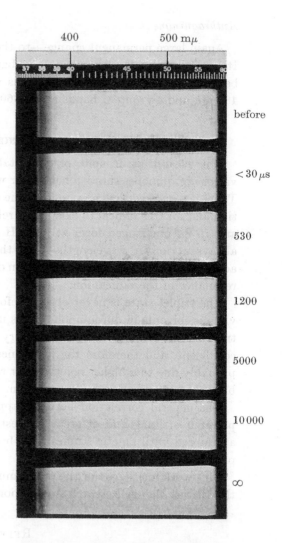

FIGURE 9. 2-SO₃Na-anthraquinone; transients in ethanol.

FIGURE 10. 2-6 diOH-anthraquinone; absence of transient in ethanol.

Bridge & Porter
Proc. Roy. Soc. A, volume 244, *plate* 5

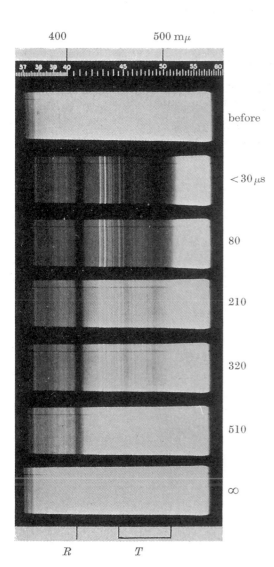

FIGURE 11. Duroquinone transients in
liquid paraffin.

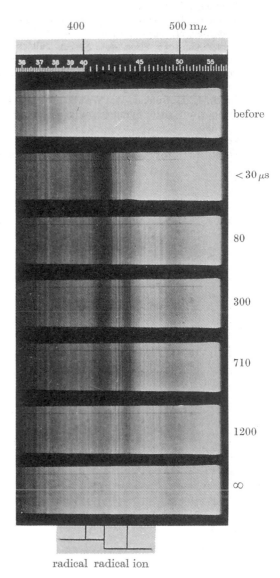

FIGURE 12. Duroquinone transients in
96 % ethanol.

Lewis, G. N. & Kasha, M. 1944 *J. Amer. Chem. Soc.* **66**, 2100.

Lewis, G. N. & Kasha, M. 1945 *J. Amer. Chem. Soc.* **67**, 1232.

Lewis, G. N. & Bigeleisen, J. 1943 *J. Amer. Chem. Soc.* **65**, 2419.

Linschitz, H., Berry, M. G. & Schweitzer, D. 1954 *J. Amer. Chem. Soc.* **76**, 5833.

Michaelis, L. 1938 *Chem. Rev.* **22**, 437.

Michaelis, L. & Granick, S. 1944 *J. Amer. Chem. Soc.* **66**, 1023.

Michaelis, L. & Woolman, S. H. 1950 *Biochem. Biophys. Acta*, **4**, 156.

Pestemer, M. 1951 *Angew. Chem.* **63**, 118.

Porter, G. 1950 *Proc. Roy. Soc.* A, **200**, 284.

Porter, G. & Windsor, M. W. 1954 *Disc. Faraday Soc.* **17**, 178.

Porter, G. & Windsor, M. W. 1957 *Nature, Lond.* **180**, 187.

Smith, L. I. 1930 *Organic syntheses*, **10**, 40.

Smith, L. I., Opie, J. W. Wawzonek, S. & Pritchard, W. W. 1939 *J. Org. Chem.* **4**, 320.

Symposium 1949 Symposium on photochemistry in relation to textiles. *J. Soc. Dy Col.* **65**, 585.

Venkataraman, B. & Fraenkel. G. K. 1955 *J. Chem. Phys.* **23**, 588.

Vogel, A. I. 1954 *A textbook of practical organic chemistry*, p. 174. London: Longmans.

Proc. R. Soc. Lond. A. **326**, 117–130 (1971)

Photochemistry of methylated *p*-benzoquinones

By D. R. Kemp and G. Porter, F.R.S.

Davy Faraday Research Laboratory of The Royal Institution,
21, Albemarle Street, London W 1X 4BS

(*Received* 19 *July* 1971)

The transient absorption at 490 nm, observed on flash photolysis of duroquinone in a variety of solvents, has been confirmed as the triplet state rather than the isomer which has recently been proposed. In ethanolic solution, the triplet state decays to give the semiquinone radical, while in liquid paraffin the evidence is not conclusive owing to an overlap of the radical absorption by the triplet at 410 nm.

The short-lived triplet absorption is observed at 490 nm on photolysis of a number of methylated *p*-benzoquinones, previous failures to observe this absorption being attributed to inadequate time resolution. The lifetime of this species is increased by methyl substitution into the quinone ring and/or the addition of water to an ethanolic solution. This phenomenon is attributed to a stabilization of the (π, π^*) triplet state relative to the (n, π^*) triplet state, and a lowest triplet state of (π, π^*) character in duroquinone. Deprotonation of this state is proposed as a primary photochemical process in polar solvents.

Introduction

Several alkyl-substituted quinones have been shown to be associated with chlorophyll in the chloroplast of the leaf (Crane 1968; Arnon & Crane 1965). Apart from their function as electron acceptors in the electron-transport chain, it has been proposed that they may play a part in phosphorylation (Vilkas & Lederer 1962). A clear understanding of the photochemistry of the simpler quinones, e.g. duroquinone (DQ), may help in establishing the photobiological function of the more complex, naturally occurring quinones.

In recent years, considerable interest has been shown in the photochemistry of DQ and in particular the assignment of the transient species (hereafter designated by T) with an absorption maximum at approximately 490 nm. Several authors (Keene, Kemp & Salmon 1965; Wilkinson, Seddon & Tickel 1968; Hermann & Schenck 1968) have suggested that an isomer of DQ may be responsible for the transient absorption, while the recent results of Land (1969) strongly support the original assignment by Bridge & Porter (1958) that it is due to the triplet state of DQ. Furthermore, Bridge and Porter (1958) concluded that the triplet state (T) did not decay to give the semiquinone radical (DQH·), while Wilkinson *et al.* (1968) have shown from energy transfer experiments that the latter is formed from the triplet state. In an attempt to resolve these discrepancies, we have investigated the series of methylated *p*-benzoquinones and, in particular, DQ. Restrictions caused by the limited time resolution of previous flash studies have been reduced by using the nanosecond flash photolysis technique.

[117]

Experimental

Nanosecond flash photolysis apparatus

The nanosecond flash photolysis apparatus was basically that described by Porter & Topp (1970). For kinetic spectrophotometric investigations, the spectrograph was replaced by an $f6$ Hilger D285 prism monochromator. Changes in light absorption were detected with a 1P28 photomultiplier operated at low gain. Large decoupling capacitors across the last few dynodes were used to ensure linear response, while small disk capacitors were used to maintain high-frequency response. Changes in incident light intensity on the photomultiplier were recorded on a Tektronix 585 A and 549 oscilloscope, and the traces were recorded on Polaroid film. Measurement of the traces was carried out by use of a travelling microscope.

Degassing of samples

Solutions of materials in organic solvents were thoroughly degassed using the freeze-thaw technique. Aqueous solutions were not frozen completely, but were cooled in liquid nitrogen until a layer of ice formed on the surface before exposure to the vacuum.

Materials

Duroquinone (Fluka AG and Aldrich Chemicals) was purified by sublimation and recrystallization from cyclohexane. The melting-points and absorption spectra compared well with those in the literature (Flaig, Salfeld & Baume (1958), and *Dictionary of organic compounds* 1965). Benzoquinone (B.D.H., technical grade) and toluquinone (Kodak Ltd) were purified by sublimation. Anthracene (James Hinton, zone-refined sample), 1,2,5,6-dibenzanthracene (Koch-Light), naphthalene (B.D.H., laboratory reagent), o-xyloquinone (K and K Laboratories), p-xyloquinone (Eastman Organic Chemicals), cumoquinone (K and K Laboratories) and durohydroquinone (K and K Laboratories) were used as supplied. Cyclohexane (B.D.H., Special for Spectroscopy) was passed through an activated silica gel column and fractionally distilled. Liquid paraffin (Medicinal BP grade) was passed through an activated silica gel column. Ethanol (Burrough's A.R. grade) and benzene (Hopkin and Williams, Analar grade) were used without further purification.

Unless otherwise stated, all experiments were performed in degassed solutions at room temperature, the concentration of quinone being in the region of 10^{-3} mol/l so that about 50 % of the 347 nm laser light was absorbed.

Results

Photolysis of duroquinone

Flash photolysis of the DQ in benzene, liquid paraffin and isopropanol gave rise to a broad, structureless absorption in the region of 500 to 410 nm. The uncorrected spectra observed at delays of 27 ns are shown in figure 1. These transient absorption spectra closely resemble those of transient T, which have been reported in liquid

Photochemistry of the methylated p-*benzoquinones*   119

paraffin (Bridge & Porter 1958), benzene (Keene *et al.* 1965; Land 1969), cyclohexane (Land 1969), ethanol (Hermann & Schenck 1968) and polymethyl methacrylate (Nafisi-Movaghar & Wilkinson 1970).

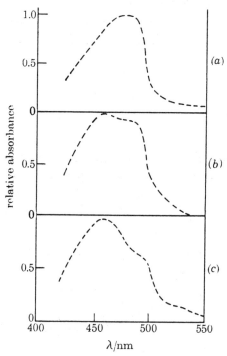

FIGURE 1. Transient spectra (uncorrected) observed 27 ns after laser photolysis of duroquinone in (*a*) benzene, (*b*) isopropanol, and (*c*) liquid paraffin.

Using the kinetic spectrophotometric technique, we observed a first-order decay of the transient absorption (T) at 490 nm in all solvents investigated (see table 1). In each case the maximum transient absorption occurred at the end of the laser pulse. The first-order decay of T is consistent with the findings of other workers (Land 1969; Bridge & Porter 1958; Nafisi-Movaghar & Wilkinson 1970). On saturating such solutions with air, the rate of decay of T was increased (see table 1). Since the concentration of oxygen in cyclohexane at room temperature is 3.0×10^{-3} mol/l (Porter & Windsor 1958), one may obtain a value for the quenching rate constant of T by oxygen in this solvent of $1.3 \times 10^9 \, \mathrm{l \, mol^{-1} \, s^{-1}}$.

The effects of several triplet donors and acceptors on the rate of decay of T in a number of solvents have been investigated. Addition of increasing amounts of anthracene ($E_T = 14\,700 \, \mathrm{cm^{-1}}$) (Lewis & Kasha 1944) to solutions of DQ in benzene, ethanol, isopropanol and liquid paraffin resulted in an increased rate of decay of T (see figure 2), and the results are summarized in table 2. Addition of 1,2,5,6-dibenzanthracene ($E_T = 18\,300 \, \mathrm{cm^{-1}}$) (Porter & Topp 1970) to solutions of DQ in benzene also resulted in quenching of the decay of T, while the presence of naphthalene ($E_T = 21\,300 \, \mathrm{cm^{-1}}$) (Lewis & Kasha 1944), in benzene and liquid paraffin

TABLE 1. FIRST-ORDER DECAY CONSTANTS OF ABSORPTION AT 490 nm FOLLOWING
LASER PHOTOLYSIS OF DQ (10^{-3} mol/l) IN A NUMBER OF SOLVENTS IN THE
ABSENCE AND THE PRESENCE OF AIR

solvent	decay constant in deaerated solution s^{-1}	decay constant in air-saturated solution s^{-1}
ethanol	1.1×10^5	3.6×10^6
benzene	1.0×10^5	1.9×10^6
cyclohexane	1.2×10^5	4.0×10^6
liquid paraffin	3.5×10^4	—

solutions, had no effect on the rate at which T decayed, in accord with previous observations (Wilkinson *et al.* 1968). From the increased rate of decay of T in the presence of anthracene, average values for the quenching rate constants (k_q) have been evaluated, and these are shown in table 2. The value in benzene is consistent with that previously reported, i.e. 6.0×10^9 l mol^{-1} s^{-1} (Land 1969).

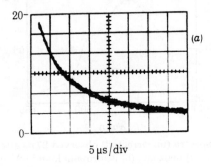

(a)

5 µs/div

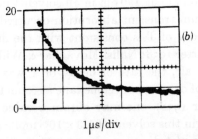

(b)

1 µs/div

FIGURE 2. Decay of absorption at 490 nm following laser photolysis of duroquinone in benzene in (a) absence of anthracene, and (b) presence of 7.5×10^{-5} mol/l anthracene.

The addition of increasing concentration of durohydroquinone (DQH$_2$) to solutions of DQ in ethanol resulted in an increased rate of decay of T (see table 3). In each case fresh solutions were prepared immediately before degassing as the hydroquinone was found to be oxidized to the quinone over a period of hours. These results lead to a rate constant for quenching of T by DQH$_2$ of 3.7×10^9 l mol^{-1} s^{-1}.

TABLE 2. FIRST ORDER DECAY CONSTANTS (k) AND SECOND ORDER QUENCHING CONSTANT (k_q) OBTAINED ON LASER PHOTOLYSIS OF DQ (10^{-3} mol/l) IN FOUR SOLVENTS AND IN THE PRESENCE OF VARIOUS CONCENTRATIONS OF ANTHRACENE

solvent	anthracene concentration/mol l^{-1}	k/s^{-1}	k_q/l mol^{-1} s^{-1}
(a) isopropanol	0	2.3×10^5	
	7.6×10^{-5}	4.3×10^5	
	1.0×10^{-4}	5.3×10^5	2.6×10^9
	1.3×10^{-4}	5.6×10^5	
	2.2×10^{-4}	7.1×10^5	
(b) ethanol	0	1.5×10^5	
	1.1×10^{-4}	4.8×10^5	3.5×10^9
	2.0×10^{-4}	9.4×10^5	
(c) liquid paraffin	0	3.5×10^4	
	1.5×10^{-4}	7.3×10^4	
	3.0×10^{-4}	9.5×10^4	2.1×10^8
	6.0×10^{-4}	1.3×10^5	
(d) benzene	0	1.0×10^5	
	5.0×10^{-5}	4.5×10^5	
	7.5×10^{-5}	6.5×10^5	6.8×10^9
	1.0×10^{-4}	7.2×10^5	

The rate of decay of T as a function of solvent composition in ethanol + water mixtures was also investigated, and the results are summarized in table 4. At low concentrations of water, the rate remained approximately constant, while percentages of water above 70 % by volume resulted in an increased rate of decay of T.

TABLE 3. FIRST-ORDER DECAY CONSTANTS, k, AT 490 nm FOLLOWING LASER PHOTOLYSIS OF DQ (10^{-3} mol/l) IN ETHANOL IN THE PRESENCE OF VARIOUS CONCENTRATIONS OF DUROHYDROQUINONE

durohydroquinone concentration/mol l^{-1}	k/s^{-1}
0	2.7×10^5
2.0×10^{-5}	3.4×10^5
1.0×10^{-4}	6.3×10^5
3.0×10^{-4}	1.6×10^6
5.0×10^{-4}	2.1×10^6

TABLE 4. EFFECTS OF INCREASING PROPORTIONS OF WATER ON THE DECAY CONSTANT, k, OF TRANSIENT T AFTER LASER PHOTOLYSIS OF SOLUTIONS OF DQ (10^{-3} mol/l) IN ETHANOL

% of water by volume	k/s^{-1}	% of water by volume	k/s^{-1}
0	1.1×10^5	70	1.3×10^5
20	1.3×10^5	80	1.8×10^5
50	1.3×10^5	90	1.9×10^5
60	1.4×10^5	100	3.5×10^5

Assignment of transient T

On flash photolysis of solutions of DQ in liquid paraffin, Bridge & Porter (1958) observed a transient (T) with an absorption maximum at 490 nm which decayed by a first-order process and was assigned to the triplet state. Wilkinson *et al.* (1968) investigated the effects of various triplet donors and acceptors on the photochemistry of DQ and showed that the rate of decay of T in liquid paraffin was unaffected by the presence of anthracene. These workers concluded that T was not the triplet state of DQ, and suggested that it might be due to an isomeric species formed from the triplet state.

From energy transfer experiments, the lifetime of DQ triplet in isopropanol has been estimated to be 10^{-5} s. (Wilkinson *et al.* 1968). The results presented here support the findings from pulse radiolysis (Land 1969) that T is formed in less than 10^{-7} s, which excludes the possibility of its being an isomer, or any other species, formed from the triplet state.

The increased rate of decay of T in the presence of anthracene and 1,2,5,6-dibenzanthracene strongly supports the assignment of T to the triplet state of DQ. Since naphthalene had no effect on the rate of decay of T, the triplet level of DQ may be estimated to be in the region of 18 300 to 21 300 cm^{-1} in accordance with the reported value of 19 300 cm^{-1} (Saltiel & Hammond 1963). However, since the anthracene and 1,2,5,6-dibenzanthracene are directly excited by the 347 nm laser light it is possible that the quenching observed may be caused by triplet–triplet interaction rather than triplet–ground state energy transfer. On the other hand, the quenching of T by anthracene in benzene has been supported by the results of Land (1969), who observed the formation of triplet anthracene absorption at the same rate as T decayed, providing strong evidence that the process was one of triplet-ground state energy transfer. The apparent discrepancy with the previous findings in liquid paraffin (Wilkinson *et al.* 1968) may perhaps be explained by the inaccuracy involved in kinetic analysis as a result of an overlap by the triplet anthracene absorption at 490 nm, and the limited time resolution of their apparatus.

The lifetimes of T (see table 1) are consistent with the estimated lifetime of triplet DQ in isopropanol (Wilkinson *et al.* 1968), while the effects of oxygen on its rate of decay are characteristic of triplet states. Further indirect evidence is derived from the effects of added DQH_2 on the rate of decay of T, since it has been shown from continuous irradiation studies (Nafisi-Movaghar & Wilkinson 1970) that DQH_2 quenches the triplet state of DQ. The effects of temperature on the lifetime of T have been investigated (Nafisi-Movaghar & Wilkinson 1970; D. J. Morantz, unpublished work) and are entirely in accordance with the assignment made here.

Origin of the semiquinone radical

Assignment of T to the triplet state of DQ introduces an inconsistency concerning the nature of the precursor of the semiquinone radical (DQH˙). From the relative changes of absorption with time at 405 and 490 nm following flash photolysis of DQ in liquid paraffin, Bridge & Porter (1958) concluded that T did not decay to give DQH˙. On the other hand, Wilkinson *et al.* (1968) have established from energy transfer experiments that DQH˙ is formed from the triplet state of DQ.

On monitoring a freshly degassed solution of DQ in ethanol at 410 nm, an instantaneous absorption (formed within the flash) was observed, a small proportion of which decayed to give a relatively long-lived species. The initial rates of decay at 410 and 490 nm were equal and, on monitoring at progressively higher wavelengths, the amount of initial absorption, and the proportion which decayed, increased. On saturating the solution with air, the same initial absorption was observed which decayed rapidly, at the same rate as that at 490 nm, to give no detectable long-lived species. These facts indicate that the initial absorption at 410 nm is associated with T, while the long-lived absorption may be assigned to DQH˙ which would not decay in the times considered. The absence of this species in the presence of oxygen is to be expected in view of the competition for the DQ triplet by the oxygen quenching process.

On pre-irradiating a freshly degassed solution before laser photolysis, the amount of long-lived absorption progressively increased while the instantaneous absorption remained approximately constant. Further pre-irradiation resulted in a build-up of the long-lived absorption with time until a constant value was attained which was approximately twice that formed before pre-irradiation. A number of workers (see Wilkinson *et al.* 1968; Hermann & Schenck 1968), have previously reported that the absorption intensity of DQH˙ increased following pre-irradiation. It has been proposed (Hermann & Schenck 1968; Nafisi-Movaghar & Wilkinson 1970), that this is due to the formation, on irradiation, of DQH_2 to form DQH˙.

$$DQ^T + DQH_2 \rightarrow 2DQH˙ \rightarrow DQ + DQH_2. \tag{1}$$

Bridge & Porter's (1958) assignment of the 410 nm absorption to DQH˙ is therefore confirmed.

On saturating the pre-irradiated solution with air, the build-up of absorption with time was removed, while the initial absorption decayed to give a long-lived transient. This is to be expected since the DQH_2 formed during irradiation will compete with oxygen for reaction with DQ triplet (equation (1)).

Monitoring a solution of DQ in liquid paraffin, in the 400 to 430 nm region, showed only the decay of an instantaneous absorption to give a long-lived species, though less than was formed in ethanol. On pre-irradiating the solution, similar results were obtained although the amount of long-lived species increased.

By analogy with the observations made in ethanol, the initial absorption is assigned to T, and the long-lived species to DQH˙. The extinction coefficient of DQH˙ in cyclohexane has been reported to be 3500 l mol^{-1} cm^{-1} (Land 1969) while in ethanol a value of 10000 l mol^{-1} has been estimated (Bridge & Porter 1958). If the extinction coefficient of DQH˙ in liquid paraffin is similar to that in cyclohexane, the reduced absorption of DQH˙ in liquid paraffin over that in ethanol may be understood. Moreover, as the optical density of DQH˙ in liquid paraffin is probably less than in ethanol, no growth of DQH˙ absorption is expected owing to the overlap of absorption by T.

The kinetic observations of Bridge & Porter (1958) may therefore be explained by an overlap of the absorption of DQH˙ by T at 405 nm. Whether T decays to give DQH˙ in liquid paraffin is not proven, but the results in ethanol suggest that this is very probably the case.

Photolysis of other methylated p-benzoquinones

(a) p-Benzoquinone

On laser photolysis of p-benzoquinone in ethanol, benzene and water, no short-lived absorption was observed in the region of 490 nm.

(b) Toluquinone (1-methyl-1,4-benzoquinone)

No transient absorption at 490 nm was observed following photolysis of solutions of toluquinone in ethanol or ethanol/water (50:50 by volume). In water and benzene, a transient absorption was observed which in water decayed by a first order process ($k_1 = 3.2 \times 10^6 \, \mathrm{s^{-1}}$). In benzene, the decay was too rapid to allow kinetic analysis.

(c) p-Xyloquinone (2,5-dimethyl-1,4-benzoquinone)

Photolysis of solutions of p-xyloquinone in ethanol gave rise to no transient absorption at 490 nm. On addition of increasing amounts of water a transient absorption appeared (see figure 3), the lifetime of which increased markedly with added water (see table 5). In ethanol + water (25:75 by volume) and water the decay was first order (see table 5) and was unaffected by the addition of oxygen due, probably, to the short lifetime.

TABLE 5. EFFECTS OF WATER ON THE DECAY OF THE TRANSIENT ABSORPTION
AT 490 nm FOLLOWING LASER PHOTOLYSIS OF p-XYLOQUINONE

solvent composition	first-order decay constant/s^{-1}
ethanol/water (50:50 by volume)	$> 3 \times 10^7$
ethanol/water (25:75 by volume)	2.0×10^7
water	3.1×10^6

(d) o-Xyloquinone (2,3-dimethyl-1,4-benzoquinone)

Solutions of o-xyloquinone in a number of solvents were investigated, and the results are given in table 6. It should be noted that dilution of the ethanol by an equal volume of benzene increased the lifetime by a factor of 2, while dilution with the same volume of water increased the lifetime by a factor of 10. Table 6 shows that in ethanol/water (50:50 by volume) and water solutions the transient is quenched by oxygen, while in the other solvents, the lifetime is probably too short for oxygen quenching to compete.

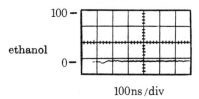

ethanol

100ns/div

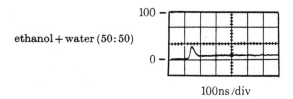

ethanol + water (50:50)

100ns/div

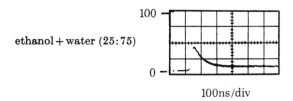

ethanol + water (25:75)

100ns/div

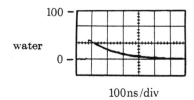

water

100ns/div

FIGURE 3. Absorption at 490 nm following laser photolysis of solutions of
p-xyloquinone in ethanol + water mixtures.

(e) *Cumoquinone* (*trimethyl-1,4-benzoquinone*)

Flash photolysis of cumoquinone in ethanol gave rise to a transient absorption at 490 nm which decayed by a first order process ($k_1 = 2.2 \times 10^6 \, \text{s}^{-1}$). At 410 nm, an instantaneous absorption was observed followed by the growth of a long-lived absorption. The latter process was first order with a rate constant of $1.8 \times 10^6 \, \text{s}^{-1}$. Typical traces showing the changes in absorption at 490 and 410 nm are given in figure 4. By analogy with similar results from DQ, the long-lived absorption at 410 nm is assigned to the semiquinone radical (Bridge & Porter 1958).

D. R. Kemp and G. Porter

The relative changes in absorption with time at 490 and 410 nm show clearly, in this case, that the species absorbing at 490 nm decays to give the semiquinone radical. This conclusion is supported by the fact that, in aerated solution, the rate of decay of absorption at 490 nm was increased ($k_1 = 5.0 \times 10^6 \, s^{-1}$), while at 410 nm the initial absorption decayed to give approximately one-third the amount of semi-quinone radical formed in degassed solution owing to competitive quenching by oxygen.

The absorption changes at 490 and 410 nm were investigated as a function of solvent composition in ethanol + water mixtures, and the results are summarized in table 7. The rate of build-up of absorption at 410 nm was decreased concurrently with the rate of decay at 490 nm, again supporting the proposal that the latter is the precursor of the semi-quinone radical. However, as the percentage of water was increased, the amount of semiquinone radical formed decreased, until in pure water no semiquinone radical was observed and only the decay of the initial absorption was seen. The initial absorption is probably due to the transient species with an absorption maximum at about 490 nm. The latter species was shown to be quenched by oxygen in aqueous solutions.

TABLE 6. FIRST-ORDER DECAY CONSTANTS OF ABSORPTION AT 490 nm FOLLOWING
PHOTOLYSIS OF SOLUTIONS OF o-XYLOQUINONE

solvent composition	decay constant in absence of air/s^{-1}	decay constant in air-saturated solution/s^{-1}
ethanol	2.2×10^7	2.3×10^7
ethanol + water (50:50 by volume)	1.7×10^6	3.1×10^6
water	8.3×10^5	1.3×10^6
ethanol + benzene (50:50 by volume)	1.1×10^7	1.1×10^7

TABLE 7. EFFECT OF WATER ON ABSORPTION CHANGES AT 490 nm AND 410 nm
FOLLOWING LASER PHOTOLYSIS OF CUMOQUINONE IN ETHANOL

percentage of water by volume	first order decay of absorption at 490 nm/s^{-1}	first order growth of absorption at 410 nm/s^{-1}
0	2.2×10^6	1.8×10^6
5	1.7×10^6	1.4×10^6
10	1.3×10^6	1.1×10^6
20	9.6×10^5	9.0×10^5
30	7.4×10^5	6.8×10^5
50	4.1×10^5	—
65	3.8×10^5	—
80	3.8×10^5	—
100	5.2×10^5	—

The similarity of the transient absorption at 490 nm in the methylated p-benzoquinones with those of DQ and chloranil (Kemp & Porter 1969), strongly suggests that these absorptions originate from the lowest triplet state of these compounds.

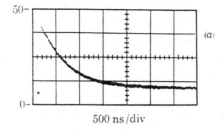

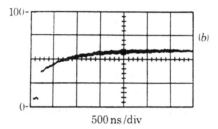

FIGURE 4. Absorption at (*a*) 490 nm and (*b*) 410 nm following laser photolysis of cumoquinone in ethanol.

DISCUSSION

An investigation of the absorption spectra of the methylated *p*-benzoquinones (Braude 1945) shows that the addition of methyl groups into the ring of *p*-benzoquinone results in a blue shift of the n–π* singlet transition and a red shift of the lowest π–π* transition. In hydroxylic solvents, such as water, these effects are enhanced, further reducing the energy gap between these transitions. Investigations on the emission spectra of the methylated *p*-benzoquinones in the vapour state (Jayswal & Singh 1965), suggests that these shifts are paralleled in the triplet manifold. Therefore, the introduction of methyl groups into the quinone ring and/or the addition of water to the system affects the relative positions of the excited states, leading to an increase in the (π, π^*) character of the lowest triplet state of these compounds.

The approximate lifetimes of the triplet states of the methylated *p*-benzoquinones (when observed) are summarized in table 8). The results in table 7 show that the lifetime of cumoquinone remains approximately constant in ethanol + water mixtures in which the water is in excess, i.e. greater than 50 % by volume. Changing the solvent from ethanol to water may cause an inversion of the (n, π^*) and (π, π^*) triplet states, resulting in the (π, π^*) state being lowest in water. The addition of a further methyl substituent to the quinone ring will tend to stabilize further the (π, π^*) state while destabilizing the (n, π^*) state. It has been shown that the triplet lifetime of DQ was approximately constant in aqueous ethanolic solutions containing less than 70 % water by volume (see table 4), and it therefore seems likely that the (π, π^*) triplet state is lowest in ethanol. A similar proposal has been made by Bryce-Smith, Gilbert & Johnson (1967) from a consideration of the photochemical reactions of these quinones with olefins and dienes.

TABLE 8. LIFETIMES OF THE TRANSIENT ABSORPTION AT 490 nm FOLLOWING LASER PHOTOLYSIS OF THE METHYLATED p-BENZOQUINONES

quinone	solvent ... ethanol	lifetime/ns ethanol + water (50:50 by volume)	water
benzoquinone	< 10	< 10	< 10
toluquinone	< 10	< 10	300
p-xyloquinone	< 10	30	320
o-xyloquinone	45	600	1200
cumoquinone	450	2400	1900
duroquinone	9000	7700	2900

< 10 denotes no transient absorption observed.

No triplet–triplet absorption was observed in p-benzoquinone, toluquinone or p-xyloquinone in ethanol. The triplet lifetimes of these compounds, under these conditions, are probably too short to be observed by our apparatus, i.e. less than 10 ns. The appreciable lifetime changes observed on varying the solvent support this proposal, while Briegleb, Herre & Wolf (1969) have shown that the lifetime of the triplet state of p-benzoquinone is short (less than $10\,\mu s$) even in the glassy solutions at $-180\,°C$.

A number of workers (see Shcheglova et al. 1969) (Barltrop & Hesp 1967), have proposed that the lowest triplet state of chloranil is (π, π^*) in character. Calculations of the electron densities in the relevant states of this compound (Barltrop & Hesp 1967) indicate that in the (π, π^*) state, the carbon atoms at positions 2 and 3 in the quinone ring are electron deficient compared to the ground state. A similar situation in DQ might be expected, and the protons on the methyl groups attached to the quinone ring will, therefore, be more acidic than in the ground state.

The base-catalysed incorporation of deuterium into DQ has been reported (Scott 1965; Clark, Hutchinson & Wilson (1968), providing evidence for the formation of an intermediate anion of the o-quinone methide (1 in equation (2)). Similar proposals have been made to explain a number of reactions, for example, the side-chain amination of DQ (Cameron, Scott & Todd 1964; Dean, Houghton & Morton 1968). These results indicate that the protons of the methyl groups are appreciably acidic even in the ground state. Excitation to the (π, π^*) triplet state would be expected to favour such a deprotonation in polar solvents, providing a route for de-activation to the ground state.

Several features of the photochemistry of DQ may now be explained following the above proposal. For example, we have shown that high concentrations of water in alcoholic media appear to quench the triplet states of DQ and cumoquinone (see tables 4 and 7). It has been suggested that hydrogen bonding may in some way lead to deactivation of the excited state (Nafisi-Movaghar & Wilkinson 1970). An alternative possibility is that proton loss in the (π, π^*) triplet state is made progressively more facile as the concentration of water increases owing to the increased basicity of the solvent.

The mechanism originally proposed for the formation of diduroquinone incorporated an isomeric species formed by intramolecular hydrogen abstraction and indeed this was one of the principal reasons for assigning the 490 nm transient to an isomer (Wilkinson *et al.* 1968; Schenck 1963). The exclusion of this species from the photochemistry of DQ requires that an alternative explanation be found. Deprotonation of the triplet state of DQ could lead directly to the *o*-quinone methide, **2**:

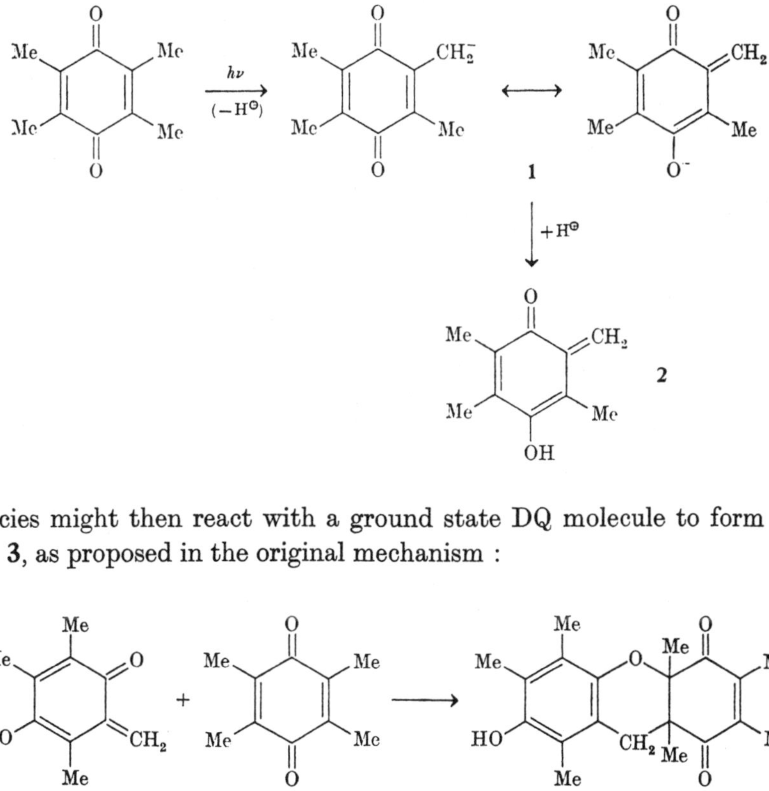

This species might then react with a ground state DQ molecule to form diduroquinone, **3**, as proposed in the original mechanism :

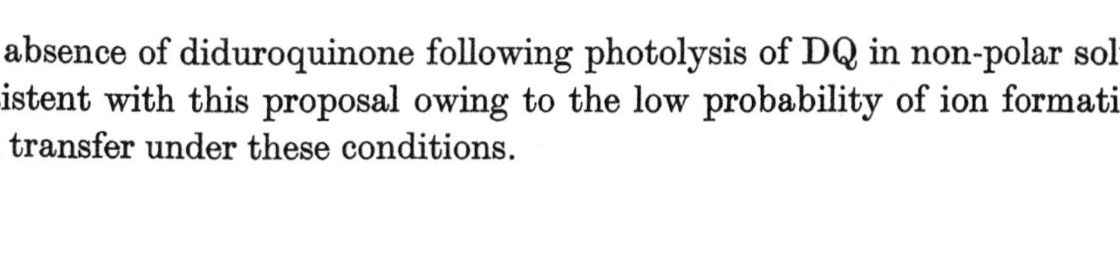

The absence of diduroquinone following photolysis of DQ in non-polar solvents is consistent with this proposal owing to the low probability of ion formation or proton transfer under these conditions.

130 D. R. Kemp and G. Porter

The authors would like to express their thanks to Dr M. B. Ledger for helpful discussions, Dr M. A. West for his technical advice and the Science Research Council for financial support (D. R. K.).

REFERENCES

Arnon, D. I. & Crane, F. L. 1965 *Biochemistry of quinones* (ed. R. A. Morton), p. 433. London: Academic Press.
Barltrop, J. A. & Hesp, B. 1967 *J. chem. Soc.* (C), 1625.
Braude, E. A. 1945 *J. chem. Soc.* p. 490.
Bridge, N. K. & Porter, G. 1958 *Proc. R. Soc. Lond.* A **244**, 259, 276.
Briegleb, G., Herre, W. & Wolf, D. 1969 *Spectrochim. Acta* **25** A, 39.
Bryce-Smith, D., Gilbert, A. & Johnson, M. G. 1967 *J. chem. Soc.* (C), p. 383.
Cameron, D. W., Scott, P. M. & Todd, Lord 1964 *J. chem. Soc.* p. 42.
Clark, V. M., Hutchinson, D. W. & Wilson, R. G. 1968 *Chem. Communs*, p. 52.
Crane, F. L. 1968 *Biological oxidations* (ed. T. P. Singer), p. 533. New York: Interscience.
Dean, F. M., Houghton, L. E. & Morton, R. B. 1968 *J. chem. Soc.* (C), p. 2065.
Dictionary of organic compounds 1965 p. 3019. London: Eyre and Spottiswoode.
Flaig, W., Salfeld, J. C. & Baume, E. 1958 *Annalen* **618**, 117.
Hermann, H. & Schenck, G. O. 1968 *Photochem. Photobiol.* **8**, 255.
Jayswal, M. G. & Singh, R. S. 1965 *J. molec. Spectrosc.* **17**, 6.
Keene, J. P., Kemp, T. J. & Salmon, G. A. 1965 *Proc. R. Soc. Lond.* A **287**, 494.
Kemp, D. R. & Porter, G. 1969 *Chem. Communs*, p. 1029.
Land, E. J. 1969 *Trans. Faraday Soc.* **65**, 2815.
Lewis, G. N. & Kasha, M. 1944 *J. Am. chem. Soc.* **66**, 2100.
Morantz, D. J. unpublished work (as quoted in Bridge & Porter 1958).
Nafisi-Movaghar, J. & Wilkinson, F. 1970 *Trans. Faraday Soc.* **66**, 2257, 2268.
Porter, G. & Topp, M. R. 1970 *Proc. R. Soc. Lond.* A **315**, 163.
Porter, G. & Windsor, M. W. 1958 *Proc. R. Soc. Lond.* A **245**, 238.
Saltiel, J. & Hammond, G. S. 1963 *J. Am. chem. Soc.* **85**, 2515.
Schenck, G. O. 1963 *Arbeitsgemeinschaft für Forschung des Landes Nordrhein-Westfalen* **120**, 27.
Scott, P. M. 1965 *J. biol. Chem.* **240**, 1374.
Shcheglova, N. A., Shigorin, D. N., Yakobson, G. G. & Tushishvili, L. Sh. 1969 *Russ. J. phys. Chem.* **43**, 1112.
Vilkas, M. & Lederer, E. 1962 *Experientia*, **18**, 546.
Wilkinson, F., Seddon, G. M. & Tickle, K. 1968 *Ber. Bunsen-Ges. phys. Chem.* **72**, 315.

Reprinted from the *Transactions of the Faraday Society*, No. 512, Vol. 61, Part 8, August, 1965

Primary Photochemical Processes in Aromatic Molecules

Part 12.—Excited States of Benzophenone Derivatives

By G. Porter and P. Suppan

Dept. of Chemistry, The University of Sheffield

Received 27th January, 1965

We have studied, in some detail, the relationship between the spectroscopic and photochemical properties of substituted benzophenones. A rather complete interpretation of these properties can now be given in terms of the electron distribution in the lowest triplet level. Three types of excited state are important, which may be classified, in decreasing order of reactivity, as $n-\pi^*$, $\pi-\pi^*$ and charge-transfer (C—T) states. The relative position of these states is changed by substitution in the ring and by change of solvent polarity. For example, p-amino- and p-hydroxy-benzophenones abstract hydrogen from paraffins with high quantum yield but not from alcohols since in the latter solvents, the C—T triplet is of lower energy than the $n-\pi^*$ triplet.

The different reactivity of hydroxy and methoxy compounds is a result of rapid deprotonation of the former in the excited state so that the equilibrium species is the base, which has a lowest triplet of C—T type. Phenyl substituted benzophenones have lowest triplet levels of $\pi-\pi^*$ type and, therefore, have intermediate reactivity.

Aromatic carbonyl compounds contain two of the features common to many photoreactive molecules, i.e. a conjugated system of π electrons and a number of non-bonding or n electrons:

In this paper we consider benzophenone molecules having a substituent X introduced to the ring which, as already discussed [1] may have a profound effect on reactivity. The substituent R will in all cases be phenyl; the properties of other substituents R were referred to [2] in part 11 and will be the subject of a later communication.

The photochemical reactions of benzophenone itself, in solution, occur through the lowest triplet level, which is of $n-\pi^*$ type, and are characteristic of attack by an electron-deficient oxygen atom. The reaction which we have studied,[3, 4] and which is further discussed in this paper, involves abstraction of a hydrogen atom from alcohols, hydrocarbons and similar solvents to form a ketyl radical and subsequently a pinacol.

$$\phi_2\overset{*}{C}O + RH \rightarrow \phi_2\dot{C}OH + \dot{R}.$$

With olefins and other unsaturated molecules this is replaced by the oxetane reaction, studied in particular by Yang.[5]

1664

The effect of substituents X, such as CH_3, in the ortho position is relatively well understood and involves an intramolecular hydrogen abstraction and formation of a transient tautomer, to the exclusion of reaction with the solvent.[1, 6] For this reason we shall not consider ortho substituents further here.

Substitution of amino, dimethyl amino or hydroxy groups into the ring at meta or para positions has a remarkable effect on the reaction with alcohols, the quantum yield being reduced from a value near unity in benzophenone itself to near zero in these substituted derivatives. That the triplet state is populated is shown by the fact that all these molecules exhibit an intense phosphorescence. Substituents such as halogens and carboxy have very little effect. The changes in reactivity are paralleled by profound changes in absorption and emission spectra and seem, therefore, to be associated with changes in the electron distribution of the lowest triplet excited state. Previous work, including that referred to in part 10, has attempted to explain these differences in terms of the inversion of two types of electronic state described as $n-\pi^*$ and $\pi-\pi^*$ and, although this gives a partial explanation of the observations there are many difficulties and exceptions. In particular, in molecules such as phenyl benzophenones where the $\pi-\pi^*$ triplet is already lower than the $n-\pi^*$ triplet, the quantum yield of reaction is much higher than when the substituent is NH_2 or OH. Furthermore, the red shift of the so-called $\pi-\pi^*$ band is greater when the substituent is OCH_3 than when it is phenyl and yet the quantum yield of the methoxy derivative is ten times greater than that of the phenyl derivative. Secondly, the $\pi-\pi^*$ bands near 250 mμ in absorption are little affected by the electron-donating substituent and the very strong new absorption at longer wavelengths appears to be a new transition. The interpretation of the strong low-energy singlet in amino and hydroxy derivatives as a red shifted first $\pi-\pi^*$ level must, therefore, be questioned. Thirdly, whilst substitution by OH has a marked effect, similar to that by NH_2, substitution by OCH_3 has no effect, the quantum yield being similar to that of benzophenone itself, although the spectra of the OH and OCH_3 substituted derivatives are nearly identical. In this paper we shall attempt to explain these observations.

EXPERIMENTAL

Steady-state quantum yields were determined using the method described by Beckett and Porter.[4] The ferrioxalate actinometer was replaced in some experiments by benzophenone + cyclohexane or benzophenone + isopropanol actinometric solutions. The concentrations of solutions were determined by a Perkin-Elmer recording spectrophotometer. Phosphorescence spectra were recorded using dilute solutions ($\sim 10^{-4}$ M) in Pyrex tubes at liquid-nitrogen temperature, by an Aminco spectrophosphorimeter. Phosphorescence lifetimes were determined by a Bequerel-type phosphorimeter, the decay curve being photographed from an oscilloscope and plotted on graph paper to determine the order of triplet decay; in all cases a satisfactory first-order decay was obtained.

The melting-points of the benzophenones were checked and the aminobenzophenones were purified by recrystallization. isoPropyl alcohol and cyclohexane were spectrograde solvents from Hopkin and Williams and B.D.H. and were used without further purification. All solutions were rigorously outgassed by the usual procedures.

RESULTS AND DISCUSSION

Since our purpose is to relate the photochemical behaviour with the electronic transition involved we have determined, in addition to the quantum yield (using 3650 Å) of photochemical reduction, the absorption spectrum, the emission (phosphorescence) spectrum in rigid media at 77°K and the triplet lifetime. Each of these parameters was determined in a polar (isopropanol) solvent and a non-polar (cyclohexane) solvent. Our results are summarized in tables 1 and 2.

TABLE 1.—SPECTROSCOPIC DATA AND QUANTUM YIELDS FOR HALOGEN AND ARYL SUBSTITUTED BENZOPHENONES

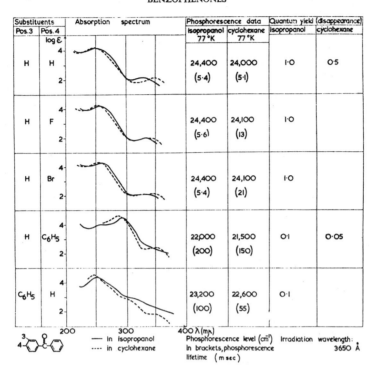

Substituents		Absorption spectrum	Phosphorescence data		Quantum yield (disappearance)	
Pos.3	Pos.4	log ε	isopropanol 77°K	cyclohexane 77°K	isopropanol	cyclohexane
H	H		24,400 (5·4)	24,000 (5·1)	1·0	0·5
H	F		24,400 (5·6)	24,100 (13)	1·0	
H	Br		24,400 (5·4)	24,100 (21)	1·0	
H	C₆H₅		22,000 (200)	21,500 (150)	0·1	0·05
C₆H₅	H		23,200 (100)	22,600 (55)	0·1	

— in isopropanol
···· in cyclohexane

Phosphorescence level (cm⁻¹)
In brackets, phosphorescence lifetime (m sec)

Irradiation wavelength:
3650 Å

TABLE 2.—SPECTROSCOPIC DATA AND QUANTUM YIELDS FOR HYDROXY, METHOXY AND AMINO SUBSTITUTED BENZOPHENONES

Substituents			Absorption spectrum	Phosphorescence data		Quantum yield (disappearance)	
Pos. 3	Pos. 4	Pos. 4'	log ε	isopropanol 77°K	cyclohexane 77°K	isopropanol	cyclohexane
H	OCH₃	OCH₃		24,400 (32)	24,100 (25)	1·0	
H	OH	H		24,400 (36) / 22,500 (72)	24,100 (34)	0·02	0·9
H	NH₂	H		22,000 (200)	23,500 (33)	0·00	0·2
NH₂	H	H		20,000 (145)	22,300 (100)	0·03	0·03
H	N(CH₃)₂	N(CH₃)₂		22,000 (106)	23,000 (41)	0·00	0·6

BENZOPHENONE AND HALOGENATED DERIVATIVES

These three molecules have a quantum yield of unity and almost identical spectra. The absorption spectra show the characteristic $n-\pi^*$ transition at 330 mμ and $\pi-\pi^*$ transition at 250 mμ, the latter showing a small red shift on halogen substitution. The $n-\pi^*$ transition shows the typical blue shift in the polar solvent and the $\pi-\pi^*$ transition the typical red shift. The triplet lifetimes of the three molecules in iso-propanol are identical but the lifetimes of the halogenated molecules are somewhat longer in the paraffinic solvent. Clearly, the lowest triplet is $n-\pi^*$ in all three cases and the observed similarity in photochemical behaviour is to be expected.

PHENYL BENZOPHENONES

Phenyl substitution has little effect on the $n-\pi^*$ transition but, as would be expected, from the extended π orbitals, has a marked effect on the $\pi-\pi^*$ transition. p and m-phenyl benzophenones have rather different absorption spectra, but their reactivities in isopropanol are about the same ($\Phi \simeq 0\cdot1$). The spectra are understandable from the observation that the $\pi-\pi^*$ band of benzophenone is a superposition of two $\pi-\pi^*$ excitations, an intense one ($\log \varepsilon > 4$) polarized parallel to the C=O axis and a weak one ($\log \varepsilon = 3$) polarized perpendicular to the C=O axis. These bands overlap so much in benzophenone that it is almost impossible to distinguish the weak $\pi-\pi^*$ band, but in benzaldehydes and alkyl-phenyl-ketones it can be seen quite plainly. Substituents such as phenyl (and halogens) shift the $\pi-\pi^*$ bands to longer wavelength in a very specific manner; para-substitution shifts only the intense $\pi-\pi^*$ component and meta- or ortho-substitution shifts only the weak one. The shifts are quite insufficient to cause an inversion of the $\pi-\pi^*$ and $n-\pi^*$ levels in the singlet state. However, the extension of the π electron system from one to two rings increases the triplet-singlet splitting (it is 8000 cm^{-1} for benzene, 10000 cm^{-1} for naphthalene, 14000 cm^{-1} for anthracene). The phosphorescence is very long-lived in both 3- and 4-ϕ-ϕ_2CO and occurs at much lower energy than in benzophenone. The assignment, therefore, is that the lowest triplet state is of $\pi-\pi^*$ nature, with a rather low quantum yield of 0·1 in virtue of its electron-distribution which has a higher electron density on the oxygen than has the $n-\pi^*$ state.

The reactivity of a lowest $\pi-\pi^*$ state has been questioned and Yang,[5] in particular, considers $\pi-\pi^*$ states to be completely unreactive ($\Phi = 0\cdot00$) and any reactivity observed to be due to the *upper* $n-\pi^*$ triplet. He gives the example of 9-anthraldehyde in which the energy separation of the $n-\pi^*$ and lowest $\pi-\pi^*$ triplets is so large that phosphorescence and reaction occur apparently from the $n-\pi^*$ level (with $\Phi = 0\cdot1$).

In this case the significant factor is the rate of conversion from the $n-\pi^*$ to the lowest unreactive level. Some phosphorescence is also to be expected from the $n-\pi^*$ level of any reactive aromatic ketone or aldehyde. In particular, all those compounds which give quantum yields of about 0·1 like 9-anthraldehyde should phosphoresce from the $n-\pi^*$ level to the same extent as 9-anthraldehyde. Yet 3-ϕ-ϕ_2CO, 4-ϕ-ϕ_2CO, α- and β-naphthaldehyde, which all have quantum yields of $\sim 0\cdot1$, show no such phosphorescence.

The intrinsic reactivity of a lowest $\pi-\pi^*$ state can be studied by energy-transfer experiments. The problem is to sensitize the $\pi-\pi^*$ triplet with the triplet level of another molecule in such a way that the $n-\pi^*$ triplet of the acceptor is effectively by-passed. α-Naphthaldehyde has been chosen as the acceptor since it has a low $\pi-\pi^*$ triplet at 19900 cm^{-1}, (and an upper $n-\pi^*$ triplet at >23000 cm^{-1}), and 4,4'-(NMe$_2$)$_2$-ϕ_2CO as the sensitizer because of its unreactive (C—T) triplet (see later) at 21700 cm^{-1}. The relevant energy levels are shown in fig. 1. Certainly, the sensitizer has also an upper $n-\pi^*$ triplet around 24000 cm^{-1}, but as it is completely unreactive ($\Phi = 0.00$) in isopropanol the conversion from the $n-\pi^*$ to the C—T state must occur almost instantaneously so that no energy transfer from the $n-\pi^*$ state of the sensitizer to that of the acceptor can be expected. At 3650 Å some absorption by the acceptor cannot be avoided, and the experiment was, therefore, designed to test what quantum yield must be attributed to α-naphthaldehyde on the assumption that every

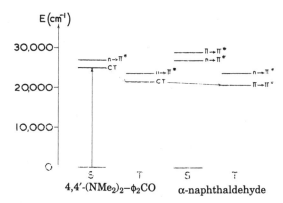

FIG. 1.—Energy levels in energy transfer from 4,4′-(NMe$_2$)$_2$ benzophenone to α-naphthaldehyde.

photon absorbed by the sensitizer was " lost " (because its energy is transferred to a completely unreactive $\pi-\pi^*$ triplet). We have found that a quantum yield of not less than 0.7 was to be attributed to α-naphthaldehyde ; this is evidently in contradiction with its known quantum yield of 0.1 which we have also checked independently. It therefore appears that the hydrogen-abstraction reaction in these compounds must proceed via the lowest $\pi-\pi^*$ state.

AMINO-BENZOPHENONES

4,NH$_2$-, 4,4′-(NMe$_2$)$_2$-, and 3-NH$_2$-benzophenones have very low quantum yields in isopropanol (0.00, 0.00, 0.03 respectively). Their absorption spectra are characterized by a relatively intense band at low energy, certainly at lower energy than the $n-\pi^*$ band of benzophenone. This band had been described as the shifted $\pi-\pi^*$ band (250 mμ) of benzophenone, but if the entire spectra are taken into account it is seen that in all these amino-benzophenones there is also a band at about 250 mμ, which corresponds both in intensity and in position to the band of benzophenone. On the other hand the intensity of the low-energy singlet in NR$_2$-benzophenones is extremely variable.

When the phosphorescence spectra are considered the singlet-triplet splitting of this excited state can be evaluated, and it is found to be less than 4000 cm^{-1}, which is far too small for a normal $\pi-\pi^*$ level. This splitting is characteristic of a state which has a small overlap of the ground-state and excited-state orbitals.

The solvent shifts are the most characteristic property of these absorption bands. When the polarity of the solvent increases, excited states which are more polar than the ground state are stabilized and the excitation energy is thereby reduced (red shift with increased polarity). On the other hand, if the excited state is less polar than the ground state a blue shift occurs for a similar reason ; this is the case with $n-\pi^*$ bands, where it may be complicated also by hydrogen-bonding effects which operate in the same direction.[7] The shifts should be compared only between the same two solvents for all molecules (in our case, cyclohexane and isopropanol). The $n-\pi^*$ band of benzophenone shows a shift of -1000 cm^{-1}, the $\pi-\pi^*$ band of $+400$ cm^{-1} (the $+$ sign will mean that the polarity increases on excitation). The lowest singlet in 4-NH$_2$-ϕ_2CO on the other hand shifts by $+3200$ cm^{-1}, and that in 3-NH$_2$-ϕ_2CO by $+2600$ cm^{-1}. These enormous shifts show that the polarity of the molecule is greatly increased in this excited state, and consequently we shall describe this excitation as an intra-molecular charge-transfer (C—T) and the state attained as a C—T state.

A valence-bond description of the charge-transfer state is simply, perhaps too simply, given as follows :

$$\overset{+}{NH_2} = \text{(ring)} = C\underset{}{\overset{O^-}{\diagup}}$$

In molecular orbital terms three charge-transfer configurations are involved :

$$\overset{+}{NH_2}-\underset{}{\text{(–)}}-C\overset{O}{\diagup} \qquad NH_2-\underset{}{\text{(+)}}-C\overset{O^-}{\diagup} \qquad \overset{+}{NH_2}-\underset{}{\text{( )}}-C\overset{O^-}{\diagup}$$

which may be described as D$^+$R$^-$A ($l\rightarrow$ring π^*), DR$^+$A$^-$ (ring $\pi\rightarrow$carbonyl π^*) and D$^+$RA$^-$ ($l\rightarrow$carbonyl π^*) respectively. The configuration D$^+$RA$^-$ interacts with D$^+$R$^-$A and DR$^+$A$^-$ but not with the locally excited states DR*A. The direct charge-transfer state D$^+$RA$^-$ is generally of too high energy to be of great importance ; it is mainly the two localized charge-transfer configurations which contribute, and of these the effect of the electron donation by lone pair NH$_2$ electrons to the ring is expected to be smaller than donation from the ring to the acceptor carbonyl π^* orbital. States of this type, in the comparable nitroanilines, have been considered by Godfrey and Murrell [8] who have calculated the relative contributions of the different electron configurations, the oscillator strengths and the dipole moments. They calculate that the energy of the C—T state should decrease in the order para, meta, ortho and that the transition of the meta compound should be of lowest intensity. The spectra of fig. 2 show that these predictions are true also for the aminobenzophenones.

This model is in agreement with the observed solvent shifts and with the small triplet-singlet splitting. It also explains immediately why those compounds which have a lowest C—T state are very unreactive, since the electron-density is greatly increased on C=O and therefore electrophilic attack is made very difficult.

In order to construct an energy-level diagram of a molecule like 4-NH$_2$-ϕ_2CO it is necessary to make certain assumptions about the levels which are not accessible to observation, i.e., the $n-\pi^*$ (S) and (T) and the $\pi-\pi^*$ (T). Since the $\pi-\pi^*$ (S) is little shifted from benzophenone to 4-NH$_2$-ϕ_2CO it is reasonable to assume that the π^* orbital, and hence the $\pi-\pi^*$ levels are also nearly the same. With these assumptions the diagrams of fig. 3 can be constructed.

It is seen that the position of the C—T and $n-\pi^*$ levels in isopropanol and the solvent shifts of the C—T and $n-\pi^*$ bands predict that inversion of the levels occurs on passing from a polar to a non-polar solvent. This is confirmed by the fact that 4-NH$_2$-ϕ_2CO and 4,4'-(NMe$_2$)$_2$-ϕ_2CO are reactive in cyclohexane with relatively high quantum yields, as would be expected if their lowest triplet is of $n-\pi^*$ nature.

In the same conditions the reactivity of 3-NH$_2$-ϕ_2CO is found to be very small in both solvents ($\Phi \simeq 0.03$), and this is understandable since its C—T state is considerably lower than that of the 4-NH$_2$- derivative, and the solvent shift is smaller (2600 cm^{-1} against 3200 cm^{-1}), so that inversion of the levels does not occur.

Additional evidence for this inversion in 4-NH$_2$-ϕ_2CO and the lack of inversion in 3-NH$_2$-ϕ_2CO is derived from phosphorescence data at 77°K, in rigid solvents. The

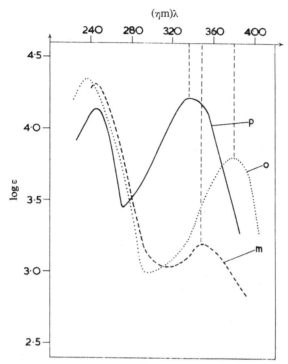

FIG. 2.—Absorption spectrum of o-, m- and p-aminobenzophenones.

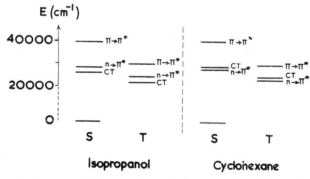

FIG. 3.—Energy levels of p-aminobenzophenone in isopropanol and in cyclohexane.

$n-\pi^*$ triplet levels show a short-lived phosphorescence near 24000 cm^{-1} ($\tau \simeq 5$-40×10^{-3} sec), while the $\pi-\pi^*$ and C—T phosphorescences are quite long-lived ($>0\cdot1$ sec). The phosphorescence lifetime of benzophenone is little changed whether the rigid solvent is isopropanol or cyclohexane at 77°K. On the contrary, the phosphorescence lifetime of 4-NH$_2$-ϕ_2CO, very long in rigid alcohol, becomes quite short in rigid paraffin (see table 2), confirming the change from a C—T to an $n-\pi^*$ level. 3-NH$_2$-

ϕ_2CO, however, has a long-lived phosphorescence in both rigid solvents, in agreement with the fact than no inversion occurs in this case. Although in the present theory we consider for the sake of simplicity that C—T, $n-\pi^*$, $\pi-\pi^*$ levels retain their completely separate identities however close they may be, the inversion of levels is not an " all or nothing " process but mixing of the states occurs when they overlap considerably. This is illustrated by the relationship of phosphorescence lifetime and position of the C—T band in different solvents given in table 3.

TABLE 3.—LIFETIME OF TRIPLET AT 77°K AND MAXIMA OF C—T ABSORPTION BAND OF 4-NH$_2$-BENZOPHENONE IN VARIOUS SOLVENTS

solvent	max. of C—T band (mμ)	lifetime of triplet (m sec)
isopropanol	336	200
benzene	313	90
cyclohexene	310	47
cyclohexane	303	33

In strongly acid solutions the absorption spectrum of amino-benzophenones shows a characteristic change which corresponds to the reaction

$$NH_2 - \phi_2CO + H^+ \rightleftharpoons \overset{+}{N}H_3 - \phi_2CO$$

(pK $4 - \overset{+}{N}H_3 - \phi_2CO \simeq 0.5$). In the protonated form the C—T band disappears completely and the $\pi-\pi^*$ band is shifted to longer wavelength. The acid form of $4 - NH_2 - \phi_2CO$ has been found photoreactive in isopropanol + water mixtures, with a quantum yield greater than 0.1; the question of reactivity involves further considerations of the pK of excited states which will be given in another publication.

HYDROXY- AND METHOXY-BENZOPHENONES

The similarity of the spectra of these compounds as against their completely different reactivities has been one of the major problems. The presence of an intense absorption band with a maximum at 2950 Å has previously been attributed to a red shift of the $\pi-\pi^*$ excitation, much in the same way as with amino-benzophenones. The nature of this absorption band can now be discussed in the same terms as for $4 - NH_2 - \phi_2CO$ and it is then clear that they have many properties in common, i.e., the solvent shift is very large and the singlet-triplet splitting is small.

This C—T singlet level, is however, much higher than in $4 - NH_2 - \phi_2CO$ and the lowest triplet in both 4—OH and $4 - OMe - \phi_2CO$ is the $n-\pi^*$ triplet. This is shown also by the fact that the phosphorescence lifetime is only about 30×10^{-3} sec in both rigid alcohol and rigid paraffin. If the lowest triplet of these compounds is the $n-\pi^*$ triplet the reactivity of $4 - OMe - \phi_2CO$ is explained but the unreactivity of $4 - OH - \phi_2CO$ seems anomalous.

The essential difference between the two compounds resides in the lability of the proton in the −OH function which has no equivalent in the −OMe group. The deprotonated form $4 - O^- - \phi_2CO$ can be prepared in alkaline solutions (pK = 6.5) and its properties are shown in table 2. The C—T absorption band is found at very low energy, nearly at the same wavelength as in $4 - NH_2 - \phi_2CO$. The phosphorescence shows that the lowest triplet is now of C—T type. The energy levels of acidic and basic forms of p-hydroxybenzophenone in isopropanol are given in fig. 4.

A study of the pK of the reaction

$$HO - \phi_2CO \rightleftharpoons {}^-O - \phi_2CO + H^+$$

in the excited singlet and triplet states shows that deprotonation must occur very efficiently in the C—T states.   The procedure of the calculation calls for some special comments and will be dealt with separately,[9] and it is sufficient here to know that the

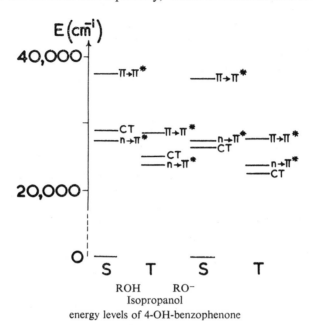

ROH   RO⁻
Isopropanol
energy levels of 4-OH-benzophenone

FIG. 4.—Energy levels of p-hydroxybenzophenone and of its basic anion in isopropanol.

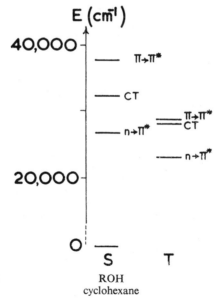

ROH
cyclohexane

FIG. 5.—Energy levels of p-hydroxybenzophenone in cyclohexane.

pKs in the excited states (C—T) singlet and triplet are < —2.   Direct evidence for deprotonation as a result of excitation is found in the observation of 4-O⁻-benzophenone as a short-lived transient produced by flash photolysis.   (In part 10 this was incorrectly attributed to a triplet absorption.) Because of this deprotonation the photochemistry of $4-OH-\phi_2CO$ is in fact that of $4-O^- -\phi_2CO$ even in strongly acid solutions.

The energy-level diagram of p-hydroxybenzophenone in cyclohexane is given in fig. 5. The $n-\pi^*$ triplet level is now easily the lowest and again this has a striking effect on the reactivity. As will be seen from table 2 the quantum yield of hydrogen abstraction rises from 0·02 in pure isopropanol to 0·9 in cyclohexane although the C—H bond in the hydrocarbon is stronger than that in the alcohol.

CONCLUSION

The photochemistry of aromatic carbonyl compounds is governed primarily by the electronic distribution of the lowest excited triplet state. This, in turn, depends on the nature and position of the substituents and on solvent effects, the importance of which cannot be overestimated. The spectroscopic datum which provides the best indication of the electron-distribution of an excited state is the solvent shift of the absorption band of this state, while other data like energy, intensity, etc., show no such unambiguous relationship. Although in principle the electron-density on the carbonyl could vary nearly continuously as the configurations interact with each other, yet the solvent shifts and the reactivities fall into three well-defined groups which correspond to the $n-\pi^*$, $\pi-\pi^*$, and C—T excitations. The properties of these states are listed in table 4.

TABLE 4.—PROPERTIES OF $n-\pi^*$, $\pi-\pi^*$ AND C—T STATES OF BENZOPHENONE DERIVATIVES

state	intensity	energy range	solvent shift	T—S splitting	reactivity	triplet lifetime 77°K	electron distribution
$n-\pi^*$	2	30,000	−800	<3,000	1	<20	$>C=O^{\delta+}$
$\pi-\pi^*$	4	40,000	+600	10,000	0·1	>100	$>C=O$
C—T	3-4	30,000	+2,500	>3,000	0·01	>100	$>C=O^{\delta-}$
	log ε	cm⁻¹ (in p-subst)	cm⁻¹	cm⁻¹	Φ	msec (in p-subst)	

The authors thank Dr. J. N. Murrell for many helpful discussions of the absorption spectra, Mr. M. Walker for his assistance in measuring the phosphorescence lifetimes and Dr. J. N. Pitts for a sample of 3-NH$_2$-benzophenone.

[1] Beckett and Porter, *Trans. Faraday Soc.*, 1963, **59**, 2051.
[2] Beckett, Osborne and Porter, *Trans. Faraday Soc.*, 1964, **60**, 873.
[3] Porter and Wilkinson, *Trans. Faraday Soc.*, 1961, **57**, 1686.
[4] Beckett and Porter, *Trans. Faraday Soc.*, 1963, **59**, 2038.
[5] Yang, Proc. *Int. Symp. Org. Photochemistry*, (Strasbourg, 1964).
[6] Yang and Rivas, *J. Amer. Chem. Soc.*, 1961, **83**, 2213.
[7] Kasha, *Disc. Faraday Soc.*, 1950, **9**, 14.
[8] Godfrey and Murrell, *Proc. Roy. Soc. A*, 1964, **278**, 71.
[9] Godfrey, Porter and Suppan, *Disc. Faraday Soc.*, 1965.

Further publications

PRIMARY PHOTOPROCESSES IN QUINONES AND DYES.
II. KINETIC STUDIES.
(The reaction was later shown to proceed via the triplet state.)
N. K. Bridge and G. Porter, *Proc. Roy. Soc.*, 1958, **A244**, 276.

ACIDIC SEMIQUINONES.
E. J. Land and G. Porter, *Proc. Chem. Soc.*, 1960, 84.

RADICALS AND RADICAL ANIONS DERIVED FROM
BENZALDEHYDE, ACETOPHENONE AND BENZIL.
A. Beckett, A. D. Osborne and G. Porter, *Trans. Faraday Soc.*, 1964, **60**, 873.

PHOTOCHEMISTRY OF BENZOPHENONE DERIVATIVES.
A. Beckett and G. Porter, *Trans. Faraday Soc.*, 1963, **59**, 2038.

FLASH PHOTOLYSIS OF BENZOPHENONE IN SOLUTION.
G. Porter and F. Wilkinson, *Trans. Faraday Soc.*, 1961, **57**, 1686.

PHOTOCHEMISTRY OF SUBSTITUTED BENZOPHENONES.
A. Beckett and G. Porter, *Trans. Faraday Soc.*, 1963, **59**, 2051.

REACTIVITIES OF n-π* AND CHARGE-TRANSFER IN THE
PHOTO-PINACOLISATION OF KETONES.
G. Porter and P. Suppan, *Proc. Chem. Soc.*, 1964, 191.

COMPARATIVE PHOTOCHEMISTRY OF AROMATIC CARBONYL
COMPOUNDS.
G. Porter and P. Suppan, *Trans. Faraday Soc.*, 1966, **62**, 3375.

REACTIVITY OF EXCITED STATES OF AROMATIC KETONES.
G. Porter and P. Suppan, International Symposium on Organic Photochemistry,
Strasbourg, 1964, *Pure App. Chem.*, 1964, **9**, 499.

TRIPLET-TRIPLET ABSORPTION SPECTRA OF
BENZOPHENONE AND ITS DERIVATIVES.
T. S. Godfrey, J. W. Hilpern and G. Porter, *Chem. Phys. Letters*, 1967, **1**, 490.

ACIDITY CONSTANTS OF AROMATIC CARBOXYLIC ACIDS IN THE S_1 STATE.
E. Vander Donckt and G. Porter, *Trans. Faraday Soc.*, 1968, **64**, 3215.

THE PHOTOCHEMISTRY OF TWO PHYTYL QUINONES TOCOPHERYLQUINONE AND VITAMIN K_1.
G. Leary and G. Porter, *J. Chem. Soc., A*, 1970, 2273.

FLASH PHOTOLYSIS OF AN ORTHO-ALKYL-BENZOPHENONE.
G. Porter and M. F. Tchir, *Chem. Comm.*, 1970, 1372.

PHOTOENOLIZATION OF ORTHO-SUBSTITUTED BENZOPHENONES.
G. Porter and M.F. Tchir, *J. Chem. Soc., A*, 1971, 3772.

THE PHOTOCHEMISTRY OF AROMATIC CARBONYL COMPOUNDS IN AQUEOUS SOLUTION.
M. B. Ledger and G. Porter, *J.C.S. Faraday I*, 1972, **68**, 539.

Chapter 9

MODELS FOR *IN-VITRO* PHOTOSYNTHESIS

The oil crisis of the 1970s stimulated a renewed interest in all alternative forms of energy storage, especially from solar energy and other renewable sources. The chemists and biologists naturally looked to photosynthesis as the source of nearly all man's energy, and noted that most of it was derived from fossil fuels laid down more than 100 million years ago. Replacement of these fuels, when they became depleted, might be addressed through natural photosynthesis itself, as biomass for example, but the efficiency of solar energy storage by plants was very low, rarely exceeding 1%.

In vitro photosynthesis was certainly possible; a simple and quite efficient example was photovoltaic generation of electric potential followed by electrolysis of water. But more economic systems were needed and much attention was paid to designing in-vitro models of plant photosynthesis. Model systems of this kind may also contrubute to the understanding of the primary processes occuring in plant photosynthesis such as light harvesting and the redox processes following electron transfer.

The first paper below is a study of the theoretical efficiencies of solar energy collection and storage in general and the second reviews the methods that are available. The papers that follow describe model systems based on chlorophyll, porphyrins and related pigments and their use to elucidate problems of energy transfer (especially to understand the mechanism of concentration quenching and how this process, so ubiquitous *in vitro*, is avoided in vivo).

Other papers describe studies of electron transfer, first the simplest model of photosynthesis, the electron transfer from chlorophyll to a quinone, and then the use of inorganic redox systems and metal catalysts like platinum to split water to hydrogen and oxygen. Usually two steps are necessary, using, in turn, a sacrificial oxidant and reductant. The complexities of carrying out oxidation and reduction in the same system and of transporting the oxidants and reductants over the large distances involved in solar energy collection have led the present author to conclude that the most viable transport process will be that of electrons and that the photovoltaic conversion of solar energy followed by electrolysis (probably of water) holds more promise than the biomimetic processes based on organic molecules.

The model systems used include solutions of pigments in lipids and liposomes whose lamellar structure has many similarities with those of the chloroplast. There is also an extensive series of investigations, (references only) carried out mainly by Dr Tony Harriman, on the photochemistry of manganese porphyrins. These are related to the process of oxygen evolution from photosystem 2 of photosythesis, where the four manganese atoms present in each photosynthetic unit are near to the site of oxidation of water.

Although this work has not led to any important practical developments for solar energy storage it has contributed to a better understanding of chemistry and physics of energy and electron transfer, both in homogeneous solution and in heterogeneous systems related to the *in vivo* photosynthetic system such as monolayers, vesicles and micelles.

J. Chem. Soc., Faraday Trans. 2, 1983, **79**, 473–482

Transfer and Storage of Chemical and Radiation Potential

BY GEORGE PORTER

Davy Faraday Research Laboratory of The Royal Institution, 21 Albemarle Street, London W1X 4BS

Received 23rd August, 1982

The change in potential which accompanies any spontaneous process is given by $RT \ln [1-(j/J)]$, where j is the net forward flux and J is the forward flux at the same reactant potential but in the absence of products. This equation is applied to thermal and photochemical reactions. For the latter, we introduce and evaluate the quantities 'radiation potential' and 'scattered radiation potential' in analogy with 'chemical potential'.

'Storage' of energy is defined and shown to be a two-stage process. Thermodynamic losses are relatively small provided the rather stringent requirements of geometry and transport kinetics can be met. The influences of these several factors on the efficiency of solar-energy collection and storage are evaluated.

The primary consideration in carrying out a chemical reaction is usually the yield of product, but the chemical potential at which this product is finally recovered is sometimes of comparable importance. This is especially true of those reactions which are used for the conversion and storage of energy; reactions such as the preparation of a chemical fuel, the charging of an electrochemical cell and the photochemical processes used for solar-energy conversion.

Consider a chemical reaction step in which products B, at chemical potential μ_B, are formed from reactants A, at chemical potential μ_A. The chemical potentials are referred to the potential of the appropriate final state of the reaction sequence, *e.g.* the products of combustion, as zero. The rate of storage of chemical potential per unit volume in the product B is $j\mu_B$, where $j = -d[A]/dt = d[B]/dt$ is the net flux per unit volume, *i.e.* the rate of change of concentration in the forward reaction $A \rightarrow B$, and [A] and [B] are concentrations. If A and B are at equilibrium, the rates of the forward and back reactions are equal and $j = 0$. If the forward reaction proceeds at a finite rate $j > 0$, there is a net overall entropy increase, as in all spontaneous reactions, and $\mu_B < \mu_A$. In the limit when j takes its maximum value and there is no back reaction $\mu_B = 0$. In both of these extreme cases $j\mu_B = 0$ and no chemical potential is stored. The product $j\mu_B$ is the rate of free-energy accumulation in the product B or the power carried by the chemical reaction, and there must be some intermediate finite rate of reaction at which the power has a maximum value. We will derive expressions relating this optimum power condition to the flux j for several cases of interest.

ENERGY TRANSFER WITHOUT LEAKAGE

$$\xrightarrow{j} A \underset{J_b}{\overset{J}{\rightleftharpoons}} B \xrightarrow{j}$$

$$\mu_A \qquad \mu_B$$

This is the steady-state case of the reaction discussed above. Mass is conserved, the injection of A into the system exactly equals the removal of B from the system. No external work is done.

474   CHEMICAL AND RADIATION POTENTIAL

The following terms will be used: $J(\equiv J_f) = k_f[A]$, where k_f is the rate constant (the specific rate) of the forward reaction $A \rightarrow B$. J is, therefore, the flux per unit volume of the forward reaction when $[B] = 0$. $J_b = k_b[B]$, where k_b is the rate constant of the reaction $B \rightarrow A$. $j = J - J_b$ is the net rate of reaction $A \rightarrow B$. K is the equilibrium constant for the reaction $A \rightleftharpoons B$. μ_A and μ_B are the chemical potentials of A and B, respectively, and are constant since the system is in a steady state.

The change in chemical potential is given by the Van't Hoff isotherm

$$\Delta\mu = \mu_B - \mu_A = -RT \ln K + RT \ln \frac{[B]}{[A]}$$

where [A] and [B] are concentrations if A and B behave ideally. For non-ideal behaviour, [A] and [B] are the activities of A and B and the concentrations in the rate equations must be multiplied by a kinetic activity factor which is usually very nearly equal to the activity coefficient. Putting $K = k_f/k_b$ and substituting the fluxes defined above, we obtain

$$\Delta\mu = RT \ln [1 - (j/J)]. \tag{1}$$

For a sequence l of reactions of this type, with a common flux j, there will be an overall loss of chemical potential given by

$$\Delta\mu = RT \sum_l \ln [1 - (j/J_l)]. \tag{2}$$

This is the very useful equation on which we shall base the rest of our discussion. It gives explicitly the free-energy change in a spontaneous reaction although it has been derived from equilibrium thermodynamics and we should therefore enquire where the approximation has been made. The assumption relating equilibrium to spontaneous change is introduced in the equation $K = k_f/k_b$, which, although widely used in chemical kinetics, is not true under all conditions, even in ideal solutions. At equilibrium, $J = J_b$ and therefore $k_f[A] = k_b[B]$. But $k_f(A) = \sum_i k_{fi}[A_i]$, where the A_i refer to species of A with differing energy content and the k_{fi} are the corresponding rate constants for this energy. Each individual k_i is constant but the concentrations $[A_i]$ are only constant provided Boltzmann equilibrium is maintained in the steady state and only under those conditions can we use a single rate 'constant' which is independent of rate.

In practical terms this does not introduce significant limitations on the use of eqn (1) except in the case of very fast reactions, or in gases at low pressure. In solutions, vibrational relaxation times are usually less than ten picoseconds at normal temperature (except in special cases such as diatomic molecules and monatomic solvents) and we shall not expect deviation from Boltzmann equilibrium or from eqn (1), when it is applied to liquid-phase reactions, unless reaction times are less than 10^{-11} s.

Before proceeding it is interesting to look at the implications of eqn (1). The loss of chemical potential in a chemical reaction can be reduced without limit by reducing j/J, provided the rate of product formation is not of consequence or is adequate. Only when the net flux j approaches J to within a few per cent do the losses in chemical potential become important in comparison with the overall energy change of chemical reactions, as is shown by the figures below, which are calculated for $T = 300$ K:

j/J	0.99	0.9	0.5	0.1
$-\Delta\mu$/eV	0.118	0.059	0.018	0.0027

CONDITION FOR MAXIMUM POWER

The rate P of production of chemical potential (the power) in the form of product B at chemical potential μ_B is given by

$$P = j\mu_B = j(\mu_A + \Delta\mu) \qquad (3)$$

$$= j\{\mu_A + RT \ln[1 - (j/J)]\}. \qquad (4)$$

When the flux j is chosen to give maximum power, $dP/dj|_J = 0$. Differentiating eqn (4) with respect to j and setting the derivative equal to zero gives

$$\frac{j_m/J}{1 - (j_m/J)} - \ln[1 - (j_m/J)] = \frac{\mu_A}{RT} \qquad (5)$$

where j_m is the flux at maximum power. The free-energy transfer efficiency $\eta = (\mu_A + \Delta\mu)/\mu_A$.

The chemical potentials with which we are concerned, *e.g.* those in combustion, electrochemical and photochemical reactions, are of the order of one electron volt. At a temperature of 300 K, we find from eqn (5) that for $\mu_A = 1$ eV,

$$\frac{\mu_A}{RT} = 38.7, \quad \frac{j_m}{J} = 0.972, \quad -\Delta\mu = 0.093 \text{ eV} \quad \text{and} \quad \eta_m = 0.907$$

and for $\mu_A = 2$ eV,

$$\frac{\mu_A}{RT} = 77.4, \quad \frac{j_m}{J} = 0.987, \quad -\Delta\mu = 0.111 \text{ eV} \quad \text{and} \quad \eta_m = 0.944.$$

Here η_m are the efficiencies of transfer of chemical potential from A to B under conditions of maximum power, *i.e.* $j = j_m$. The maximum possible transfer efficiency is, of course, unity but this occurs when $j = 0$ and hence at zero power.

It will be noted that, under conditions of maximum power, the energy losses are considerable and amount to 10% when the chemical potential to be transferred is 1 eV.

ENERGY TRANSFER WITH LEAKAGE

$$\xrightarrow{j} A \underset{J_b}{\overset{J}{\rightleftharpoons}} B \xrightarrow{j-J_x}$$

$$\downarrow J_x$$

$$X$$

There is now an additional reaction which gives the unwanted product X, with a flux J_x and rate constant k_x. The power P is now given by

$$P = (j - J_x)\{\mu_A + RT \ln[1 - (j/J)]\}. \qquad (6)$$

Making the substitution $J_x = (k_x/k_b)(J - j)$ and putting $dP/dj|_J = 0$ as before, we obtain

$$\frac{(j_m/J) - k_x/(k_b + k_x)}{1 - (j_m/J)} - \ln[1 - (j_m/J)] = \frac{\mu_A}{RT}. \qquad (7)$$

The efficiency at maximum power

$$\eta_m = \frac{(j_m - J_x)}{j_m} \frac{(\mu_A + \Delta\mu)}{\mu_A} = \{1 - (k_x/k_b)[(J/j_m) - 1]\}(\mu_A + \Delta\mu)/\mu_A. \qquad (8)$$

476   CHEMICAL AND RADIATION POTENTIAL

Example: For $\mu_A = 1$ eV,

$$\frac{\mu_A}{RT} = 38.7$$

and for

$\dfrac{k_x}{k_x + k_b}$	0	0.5	0.91

we obtain

$\dfrac{j_m}{J}$	0.972	0.986	0.997
$-\Delta\mu/$eV	0.093	0.110	0.150
η_m	0.907	0.890	0.850.

The high values of efficiency, even when $k_x = 10\,k_b$, are a result of the high value of the relative flux j_m/J for optimum power under these conditions so that the back reaction flux is only 0.3% of the forward flux.

APPLICATION TO PHOTOCHEMICAL REACTIONS

$$\xrightarrow{\ j\ } A + h\nu \underset{J_b}{\overset{J}{\rightleftharpoons}} A^* \xrightarrow{\ j\ }$$
$$\qquad\quad \mu_A\ \mu_R \qquad \mu_A$$

For a photochemical reaction without leakage, when the only fates of the product A^* are reaction (with flux j) to give the required product and reverse reaction (with flux J_b), the first product (B in the last section) becomes the excited state A^*. The potential of the 'reactants' on the l.h.s. is composed of the chemical potential of A and the potential of the light quanta absorbed. We will call μ_R the radiation potential or the partial molar quantal free energy of the radiation, by analogy with the chemical potentials, and define it as the maximum work which can be derived from one einstein of the quanta when they are absorbed at the ambient temperature T_A.

The change (loss) in potential in the light absorption process is given, from eqn (1), as

$$\Delta\mu = \mu_{A^*} - \mu_A - \mu_R = RT \ln\left[1 - (j/J)\right]. \tag{9}$$

To evaluate this equation we need to know μ_R.

RADIATION TEMPERATURE AND POTENTIAL (T_R AND μ_R)

Let the radiation be 'monochromatic', consisting of a narrow band centred at wavelength λ. The total energy of this radiation, $E = Nhc/\lambda$ J ein^{-1}, where N is the Avogadro number, but it has entropy associated with it of an amount E/T_R e.u. ein^{-1}, where T_R is the effective temperature of the radiation. This entropy is lost when the radiation is destroyed in the absorption process and at least an equivalent amount of entropy must be created in the absorbing mixture at ambient temperature T_A. The energy available to do work at temperature T_A is therefore given by

$$\mu_R = E(T_R - T_A)/T_R \tag{10}$$

and the maximum possible efficiency

$$\eta_m = \mu_R/E = (T_R - T_A)/T_R. \tag{11}$$

This is merely the Carnot theorem applied to radiation.

It remains to evaluate the effective radiation temperature T_R. The temperature of a monochromatic beam of radiation is equal to the temperature of the black body which gives the same irradiance, E_λ, at the same wavelength λ for unit wavelength interval and unit solid angle, Ω. This temperature is given directly by the Planck radiation formula as

$$T_R = \frac{hc}{k\lambda \ln\left[1 + (2hc^2\Omega/\lambda^5 E_\lambda)\right]} \tag{12}$$

where Ω is the solid angle subtended by the source (including any optical concentrators) at the receiver and the other symbols have their usual meaning. It is assumed that the refractive index is unity.

Example: The sun at A.M.1 (air mass) and without concentrators gives an irradiance at the earth's surface of $1.25\ \text{W m}^{-2}\ \text{nm}^{-1}$ at 700 nm.[1] The solid angle subtended by the sun on earth is 6.8×10^{-5} sr. From eqn (12) we obtain $T_R = 5590$ K and from eqn (11) we obtain $\eta_{max} = 0.946$. If the sun were a perfect black body all other wavelengths would give the same value of T_R and η_{max}.

TEMPERATURE AND POTENTIAL OF SCATTERED RADIATION (T_{RS} AND μ_{RS})

The temperature, potential and efficiency given by eqn (12) with eqn (10) and (11) are not appropriate to a photochemical reaction because the direction of the radiation beam is lost in the absorption process, the entropy increases further and the effective temperature decreases. The process is equivalent to scattering the radiation from angle Ω to 4π and the change in temperature [assuming the last term in the denominator of eqn (12) is much greater than unity] is given by

$$\frac{1}{T_{RS}} - \frac{1}{T_R} = \frac{k\lambda}{hc} \ln \frac{4\pi}{\Omega} \tag{13}$$

and

$$\Delta\mu \text{ (scattering)} = \mu_{RS} - \mu_R = -RT_A \ln \frac{4\pi}{\Omega} \tag{14}$$

from eqn (10) and (13).

Specifically, the temperature of the scattered radiation

$$T_{RS} = \frac{hc}{k\lambda} \frac{1}{\ln\left[1 + (8\pi hc^2/\lambda^5 E_\lambda)\right]} \tag{15}$$

and

$$\mu_{RS} = \frac{Nhc}{\lambda}\left[1 - (kT_A\lambda/hc) \ln\left[1 + (8\pi hc^2/\lambda^5 E_\lambda)\right]\right\}. \tag{16}$$

This is the appropriate expression for photochemical processes and it allows the direct calculation of the potential of scattered monochromatic radiation from a knowledge of its irradiance $E_\lambda (\text{J m}^{-2}\ \text{s}^{-1}\ \Delta\lambda^{-1})$. The solid angle is not required;

if the solid angle at the receiver is increased by concentrators the measured intensity increases in proportion and hence the temperature, potential and efficiency calculated from eqn (15) and (16) also increase. Note that radiation potential is a function of the ambient temperature.

Example. If the sun's radiation at 700 nm is scattered over 4π or, what is equivalent, absorbed and used in a photochemical reaction, its effective temperature is reduced from 5590 to 1303 K and η_{max} from 0.946 to 0.77.

PHOTOCHEMICAL POTENTIAL AT MAXIMUM POWER

If the radiation and the photochemical system A/A* were in equilibrium and there were no losses other than fluorescence, eqn (16) would also apply to the maximum chemical potential of μ_{A*} in an isotropic system. There could then be no photochemical reaction of A*. When A* is removed by reaction its chemical potential falls further and the optimum flux j_m for maximum power and the corresponding potential and efficiency are calculated as before using the flux equation

$$\Delta\mu = \mu_{A*} - \mu_A - \mu_{RS} = RT_A \ln[1-(j/J)].$$

Under optimum conditions of maximum power $j = j_m$ is given by eqn (5) or, more generally, if there are leakage processes, by eqn (7).

Example: For $\lambda = 700$ nm, and for the sun at A.M.1, $\mu_{RS} = 1.36$ eV. Substitution of $\mu/RT = 52.7$ in eqn (5) gives $j_m/J = 0.98$ from which $\Delta\mu$ (at maximum power) $= -0.101$ eV for the sun at A.M.1. The overall efficiency of conversion of the energy of light quanta of 700 nm wavelength into chemical potential at maximum power is therefore given as follows:

$$\eta_m = \frac{(\mu_{A*} - \mu_A)\lambda}{Nhc} = 1 - \frac{T_A}{T_R} - \frac{RT_A\lambda}{Nhc}\ln\frac{4\pi}{\Omega} + \frac{RT_A\lambda}{Nhc}\ln[1-(j_m/J)] \quad (17)$$

$$= 1 - 0.054 - 0.177 - 0.057 = 0.71.$$

The last three terms give the fractional losses due to (a) entropy of the original radiation, (b) entropy increases on or before absorption due to scattering and (c) losses (minimum) due to power extraction. The second, and largest, term can be reduced by the use of concentrators.

If there are leakages from the excited state due to internal conversion or other radiationless deactivation processes, eqn (7) must be used with the relation

$$\frac{k_x}{k_x + k_b} = 1 - \phi_f \quad (18)$$

where ϕ_f is the quantum yield of fluorescence in the absence of reaction. This will increase the last term in eqn (17). In the above example, if $\phi_f = 0.5$, j_m/J is increased from 0.98 to 0.99 and the overall efficiency at maximum power is reduced very little, from 0.71 to 0.70.

POLYCHROMATIC RADIATION

Although the other wavelengths in the sun's spectrum will, because the sun approximates to a black body, give similar temperatures and therefore similar efficiencies on the basis of the above calculation, this calculation has assumed that the energy of the absorbed quantum is exactly equal to the energy E of the excited state. This cannot be true of polychromatic radiation when a single absorber is used and allowances must be made for losses due to non-absorption or the degradation of energy in excess of the zero-point excitation.

The most important case is that of a single absorber with a threshold excitation wavelength λ_0 such that the yield

$$\phi_{A^*} = 1 \text{ for } \lambda < \lambda_0 \quad \text{and} \quad \phi_{A^*} = 0 \text{ for } \lambda > \lambda_0.$$

The energy of all photons absorbed, after degradation of excess vibrational energy, is hc/λ_0 and the fraction θ of the energy absorbed from a polychromatic source which is available in photochemical change is given by

$$\theta = \frac{\displaystyle\int_{\lambda=0}^{\lambda=\lambda_0} E_\lambda \frac{\lambda}{\lambda_0} \, d\lambda}{\displaystyle\int_{\lambda=0}^{\lambda=\infty} E_\lambda \, d\lambda} \tag{19}$$

where E_λ is the energy of the radiation in wavelength interval $d\lambda$. For a threshold wavelength of 700 nm and A.M.1, the fraction $\phi = 0.38$,[2] giving an overall solar efficiency at this wavelength of $0.38 \times 0.71 = 0.27$.

Other approaches, which will be reviewed elsewhere, have been used to evaluate the maximum efficiency of solar energy and have given very similar values.[2-12] The approach used here, by evaluating the four loss processes separately, is more readily applicable to photochemical processes in general as well as to thermal and electrochemical processes. In the latter, the flux $j = I/F$ and the potential $\mu = nFE$, where I is the current, F is the Faraday constant, E is the electric potential and n is the number of electrons transferred per mole. $j = j_m$ then becomes the condition for impedance matching at maximum power.

STORAGE OF CHEMICAL POTENTIAL

We wish to convert a reactant A which is initially in store (a) to a product B which is finally in store (b) and which is indefinitely stable when in that store. Since the overall reaction is A_s (in store) $\rightarrow B_s$ (in store) and since all reactions are reversible, this would, at first sight, appear to be an impossible task. Since the back reaction can only be prevented by an activation energy barrier (infinitely large for indefinite stability) it would seem to follow either that there is a similar large barrier preventing the forward reaction or that the energy of the reactants is infinitely greater than that of the products.

Indefinite storage is, however, possible without any significant further losses of chemical potential other than those already discussed. We must first define the difference between a store and a reaction vessel and, second, we must distinguish between open and closed stores.

We define a store of chemical potential as a volume in which (a) the chemical contents are indefinitely stable when the store is closed because there is then a very high barrier to reaction and (b) the chemical potential of the contents is constant and independent of the quantity of contents in the store. It follows from (b) that the store has a variable volume which is proportional to the mass of the contents. Its contents are not in a steady state.

The reaction vessel (r.v.) is a volume where reaction occurs at finite rate in both directions and the contents are in a steady state. Reactants or products may be transferred selectively between r.v. and store only when the store 'door' is open.

480   CHEMICAL AND RADIATION POTENTIAL

The isolation of the r.v. from the store (closing the door) may be achieved in a number of ways, for example: (1) store and reaction vessel are connected by a physical mechanism, such as tap or diaphragm, which may be opened and closed without expenditure of energy; (2) in an electrochemical reaction the electrical circuit may be made or broken; (3) the isolation may occur spontaneously by phase separation, e.g. the elimination of a gas or precipitation of a solid, which moves out of the reaction zone.

Finally, the requirement that the reaction can proceed in the reaction vessel but not in the store may also be satisfied in a variety of ways. For example: (1) The reaction has a very high activation energy except in the presence of a catalyst which is present in the r.v. but not in the store; (2) The chemical reaction occurs between two components which are present together in the reaction vessel but are removed to separate stores.

A schematic representation of a reaction with storage is shown in fig. 1.

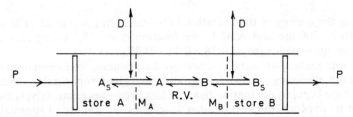

FIG. 1.—Schematic representation of a reaction with storage.

The r.v. is separated from the reactant store by a membrane M_A permeable only to A and from the product store by a membrane M_B permeable only to B. The doors (D) (taps etc.) may be open or closed and the volumes of the stores are varied by pistons P so as to maintain a constant chemical potential of A_s and B_s in the stores. The conversion of stored A into stored B proceeds in two stages, an open stage and a closed stage.

OPEN STAGE

$$A_s \underset{\mu_{A_s}}{\overset{j\rightarrow}{\rightleftharpoons}} A \underset{\mu_A}{\overset{j\rightarrow}{\rightleftharpoons}} B \underset{\mu_B}{\overset{j\rightarrow}{\rightleftharpoons}} B_s$$
$$\qquad\qquad\qquad\qquad\qquad\quad \mu_{B_s}$$

There is a net flux j converting A_s into B_s which is the same for each of the three steps shown unless there are leakage reactions. Each of the three steps i has its own forward and backward fluxes J_{fi} and J_{bi}. The loss in chemical potential for maximum power is, for each step, given by the flux eqn (1).

The additional losses due to storage occur in the transport processes between reaction vessel and stores and are caused by the departures from equilibrium which are necessary to ensure transport. The problem is identical for the two transport processes and we need consider only transport of the product B. The change of chemical potential in this process

$$\Delta\mu_t = \mu_{B_s} - \mu_B = \boldsymbol{RT} \ln\left[1 - (j/J)\right]$$

where j and J are now the fluxes due to diffusion. The net flux j is already determined as $j = j_m$ by the requirements for maximum power in the preceding photochemical reaction but J is now a variable depending on the design of the reaction vessel and store. We wish to make J as large as possible so as to make $\Delta\mu_t$ negligible; usually it will be adequate to make J a factor of two greater than j in order to ensure that $\Delta\mu_t$ is < 0.02 eV.

There are, therefore, no unavoidable thermodynamic limitations on the efficiency of the transport process but the kinetic criteria may be difficult to meet. Since the diffusion coefficients are the same in the forward and back directions

$$1 - (j/J) = J_b/J = [B_s]/[B]$$

and if $j/J = 0.5$, the value chosen in the last paragraph, $[B_s]/[B] = 0.5$ also. The net forward flux through unit area due to diffusion is

$$j'_m = \frac{([B] - [B_s])D}{l}$$

where D is the diffusion coefficient, l is the diffusion length (*i.e.* the diffusion distance from r.v. to store) and j'_m is the flux across unit area for maximum power (j' is related to the volume flux j by $j' = jl$ for an isotropic distribution in a vessel of length l).

It is instructive to evaluate this expression numerically for the case of solar radiation storage. The total solar flux at A.M.1 is 8 W dm^{-2}, which, using the efficiencies of an earlier section (0.38), is equivalent to 1.7×10^{-5} ein s^{-1} for a threshold wavelength of 700 nm. This gives a flux $j'_m = 1.7 \times 10^{-5}$ mol dm^{-2} s^{-1} if the quantum yield is unity. The geometry is assumed to be the most favourable one, where the solar and material fluxes are through the same plane, *i.e.* in fig. 1 the direction of the incident light is perpendicular to the membranes. Using the typical value $D = 10^{-7}$ dm^2 s^{-1} and $[B_s]/[B] = 0.5$ we obtain

$$[B]/l = 340 \text{ mol dm}^{-4}.$$

Since the concentrations cannot greatly exceed 1 mol dm^{-3} it follows that diffusion distances from r.v. to store in a solar-energy storage reaction must usually be < 0.3 mm. The requirement for such small diffusion distances suggests that phase separation in heterogeneous systems may prove more successful than diffusion across a membrane.

CLOSED STAGE

After closing the connections between the r.v. and the store there is a further loss of energy in the reaction vessel as the mixture returns to equilibrium. In a thermal reaction this does not produce an overall loss since it results in an increase in the potential of the product B which is utilised when the door is opened again on the next cycle. In a photochemical reaction, however, the back reaction results in an irrecoverable loss as the excited state and its products return to the ground state in the dark. The fractional energy loss ϕ_c in one open/close cycle is given by

$$\phi_c = \frac{\text{energy loss per dm}^2 \text{ of r.v.}}{\text{energy stored per cycle per dm}^2} = \frac{l[B]}{j'_m t} = \frac{[B]}{j_m t} \tag{20}$$

where l is the length of the reaction vessel in the flux direction and t is the open period of one cycle; it is assumed that the door is open during the light period and closed during the dark period and that both periods are longer than the time required to reach the steady state.

482   CHEMICAL AND RADIATION POTENTIAL

The fractional energy loss for a 0.3 mm long absorber and other conditions as above is then given by $\phi_c t = 176$ s. For a twelve hour cycle $\phi_c = 4 \times 10^{-3}$, which is tolerable, but cyclic losses in storage would become serious if the cycle were one hour or less in duration or if the capacity of the reaction vessel [B]l were increased.

Indefinite storage of the products of chemical and photochemical change is therefore possible, without significant additional losses, but two additional requirements are imposed: a relatively fast diffusional transport process and, in photochemical reactions, a long light period coupled with a small capacity per unit area of the reaction vessel. These requirements are more than adequately satisfied by the diurnal cycle and the fine lamellar structure of the photosynthetic apparatus.

I am grateful to Dr Mary Archer for her detailed comments on this paper and for several helpful suggestions.

[1] *Handbook of Geophysics and Space Environments*, ed. S. L. Vallery (McGraw-Hill, New York, 1965).
[2] J. R. Bolton, A. F. Haught and R. T. Ross, in *Photochemical Conversion and Storage of Solar Energy*, ed. J. S. Connolly (Academic Press, New York, 1981), pp. 297–339.
[3] L. N. M. Duysens, *Brookhaven Symp. Biol.*, 1958, **11**, 10.
[4] W. Shockley and H. J. Queisser, *J. Appl. Phys.*, 1961, **32**, 510.
[5] R. T. Ross, *J. Chem. Phys.*, 1966, **45**, 1.
[6] R. T. Ross and M. Calvin, *Biophys. J.*, 1967, **7**, 595.
[7] R. T. Ross and T. L. Hsiao, *J. Appl. Phys.*, 1977, **48**, 4783.
[8] R. Hill, *Plant Cell Physiol. (Special Issue)*, 1977, 47.
[9] W. W. Parson, *Photochem. Photobiol.*, 1978, **28**, 389.
[10] R. S. Knox, in *Topics in Photosynthesis*, ed. J. Barber (Elsevier, Amsterdam, 1978), vol. II, p. 55.
[11] P. T. Landsberg and G. Tonge, *J. Appl. Phys.*, 1980, **51**, R1–R20, no. 7.
[12] D. C. Spanner, *Thermodynamics* (Academic Press, New York, 1964), p. 213.

Proc. Roy. Soc. Lond. A. **315**, 149–161 (1970)

Model systems for photosynthesis
I. Energy transfer and light harvesting mechanisms

By Angela R. Kelly and G. Porter, F.R.S.

Davy Faraday Research Laboratory of the Royal Institution,
21 Albemarle Street, London W1X 4BS

(*Received 7 July* 1969)

Absorption spectra, triplet state and fluorescence properties of solutions of chlorophylls b and a and pheophytins b and a in lecithin have been studied over a range of concentration from 10^{-4} to 5×10^{-2} mol l^{-1}.

Energy transfer has been investigated for the pairs chlorophyll b to a, pheophytins b to a, and is adequately described by the Förster inductive resonance mechanism.

We have also examined the self-quenching of fluorescence and triplet state formation for each of the four molecules. With the exception of pheophytin b, which forms aggregates at high concentrations, there were no observable quenchers and, in each case, the half quenching concentration is close to that at which energy transfer between like molecules occurs.

Mechanisms for this self-quenching and the possible relevance of these models to photosystem II in photosynthesis are discussed.

Current concepts in photosynthesis (see, for example, a recent review by Boardman (1968)) suggest the participation of two photochemical systems. Many workers believe that the two photosystems represent physically separated pigment arrays each with characteristic absorption spectra. Further evidence for this has been supplied recently by experiments in which chloroplasts are treated with a mild detergent and then fractionated by differential centrifugation. Two particles are separated: the heavier one contains chlorophyll embedded in a lipid matrix and has photochemical properties resembling the function of photosystem II. Properties of interest are its absorption spectrum $\lambda_{max} = 650$ nm (chlorophyll b) and 670 nm (chlorophyll a) and a fluorescence quantum yield of 0.016 (higher than the 0.003 found for the intact chloroplast). In order to establish a suitable model for photosystem II in photosynthesis it is desirable to reproduce these properties as closely as possible.

Apart from studies of the spectral properties and aggregation effects in various solvents (Brody & Brody 1967; Broyde, Brody & Brody 1968; Amster & Porter 1966) and energy transfer in ether solutions (Watson & Livingston 1950) the principal models for reproducing chloroplast-like conditions have been monomolecular layers. Classical work by Bellamy, Tweet & Gaines (1964) established that energy transfer between like and unlike molecules in diluted monolayers could be described by a Förster mechanism (Förster 1948, 1957). They observed a marked decrease in the fluorescence yield as the separation of molecules within a layer was decreased but suggested no explanation. More recently, Trosper, Park & Sauer

Angela R. Kelly and G. Porter

(1968) studied monolayers of chlorophyll *a* diluted by various chloroplast lipids. They found fluorescence depolarization brought about by a single 'Förster' type energy transfer, and also observed concentration quenching. Absorption spectra of monolayers which have been reported have a red absorption maximum at 685 nm suggesting a possible resemblance to system I particles.

There is an intermediate range of both state and concentration which has not been studied, making it difficult to relate results in monolayers or in the chloroplast to the established properties of dilute solutions. We have, therefore, used another simple model system—a solid solution of the pigments of both chlorophylls and their derivatives, pheophytins—in a lecithin matrix. These studies represent an extension of preliminary work on solid solutions of chlorophyll in cholesterol (Porter & Strauss 1966). The light harvesting process is thought to involve a large number of 'bulk' pigment molecules before the energy is finally trapped. Our experiments have been designed to clarify the conditions for collection of light energy by these traps.

EXPERIMENTAL

Preparation of samples

Lecithin was chosen as the matrix because of its occurrence in the leaf and because it was found to form transparent solid solutions.

Both chlorophyll and lecithin decompose when heated in air, consequently the melting and rapid cooling techniques previously used for preparing glasses were not applicable. Instead, appropriate amounts of the two components were dissolved in chloroform, the volume of this solution was then reduced until it was tacky and a few drops of this mixture were then spread on a microscope slide. The remaining chloroform was evaporated off in a stream of nitrogen and samples allowed to stand in this atmosphere for at least half an hour. This ensured removal of oxygen from the matrix and also allowed equilibration of the solution (important in the cases where aggregation occurred). Measured triplet yields and lifetimes indicated that rediffusion of oxygen into the sample is relatively slow and transients were still readily observed in samples which had been kept in the dark for several days. In order to compare relative yields of triplet formation and of fluorescence over a wide range of concentrations, the path lengths of the solid solutions were varied so that samples having the same optical density were compared in each experiment. Path lengths ranged from $200 \mu m$ to less than $5 \mu m$ and were measured by focusing methods using a Zeiss microscope with a vernier scale on its fine-focus adjustment.

Flash photolysis

Samples were examined by means of a modified version of the microbeam flash photolysis apparatus previously described (Porter & Strauss 1966). The only significant modification was in the method of excitation; the capillary flash lamp was mounted in a vertical position to one side of the apparatus and the exciting light was reflected vertically on to the sample by means of a prism. A lens mounted 10 cm in front of the capillary enabled a sharp one-third size image to be projected onto the plane of the sample. Vertical illumination makes possible the examination of a wider range of samples, including liquid samples in short path length cells, and ensures more uniform and reproducible illumination of the selected area.

Fluorescence

Fluorescence spectra were recorded on an Aminco–Keirs recording spectro-fluorimeter using an HTV R 136 photomultiplier tube. The wavelength response of the instrument was calibrated using, as secondary fluorescence standards, quinine sulphate (10^{-4} mol l^{-1} in 0.1 N H_2SO_4) and a chelate of $AlCl_3$ and pontachrome blue black red (P.B.B.R. reagent, Eastman-Kodak) as described by Augauer & White (1963).

Despite the short path lengths used, sample optical densities were still fairly high and it was found that reproducible measurements of fluorescence spectra of the solid solutions on microscope slides could best be achieved by mounting the sample in the centre of the cavity at an angle of 45° to both exciting and emitted light paths.

Absolute fluorescence yields were determined by plotting the spectra (corrected for instrument response) as relative quanta/wavenumber and determining graphically the area under the curve. Secondary standards used for comparison were quinine sulphate, rhodamine *b* and a solution of chlorophyll *a* in ether. Relative fluorescence yields were determined from the peak heights and comparisons of the unknown with a standard were made in every experiment.

Absorption spectra

Absorption spectra were recorded on the Unicam SP 800 spectrophotometer. In the visible region, scatter by lecithin is fairly small and no sample was placed in the reference beam.

Materials

A mixture of pheophytins *a* and *b* was kindly supplied by Professor R. B. Woodward. The two components were separated on a Zeokarb resin column by the method of Wilson & Nutting (1963). Some difficulty was experienced with this method; as an alternative, eluting the pheophytin *a* from icing sugar columns in 5 % benzene/heptane was also tried (J. M. Kelly, private communication). Pheophytin *a* samples from both sources were used with identical results.

Samples of pure crystalline chlorophylls were obtained from three sources—Koch Light Limited, Sigma Chemical Company and Sandoz Weidemann Limited. Although identical results were obtained with all three, absorption spectral studies showed that the samples from Sigma were most free of pheophytins. Lecithin, obtained from the Nutritional Biochemical Corporation, New York, was 95 % pure. It was further purified, prior to use, by dissolving in a minimum of chloroform and reprecipitating in excess acetone. It was important during this preparation not to expose the solid to air. The purified lecithin was stored in solution in chloroform in stoppered flasks. A film of lecithin prepared from such a solution is optically clear (transmission of a $50\,\mu$m sample was $> 80\,\%$) down to 350 nm.

RESULTS

Absorption spectra

Absorption spectra of the four compounds in dilute solution in lecithin closely resemble the spectra in ether (Seeley & Jensen 1965), but both Soret and red bands are shifted by 6–10 nm to the red. This is in accordance with an observation of Chapman & Fast (1968) on chlorophyll–lecithin–water dispersions.

The spectra of both chlorophylls and of pheophytin a are quite unaffected by increases in concentration over the range 10^{-4} to $5 \times 10^{-2}\,\mathrm{mol\,l^{-1}}$, indeed both the chlorophylls are extremely soluble in lecithin ($> 10^{-1}\,\mathrm{mol\,l^{-1}}$). The solubility limit for pheophytin a occurred at $\sim 8 \times 10^{-2}\,\mathrm{mol\,l^{-1}}$. In contrast, marked changes occur in the absorption spectra of pheophytin b at relatively lower concentrations. At concentrations greater than $10^{-3}\,\mathrm{mol\,l^{-1}}$ new bands appear in the spectrum at 476 nm which are attributed to the formation of some aggregate. At concentrations up to $4.5 \times 10^{-3}\,\mathrm{mol\,l^{-1}}$ both monomer and aggregate bands increased with increasing concentration but at still higher concentrations only the aggregate bands increased, suggesting a limiting solubility of monomeric pheophytin b in lecithin of 2.3×10^{-3} $\mathrm{mol\,l^{-1}}$. That no further increase in absorption at 436 nm occurred indicated that there was no absorbance by the aggregate at this wavelength and allowed a difference spectrum to be deduced for this species. This is shown in figure 1.

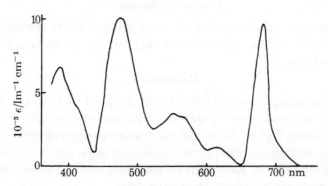

FIGURE 1. Spectrum attributed to aggregate of pheophytin b in lecithin.

Fluorescence spectra

Fluorescence spectra for the four compounds in dilute solution in lecithin show good mirror image relationships with the absorption spectra and are given in figure 2. No changes in fluorescence spectral shapes were detected even at the highest concentrations and the pheophytin b dimer appeared to be non-fluorescent.

However, the emissions occur nearly at the limit of detection of our photocell and the possibility of an aggregate emission further to the red cannot be discounted.

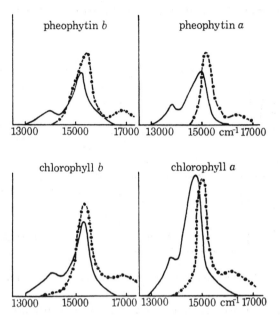

FIGURE 2. Absorption and fluorescence spectra. ———, Fluorescence spectra as relative quantum intensity; —●—, absorption spectra 10^{-4} ϵ/lm^{-1} cm^{-1}.

Triplet states

Triplet states of the four compounds were characterized in dilute solution by their difference spectra. These were determined photoelectrically point-by-point over a range of wavelengths 400 to 700 nm. At a number of wavelengths the rate constant for triplet decay was calculated and found to be first order over two to three lifetimes. The same lifetime was found at every wavelength, indicating the presence of only one transient species—the triplet state. The spectra, which are shown in figure 3, resemble closely those reported by Linschitz & Sarkanen (1958) for fluid solutions. Table 1 summarizes the spectral properties of the four pigments in dilute solution in lecithin.

Energy transfer

The fluorescence spectra of chlorophylls *a* and *b* overlap to a considerable extent which makes experimental detection of energy transfer difficult. We have found that this can be overcome conveniently by using flash photolysis to study relative triplet yields of donor and acceptor. No change in the triplet lifetimes of these components is observed under any of our conditions showing that any changes in relative triplet yields are due solely to processes affecting the singlet state.

154   Angela R. Kelly and G. Porter

A filter combination consisting of 1 cm of a solution of *p*-nitroaniline in propylene glycol and a blue gelatine Wratten 47 B enabled us to excite selectively in the Soret band of the donors in each of the pairs (donor named first) chlorophyll *b*–chlorophyll *a* and pheophytin *b*–pheophytin *a*. Figure 4 illustrates this for the chlorophylls.

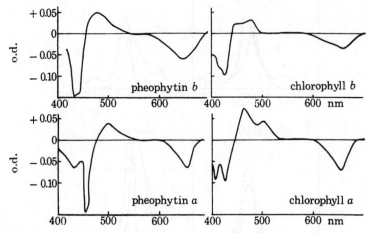

FIGURE 3. Difference spectra observed 30 μs after flash photolysis.

TABLE 1. PHOTOPHYSICAL PROPERTIES OF CHLOROPHYLLS
AND PHEOPHYTINS IN LECITHIN

	absorption spectra		fluorescence		triplet state		1st order decay constant
	λ_{max} nm	$10^{-4}\,\epsilon$ lm^{-1} cm^{-1}	ν_{max} cm^{-1}	quantum yield at ∞ dilution	λ_{max} nm	$10^{-4}\,\epsilon$ lm^{-1} cm^{-1}	s^{-1}
pheophytin *b*	436	19	15040	0.14 ± 0.02	480	5.15	1980 ± 50
	658	5.5					
pheophytin *a*	416	10.5	14750	0.11 ± 0.02	470	1.05	2090 ± 40
	666	5.5					
chlorophyll *b*	460	14.0	15380	0.10 ± 0.03	480	7.2	535 ± 30
	660	4.5					
chlorophyll *a*	436	12.0	14710	0.26 ± 0.04	470	5.7	1080 ± 40
	676	8.8					

Furthermore, an examination of the triplet difference spectra reveals that for each pair Δ o.d. (acceptor) is zero just where the singlet depletion of the donor is greatest. This point can then be used as a reference which enables the contribution of the two components to the net optical density change to be calculated at any other wavelength.

A donor concentration of 7×10^{-4} mol l^{-1} was used throughout and path lengths of the samples were kept approximately the same. This meant that the fraction of exciting light absorbed was similar from one experiment to the next. Relative triplet yields of donor and acceptor were calculated from the appropriate optical density change. Small variations in the fraction of exciting light absorbed had to be allowed for and this is done by calculating

$$\text{relative triplet yield} = \frac{\Delta \text{ o.d.}}{\text{fraction of exciting light absorbed by donor}}.$$

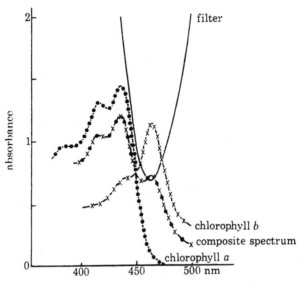

FIGURE 4. Spectra of donor and acceptor and filter used for selective excitation in the energy transfer experiments.

Results for the pheophytins and the chlorophylls are shown in figures 5 and 6 respectively. The yields have been compared with those in the absence of acceptor. For the chlorophylls it was just possible to resolve the components of the composite fluorescence spectra. Relative fluorescence yields fall on the same curve. In table 2 the concentrations of acceptor needed to half-quench the donor are listed along with the corresponding mean separations of the donor–acceptor pair.

These experimental disturbances may be compared with those predicted by the inductive resonance theory of Förster (1948). For weakly interacting chromophores, such as those in the relatively dilute solutions studied here this predicts that

$$R_0^6 = \frac{9000 \ln 10 K^2 \phi_D}{128\pi^5 n^4} \int \frac{f(\nu)\,\epsilon(\nu)\,d\nu}{\nu^4}. \tag{1}$$

where n is the refractive index of the medium $= 1.434$. ϕ_D is the quantum yield of donor fluorescence, and K^2 is a factor which takes account of the orientation of transition dipoles. Like Bennett & Kellog (1964) we have assumed our solutions to be random three dimensional arrays and have taken a value of $K^2 = \frac{2}{3}$. The integral

$$\Omega = \int \frac{f_D(\nu)\,\epsilon_A(\nu)\,d\nu}{\nu^4}$$

represents the overlap of donor emission and acceptor absorption. $f(\nu)$ is the fluorescence spectral distribution of the donor and must be normalized to unity on a wave number scale.

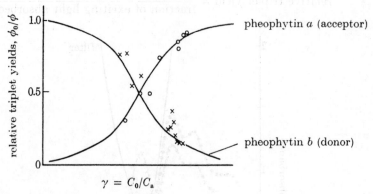

FIGURE 5. Energy transfer from pheophytin b to a. The heavy lines are theoretical curves (Förster theory).

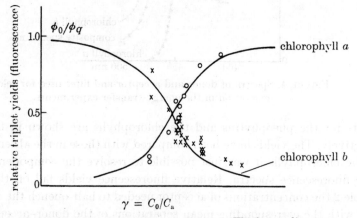

FIGURE 6. Energy transfer from chlorophyll b to a. The heavy lines are theoretical curves (Förster theory).

TABLE 2

pair	$\dfrac{C_0}{\text{mol l}^{-1}}$	R_0/nm	
		expt.	theory
chlorophyll b (donor)–chlorophyll a	1.53×10^{-3}	6.5	5.45
pheophytin b (donor)–pheophytin a	1.8×10^{-3}	5.5	5.7

This overlap integral was evaluated graphically, calculating the value at suitable wavenumber intervals and determining the area under the curve by a weighing procedure. The calculations have since been checked numerically on our Elliott 803 computer by Dr L. K. Patterson. The agreement is excellent.

Förster theory also predicts the shape of the quenching curves (Förster 1946):

$$\left.\begin{array}{l} \phi_D/\phi' = 1 - \sqrt{\pi}\,\gamma\exp{(\gamma^2)}\,(1-\mathrm{erf}\,\gamma), \\ \phi_A/\phi'' = \sqrt{\pi}\,\gamma\exp{(\gamma^2)}\,\{1-\mathrm{erf}\,\gamma\}, \end{array}\right\} \tag{2}$$

where ϕ_D is the yield of any process emanating from $^1D*$ in presence of quencher, ϕ' the yield in absence of quencher, ϕ_A the yield of any process from $^1A*$ with quenching, ϕ'' the limiting yield after complete transfer and γ is a linear function of concentration $= C_a/C_0$ where C_a is the acceptor concentration and C_0 the critical concentration corresponding to an average of one quencher molecule in a sphere of radius R_0.

By fitting the experimental data to these curves it is possible to make a more accurate estimate of C_0. Good fits were obtained for the pheophytins for both sensitization and quenching (figure 5). Similar plots for the chlorophylls are shown in figure 6. At high concentrations of the acceptor the total yield of both triplet and fluorescence falls as a result of self-quenching which is discussed in the next section. In the low concentration region, where self-quenching is unimportant, the energy transfer between dissimilar molecules is well accounted for by the Förster theory.

Concentration quenching

At concentrations similar to those of the acceptor at which energy transfer occurs from b to a molecules, concentration quenching is observed in solutions of each of the four species alone, resulting in a decrease in both fluorescence and triplet formation. Triplet yields as a function of concentration for chlorophylls a and b and pheophytins a and b are shown in figure 7. The results for both chlorophylls fall on the same curve, that for pheophytin a is displaced to higher concentrations, a result which could be predicted because of the smaller value of the overlap integral for absorption and fluorescence of this molecule.

The Förster expression relating relative yield to concentration which gave a satisfactory account of b to a transfer gives a less satisfactory description of self-quenching as is shown by the broken line graph of figure 8 for chlorophyll b. The experimental yields fall more sharply with concentration than the theoretical curves suggesting a higher power dependence on concentration. The full line in figure 8, which is calculated from the function

$$\phi/\phi_0 = 1/(1+\gamma^2) \tag{3}$$

with $\gamma = C/C_{\frac{1}{2}}$ where $C_{\frac{1}{2}}$ is the half quenching concentration, accounts for the data within the limits of experimental error. Similarly, for chlorophyll a and pheophytin a, the self-quenching data are better described by function (3) than by the Förster expression (equation (2)). The formation of aggregates at higher concentrations of pheophytin b could account for the different shape of quenching curve found in this case, the quenching process probably being a mixture or self-quenching and direct energy transfer to the aggregate.

Angela R. Kelly and G. Porter

Simultaneous b to a transfer and concentration quenching

The relationship between energy transfer from chlorophyll b to a and the self-quenching of chlorophyll b is shown in figure 8. The steepest curve (full line), which is best described by the concentration squared dependence, is a pure self-quenching curve for chlorophyll b. The broken curve, which is best described by the Förster

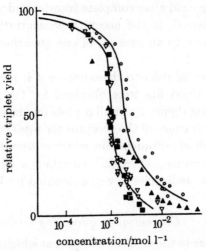

FIGURE 7. Self-quenching of triplet state formation for the four pigments:
∇, chlorophyll a; $\blacksquare$, chlorophyll b; $\bigcirc$, pheophytin a; $\blacktriangle$, pheophytin b.

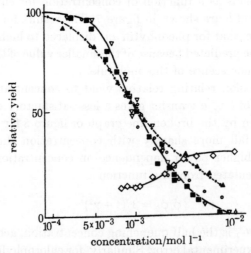

FIGURE 8. Self-quenching and energy transfer for chlorophyll b. —$\bullet$—, Förster function
(equation (2)) with $C_0 = 1.8 \times 10^{-3}$ mol l^{-1}. ——, Empirical function $\phi_0/\phi = 1/(1+\gamma^2)$
with $\gamma = C/C_0$ and $C_0 = 1.3 \times 10^{-3}$ mol l^{-1}. $\blacktriangle$, 6×10^{-4} mol l^{-1} chlorophyll b, increasing
amounts of chlorophyll a; $\bigcirc$, 6×10^{-4} mol l^{-1} chlorophyll a, increasing amounts of
chlorophyll b; ∇, 2×10^{-4} mol l^{-1} chlorophyll a; $\blacksquare$, pure self-quenching of chlorophyll b;
$\diamondsuit$, relative yield of excitation reaching chlorophyll a (6×10^{-4} mol l^{-4}) with increasing
concentrations of chlorophyll b.

expression (2), shows the quenching of a 7×10^{-4} mol l^{-1} solution of chlorophyll b by varying concentrations of chlorophyll a. The two sets of intermediate points are for cases where both energy transfer to a and self-quenching of b were operative.

TABLE 3. SELF-QUENCHING OF THE SINGLET STATE OF PHEOPHYTINS
AND CHLOROPHYLLS

	half quenching separations/nm		
system	experimental	theory	reference
(a) *solid solutions in lecithin*			
pheophytin b	7.2	5.5	this work
pheophytin a	5.0	4.8	this work
chlorophyll b	7.0	5.55	this work
chlorophyll a	7.3	5.45	this work
(b) *other systems*			
chlorophyll a–cholesterol	6.7	—	Porter & Strauss (1966)
chlorophyll a–ether	6.0–7.0	5.17	Watson & Livingston (1950)†
chlorophyll a monolayers stearicacid diluent	5.0	6.7	Tweet, Gaines & Bellamy (1964)
chlorophyll a monolayer various lipid diluents	6.0–8.0	6.7	Trosper, Park & Sauer (1968)

† Our calculations.

For one of these intermediate cases, that where the chlorophyll a concentration was 6×10^{-4} mol l^{-1}, the yield of triplet chlorophyll a as a function of the concentration of chlorophyll b is also shown. It is seen that the concentration quenching of chlorophyll b is accompanied by a small increase in the energy transferred to chlorophyll a.

DISCUSSION

Although we have worked in a medium in which diffusion is excluded and have mainly studied triplet yields rather than fluorescence, our results are very similar to those described many years ago by Watson & Livingston (1950) for fluid solutions of chlorophylls. These authors found that b to a transfer first became observable at concentrations lower than those required for the onset of self-quenching. Although we have found that the half-quenching concentrations are similar for b to a transfer and for self-quenching, reference to our figure 8 shows that the onset of b to a transfer appears at a lower concentration than that for self-quenching owing to the lower power concentration dependence of the b to a transfer.

Watson & Livingston gave several arguments to support their view that concentration quenching was not a result of ground state dimer formation. We have found no evidence for dimers or other aggregates in our solutions except those of pheophytin b at relatively high concentrations. Furthermore, quenching by collisions of the second kind is excluded in the rigid lecithin matrix of our experiments.

The higher power of concentration dependence of self-quenching over b to a transfer which we have found was also clearly described by Watson and Livingston. These authors ascribed the quenching to some form of energy degradation during an energy transfer process, following earlier suggestions of Vavilov (1943). Our results lend further support to this view; collisional processes are excluded in our systems and the near identity of half quenching concentrations with those for b to a transfer argues strongly for a mechanism of self-quenching in which Förster type long distance transfers play a part.

Concentration quenching of this type occurs in other molecules, and we have observed concentration quenching of β-naphthol in fructose or glucose glasses and in cholesterol using the same method of triplet state yield measurement. The half-quenching concentrations (10^{-1} mol l^{-1}) are much higher than for the chlorophylls, but correspond to critical transfer distances (1.7 nm) which are again close to those calculated by Förster theory.

The dependence of concentration quenching on the second power of the concentration would be explained if an energy transfer process dependent on the first power of the concentration was followed by quenching at a trap whose concentration was also dependent on the first power of concentration. Such a trap could be an impurity or a dimer (although the concentration of dimer is proportional to the second power of monomer concentration, the probability that a molecule to which a transfer has taken place is a dimer is proportional only to the first power of monomer concentration). The dimer, in the sense referred to here, does not necessarily imply a chemically bonded complex but may be merely a statistical association of two molecules close enough to cause quenching when one or other of them is excited. Impurities, which would have to be in the solute in order to explain the concentration dependence, seem to be an unsatisfactory explanation in view of the generality of the phenomenon, the constancy of self-quenching rates when solutes are derived from several sources and the independence of half-quenching concentrations on added impurity.

The simplest explanation of the second power concentration dependence of quenching would be given in terms of a ground state equilibrium of monomers and dimers, the directly excited dimers being rapidly quenched. Although this would not account for the close similarity between concentration quenching and chlorophyll b to a energy transfer, it is impossible to exclude it completely on present evidence. Fluorescence lifetime measurements should give an unequivocal answer on this point. Such measurements will be described in part two of this paper where the mechanism of concentration quenching will be considered further. It is, however, clear from the present work that, whatever the mechanism of radiationless conversion in concentration quenching, it does not result in formation of the triplet state and, therefore, presumably proceeds directly to the ground state.

Finally, we may comment briefly on the relevance of our results to the light harvesting mechanism of photosystem II. Whilst it can be shown that because of the longer lifetime of the triplet state, triplet exciton migration may result in more jumps during the excited state lifetime than the corresponding process for singlets, such a process cannot make a contribution to the overall energy harvesting if the triplet state is not significantly populated; this is just what we have found in systems having chlorophyll concentration comparable with those of the chloroplast. Instead, it seems likely that a process of singlet energy transfer among the bulk pigment molecules would lead to collection of light by the traps. Within the trap, the triplet state of chlorophyll may still have an important role.

One of us (A. R. K.) acknowledges a Science Research Council studentship during the tenure of which most of this work was carried out.

REFERENCES

Amster, R. L. & Porter, G. 1966 *Proc. Roy. Soc. Lond.* A **296**, 38.

Augauer, R. J. & White, C. E. 1963 *Anal. Chem.* **35**, 144.

Bennett, R. G. & Kellog, R. E. 1964 *J. chem. Phys.* **41**, 3040.

Bellamy, W. D., Tweet, A. G. & Gaines, G. L. 1964 *J. chem. Phys.* **41**, 2068.

Boardman, N. K. 1968 *Adv. Enzymol.* (ed. Nord), p. 1.

Brody, M. & Brody, S. S. 1967 *Biochim. biophys. Acta* **112**, 54.

Broyde, S. B., Brody, S. S. & Brody, M. 1968 *Biochim. biophys. Acta* **153**, 186.

Chapman, J. & Fast, G. 1968 *Science, N.Y.* **160**, 188.

Förster, T. 1946 *Naturwissenschaften* **33**, 166.

Förster, T. 1948 *Annln. Phys.* **2**, 55.

Förster, T. 1957 *Disc. Faraday Soc.* **17**, 1.

Linschitz, H. & Sarkanen, K. 1958 *J. Am. chem. Soc.* **80**, 4826.

Porter, G. & Strauss, G. 1966 *Proc. Roy. Soc. Lond.* A **295**, 1.

Seeley, G. R. & Jensen, R. G. 1965 *Spectrochim. Acta* **21**, 1835.

Trosper, T., Park, R. B. & Sauer, K. 1968 *Photochem. photobiol.* **7**, 451.

Tweet, A. G., Gaines, G. L. & Bellamy, W. D. 1964 *J. chem. Phys.* **40**, 2596.

Vavilov, S. 1943 *J. Phys. U.S.S.R.* **7**, 141.

Watson, J. & Livingston, R. 1950 *J. chem. Phys.* **18**, 802.

Wilson, G. R. & Nutting, M. D. 1963 *Anal. Chem.* **35**, 144.

CONCENTRATION QUENCHING OF CHLOROPHYLL FLUORESCENCE IN BILAYER LIPID VESICLES AND LIPOSOMES

G.S. BEDDARD, S.E. CARLIN and G. PORTER

Davy Faraday Research Laboratory of the Royal Institution, London W1X 4BS, UK

Received 14 June 1976

The fluorescence yields and lifetimes of chlorphyll-a in lipid liposomes and vesicles have been measured in an attempt to understand the light harvesting mechanism of photosynthesis. Concentration quenching of the fluorescence was observed in all systems, the extent depending on the lipid used. The system having the highest half-quenching concentration $(7.0 \times 10^{-2}$ molal) was a 3:1 mole to mole mixture of monogalactosyl diglyceride and digalactosyl diglyceride.

1. Introduction

In the chloroplast thylakoid membrane the light harvesting chlorophyll molecules are at an average concentration of approximately 0.1 molal, yet energy migration to the reaction centres is very efficient [1]. However, in all in vitro systems studied so far, e.g. in solution [2]; lipid monolayers [3]; multilayers [4]; black lipid membranes [5]; detergent micelles [6] and polymethylmethacrylate [7], concentration quenching of the fluorescence of the chlorophyll molecules is observed at relatively low concentrations. This suggests that there may be some major structural difference between the molecular environment of the chlorophyll in all these model systems and in the chloroplast thylakoid membrane. One such possibility is that the chlorophyll molecules are separated from each other by lipid molecules, perhaps involving coordination to the magnesium atom [8].

Chlorophyll molecules in green plant cells are contained in the lamellae of the chloroplasts [9] and it is considered [10] that vesicles and liposomes are good models for the study of the biophysical properties of membrane lipids. These lipid systems are convenient as they are more stable than the black lipid membranes, and suspensions of low optical density and high concentration may easily be prepared.

In an attempt to explain the apparent absence of concentration quenching in vivo we have measured the relative fluorescence yields and lifetimes of chlorophyll-a in lipid vesicles and liposomes over a 100 fold concentration range.

2. Experimental

BDH spectroscopic chloroform and Hopkins and Williams Analar diethyl ether and Analar water were used as received. All other compounds were of the purest grades available and were stored below $0°C$ in the dark. Grade 1 egg lecithin was obtained from Lipid Products, mono- and digalactosyl diglycerides were obtained from P-L Biochemicals Limited, chromatographically pure cholesterol and the chlorophyll-a were obtained from Sigma Chemical Company. The ampules of chlorophyll were dissolved in 10 ml chloroform and stored as above.

The liposomes were prepared by mixing a solution of lipid and chlorophyll-a in chloroform which was evaporated to dryness on a rotary evaporator. When completely dry, 15 ml of Analar water were added to the mixture which was then shaken vigorously for about 10 minutes until all frothing had stopped.

The vesicles were prepared in a similar manner except that, after adding the water, the mixture was sonicated in an ice bath in the dark and under nitrogen using a Dawe 7532A Sonoprobe (90 W incident power) until the cloudiness disappeared. The sonication was

Volume 43, number 1   CHEMICAL PHYSICS LETTERS   1 October 1976

performed in half minute intervals, separated by approximately five minutes. When prepared in this way the lipid/chlorophyll mixture forms single spherical bilayers (vesicles) a high percentage of which are of uniform size [11]. The amount of chlorophyll was the same in all samples used and the concentration was changed by altering the quantity of lipid.

Steady state fluorescence measurements were made in aerated solution on a Perkin Elmer MPF4 spectrofluorometer with an HTV R446 photomultiplier. Fluorescence yields, on excitation at 430 nm, were calculated relative to a dilute aerated 10^{-6} M chlorophyll-a solution in diethyl ether. Absorption spectra of the liposomes were recorded on a Unicam SP800 with a scattering attachment and, for the vesicles, a Perkin Elmer 124 double beam spectro-

photometer was used. To prevent distortion of the fluorescence spectra the maximum optical density used was 0.15 at 430 nm.

The single photon counting apparatus used to measure the fluorescence lifetimes was that described by Beddard et al. [12] except that the emission photomultiplier was a Phillips 56 TVP/03, which was cooled to $4°C$ in a stream of dry nitrogen. The picosecond fluorescence lifetimes were measured using the apparatus previously described [13].

3. Results and discussion

The fluorescence concentration quenching curves of chlorophyll-a in egg lecithin vesicles and liposomes

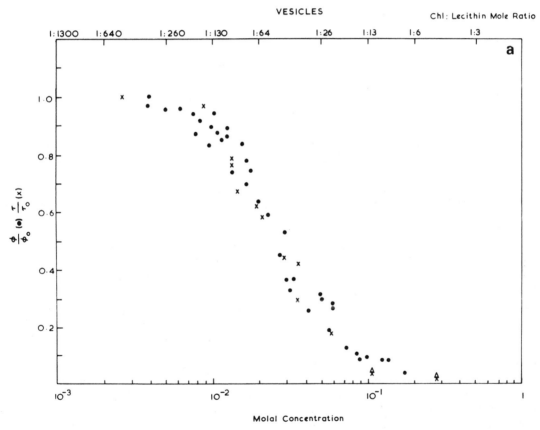

Fig. 1a.

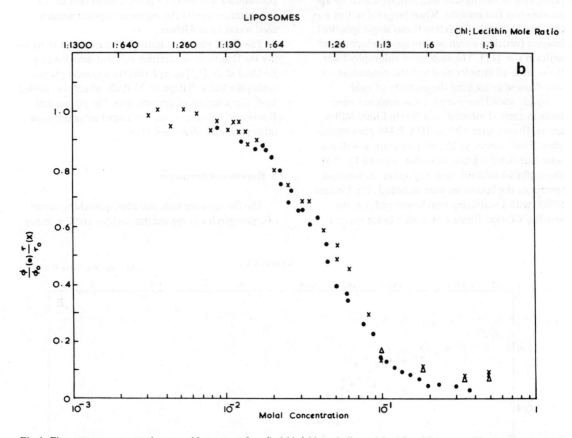

Fig. 1. Fluorescence concentration quenching curves of egg/lecithin/chlorophyll-a vesicles (a) and liposomes (b) plotted against total molal concentration where ($\bullet$) represents relative yields, ($\times$) relative lifetimes, and ($\triangle$) relative mean lifetimes. $\phi_0 = 0.26$ and $\tau_0 = 5.7$ ns all in aerated solution.

are shown in figs. 1a and 1b, respectively, and the half-quenching values for all systems studied are given in table 1.

The quantum yield curves may be fitted to an empirical equation of the form

$$\log(\phi_0/\phi - 1) = \log k + M \log c, \qquad (1)$$

where $M = 1.94 \pm 0.1$, $k = 1.27 \times 10^3$ for the egg lecithin vesicles, and $M = 1.94 \pm 0.20$, $k = 4.7 \times 10^2$ for the liposomes. c is the chlorophyll concentration in moles/1000 g and k is given by $c_{1/2}^{-2}$. This empirical equation describes several systems showing concentration quenching and has been interpreted by Beddard and Porter [8].

The effect of oxygen quenching of the fluorescence on the ϕ/ϕ_0 curves was measured at several concentrations by comparison with samples degassed by four freeze−pump−thaw cycles. At low chlorophyll concentrations the degassed sample yields were ≈ 10% greater than the aerated ones, but no significant change in yield was detected at concentrations greater than half-quenching concentrations.

The fluorescence decays, when longer than about 1 ns, were measured on the nanosecond single-photon counting instrument; those of shorter duration were measured by Mr. C.J. Tredwell and were excited at 530 nm by a 6 ps pulse from a neodymium glass laser and observed by a streak camera and optical multi-

Table 1

	$C_{1/2}$ molal	Chl-a lipid mole ratio	Ref.
lecithin (egg) vesicles	2.8×10^{-2}	1 : 51	this work
lecithin (egg) liposomes	4.6×10^{-2}	1 : 28	this work
Mgdg–Dgdg vesicles [a] 3 : 1 mole ratio	3.9×10^{-2}	1 : 30	this work
Mgdg–Dgdg liposomes 3 : 1 mole ratio	7.0×10^{-2}	1 : 17	this work
lecithin (egg) + 2% cholesterol vesicles	2.3×10^{-2}	1 : 54	this work
Mgdg monolayers	2.0×10^{-2}		[3]
multilayers (lecithin)	1.4×10^{-3}		[4]
PMMA [b]	3.3×10^{-3}		[7]
ether solution	1.6×10^{-2} molar		[2]

[a] Mgdg, Dgdg = mono- and di-galactosyl diglyceride respectively.

[b] PMMA = polymethylmethacrylate.

channel analyser [13]. The decays were interpreted by an equation of the form

$$I(t) = \exp[-t/\tau_1 - (t/\tau_2)^{1/2}] . \qquad (2)$$

Where the decays were measured by the photon counting instrument an iterative convolution technique, using a gradient expansion algorithm, was used to analyse the data. The fluorescence decay of dilute chlorophyll vesicle suspensions was exponential with a lifetime of 5.7 ns. The decays up to 60% quenching were also close to exponential for $2\frac{1}{2}$ decades decrease in intensity past the maximum, the contribution of the τ_2 term being very small in comparison to τ_1. At higher concentrations the decays depart from exponential; for example, at 0.1 M chlorophyll-a $\tau_1 = 761$ ps and $\tau_2 = 132$ ps in egg lecithin liposomes. Both mean lifetime τ_M and the lifetime τ_1 are plotted in fig. 1. For exponential decays these points are the same. Two types of process are described by eq. (2); these are diffusional quenching of fluorescence in solution [14] and quenching by single step resonance energy transfer to a random distribution of quenching molecules [15].

The mechanism proposed for the type of concentration quenching studied here is energy migration by Förster type resonance transfer between like molecules, followed by trapping at a pair of molecules closer than a critical distance, which we call a statistical pair trap [8]. The statistical pair is any two chlorophyll molecules which, in the random distribution, are sep-

arated by less than 10 Å and which do not interact appreciably in their ground states but interact strongly when one of the pair becomes excited so as to form an excimer or other complex which does not fluoresce in the monomer region.

If the quenching were occurring in a single step, as described by Förster [15], then eq. (2) describes the process with τ_1 given by the natural decay time of chlorophyll. None of the decays could be fitted to eq. (2), even for the highest concentrations, with τ_1 at 5.7 ns and by varying τ_2 over a wide range. A fit was only obtained by varying τ_1 and τ_2 simultaneously, which implies that quenching is not occurring by single step transfer after excitation.

At low concentrations, where the mean distance between fluorophores is large, the energy migration is slow and, as the trap concentration is low, we observe little quenching. At intermediate concentrations, up to 75% quenching, trapping occurs after a number of transfers and competes with fluorescence. Because the same fluorophore can be visited several times during the migration, and because the traps are at various distances from the initially-excited molecule, we expect the rate of fluorescence quenching to be time dependent. This is analogous to the situation of molecular diffusion quenching in solution [14] where the quenching rate has one term independent of time (t) and one dependent on $t^{-1/2}$. In the mechanism under consideration here, detailed Monte Carlo calculations of the rate of quenching have been performed and we find the

Volume 43, number 1   CHEMICAL PHYSICS LETTERS   1 October 1976

quenching rate again has time dependent and time independent terms, but the time dependent term has a different exponent from that in the diffusional quenching case. The details of these calculations will be given elsewhere [16].

The larger half quenching concentrations in liposomes over vesicles (table 1) may be due to the small radius of curvature (radius 6 nm) of the inner layer of the bilayer in vesicles forcing the porphyrin rings of the chlorophyll molecules much closer together than they are in the outside layer (radius 10 nm), thus resulting in an increased fraction of statistical pair quenchers at a given concentration. In the liposomes the large number of bilayers, and their large radii of curvature, causes this effect to be negligible.

Concentration quenching curves of chlorophyll in dipalmitoyl-lecithin liposomes have recently been published [17]. Quenching was attributed to the formation of ground state dimers and half quenching of the fluorescence yield occurred at much lower concentrations than those reported here. Using our method of preparation, it is possible to form liposomes and vesicles without causing aggregation of the chlorophyll molecules [18] indicating that quenching must occur by another mechanism.

The absorption and fluorescence spectra in the lipid were measured at each concentration. The normalised absorption spectra at 5×10^{-3} molal and 0.3 molal concentrations were superimposable, as were the fluorescence spectra. This indicates that in our preparations no chlorophyll aggregation occurs, which agrees with the ESR measurements of Tomkiewicz et al. [18]. At the highest concentrations used, where the quencher concentration is also high, the presence of aggregate absorption would be easily observable. In egg lecithin, the absorption maxima of chlorophyll-a are at 670 nm and 436 nm and the emission maximum is at 677 nm. The wavelength of the red absorption band agrees well with one of the principal absorption peaks in spinach chloroplast membranes [19] suggesting that this band is due to chlorophyll in a similar environment to that in the liposomes and vesicles.

Further evidence for the lack of aggregate formation is provided by the close similarity of the lifetime and yield curves. If quenching were due to non-fluorescent aggregates, the yield would fall rapidly but with no change in lifetime and the identity $\phi/\phi_0 =$

τ/τ_0, which is followed closely in our work, would no longer hold.

It is known that energy migration is very efficient in the photosynthetic membranes of the chloroplasts of green plants, which implies that concentration quenching does not occur to any great extent. Furthermore, when the photochemical reaction centres are closed, the fluorescence yield rises to 0.024 and the lifetime increases to 1.5 ns. These values are larger than those observed at similar concentrations in the in vitro systems.

If the mechanism of concentration quenching is as we have suggested, then it should be reduced by holding the chlorophyll molecules further apart than the trap forming distance by choice of a suitable lipid or lipoprotein membrane. The lipids could have bulky hydrophilic groups which keep the chlorophyll molecules apart by specific interactions with the porphyrin ring.

In the thylakoid membrane, the most abundant type of lipids are the galactolipids (mono- and di-) the composition mole ratio being approximately 2:2:1 galactolipid:other lipids:chlorophyll [20]. As the galactose groups on the galactolipids are bulkier than the choline groups in the lecithin and are particularly well adapted for hydrogen bonding to the chlorophyll carbonyl groups or coordination with the magnesium atom, we expect to see less quenching with the galactolipids than with lecithin.

Concentration quenching curves have been obtained for a 3:1 mole to mole ratio of monogalactosyl diglyceride:digalactosyl diglyceride and the results can be seen in table 1. As anticipated, the values of the half quenching concentrations in both galactolipid liposomes and vesicles are approximately a factor of two greater than the corresponding lecithin systems. If it is possible to increase the value of the half quenching concentration by altering the lipid composition in these systems by another factor of two or three, we shall have a viable model for the light harvesting membrane of photosynthetic organisms.

Altering the fluidity of the hydrocarbon chain region of the bilayers is another possibility for preventing concentration quenching and this has been investigated by the use of cholesterol, which is found in small quantities in chloroplasts and is known [21] to have a marked effect on the fluidity of the hydrocarbon region of egg lecithin vesicles and liposomes. It was found

Volume 43, number 1   CHEMICAL PHYSICS LETTERS   1 October 1976

that adding 2% cholesterol to egg lecithin vesicles has a negligible effect on the half quenching value of pure egg lecithin, indicating that the electrostatic interactions in the polar head group region are more important in determining the position of the chlorophyll molecules in the bilayer.

In addition to the chlorophyll lipid interactions, some of the chlorophyll in the chloroplast thylakoid membranes may be associated directly with the intrinsic protein, known to be present in large quantities in the membrane. These proteins could have a structure such that the chlorophylls are held far enough apart to prevent quenching and still allow energy transfer to a photochemical reaction centre. These chlorophyll protein complexes may account for the other main absorption maximum at 678–682 nm which is observed in the chloroplast.

Acknowledgement

The authors wish to thank C.J. Tredwell for measuring the picosecond decays, and Queen Mary College for computer facilities. G.S.B. thanks the Royal Society for the Mr. and Mrs. John Jaffé Research Fellowship, and S.E.C. thanks the Science Research Council for a Studentship.

References

[1] K. Sauer, in: Bioenergetics of photosynthesis, ed. Govindjee (Academic Press, New York, 1975).

[2] W.E. Watson and R. Livingston, J. Chem. Phys. 18 (1950) 802.

[3] W.D. Bellamy, A.G. Tweet and G.L. Gaines, J. Chem. Phys. 41 (1964) 2068;
T. Trosper, R.B. Park and K. Sauer, Photochem. Photobiol. 7 (1968) 451.

[4] A.R. Kelly and G. Porter, Proc. Roy. Soc. A315 (1970) 149;
A.R. Kelly and L.K. Paterson, Proc. Roy. Soc. A324 (1971) 117.

[5] N. Alamuti and P. Lauger, Biochem. Biophys. Acta 362 (1970) 211.

[6] E. Lehoczki and K. Csatorday, Biochem. Biophys. Acta 396 (1975) 86.

[7] P. Avis, J. Ferreira and G. Porter, unpublished results.

[8] G.S. Beddard and G. Porter, Nature 260 (1976) 366.

[9] P.M. Lintilhac and R.B. Park, J. Cell. Biol. 28 (1966) 582.

[10] A.D. Bangham, Progr. Biophys. Mol. Biol. 18 (1968) 29.

[11] E. Korn, ed., Methods in membrane biology, Vol. 1 (Plenum Press, New York, 1974).

[12] G.S. Beddard, S.E. Carlin and C. Lewis, Faraday Trans. II 71 (1975) 1894.

[13] G.S. Beddard, G. Porter and C.J. Tredwell, Nature 258 (1975) 166.

[14] R.M. Noyes, J. Chem. Phys. 22 (1954) 1349.

[15] T. Förster, Naturwissenschaften 33 (1946) 166.

[16] G.S. Beddard and G. Porter, in preparation.

[17] A.G. Lee, Biochemistry 14 (1975) 4397.

[18] M. Tomkiewicz and G.A. Corker, Photochem. Photobiol. 22 (1975) 249.

[19] P. Nicholls, J. West and A.D. Bangham, Biochem. Biophys. Acta 363 (1974) 190.

[20] R.B. Park, in: The chlorophylls, eds. L. Vernon and G. Seely (Academic Press, New York, 1966).

[21] D. Chapman, in: Biological membranes, Vol. 2, eds. D. Chapman and D.F.H. Wallach (Academic Press, New York, 1973) p. 91.

Proc. Roy. Soc. Lond. A. **319**, 319–329 (1970)

The interaction of photo-excited chlorophyll a with duroquinone, α-tocopherylquinone and vitamin K_1

By J. M. Kelly and G. Porter, F.R.S.

Davy Faraday Research Laboratory of The Royal Institution,
21 Albemarle Street, London, W 1X 4BS

(*Received 8 May* 1970)

The reversible electron transfer between photoexcited chlorophyll a and duroquinone has been studied in neutral, acid and alkaline ethanolic solutions. Rates for the reaction of chlorophyll a singlet and triplet states with duroquinone have been determined, and the kinetics for the decay of the radicals so formed have been investigated. In neutral and alkaline solution, evidence has been found for the recombination of the chlorophyll radical cation with the semiquinone radical anion of duroquinone at a rate close to that for a diffusion controlled reaction. In acidic solutions, the chlorophyll radical cation decays by two processes: 60% reacts by second-order process within milliseconds, and the residual 40% decays by the first-order process having a lifetime of approximately 1 s. The decay rate of these oxidized chlorophyll a species is greatly increased by durohydroquinone.

Analogous reactions have been found for triplet chlorophyll a with α-tocopherylquinone and with vitamin K_1.

Introduction

The importance of substituted p-benzoquinones in the photosynthetic electron transport chain has been shown by several authors (for reviews see Witt (1967) and Boardman (1968)) and the recognition of this has led to extensive investigations of the '*in vitro*' properties of solutions of chlorophyll a with a number of quinones. However, such a variety of experimental techniques has been employed under different conditions of temperature, reactant concentrations and solvent, that it is difficult to correlate the results into a consistent scheme.

The interaction of the excited singlet state of chlorophyll a with quinones has been shown in studies of the fluorescence quenching by these and other oxidizing agents (Livingston 1950; Dilung & Chernyuk 1961; Krasnovskii & Drozdova 1966). The reaction of the triplet state of chlorophyll a with quinones has been investigated by flash photolysis and other techniques. Tollin & Green (1962), by observing the formation of the semiquinone radical anion, using an e.s.r. technique, have shown that such interactions may lead to charge separation, and Evstigneev (1968) has observed the production of an electrode-active oxidized form of chlorophyll a under similar conditions. It has also been shown that this electron transfer is accompanied by proton ejection (Quinlan & Fujimori 1967). In viscous alcoholic media at low temperatures is is possible to record the visible absorption spectrum of this oxidized chlorophyll species (Krasnovskii & Drozdova 1964; Evstigneev 1968).

The 'secondary' reactions of the system, that is the reactions of the oxidized

J. M. Kelly and G. Porter

chlorophyll a species, have not been unambiguously assigned. However, two principal pathways have been proposed (Mukherjee, Cho & Tollin 1969):

(1) disproportionation of the oxidized chlorophyll a species

$$Chl^{ox} + Chl^{ox} \rightarrow Chl^{2ox} + Chl, \tag{1}$$

or

(2) reaction of the oxidized chlorophyll species with the semiquinone radical anion

$$Chl^{ox} + Q^{\cdot -} \rightarrow Chl + Q, \tag{2}$$

where Chl^{ox} is $Chl^{\cdot +}$ or a deprotonated form and Chl^{2ox} is Chl^{2+} or a deprotonated form.

Flash photolysis studies by Chibisov, Karyakin, Drozdova & Krashovskii (1967) and by Seifert & Witt (1968) favour the second course, although the Russian workers suggested that a first-order decay process for the oxidized chlorophyll species was also important. Ke, Vernon & Shaw (1965) have also shown that in acidic ethanolic solutions of chlorophyll a and various quinones (for example, trimethyl quinone and ubiquinione-30) transients may be formed which decay with lifetimes of milliseconds.

We have investigated the interactions of photo-excited chlorophyll a with duroquinone in ethanolic solutions of varying pH, and have extended this study to α-tocopherylquinone (3-hydroxy:3,7,11,15 tetramethylhexadecyl trimethyl p-benzoquinone), and to vitamin K_1 (2-methyl 1,3 phytyl naphthoquinone). Duroquinone has been used because it is stable in aerated ethanolic solutions (unlike p-benzoquinone which forms orange polymeric suspensions), it may readily be purified and, having a lower oxidation potential than less substituted benzoquinones, it is less likely to contain traces of hydroquinone.

EXPERIMENTAL

Absorption spectra

These were recorded in the range 200 to 750 nm on a Unicam SP 800 ultraviolet/visible spectrophotometer. As a cell of pathlength 10 cm was used for the flash photolysis studies, it was possible to record the spectra of these samples both before and after degassing, and after an experiment.

Fluorescence spectra

An Aminco-Keirs spectrofluorimeter was utilized, with an HTV R 213 tube (cathode material S-10). With this system, it was not possible to record spectra at wavelengths longer than 750 nm. Samples were contained in 1 cm square cells. In the experiments reported here, the sample was excited at 616 nm and the fluorescence observed at 672 nm. The concentration of chlorophyll a was approximately $3 \times 10^{-6} \, \mathrm{mol \, l^{-1}}$.

Flash photolysis

The apparatus was a conventional kinetic flash spectrophotometer. The flash, of lifetime $30\,\mu s$ was usually $200\,J$ in energy. An HTV R 213 photomultiplier was used for monitoring the system.

In experiments with chlorophyll a, light from the flash lamps was filtered using 1 cm of 10% potassium dichromate aqueous solution. This effectively protects duroquinone from direct excitation, and this was verified experimentally with a degassed $2 \times 10^{-3}\,mol^{-1}$ solution of duroquinone in ethanol, where no transient could be observed. The monitoring beam was filtered using one of the following as suitable: Wratten no. 29 (1 % transmission at 610 nm), no. 3 (440 nm), and no. 2B (400 nm). All experiments were performed at room temperature.

Outgassing of samples

This was performed in a room illuminated only by diffuse green light. Air was removed by subjecting the sample to a series (6 or 7) of freeze-pump-thaw-shake cycles, until no oxygen discharge could be noticed with an electric discharge high vacuum tester.

Materials

(a) Chlorophyll a was prepared from spinach using the method of Strain *et al.* (1960). The purity of the materials was checked by visible absorption spectroscopy. Material supplied by Sigma Chemicals was also used, and the results obtained were identical. (b) Duroquinone was supplied by Fluka A. G. and purified by sublimation. The extinction coefficients (Flaig, Salfeld & Baume 1958) and melting point (*Dictionary of organic compounds* 1965) were in good agreement with the literature. (c) α-Tocopherylquinone and α-tocopherol were supplied by Eastman Kodak and used without further purification. (d) Vitamin K_1. Natural phytol material, supplied by Mann Research Laboratories. (e) Durohydroquinone was supplied by K. and K. Laboratories. This contained approximately 2 % duroquinone, but was not further purified as recrystallization is not a satisfactory method of separating these compounds. (f) Ethanol. A. R. Absolute alcohol as supplied by Burroughs Limited.

RESULTS

Fluorescence quenching of chlorophyll a

The reaction of the lowest excited singlet state of chlorophyll a with duroquinone has been examined by recording the intensity of fluorescence of chlorophyll a in aerated ethanol solutions of duroquinone (0 to $3 \times 10^{-2}\,mol\,l^{-1}$), under conditions where the chlorophyll a absorbs the same amount of light in each solution. The half quenching concentration of duroquinone is $1.6 \pm 0.2 \times 10^{-2}\,mol\,l^{-1}$, that is the concentration of duroquinone where the fluorescence of chlorophyll a is reduced to half that of a solution containing no duroquinone. The wavelength of maximum emission and also the shape of the fluorescence is unchanged in these solutions. A

J. M. Kelly and G. Porter

Stern–Volmer relation is not obeyed, but to a first approximation the following relation holds:

$$I_0/I = 1 + C_1(Q) + C_2(Q)^2,$$

where

$$C_1 = 48 \pm 3\,\mathrm{l\,mol^{-1}},$$

and

$$C_2 = 750 \pm 200\,\mathrm{l^2\,mol^{-2}}.$$

In 0.1 mol l^{-1} acetic acid/ethanol, the results obtained are identical with the above

In our laboratory, Patterson has found, for aerated solutions of chlorophyll a in ethanol, a lifetime (τ) of $5.3 \pm 0.2 \times 10^{-9}$ s. This is longer than that obtained from the product of the natural radiative lifetime and the quantum yield of fluorescence (Weber & Teale 1957) previously reported for these solutions. Thus for the reaction

$$\mathrm{Chl^s + DQ \rightarrow Chl...DQ,} \tag{3}$$

we may derive that

$$k_3 = C_1\tau$$
$$= 9 \pm 2 \times 10^9\,\mathrm{l\,mol^{-1}\,s^{-1}}.$$

This rate is somewhat higher than that predicted for a diffusion-controlled reaction ($5.5 \times 10^9\,\mathrm{l\,mol^{-1}\,s^{-1}}$).

These observations are comparable to previously reported work for other similar systems, Livingston & Ke (1950), Dilung & Chernyuk (1961) and Krasnovskii & Drozdova (1966).

Flash photolysis studies of solutions of chlorophyll a and duroquinone

All work reported in this section was performed with outgassed solutions. As chlorophyll a is decomposed in aerated solutions, particularly if this is acidified, outgassing operations were carried out using diffuse green lighting. Under such conditions no decomposition of the chlorophyll a was observed. For reproducible kinetics, it is necessary to protect the solution from white light as the duroquinone undergoes an efficient reaction with the solvent to form the hydroquinone

$$\mathrm{Q^* \xrightarrow[h_\nu]{solvent} QH_2.} \tag{4}$$

On flashing ethanolic solutions of chlorophyll a (1 to 2×10^{-6} mol l^{-1}) with orange light ($\lambda > 550$ nm), chlorophyll a triplet is observed. This species decays by both first and second-order processes:

$$-\mathrm{d}(T)/\mathrm{d}t = k_i(T) + k_{ii}(T)^2,$$

where

$$k_i = 1300 \pm 300\,\mathrm{s^{-1}},$$

$$k_{ii} = 2 \pm 1 \times 10^9\,\mathrm{l\,mol^{-1}\,s^{-1}}.$$

(In deriving k_{ii}, we have assumed that there is negligible absorption of the triplet state at 664 nm introducing a possible error of 10 %.)

The decay kinetics and difference spectrum obtained with solutions in 0.1 mol l^{-1} acetic acid/ethanol are identical with the above. These results are in agreement with those found in other solvents, Seely (1966), Parker & Joyce (1967) and Ke *et al.* (1965).

On addition of small quantities of duroquinone (0 to 4×10^{-6} mol l^{-1}) to chlorophyll solutions in neutral ethanol, two transients may be observed at 664 nm. The earlier species has a photoelectric spectrum identical to the chlorophyll a triplet state and the second, which decays by second-order kinetics to reform the ground state of chlorophyll a, has the difference spectrum shown in figure 1. On increasing the concentration of duroquinone in the solution, the lifetime of the earlier species is so reduced that it could not be observed.

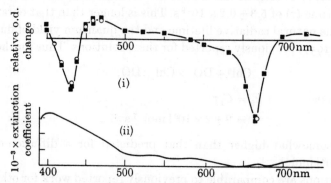

FIGURE 1. Difference spectrum (i) and the derived absorption spectrum (ii) for the transients formed on flashing an ethanolic solution of chlorophyll a and duroquinone. ■, Neutral ethanol; ○, 0.1 mol l^{-1} pyridine/ethanol.

With a duroquinone concentration of 1×10^{-4} mol l^{-1}, only one transient could be observed at all wavelengths and the derived rate constants (second-order plots are linear over three half-lives) are shown below:

$$k = 4.4 \pm 0.5 \times 10^4 \, (-\Delta\epsilon_{664}) \, l \, mol^{-1} s^{-1}$$

$$= 9.1 \pm 0.9 \times 10^4 \, (-\Delta\epsilon_{430}) \, l \, mol^{-1} s^{-1}$$

$$= 2.5 \pm 0.3 \times 10^5 \, (\Delta\epsilon_{460}) \, l \, mol^{-1} s^{-1}.$$

($\Delta\epsilon_\lambda = \epsilon_{tr}^\lambda - \epsilon_{g.s.}^\lambda$, where ϵ_{tr}^λ and $\epsilon_{g.s.}^\lambda$ are the extinction coefficients of transient and ground state respectively at a wavelength λ.) On inserting the relative values for the change in optical density at these wavelengths we obtain a value of

$$k = 4.7 \pm 0.7 \times 10^4 \, (-\Delta\epsilon_{664}) \, l \, mol^{-1} s^{-1}$$

$$= 3.2 \pm 0.5 \times 10^9 \, l \, mol^{-1} s^{-1},$$

(assuming $\epsilon_{g.s.}^{664} \gg \epsilon_{tr}^{664}$ and $\epsilon_{g.s.} = 69000$).

After three half-lives, deviations from second-order kinetics were noted, the experimental rate being faster than that predicted for a second-order process. These are partially attributable to reaction of the photoproducts with traces of duro-hydroquinone.

These observations are consistent with the following scheme

$$Chl^T + DQ \rightarrow Chl...DQ, \tag{5}$$

$$\text{Chl...DQ} \rightarrow \text{Chl}^{\cdot+} + \text{DQ}^{\cdot-}, \tag{6}$$

$$\text{Chl}^{\cdot+} + \text{DQ}^{\cdot-} \rightarrow \text{Chl} + \text{DQ}, \tag{7}$$

where $k_7 = 3.2 \pm 0.5 \times 10^9 \, \text{l mol}^{-1} \text{s}^{-1}$.

In alkaline solution ($0.1 \, \text{mol l}^{-1}$) pyridine/ethanol), a similar decay was observed with a second-order rate constant for reaction (7) of $3.0 \pm 0.3 \times 10^9 \, \text{l mol}^{-1} \text{s}^{-1}$.

The rate constant k_5 has been determined by measuring the rate of decay of the chlorophyll triplet in ethanolic solution containing concentrations of duroquinone (5×10^{-7} to $4 \times 10^{-6} \, \text{mol l}^{-1}$). This is most easily and accurately performed at a monitoring wavelength of 406 nm, where the transients derived from chlorophyll and quinone reaction show neither absorption nor depletion (see figure 1). A linear plot of rate constant for triplet decay versus quinone concentration was observed giving $k_5 = 1.4 \pm 4 \times 10^9 \, \text{l mol}^{-1} \text{s}^{-1}$.

In acidic solution ($0.1 \, \text{mol l}^{-1}$ acetic acid/ethanol, unless otherwise stated), a rate constant of $k_5 = 1.5 \pm 0.5 \times 10^9 \, \text{l mol}^{-1} \text{s}^{-1}$ was obtained. However, the decay of the $\text{Chl}^{\cdot+}$ species, presumably formed, is quite different and two processes must be involved. On monitoring at 664 nm, 60% of the oxidized chlorophyll was observed to decay within a few milliseconds, while the remainder decays by a first-order process in approximately 1 s. It is proposed that this phenomenon may be explained by the following scheme:

$$\text{Chl}^T + \text{DQ} \rightarrow \text{Chl}^{\cdot+} + \text{DQ}^{\cdot-}, \tag{5) and (6}$$

$$\text{DQ}^{\cdot-} + \text{H}^+ \rightarrow \text{DQH}^{\cdot}, \tag{8}$$

$$\text{Chl}^{\cdot+} + \text{DQH}^{\cdot} \rightarrow \text{Chl} + \text{DQ} + \text{H}^+, \tag{9}$$

$$\text{DQH}^{\cdot} + \text{DQH}^{\cdot} \rightarrow \text{DQ} + \text{DQH}_2, \tag{10}$$

$$\text{Chl}^{\cdot+} \rightarrow \text{Chl}, \tag{11}$$

$$\text{Chl}^{\cdot+} + \text{DQH}_2 \rightarrow \text{Chl} + \text{DQH}^{\cdot} + \text{H}^+, \tag{12}$$

where $k_9(\text{DQH}^{\cdot}) \gg k_{11} + k_{12}(\text{DQH}_2)$.

Experimentally, it has been shown that the initial second-order rate constant for decay of $\text{Chl}^{\cdot+}$ is $2.7 \pm 0.3 \times 10^3 \, (-\Delta\epsilon_{664}) \, \text{l mol}^{-1} \text{s}^{-1}$, i.e. $1.9 \pm 0.2 \times 10^8 \, \text{l mol}^{-1} \text{s}^{-1}$, and that the long-lived transient decays by a first-order process with a rate constant of $6 \pm 2 \times 10^{-1} \text{s}^{-1}$. The ratio of the optical density at 664 nm of the long lived transient (extrapolated to $t = 0$) to that of the initial optical density at the same wavelength ($\Delta(\text{o.d.})^{664}_{\text{l.l.t.}}/\Delta(\text{o.d.})^{664}_0$) is 0.40 ± 0.06.

If the above scheme is assumed, it is possible to derive theoretically an expression for this latter ratio:

$$\text{conc}_{\text{l.l.t.}}/\text{conc}_0 = \Delta(\text{o.d.})^{664}_{\text{l.l.t.}}/\Delta(\text{o.d.})^{664}_0$$

$$= 2B^{1/(1-2B)},$$

where $B = k_{10}/k_9$.

A plot of the variation of the optical density ratio versus k_{10}/k_9 was constructed, and our experimental results lead to a value of $k_{10}/k_9 = 0.60 \pm 0.15$. Thus we may assign

$$k_{10} = 1.1 \pm 0.4 \times 10^8 \, \text{l mol}^{-1} \text{s}^{-1}.$$

The theoretically predicted decay for the Chl^{+} derived by computer program assuming the above mechanism, agrees well with that found experimentally.

The rate constant k_{10} may be determined independently by flash photolysis studies of the semiquinone radical produced by photolysis of solutions of duroquinone in $0.1 \, mol \, l^{-1}$ acetic acid/ethanol. Although the second-order kinetics are not obeyed after one half-life, owing to the formation of a long-lived metastable species (proposed by other workers, Michaelis & Granick (1944) and Bridge & Porter (1958), as the quinhydrone), the initial second-order rate constant

$$k_{10} = 1.3 \pm 0.2 \times 10^4 \, \Delta\epsilon_{418} \, l \, mol^{-1} s^{-1}$$

(i.e. $k_{10} = 1.3 \times 10^8 \, l \, mol^{-1} s^{-1}$, assuming $\Delta\epsilon_{418} = 10\,000$). This agrees with our independently estimated value, but is lower than that found by other workers $(4 \times 10^8 \, l \, mol^{-1} s^{-1})$ for a solvent system of 50:50 ethanol/water (Bridge & Porter 1958; Land & Porter 1960). This decrease in the rate of radical–radical reaction in pure ethanolic solution compared to that in aqueous solution has also been noticed for the dimerization of benzophenone ketyl radicals (M. B. Ledger, private communication).

The experimental values for the replenishment of chlorophyll *a* ground state are independent of the concentration of acetic acid in solution within the range 3×10^{-2} to $3 \times 10^{-1} \, mol \, l^{-1}$, indicating that the semiquinone is reacting in these solutions in its protonated form and not as the radical anion at a concentration controlled by the following equilibrium:

$$Q^{\cdot -} + H^+ \rightleftharpoons QH^{\cdot}. \tag{13}$$

The mechanism proposed above would lead us to expect the difference spectrum of the long-lived species to be unlike that observed at short time delays after the flash, especially in the wavelength region below 450 nm. Thus

$$\Delta(o.d.)_0^\lambda = (\epsilon_{Chl^{\cdot +}} - \epsilon_{Chl} + \epsilon_{QH^{\cdot}} - \epsilon_Q) \, cl,$$

$$\Delta(o.d.)_{l.l.t.}^\lambda = (\epsilon_{Chl^{\cdot +}} + - \epsilon_{Chl} + \tfrac{1}{2}\epsilon_{QH_2} - \epsilon_Q) \, cl,$$

where c is the concentration of transient, l the pathlength and ϵ extinction coefficient. As ϵ_Q and ϵ_{QH_2} are comparatively negligible at wavelengths (λ) above 400 nm, we may derive the expression

$$\Delta(o.d.)_0^\lambda - \Delta(o.d.)_{l.l.t.}^\lambda = \epsilon_{QH^{\cdot}} \cdot cl.$$

The experimental results are shown in figure 2 (normalized to

$$(o.d.)_0^{664} = (o.d.)_{l.l.t.}^{664} = 100).$$

A marked difference may be observed and this is consistent within error to the absorption spectrum of the semiquinone radical.

It was found that if adequate precautions were not taken to filter out light of wavelength less than 500 nm, then the rate of decay of the long-lived transient increased with the duration of the experiment. Concurrent with this was an increase in the permanent absorption in the wavelength region, 280 to 310 nm, (the region of the hydroquinone absorption). When quantities of durohydroquinone (0 to

326   J. M. Kelly and G. Porter

$2 \times 10^{-4}\,\mathrm{mol\,l^{-1}}$) were added to a solution of chlorophyll a ($1 \times 10^{-6}\,\mathrm{mol\,l^{-1}}$) and duroquinone ($1 \times 10^{-4}$) in $0.1\,\mathrm{mol\,l^{-1}}$ acetic acid/ethanol, it was found that the first-order rate constant for decay of the $\mathrm{Chl}^{\cdot+}$ was directly proportional to the $\mathrm{QH_2}$ concentration. Thus we may assign a value for the rate constant for reaction (12) of $5.5 \pm 2 \times 10^{5}\,\mathrm{l\,mol^{-1}\,s^{-1}}$ (table 1).

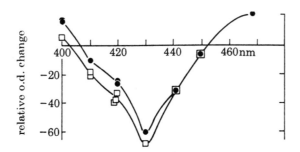

FIGURE 2. Spectra of the short- and long-lived transients formed from chlorophyll a and duroquinone in acidic ethanol. ●, $(\mathrm{o.d.})_0$; □, $(\mathrm{o.d.})_{\mathrm{l.l.t.}}$

TABLE 1. PROPOSED MECHANISM FOR THE INTERACTION OF EXCITED
CHLOROPHYLL AND DUROQUINONE IN ETHANOL

$\mathrm{Chl}^{s} + \mathrm{DQ} \to \mathrm{Chl...DQ}$	$k_3 = 9 \pm 2 \times 10^{9}\,\mathrm{l\,mol^{-1}\,s^{-1}}$
$\mathrm{Chl}^{T} + \mathrm{DQ} \to \mathrm{Chl...DQ}$	$k_5 = 1.4 \pm 0.4 \times 10^{9}\,\mathrm{l\,mol^{-1}\,s^{-1}}$
$\mathrm{Chl...DQ} \to \mathrm{Chl} + \mathrm{DQ}$	$1.1 \times k_6$
$\mathrm{Chl...DQ} \to \mathrm{Chl}^{\cdot+} + \mathrm{DQ}^{\cdot-}$	k_6
$\mathrm{Chl}^{\cdot+} + \mathrm{DQ}^{\cdot-} \to \mathrm{Chl} + \mathrm{DQ}$	$k_7 = 3.2 \pm 0.5 \times 10^{9}\,\mathrm{l\,mol^{-1}\,s^{-1}}$
$\mathrm{DQ}^{\cdot-} + \mathrm{H^+} \to \mathrm{DQH}^{\cdot}$	$(k_8 = 7 \times 10^{9}\,\mathrm{l\,mol^{-1}\,s^{-1}})$†
$\mathrm{Chl}^{\cdot+} + \mathrm{DQH}^{\cdot} \to \mathrm{Chl} + \mathrm{DQ} + \mathrm{H^+}$	$k_9 = 1.9 \pm 0.2 \times 10^{8}\,\mathrm{l\,mol^{-1}\,s^{-1}}$
$\mathrm{DQH}^{\cdot} + \mathrm{DQH}^{\cdot} \to \mathrm{DQH_2} + \mathrm{DQ}$	$k_{10} = 1.1 \pm 0.4 \times 10^{8}\,\mathrm{l\,mol^{-1}\,s^{-1}}$
$\mathrm{Chl}^{\cdot+} \to \mathrm{Chl}$	$k_{11} = 6 \pm 2 \times 10^{-1}\,\mathrm{s^{-1}}$
$\mathrm{Chl}^{\cdot+} + \mathrm{DQH_2} \to \mathrm{Chl} + \mathrm{DQH}^{\cdot} + \mathrm{H^+}$	$k_{12} = 5.5 \pm 2 \times 10^{5}\,\mathrm{l\,mol^{-1}\,s^{-1}}$

† Bridge & Porter (1958).

Studies of solutions of chlorophyll a with α-tocopherylquinone and vitamin K_1

Similar experiments have been performed with α-tocopherylquinone and vitamin K_1. The results obtained in these cases are consistent with an analogous scheme to that of table 1 and are shown in table 2.

As with solutions of duroquinone, it is necessary to exclude light of wavelength less than $5000\,\mathrm{nm}$, which causes the quinones to form reduced and isomeric species (Leary & Porter 1970) as these compounds seriously modify the decay kinetics of the oxidized chlorophyll. No evidence has been found for the production of long lived isomeric or other derivatives of the quinone similar to those observed on direct irradiation of the quinone (Leary & Porter 1970).

TABLE 2. COMPARISON OF THE RATE CONSTANTS FOR REACTIONS OF CHLOROPHYLL
a WITH DUROQUINONE, α-TOCOPHERYLQUINONE AND VITAMIN K_1 RESPEC-
TIVELY

reaction step	duroquinone	α-tocopheryl-quinone	vitamin K_1	units and factor
fluorescence half-quenching concentration	1.6 ± 0.2	2.3 ± 0.2	1.5 ± 0.2	$\times 10^{-2}$ mol l^{-1}
$Chl^T + Q$	1.4 ± 0.4	0.7 ± 0.2	1.2 ± 0.3	$\times 10^9$ l mol^{-1} s^{-1}
$Chl^{\cdot+} + Q^{\cdot-}$	3.2 ± 0.5	3.3 ± 0.4	4.5 ± 0.6	$\times 10^9$ l mol^{-1} s^{-1}
$Chl^{\cdot+} + QH^{\cdot}$	1.9 ± 0.2	1.4 ± 0.2	3.3 ± 0.3	$\times 10^8$ l mol^{-1} s^{-1}
$QH^{\cdot} + QH^{\cdot}$	1.1 ± 0.4	0.4 ± 0.1	0.7 ± 0.2	$\times 10^8$ l mol^{-1} s^{-1}
$Chl^{\cdot+} \rightarrow Chl$ (acidic solution) probably reaction with solvent impurities	0.6 ± 0.2	1.7 ± 0.2	1.8 ± 0.2	s^{-1}

The reaction of chlorophyll radical cation with α-tocopherol

When quantities of α-tocopherol (TPH), (0 to 2×10^{-3} mol l^{-1}) were added to an
acidic solution of chlorophyll *a* and α-tocopherylquinone (1×10^{-4} mol l^{-1}), the
chlorophyll *a* radical cation was found to decay by a first-order process with a rate
constant directly proportional to the concentration of α-tocopherol. The gradient
of this plot is $1.8 \pm 0.3 \times 10^5$ l mol^{-1} s^{-1}. As the decay of the chlorophyll *a* triplet in
ethanol is unaffected by α-tocopherol in concentrations as high as 1×10^{-3} mol l^{-1}
(its spectrum and decay kinetics are quite unchanged) we may assume that the
following mechanism holds

$$Chl^T + TQ \rightarrow Chl^{\cdot+} + TQ^{\cdot-} \qquad (k = 7 \pm 2 \times 10^8 \text{ l mol}^{-1} \text{ s}^{-1}), \qquad (14)$$

$$Chl^{\cdot+} + TPH \rightarrow Chl + TP^{\cdot} + H^+ \qquad (k = 1.8 \pm 0.3 \times 10^5 \text{ l mol}^{-1} \text{ s}^{-1}). \qquad (15)$$

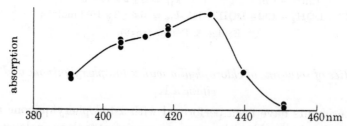

FIGURE 3. Long-lived transient observed in acid solutions of chlorophyll *a*
d-α-tocopherylquinone and α-tocopherol.

At wavelengths between 390 and 450 nm, a long-lived species was noted (first-
order decay rate constant $= 1.6 \pm 0.2$ s^{-1}) and the absorption spectrum of this is
shown in figure 3. The identity of this species is not known, but it is certainly an
oxidized form of α-tocopherol, as the TQH$^{\cdot}$ radicals are much shorter lived under
these conditions.

Discussion

As both the lowest singlet and triplet states of duroquinone lie well above those of chlorophyll a, processes involving energy transfer from chlorophyll a to the quinone must be considered impossible. The phenomena observed here must be caused by electron transfer as indicated in our scheme.

It is possible that the Chl$^{\cdot+}$ species may exist in these solutions as the deprotonated form but no specific evidence has been found for this. As the rate constant of reaction (7) is close to that for which a diffusion controlled reaction, it seems probable that the oxidized chlorophyll a is indeed in its cationic form.

The slow first-order decay of Chl$^{\cdot+}$ in acidic media is probably caused by reaction of this oxidized chlorophyll species with the solvent or more probably with impurities therein.

The results of this study are similar to those obtained for chlorophyll a and benzoquinone by Seifert & Witt (1968) and by Chibisov et al. (1967) in neutral alcoholic solutions, and they resolve the discrepancy in rates observed by these workers and by Ke et al. (1965) who used acid conditions. The transient spectra observed here resemble those recorded by Krasnovskii & Drozdova (1964) in viscous alcoholic solutions at low temperatures.

It is of interest in the light of our conclusions to re-examine the results of Tollin & Green (1962) and Mukherjee, Cho & Tollin (1969) (e.s.r. studies at low temperatures) and Evstigneev (1968) (observations of electrode-active species). These workers have studied principally benzoquinone, so that an exact parallel does not exist.

Evstigneev's measurements of photo-induced electrode potential changes require the formation of a substantial steady-state concentration of intermediate species. Two interesting conclusions were reached: (a) that the concentration of active species was dependent on pH (decreasing sharply at pH > 6), and (b) that it was at a maximum at a quinone concentration of approximately $10^{-4}\,\mathrm{mol\,l^{-1}}$. Clearly from our results, a steady state concentration of Chl$^{\cdot+}$ would be several orders of magnitude smaller in neutral ethanol than in acid solution where a substantial proportion (40 %) for duroquinone) has a half-life of approximately 1 s. Similarly, with concentrations of quinone greater than $10^{-4}\,\mathrm{mol\,l^{-1}}$ we might expect that reaction of Chl$^{\cdot+}$ with hydroquinone present as impurity would reduce this steady state concentration and therefore the magnitude of signal recorded.

Several discrepancies exist between the present work and the study reported by Tollin & Green (1962) and Mukherjee et al. (1969), from steady state e.s.r. measurements. First, these workers were able to detect only signals from the semiquinone and not from the chlorophyll transient, and this led them to suggest that two chlorophyll transients disproportionate rapidly. We have no evidence for such a reaction. Secondly, they propose a rate constant for reaction of the chlorophyll triplet state with quinone of $5 \times 10^4\,\mathrm{l\,mol^{-1}\,s^{-1}}$ in ethanol at $-40\,^{\circ}\mathrm{C}$. Even allowing for differences in diffusion rate in such a medium because of the lower temperature ($\eta_{20\,^{\circ}\mathrm{C}} = 1.2\,\mathrm{cP}$, $\eta_{-40\,^{\circ}\mathrm{C}} = 4.3\,\mathrm{cP}$), this rate is much slower than that reported here and by other workers. A possible explanation for the lack of observation of the chlorophyll radical is that sufficient quantities of QH_2 were present (the quinone concentration used is 10^{-2} to 4×10^{-2} mol l^{-1}) to quench the Chl$^{\cdot+}$ formed, particularly in alkaline solutions and that the reaction

$$Chl^{\cdot+} + QH^- \rightarrow Chl + Q^{\cdot-} + H^+$$

may cause the rapid decrease of Chl$^{\cdot+}$ and formation of the (observed) semiquinone radical anion.

It seems, therefore, that our work may be consistent with most of the observations of these workers, if allowance is made for the efficient reaction of oxidized chlorophyll with hydroquinone present as impurity in the quinone.

The results from the studies with α-tocopherylquinone and vitamin K_1 indicate that the long-alkyl chain of this species modifies the rate of some of the reaction steps (particularly that of semiquinone radical disproportionation), but that the mechanism is basically unaffected. There is no evidence for production of transient isomeric species of either α-tocopherylquinone or vitamin K_1 which have been found on direct photolysis of the natural quinones (Leary & Porter 1970).

It is hoped that the extension of this work to the solid phase (e.g. lecithin solutions (Kelly & Porter 1970) and monolayers) may provide some insight into the primary processes of photosynthesis.

References

Boardman, N. K. 1968 *Adv. Enzymol.* **30**, 1.
Bridge, N. K. & Porter, G. 1958 *Proc. Roy. Soc. Lond.* A **244**, 276.
Chibisov, A. K., Karyakin, A. V., Drozdova, N. N. & Krasnovskii, A. A. 1967 *Dokl. Akad. Nauk SSSR* **175**, 737.
Dictionary of organic compounds 1965 p. 3019. London: Eyre and Spottiswoode.
Dilung, I. I. & Chernyuk, I. N. 1961 *Dokl. Akad. Nauk SSSR* **140**, 162.
Evstigneev, V. B. 1968 *J. Chim. phys.* **65**, 1447. and references therein.
Flaig, W., Salfeld, J. C. & Baume, E. 1958 *Justus Liebigs Annln Chem.* **618**, 117.
Ke, B., Vernon, L. P. & Shaw, E. R. 1965 *Biochemistry, N,Y.* **4**, 137.
Kelly, A. R. & Porter, G. 1970 *Proc. Roy. Soc. Lond.* A **315**, 149.
Krasnovskii, A. A. & Drozdova, N. N. 1964 *Dokl. Akad. Nauk. SSSR* **158**, 730.
Krazsnovskii, A. A. & Drozdova, N. N. 1966 *Dokl. Akad. Nauk. SSSR* **166**, 223.
Land, E. J. & Porter, G. 1960 *Proc. Chem. Soc.* p. 84.
Leary, G. & Porter, G. 1970 *J. Chem. Soc.* (in the Press).
Livingston, R. & Ke, C-L. 1950 *J. Am. Chem. Soc.* **72**, 909.
Michaelis, L. & Granick, S. 1944 *J. Am. Chem. Soc.* **66**, 1023.
Mukherjee, D. C., Cho, D. H. & Tollin, G. 1969 *Photochem. Photobiol.* **9**, 273.
Parker, C. A. & Joyce, T. A. 1967 *Photochem. Photobiol.* **6**, 395.
Quinlan, K. P. & Fujimori, E. 1967 *J. Phys. Chem. Ithaca* **71**, 4154.
Seely, G. R. 1966 *The Chlorophylls* (Eds. L. P. Vernon and G. R. Seely), p. 530. New York: Academic Press.
Seifert, K. & Witt, H. T. 1968 *Naturwissenschaften* **55**, 222.
Strain, H. H., Thomas, M. R., Crespi, H. L., Blake, M. I. & Katz, J. J. 1960 *Ann. N.Y. Acad. Sci.* **84**, 617.
Tollin, G. & Green, G. 1962 *Biochim. biophys. Acta* **60**, 524.
Weber, G. & Teale, F. W. J. 1957 *Trans. Faraday Soc.* **53**, 646.
Witt, H. T. 1967 *Nobel Symposium 5-Fast Reactions and Primary Processes in Chemical Reactions* (ed. S. Claesson), p. 81. New York: Interscience.

Proc. R. Soc. Lond. A. **342**, 317–325 (1975)

Model systems for photosynthesis
V. Electron transfer between chlorophyll and quinones in a lecithin matrix

By G. S. Beddard, Sir George Porter, F.R.S. and G. M. Weese†

Davy Faraday Research Laboratory of The Royal Institution,
21 Albemarle Street, London W1X 4BS)

(*Received* 31 *July* 1974)

The relative fluorescence yield, the lifetime and the triplet yield have been investigated, for solutions in lecithin, of chlorophyll *a* in the presence of a series of quinones.

The singlet and triplet yields show the same dependence on quinone concentration; the lifetimes show a slightly weaker dependence. It is proposed that the quenching proceeds via electron transfer from the excited singlet state of the chlorophyll *a* molecule to its nearest neighbour quinone.

A model, based upon a random distribution of molecules, is used to derive quenching parameters. The model gives a good account of the experimental data and it is found that the quenching ability is directly related to the oxidation potential of the quinone.

Introduction

The primary photochemical act of photosynthesis is considered to be the donation of an electron from chlorophyll to an acceptor molecule (Clayton 1971; Butler 1973). From the shape of the redox titration curve of the initial fluorescence, Reed, Zankel & Clayton (1969) concluded that the primary electron acceptor is not a quinone. Quinones are present in photosystem II, however, and possibly very close to the chlorophyll. They interact in a secondary electron pool, closely coupled to, but not identical with the primary electron acceptor. Several quinone molecules are present for each chlorophyll molecule in the reaction center and some reduction of the quinone does occur. It was therefore of interest to investigate the quenching of chlorophyll *a* by quinones in lecithin because of the possible mechanistic similarities between this system and those responsible for the low fluorescence yield in the reaction centres.

Experimental

Chlorophyll *a*, quinone and lecithin were dissolved in chloroform and deposited on to quartz slides by evaporation of the chloroform in a current of oxygen free nitrogen (Weese 1972). The homogeneity and uniform thickness of the sample were

† Present address: Department of Chemistry, University of Ottawa, Ottawa, K1N 6NS, Ontario, Canada.

checked by recording the absorption spectrum at different positions on the slide. The absorption spectrum over the range 350–750 nm was recorded by a Perkin Elmer 124 spectrophotometer, the thickness of the sample was adjusted to give an absorbance of approximately 0.7 at 438 nm. Concentrations were determined from the absorption spectrum and the sample path length, measured by means of a Zeiss microscope. Because of the high sample absorbance, fluorescence spectra were recorded by front face reflexion on an Hitachi MPF 2A fluorimeter. Excitation at 435 nm was used for all the samples, fluorescence being measured from 600 to 700 nm. A Wratten 47B filter was placed in the exciting light beam and a Corning 2412 filter in front of the red-sensitive photomultiplier (H.T.V. R106) to reduce the scattered light. Relative quantum yields were calculated from the area under the fluorescence spectrum. Absolute quantum yields were determined by reference to a quinine sulphate standard (Parker 1968).

Both a pulse sampling method (Kelly & Patterson 1971) and a photon counting method (Beddard 1973) were used to measure the fluorescence decay profiles. Results from the two methods were compatible. Convolution of the exciting pulse time profile with an exponential decay function was used to determine the fluorescence decay profile. A Mullard 56 TVP photomultiplier was used to observe the fluorescence.

Time-resolved triplet state absorption spectra were recorded on the microbeam flash photolysis apparatus described in part I (Kelly & Porter 1970). A flash duration of 40 µs and energy of 70 J were used. An E.M.I. 6250B photomultiplier was used to monitor the absorption.

Materials

Crystalline chlorophyll *a* was obtained from Sigma Ltd. The purity of this material was checked by comparing the ratio of the absorbance of the red and blue peaks with literature values (Seely & Jensen 1965). Lecithin was obtained from the Nutritional Biochemical Corporation and was purified as described previously (Kelly 1968). Duroquinone was obtained from Fluka A.G. 2,5-dichloro and 2,5-dimethyl-*p*-benzoquinone and α-tocopherylquinone from Eastman Kodak, plastoquinone-9 from La Roche and Company. The A.R. solvents and *p*-benzoquinone were obtained from B.D.H. Ltd.

Results

The chlorophyll *a*–lecithin–quinone system was examined in detail for six quinones over the concentration range 10^{-4} to 10^{-1} M. The chlorophyll *a* concentration was kept at 4×10^{-4} M, a low enough concentration to prevent any significant self-quenching (Kelly & Patterson 1971). The addition of quinone did not change the position of the chlorophyll *a* fluorescence, and no new emissions were observed. The absolute fluorescence quantum yield of chlorophyll in lecithin was measured as 0.26 at 4×10^{-4} M in chlorophyll. This value is close to that obtained by dividing the natural fluorescence decay time, measured from the absorption spectrum, into the measured decay time of 5.1 ns, measured under the same conditions.

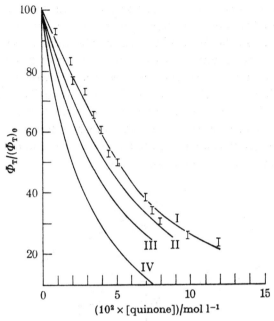

FIGURE 1. Effect of quinone concentration on the relative triplet yield of chlorophyll *a* for four *p*-benzoquinones. I, 2,5-Dichloro-*p*-benzoquinone; II, *p*-benzoquinone; III, 2,5-dimethyl-*p*-benzoquinone; IV, duroquinone. (Experimental points are included for duroquinone only.)

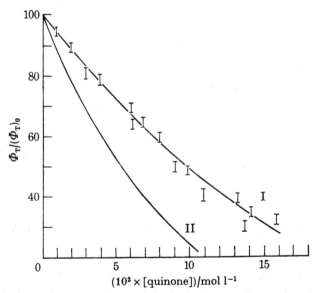

FIGURE 2. Effect of quinone concentration on the relative triplet yield of chlorophyll *a* for two natural quinones. I, Plastoquinone; II, α-tocopherylquinone. (Experimental points are included for plastoquinone only.)

320   G. S. Beddard and others

The variations in the three parameters, fluorescence lifetime, yield and triplet yield, measured as a function of quinone concentration are shown in figures 1, 2, 3 and 4. Each is expressed in the form X/X_0 with X_0 the value at zero quinone concentration. The triplet was found to decay with first order kinetics with a rate constant of $1.07 \pm 0.03 \times 10^3\,\mathrm{s^{-1}}$ over the whole wavelength range studied and was independent of the quinone over the concentration range used. The triplet difference spectrum is shown in figure 5. The triplet yield and singlet yield show the same dependence on quinone concentration, the singlet lifetime showing a smaller rate of change. The concentration dependence of the triplet yield was found to be the same at 430, 470 and 600 nm in the triplet absorption spectrum.

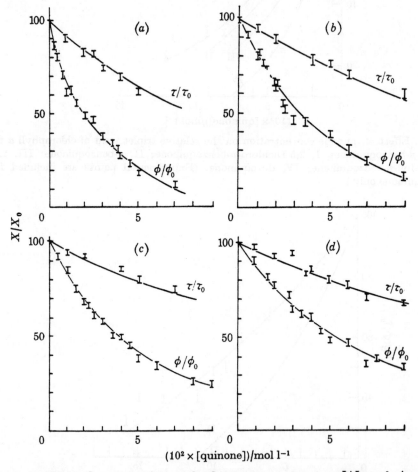

FIGURE 3. Effect of concentration on the fluorescence parameters Φ/Φ_0 and τ/τ_0 of chlorophyll a for four p-benzoquinones. (a) 2,5-Dichloro-p-benzoquinones; (b) benzoquinone; (c) 2,5-dimethyl-p-benzoquinone; (d) duroquinone. ———, Theoretical curves; I, experimental points.

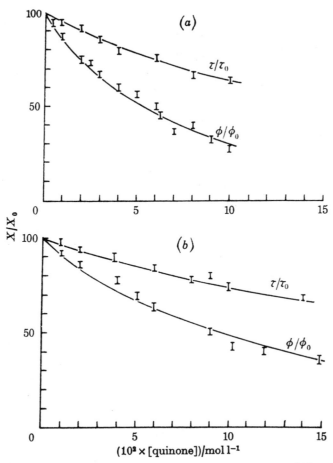

FIGURE 4. Effect of concentration on the fluorescence parameters Φ/Φ_0 and τ/τ_0 of chlorophyll a for the natural quinones. (a) Plastoquinone-q; (b) α-tocophenylquinone. ——, Theoretical curves; I, experimental points.

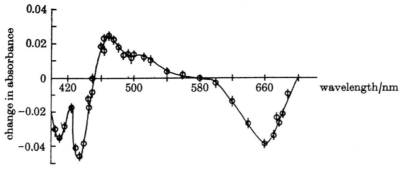

FIGURE 5. The chlorophyll a triplet difference spectrum in lecithin.

322   G. S. Beddard and others

DISCUSSION

Our measurements on quinone quenching of chlorophyll a reveal a concentration dependence very different from those on self-quenching of chlorophyll b in the absence of quinone made by Kelly & Patterson (1971). In self-quenching of chlorophyll b, the reduction in the fluorescence yield occurred at a lower concentration and with a higher power dependence upon concentration than the reduction in lifetime. These results were interpreted in terms of the presence of non-fluorescent pairs of molecules, statistically distributed, which led to *in situ* quenching. The fluorescence yield fell rapidly with an increase in concentration and the reduction was almost complete before a change in lifetime was observed. In the results presented here the lifetime and yield decrease in sympathy indicating that *in situ* quenching is not significant. The constant triplet lifetime indicates that the quenching process occurs from the chlorophyll a, S_1 singlet state only.

The lack of any spectral change in the absorption or fluorescence spectral distribution by adding quinone does not eliminate the possible presence of quinone–chlorophyll complexes. White & Tollin (1971) and Raman & Tollin (1971) have proposed chlorophyll–quinone complexes to explain the photochemistry of their systems. They propose quinone interactions with the chlorophyll triplet state in fluid polar solutions.

In the experiments described here, if chlorophyll–quinone complexes are important, a reduction in the fluorescence quantum yield with no corresponding reduction in the lifetime would be expected, as is observed in chlorophyll self-quenching (Kelly & Patterson 1971).

The quinone concentration required to reduce the fluorescence to half of its value in the absence of quinone (the half quenching concentration, $C_{\frac{1}{2}}$) is seen to be dependent upon the oxidation potential of the quinone (table 1). This provides powerful, if indirect, evidence that the quenching process is an electron transfer from photoexcited chlorophyll a to the quinone. In the flash kinetic experiments the only transient species observed was the chlorophyll triplet. No direct evidence for an electron transfer reaction, in the form of $Chl^{\cdot+}$ and $Chl^{\cdot-}$ species, was found. In this static medium, however, the $Chl^{\cdot+}$ and $Chl^{\cdot-}$, if formed, could not diffuse apart and the back transfer of an electron would be a very rapid and efficient process. The chlorophyll triplet has a higher ionization potential than the singlet (Witt 1967), and this nullifies the effect of its longer lifetime in its possible reaction with the quinone.

The concentration dependences of the yields and lifetimes indicate that the separation between donor and acceptor is important in determining the degree of quenching. At high concentrations an increasing fraction of molecules will find themselves close together. A model based upon the random distribution within the matrix was used to interpret the results. The transfer was considered to occur only between nearest neighbours and the equations derived by Minn & Filipescu (1970) were used to describe the electron transfer process:

$$\Phi/\Phi_0 = \tau_0^{-1} C_a \int_0^\infty F_1(R)\, \mathrm{d}R \tag{1}$$

and

$$\tau/\tau_0 = \frac{\tau_0^{-1} \int_0^\infty F_1(R)\, \mathrm{d}R}{\int_0^\infty F_0(R)\, \mathrm{d}R}, \tag{2}$$

where

$$F_n(R) = 4\pi R^2 n!\, [\tau_0^{-1} + k(R)]^{-1-n} \exp\left(-\frac{4\pi}{3} C_a R^3\right),$$

and $k(R)$ is the angle-averaged rate constant for transfer to a nearest neighbour molecule at a distance R. The exponential term represents the probability of finding an excited donor and its closest acceptor at a distance R (Chandrasekhar 1943) and is based on a random distribution of molecules. C_a is the concentration of the acceptor and τ_0 the natural lifetime of the donor. Because of the cubic components in the exponential term, the integrals have sharp cut-off regions beyond which no significant transfer occurs. This effect explains the partial success of the sphere of quenching model (Perrin 1924). The maximum value of the function $F_0(R)$ is found to represent the most probable transfer distance. A crude rate expression in $k(R)$ was used, assuming that an exchange interaction would be sufficient to account for the electron transfer. Electron transfer by exchange required spatial overlap of the electron orbitals of the donor and acceptor molecules and the rate expression for electron transfer at a distance R was defined as

$$k(R) = A \exp(-2R/L). \tag{3}$$

The exponential dependence arises from the exponential decline of the electronic wave function with distance. A corresponds to the rate of transfer (s^{-1}) at 'zero' distance. By comparison with the Dexter exchange theory (Dexter 1953), L may be considered as the average of the radii of the donor orbital of chlorophyll a and the acceptor orbital of the quinone. In the crude model this definition is too exact; the value of L determines the overlap between donor and acceptor orbitals at any particular separation distance. Since the chlorophyll a orbital remains the same in every chlorophyll–quinone pair, changes in L reflect changes in the quinone acceptor orbital. By substituting the rate expression (3) into the expressions (1) and (2) for the relative fluorescence lifetimes and yields, quenching curves were generated by the numerical evaluation of the integrals involved. The calculated quenching curves were simultaneously fitted to the experimental curves at five concentrations and at both Φ/Φ_0 and τ/τ_0 for each chlorophyll–quinone pair. The values of A and L were varied in order to minimize the square of the differences between the calculated and experimental points.

The function $F_0(R)$ shows a well defined maximum, at the most probable transfer distance, the function value falling steeply on either side of the maximum. Table 2 shows this transfer distance together with the half quenching concentration. It can be seen that the transfer distance does not decrease appreciably after the half quenching concentration is reached, suggesting that the transfer efficiency approximates its maximum value at the half quenching concentration. As expected, the maximum transfer probability occurs at larger distances for the more strongly oxidizing quinones than for those with smaller oxidation potentials. The predicted transfer distances appear very large for electron transfer because the model assumes point molecules. The porphyrin ring is *ca.* 1.7 nm in diameter, however, and the actual electron transfer distances are therefore much smaller than the theoretically

TABLE 1. EXPERIMENTAL AND THEORETICAL DATA FOR THE QUENCHING
OF CHLOROPHYLL *A* FLUORESCENCE BY QUINONES

quinone	oxidation potential/V	$C_{\frac{1}{2}}/10^2$ M	$A/10^{-11}$ s^{-1}	L/nm
2,5-dichloro-*p*-benzoquinone	0.74	2.0	179	0.379
p-benzoquinone	0.71	2.8	64	0.379
2,5-dimethyl-*p*-benzoquinone	0.60	3.8	39.4	0.342
duroquinone	0.47	4.9	14.3	0.342
plastoquinone-9	0.53	5.8	3.44	0.397
α-tocopherylquinone	0.47	9.4	1.04	0.379

TABLE 2. DISTANCE OF MOST PROBABLE TRANSFER AT
VARIOUS QUINONE CONCENTRATIONS†

(*C*, Concentration; *D*, most probable transfer distance.)

quinone

2,5-dichloro-*p*-benzoquinone

$C/10^2$ M	1	2†	4	5	6
D/nm	3.1	2.7	2.6	2.5	2.5

p-benzoquinone

$C/10^2$ M	1	2.8†	4	6	9
D/nm	3.0	2.4	2.3	2.2	2.1

2,5-dimethyl-*p*-benzoquinone

$C/10^2$ M	1	2	3.8†	6	10
D/nm	3.0	2.5	2.3	2.1	2.0

duroquinone

$C/10^2$ M	1	2	4.9†	6	9
D/nm	3.0	2.4	2.0	1.9	1.9

plastoquinone-9

$C/10^2$ M	1	3	5.8†	8	10
D/nm	3.0	2.3	2.0	1.9	1.9

α-tocopherylquinone

$C/10^2$ M	1	3	6	9.4†	14
D/nm	3.0	2.1	1.6	1.7	1.6

† These concentrations represent $C_{\frac{1}{2}}$.

predicted values. Table 1 also shows the calculated values of the maximum transfer rate A and the overlap parameter L for each of the quinones. It can be seen that, for both the simple and natural benzoquinones, the value of A increases with the oxidation potential of the quinone, again indicative of an electron transfer process. The values of L also vary in a similar manner, consistent with our definition of L as a factor which monitors the overlap of the orbitals of the two molecules.

The bulky side groups on plastoquinone-9 and α-tocopherylquinone make the electron transfer less efficient than the oxidation potential would indicate, when compared to the simple quinones. This inhibitory effect would not be expected to operate in the ordered structure of the chloroplast, where the phytyl sidechains of the quinone and chlorophyll facilitate their close proximity by orientation at the interface between phases of different polarity.

G. S. B. and G. M. W. wish to thank the Science Research Council for the award of Studentships.

REFERENCES

Beddard, G. S. 1973 Thesis, University of London.

Butler, W. L. 1973 *Accnts chem. Res.* **6**, 177.

Chandraskhar, S. 1943 *Rev. Mod. Phys.* **15**, 1.

Clayton, R. K. 1971 *Adv. chem. Phy.* **19**, 353.

Dexter, D. L. 1953 *J. chem. Phys.* **21**, 836.

Kelly, A. R. 1968 Thesis, University of Sheffield.

Kelly, A. R. & Porter, G. 1970 *Proc. R. Soc. Lond.* A **315**, 149.

Kelly, A. R. & Patterson, L. 1971 *Proc. R. Soc. Lond.* A **324**, 117.

Minn, F. & Filipescu, N. 1970 *J. chem. Soc. Lond.* A 1016.

Parker, C. A. 1968 *Photoluminescence from solutions.* Amsterdam: Elsevier.

Perrin, F. 1924 *C. r. hebd. Séanc. Acad. Sci., Paris* **178**, 1978.

Raman, R. & Tollin, G. 1971 *Photochem. Photobiol.* **13**, 135.

Reed, D. W., Zankel, R. L. & Clayton, R. K. 1969 *Proc. natn. Acad. Sci. U.S.A.* **63**, 42.

Seely, G. R. & Jensen, R. G. 1965 *Spectrochim. Acta.* **21**, 1835.

Weese, G. M. 1972 Thesis, University of London.

White, R. A. & Tollin, G. 1971 *Photochem. Photobiol.* **14**, 15, 43.

Witt, H. T. 1967 Nobel Symposium 5. *Fast reactions and primary processes* (ed. S. Claesson), p. 81.

J. Chem. Soc., Faraday Trans. 2, 1981, **77**, 1939–1948

Colloidal Platinum Catalysts for the Reduction of Water to Hydrogen, Photosensitised by Reductive Quenching of Zinc Porphyrins

By Anthony Harriman,† George Porter
and Marie-Claude Richoux

Davy Faraday Research Laboratory, The Royal Institution, 21 Albemarle Street, London W1X 4BS

Received 11th March, 1931

EDTA transfers an electron to the triplet excited state of $ZnTMPyP^{4+}$ in aqueous solution and the one-electron reduction product of the zinc porphyrin reduces water to H_2. A Pt catalyst is required to promote this latter reaction and it was found that the nature of the support used to stabilise the colloidal Pt particle against flocculation had a great effect upon the efficiency of the Pt catalyst. The most effective support was Carbowax 20M and irradiation of $ZnTMPyP^{4+}$ in the presence of EDTA (0.1 mol dm^{-3}) and Pt–Carbo ($10^{-4} \text{ mol dm}^{-3}$) resulted in formation of H_2 with an optimum quantum yield ($\phi_{\frac{1}{2}H_2}$) of 0.07.

The average lifetime of the reduced zinc porphyrin was increased by binding the porphyrin to a polymer and by using a surfactant derivative of the porphyrin. In the latter case, the surfactant porphyrin was used as a support for colloidal Pt particles and upon irradiation in the presence of EDTA H_2 was produced with $\phi_{\frac{1}{2}H_2}$ of *ca.* 0.004; this was increased significantly when methyl viologen was present in the aqueous solution.

In recent years there have been considerable advances made towards the overall storage of sunlight in the form of hydrogen fuel. The most successful approach to this ideal has been the photoreduction of water, *via* an electron relay, at the expense of a sacrificial electron donor (such as EDTA, H_2S or cysteine).[1-4] Such systems have been refined to a high degree and the most recently proposed system,[5,6] using a water-soluble zinc porphyrin as chromophore and EDTA as sacrificial donor, produced hydrogen with an optimum quantum yield ($\phi_{\frac{1}{2}H_2}$) of 0.6 upon irradiation with 550 nm light. With few exceptions, the electron relay used in these systems has been methyl viologen (MV^{2+}); the popularity of the $MV^{2+/+}$ couple as an electron relay lies in the facile reduction of water to H_2 on the surface of a Pt catalyst

$$2MV^+ + 2H_2O \rightleftharpoons 2MV^{2+} + H_2 + 2OH^- \qquad (1)$$

and also in the fact that the reduced form of MV^{2+} is highly coloured and can be monitored easily.

However, if we attempt to construct a practical device[7] for photoreduction of water to H_2 then MV^{2+} presents several problems. It is a classified poison, it can be hydrogenated in the presence of a Pt catalyst, the reduced form is extremely reactive towards oxygen and it is expensive. In addition, the intense absorption in the visible region shown by the reduced form (MV^+) acts as an inner filter and can inhibit the primary photochemical processes. Thus, we need to turn our attention away from MV^{2+} and consider some alternative cycles for production of H_2 from water.

Amongst possible alternative schemes, one that is worthy of special attention involves reduction of water by the reduced form of the chromophore. This scheme is illustrated below

$$S^* + D \rightarrow S^- + D^+ \qquad (2)$$

$$D^+ \rightarrow \text{products} \qquad (3)$$

$$2S^- + 2H_2O \rightleftharpoons 2S + H_2 + 2OH^- \qquad (4)$$

where S is the chromophore and D is an irreversible electron donor. This scheme has been realised using acridine dyes[8,9] and phthalocyanines[10,11] as chromophores, although in the latter case the yields of H_2 were very low. In addition, recent work[5,6] has shown that water-soluble zinc porphyrins will function in reductive cycles using EDTA as the irreversible electron donor. However, as we have reported in a short communication,[12] the quantum yield for production of H_2 from the above reductive cycle was only ca. 5% of that found in the presence of MV^{2+}. In this paper we explore several methods for improving the reductive system and, in particular, we have attempted to understand the role of the Pt catalyst in the H_2 production step.

EXPERIMENTAL

All experimental operations were carried out as reported previously.[6,12] All solutions used water, redistilled from alkaline permanganate, at pH 5.0 outgassed by the freeze–pump–thaw method, and the concentration of the porphyrin chromophore was approximately 10^{-5} mol dm^{-3}. Quantitative estimation of evolved H_2 was made by g.c. using the conditions suggested by Valenty.[13]

EDTA (sodium salt) (B.D.H. Chemicals) was used as received. ZnTPyP and ZnTMPyP^{4+} were prepared as described in a previous paper.[6] The zinc complex of 5,10,15,20-tetra(N-octadecyl-4-pyridyl)porphine (ZnTOPyP^{4+}) was prepared by alkylation of ZnTPyP using a literature method.[14] The mono(4-carboxyphenyl) tri(4-methylphenyl)porphine was prepared by a literature method and purified by repeated chromatography on silica.[15] The compound was converted to the acid chloride by treatment with thionyl chloride which, after drying under vacuum, was refluxed with a suspension of poly(vinyl alcohol) (MW = 16 000) in benzene to afford esterification. The polymer-bound porphyrin was heated at 80 °C in aqueous solution containing an excess of $ZnCl_2$ until complete metallation had occurred (as monitored by absorption spectroscopy) and the excess salt was removed by ion-exchange chromatography. The final polymer, after purification by gel permeation chromatography, contained approximately 2% w/w zinc porphyrin.

Colloidal Pt supported on PVA (MW = 16 000) was prepared as before;[6] the average particle size was measured by Dr. J. Kiwi (EPFL, Lausanne, Switzerland) using the light-scattering technique. The other colloidal Pt catalysts were prepared by citrate reduction of H_2PtCl_6 in dilute aqueous solution following the method of Bond.[16] Particular attention was given to the cleanliness of the glassware used for such preparations since this appeared to influence the properties of the Pt–sol produced.[17] After removal of excess inorganic residues on an ion-exchange column, the colloidal Pt solution was stirred vigourously whilst the supporting agent was added slowly and the solution was then stirred at 45 °C for 48 h. The supporting agents used were Carbowax 20M (Carbo) (B.D.H. Chemicals), sodium lauryl sulphate (LS) (B.D.H. Chemicals) and cetyltrimethylammonium chloride (CTAC) (Koch-Light Chemicals) and, in all cases, a molar ratio of support : Pt = 2 : 1 was used. The total concentration of Pt in the final solution was measured by atomic absorption. We are greatly indebted to Prof. M. Grätzel (EPFL, Lausanne, Switzerland) for supplying full experimental details for preparation of these supported Pt catalysts.

Careful adherence to the conditions recommended by previous workers[16-18] should result in stabilised Pt particles having an average particle radius of 3–4 nm.

STRUCTURE OF THE PORPHYRINS

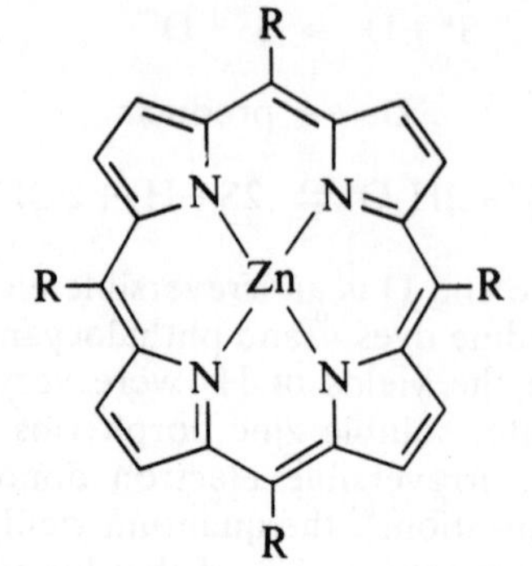

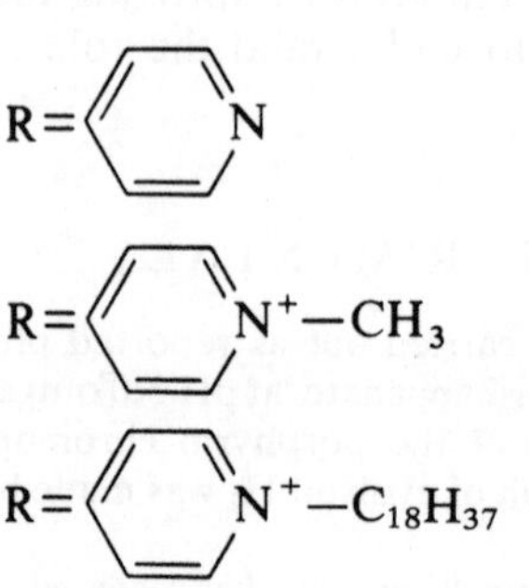

$R = $ [pyridine structure]   ZnTPyP

$R = $ [N-methylpyridinium structure] $N^+ - CH_3$ ZnTMPyP^{4+}

$R = $ [N-octadecylpyridinium structure] $N^+ - C_{18}H_{37}$ ZnTOPyP^{4+}

RESULTS AND DISCUSSION

It was shown previously that the triplet excited state of ZnTMPyP^{4+} was quenched by EDTA and that the quenching reaction led to irreversible reduction of the porphyrin ring.[6] Thus, irradiation of ZnTMPyP^{4+} in outgassed aqueous solution containing EDTA $(0.1 \, \mathrm{mol \, dm^{-3}})$ resulted in bleaching of the chromophore, the quantum yield for the bleaching process (ϕ_{BL}) was found to be 0.08, but no H$_2$ was produced during this reaction.[12] However, when irradiation was carried out in the presence of a Pt–PVA catalyst, the bleaching process was inhibited and H$_2$ was produced as a reaction product.[12] The observed results were consistent with the following scheme

$$*\mathrm{ZnTMPyP^{4+}} + \mathrm{EDTA} \rightarrow \mathrm{ZnTMPyP^{3+}} + \mathrm{EDTA^+} \tag{5}$$

$$\mathrm{EDTA^+} \rightarrow \mathrm{products} \tag{6}$$

$$2\mathrm{ZnTMPyP^{3+}} \rightarrow \mathrm{ZnTMPyP^{4+}} + \mathrm{ZnTMPyP^{2+}} \tag{7}$$

$$2\mathrm{ZnTMPyP^{3+}} + 2\mathrm{H_2O} \rightarrow 2\mathrm{ZnTMPyP^{4+}} + \mathrm{H_2} + 2\mathrm{OH^-} \tag{8}$$

and the effect of the Pt–PVA concentration on ϕ_{BL} and on the quantum yield for formation of H$_2$ $(\phi_{\frac{1}{2}H_2})$ is given in fig. 1. It was shown by flash-photolysis experiments that the species responsible for H$_2$ production was the one-electron reduction product of the zinc porphyrin (ZnTMPyP^{3+}), and the effect of Pt–PVA concentration on the lifetime of this species is given in fig. 2. As seen from fig. 1 and 2, very high concentrations of the Pt catalyst were required before the bleaching process

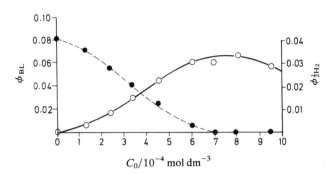

FIG. 1—Effect of Pt–PVA concentration on the quantum yields for the bleaching process (●) and for formation of H_2 (○).

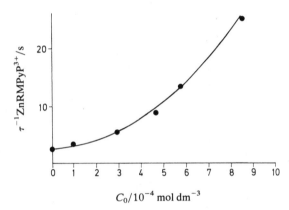

FIG. 2. Effect of Pt–PVA concentration on the half-life of $ZnTMPyP^{3+}$.

was inhibited completely and it is this observation that restricts the application of such systems to practical solar energy storage devices.

The active reducing species ($ZnTMPyP^{3+}$) was unstable towards disproportionation, and the half-life of the radical, in the absence of Pt–PVA, was only *ca.* 0.3 s under the conditions used for the flash-photolysis experiments. Thus, there was a limited time available for the radical to encounter a Pt particle and in order to ensure that this process could compete with disproportionation it was necessary to use a high concentration of Pt–PVA particles.

In all colloidal catalysts, Pt is present in an aggregated form and, if the particle has a spherical shape, the aggregation number (n) is related to the radius (r) of the colloidal particle by

$$n = \tfrac{4}{3}\pi r^3 \rho N / M \tag{9}$$

where ρ is the density of Pt, M is the atomic weight and N is Avogadro's constant. (With PVA as supporting agent, there must be some doubt about the spherical nature of the particle.) For the Pt–PVA catalyst that we have used here, the average radius of a supported particle was *ca.* 30 nm so that $n = 7.5 \times 10^6$. Now, the total concentration of Pt present in solution (C_0), as measured by atomic absorption, is related to the molar concentration of Pt–PVA particles by

$$[\text{Pt–PVA}] = C_0/n \tag{10}$$

and, hence, for a given C_0, the larger the radius of the particle the lower the concentration of particles. In addition, the rate of diffusional encounter (k_D) between an organic radical, such as ZnTMPyP^{3+}, and a Pt–PVA particle depends[19] upon the radius of the particle and for $r = 30$ nm k_D is *ca.* 2.8×10^{11} (particle moles)$^{-1}$ s^{-1} where a particle mole is defined as n/N. Therefore, in order for the Pt–PVA particle to intercept the disproportionation reaction, it is necessary that $k_D[\text{Pt–PVA}]$ is much greater than $k_{DIS}[\text{ZnTMPyP}^{3+}]$, where k_{DISS} is the rate constant for the disproportionation reaction. The diffusional controlled pseudo-first-order rate constant $k_D[\text{Pt–PVA}]$ is given by

$$k_D[\text{Pt–PVA}] = \frac{3 \times 10^3 \times M \times D}{\rho} \frac{C_0}{r^2} \tag{11}$$

where the diffusion coefficient (D) of ZnTMPyP^{3+} is given a typical value of 10^{-5} cm^2 s^{-1}. The equation predicts a linear relationship between the pseudo-first-order rate constant and the total concentration of Pt and it also requires that the rate of reaction be proportional to $1/r^2$.

In the absence of Pt, the lifetime of ZnTMPyP^{3+} was 0.3 s and for this to be decreased by a factor of ten the calculated concentration of Pt–PVA required is *ca.* 8×10^{-4} mol dm^{-3}. As shown in fig. 2, the actual concentration of Pt–PVA needed to achieve a tenfold reduction in the lifetime of the radical was in very good agreement with this calculated value showing that almost every collision between a radical and a Pt–PVA particle was effective in finding an active site upon the particle surface. Over the C_0 range studied, the lifetime of ZnTMPyP^{3+} was not a linear function of $(C_0)^{-1}$, as is predicted by eqn (11), but this must be expected since at modest concentrations of Pt there are competitive processes for the deactivation of ZnTMPyP^{3+}. In order to test the linear relationship expressed in eqn (11), very high concentrations of Pt–PVA are required.

Before such systems can be proposed as practical devices for the photoreduction of water to H$_2$ it is essential that the concentration of Pt is reduced to a more economical level. With the above system, this may be achieved in two ways; first, k_{DIS} can be reduced to a minimum so that the lifetime of ZnTMPyP^{3+} is increased significantly (say >10 s) and, secondly, $k_D[\text{Pt}]$ can be increased, either by increasing the rate of diffusional encounters or by decreasing n so that, for a given C_0, [Pt] is increased.

In order to achieve the latter situation, several different supports were used to stabilise colloidal Pt particles in aqueous solution. It has been found by others,[18,20] and confirmed by ourselves, that medium chain length surfactant-type molecules afford good protective properties to colloidal Pt particles. In particular, alkanes with positively (CTAC) and negatively (LS) charged head groups stabilise Pt particles formed by citrate reduction of H$_2$PtCl$_6$ whilst low molecular weight, cross-linked poly(ethylene glycol) (Carbo) affords excellent protective behaviour.[18] For all three systems, the expected radii of the stabilised Pt particles are in the range 3–4 nm,[16–18] which represents a tremendous improvement over the Pt–PVA particles. For these particles, n is *ca.* 1.2×10^4 and k_D is *ca.* 3.5×10^{10} (particle moles)$^{-1}$ s^{-1}.

TABLE 1—CONCENTRATION OF Pt CATALYST REQUIRED TO REDUCE THE LIFETIME OF ZnTMPyP^{3+} TO ONE-TENTH OF ITS VALUE OBSERVED IN THE ABSENCE OF Pt

catalyst	concentration/mol dm^{-3}
Pt–PVA	9.5×10^{-4}
Pt–CTAC	3.2×10^{-4}
Pt–Carbo	9.0×10^{-5}
Pt–LS	6.2×10^{-5}

Flash-photolysis experiments showed that in the presence of the above surfactant-supported catalysts the lifetime of ZnTMPyP^{3+} decreased with increasing concentration of Pt. Table 1 gives the total concentration of Pt (C_0) required to reduce the lifetime of ZnTMPyP^{3+} to one-tenth of the value observed in the absence of Pt. Table 1 shows that all of the surfactant-supported catalysts were more efficient than Pt–PVA. Of the three catalysts, the cationic support (Pt–CTAC) was considerably less efficient than the other two, probably due to electrostatic repulsion between ZnTMPyP^{3+} and the positively charged layer around the catalyst surface. The other two catalysts gave comparable kinetic results although Pt–LS was the most efficient.

Separate steady-state irradiation experiments showed that H₂ production still occurred with Pt–Carbo and Pt–CTAC but not with Pt–LS. In the latter case, although the lifetime of ZnTMPyP^{3+} was shortened drastically by the presence of the catalyst, there was no H₂ produced even at high Pt concentrations. With both Pt–Carbo and Pt–CTAC, H₂ production was observed although at a given C_0 the Pt–Carbo catalyst was some thirty times more efficient, for H₂ production, than Pt–CTAC. Thus, the Pt–Carbo catalyst seems to be the most suitable for the present needs and, upon irradiation of ZnTMPyP^{4+} in outgassed aqueous solution containing EDTA (0.1 mol dm^{-3}) and Pt–Carbo ($C_0 = 10^{-4}$ mol dm^{-3}), H₂ was produced with a quantum yield ($\phi_{\frac{1}{2}H_2}$) of *ca.* 0.07. The longevity of this system was good and the total amount of H₂ produced upon exhaustive photolysis of the system corresponded to a turnover with respect to the chromophore of >2000. (Note that for this experiment the buffer concentration was very high and H₂ was removed at frequent intervals by purging with N₂.) This system is the best that we have been able to produce for the photogeneration of H₂ *via* a reductive cycle, and the total amount of H₂ obtained from the system was comparable with the optimum yield obtained previously[6] with the ZnTMPyP^{4+}/MV^{2+}/EDTA system, although the rate of H₂ production is much slower in the reductive cycle.

In order to increase the lifetime of the reduced zinc porphyrin, it was decided to synthesise a porphyrin derivative where the disproportionation reaction would be hindered by steric factors. The synthetic route chosen was to attach the porphyrin to a PVA polymer backbone *via* a single ester linkage, as described in the experimental section. This polymer-bound zinc porphyrin had a long triplet lifetime ($\tau_T = 0.8$ ms) although the reaction between the triplet state and EDTA was inefficient ($k = 2 \times 10^4$ dm^3 mol^{-1} s^{-1}) compared with the corresponding reaction[12] with ZnTMPyP^{4+}. This decrease in the rate of reaction may be a consequence of positioning the porphyrin in a fairly hydrophobic region so that there is little contact with EDTA, which is hydrophillic. However, flash-photolysis experiments showed that excitation of the polymer-bound zinc porphyrin in outgassed aqueous solution containing EDTA (0.1 mol dm^{-3}) resulted in formation of the one-electron reduction product of the porphyrin which, in this case, had a lifetime >1 s. Thus, attaching the porphyrin to a polymer support gave a substantial increase in the lifetime of the reduction product but, for a given concentration of EDTA, the yield of this product was much lower than that found with ZnTMPyP^{4+}.

The polymer-bound zinc porphyrin was used as a support for Pt particles so that the porphyrin was placed in close proximity to the catalyst surface. A picture of the supposed structure of the supported particle is given in fig. 3 and, for such

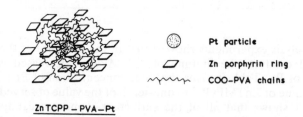

ZnTCPP − PVA − Pt

Pt particle
Zn porphyrin ring
COO−PVA chains

FIG. 3.—Idealised structure of Pt particles supported by polymer-bound zinc porphyrin.

a structure, the reduced porphyrin should be produced at a site close to the Pt surface. This should assist electron transfer from the reduced porphyrin to the Pt particle since it avoids the need for long-range mass diffusion and, in fact, the average lifetime of the reduced porphyrin was relatively short (*ca.* 0.05 s) in the presence of Pt ($C_0 = 5 \times 10^{-5}$ mol dm^{-3}). However, very little H$_2$ was observed with this system ($\phi_{\frac{1}{2}H_2} < 10^{-3}$), probably reflecting the poor efficiency with which EDTA quenched the triplet excited state of the zinc porphyrin.

A second attempt was made to prepare a functionalised Pt catalyst and in this case a surfactant zinc porphyrin was used as support for the Pt particle. The surfactant porphyrin was synthesised by alkylating each of the four N atoms of ZnTPyP with a C$_{18}$ alkyl chain so that the final porphyrin has a total electronic charge of 4+. The porphyrin was insoluble in water, at concentrations below the c.m.c., but was readily dispersed in CTAC micelles. Under such conditions, flash-photolysis measurements showed that the bimolecular rate constant for quenching the triplet excited state with EDTA was 2.4×10^5 dm^3 mol^{-1} s^{-1}, which was comparable to that observed[12] with the water-soluble ZnTMPyP^{4+} showing that, with this structure, the zinc porphyrin was freely accessible to the aqueous phase.

Again, the zinc porphyrin derivative was used as a support for colloidal Pt particles, although in some cases the surfactant porphyrin was diluted 1/1 w/w with CTAC, so that the porphyrin and Pt were held in close proximity. Steady-state irradiation experiments showed that in the presence of EDTA (0.1 mol dm^{-3}) H$_2$ was produced as a reaction product with a quantum yield ($\phi_{\frac{1}{2}H_2}$) of *ca.* 0.004 ± 0.002. The $\phi_{\frac{1}{2}H_2}$ value was measured for irradiation at $\lambda = 430 \pm 30$ nm and was found to depend upon the sample of functionalised Pt catalyst. Different batches of catalyst gave different $\phi_{\frac{1}{2}H_2}$ values, but this is not too surprising in view of the uncertain structure of the catalyst. Consequently, at this stage the reported $\phi_{\frac{1}{2}H_2}$ value should be regarded as a fairly crude estimate although it is the average of four separate batches of the Pt catalyst.

An idealised structure for the functionalised Pt catalyst, with and without added CTAC, is shown in fig. 4 and it is seen that this system can be described as a micro-photoelectrochemical cell where the chromophore is coated onto the surface of a

PHOTOREDUCTION OF H$_2$O TO H$_2$

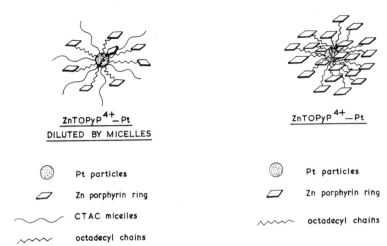

ZnTOPyP^{4+}—Pt
DILUTED BY MICELLES

ZnTOPyP^{4+}—Pt

⊙ Pt particles

▱ Zn porphyrin ring

〜〜 CTAC micelles

⋀⋀ octadecyl chains

⊙ Pt particles

▱ Zn porphyrin ring

⋀⋀ octadecyl chains

FIG. 4.—Idealised structure of Pt particles supported by ZnTOPyP^{4+} or ZnTOPyP^{4+}/CTAC (1/1 w/w).

Pt electrode. Previous work[21] has demonstrated that similar macro photoelectrochemical cells can be used for the photoreduction of water to H$_2$ but the reaction mechanism remains obscure.

With our functionalised Pt catalyst, fluorescence from the zinc porphyrin was observed although the fluorescence quantum yield was reduced relative to that found for ZnTOPyP^{4+} dispersed in CTAC micelles. (An accurate quantum yield was not possible due to light scattering.) Similarly, there was a reduction in the triplet-state lifetime when the zinc porphyrin was deposited upon the surface of the Pt particle. These quenching effects are most probably due to Förster-type energy transfer to the metal substrate but this does not present a particularly great problem since the effects are rather small. In separate steady-state irradiation experiments it was found that ZnTOPyP^{4+} deposited upon the surface of a clean Pt electrode gave rise to a photoconductivity effect[22] upon illumination when there was an electron acceptor (O$_2$ or MV^{2+}) present in the electrolyte solution. In the absence of an added electron acceptor there was practically no photocurrent generated upon illumination. Such effects have been well documented[22-25] for porphyrins and indicate the photoproduction of charge carriers from the neutral exciton state, and recent work[22,23] suggests that the charge carriers are formed at the interface between the porphyrin and the electrolyte solution. Thus, when MV^{2+} is used as electron acceptor, irradiation results in charge transfer from metalloporphyrin to MV^{2+}, probably *via* a triplet-state process,[22,23] and this is followed by migration of the oxidising equivalent from the metalloporphyrin into the bulk of the Pt metal:

$$ZnP^* + MV^{2+} \rightarrow ZnP^+ + MV^+ \tag{12}$$

$$ZnP^+ + Pt \rightarrow ZnP + Pt^+. \tag{13}$$

Similarly, irradiation of the functionalised Pt catalyst in the presence of MV^{2+} resulted in the transient formation of MV$^+$ but H$_2$ was not detected as an overall reaction product, presumably because short circuiting can occur:

$$MV^+ + Pt^+ \rightarrow MV^{2+} + Pt. \tag{14}$$

Irradiation of the functionalised Pt catalyst in the presence of EDTA did lead to formation of H_2 and the overall reaction mechanism is probably similar to that elaborated for the water-soluble zinc porphyrin system:

$$*ZnTOPyP^{4+} + EDTA \rightarrow ZnTOPyP^{3+} + EDTA^+ \tag{15}$$

$$EDTA^+ \rightarrow products \tag{16}$$

$$ZnTOPyP^{3+} + Pt \rightarrow ZnTOPyP^{4+} + Pt^- \tag{17}$$

$$Pt^- + H^+ \rightarrow Pt + \tfrac{1}{2}H_2. \tag{18}$$

Since H_2 formation occurs at very low concentrations of Pt, it seems feasible that the production of H_2 does not require interaction between particles but that each particle can function independently.

When MV^{2+} was added to the bulk aqueous solution, the observed yield of H_2 was increased significantly ($\phi_{\frac{1}{2}H_2} = 0.01$) relative to identical experiments performed in the absence of MV^{2+} ($\phi_{\frac{1}{2}H_2} = 0.004$). Thus, either MV^{2+} reacts directly with the triplet excited state of $ZnTOPyP^{4+}$ or else MV^{2+} is reduced by $ZnTOPyP^{3+}$. In the latter case, MV^{2+} functions as an electron relay thereby freeing the metalloporphyrin to undergo a second electron-transfer reaction. Note that studies with macro-photoelectrochemical cells also found that the yield of H_2 was increased when MV^{2+} was used as an electron relay.[21]

Work is now in progress aimed at establishing a reaction mechanism for H_2 production from both macro and micro photoelectrochemical cells which use metalloporphyrins as sensitisers. Note that if metal substrates with different work functions are used then the direction of the charge-transfer step can be tuned. That is, with Pt as substrate and a zinc porphyrin as chromophore the charge carriers must be produced at the electrolyte interface and charge separation is achieved by migration of the positive holes into the Pt bulk.[22,23] However, if Al or SnO_2 are used as substrate, then charge transfer occurs at the porphyrin/substrate interface and now the positive holes migrate towards the electrolyte solution.[22] Thus, by careful design of the system it may be possible to produce both H_2 and O_2 from water.

CONCLUSIONS

The work described in this paper has substantiated previous findings that careful preparation of colloidal Pt particles can result in increased efficiency for the Pt catalysed reduction of water to H_2. It is now possible to prepare stabilised Pt particles with an average radii of ca. 3–4 nm which intercept reverse electron transfer or disproportionation reactions of the reduced electron relay and lead to production of H_2. When these particles are used in conjunction with a reductive cycle, the chromophore functions as the electron relay, thus avoiding the need for an electron acceptor such as MV^{2+}. Furthermore, the electron relay can be attached directly to the support used to stabilise the Pt particle and so overcome the problems associated with long-range diffusional mass transfer. Using such systems, it was possible to photogenerate H_2 in low yield using visible light and this observation is of considerable importance because it provides a route whereby electron transfer to the Pt particle can compete with other decay pathways of the reduced relay at very low concentrations of Pt. In fact, by careful design of such systems it may be possible to construct a unit that will operate in aerated solution. This is essential if such photochemical systems are to be used in practical solar energy storage devices.

1948   PHOTOREDUCTION OF H_2O TO H_2

Due to work carried out in other laboratories and in our own, the photoreduction of water to H_2 at the expense of a sacrificial organic substrate is at an advanced stage. In previous papers,[6,12,26] we have developed photosensitisers that operate with high efficiency and that absorb at long wavelength. Now, following the work of Grätzel *et al.*, we have available several efficient Pt catalysts that facilitate reduction of water to H_2. The next step is to develop systems that do not need sacrificial substrates and that can be coupled to analogous systems capable of the oxidation of water to O_2. This may be achieved in two ways: first, by using two specific redox catalysts in a four-photon system or, secondly, by using a common redox couple to link together two separate four-photon systems. Although considerable progress has been achieved in the preparation of specific redox catalysts,[18] the latter approach seems to be the most feasible and this is our next goal.

We thank the S.R.C., the E.E.C. and G.E. (Schenectady) for financial support. We are greatly indebted to Prof. M. Grätzel for detailed discussions and for providing preprints of his work.

[1] B. V. Koryakin, T. S. Dzhabiev and A. E. Shilov, *Dokl. Akad. Nauk SSSR*, 1977, **238**, 620.
[2] J. M. Lehn and J. P. Sauvage, *Nouv. J. Chim.*, 1977, **1**, 441.
[3] K. Kalyanasundaram, J. Kiwi and M. Grätzel, *Helv. Chim. Acta*, 1978, **61**, 2720.
[4] A. Moradpour, E. Amouyal, P. Keller and H. Kagan, *Nouv. J. Chim.*, 1978, **2**, 547.
[5] K. Kalyanasundaram and M. Grätzel, *Helv. Chim. Acta*, 1980, **63**, 478, G. McLendon and D. S. Miller, *J. Chem. Soc., Chem. Commun.*, 1980, 533.
[6] A. Harriman, G. Porter and M-C. Richoux, *J. Chem. Soc., Faraday Trans. 2*, 1981, **77**, 883.
[7] A. Harriman, A. Mills and G. Porter, to be published.
[8] A. Krasna, *Photochem. Photobiol.*, 1979, **29**, 267.
[9] K. Kalyanasundaram and D. Dung, *J. Phys. Chem.*, 1980, **84**, 2551.
[10] A. Harriman and M. C. Richoux, *J. Photochem.*, 1980, **14**, 253.
[11] A. Harriman, G. Porter and M-C. Richoux, *J. Chem. Soc. Faraday Trans. 2*, 1980, **76**, 1618.
[12] A. Harriman and M-C. Richoux, *J. Photochem.*, 1981, **15**, 335.
[13] S. J. Valenty, *Anal. Chem.*, 1978, **50**, 669.
[14] T. Yamamura, *Chem. Lett.*, 1977, 773.
[15] J. L. Y. Kong and P. A. Loach, *Frontiers of Biological Energetics – Electrons to Tissues*, ed. P. L. Dutton, J. S. Leigh and A. Scarpa (Academic Press, New York, 1978), vol. 1, p. 73.
[16] G. C. Bond, *Trans. Faraday Soc.*, 1956, **52**, 1235.
[17] K. Kinoshita and P. Stonehart, *Modern Aspects of Electrochemistry* ed. J. O'M. Bockris and B. E. Conway (Plenum Press, New York, 1977), number 12, chap. 4, p. 183.
[18] P. A. Brugger, P. Cuendet and M. Grätzel, *J. Am. Chem. Soc.*, in press.
[19] A. Henglein, *J. Phys. Chem.*, 1979, **83**, 2209.
[20] M. Horisberger, *Biologie Cellulaire*, 1979, **36**, 253.
[21] M. de Backer, M. C. Richoux, F. Leclercq and G. Lepoutre, *Rev. Phys. Appl.*, 1980, **15**, 529.
[22] T. Kawai, K. Tanimura and T. Sakata, *Chem. Phys. Lett.*, 1978, **56**, 541.
[23] T. Kawai, K. Tanimura and T. Sakata, *Chem. Lett.*, 1979, 137.
[24] T. Katsu, K. Tamagake and Y. Fujita, *Chem. Lett.*, 1980, 289.
[25] K. Tanimura, T. Kawai and T. Sakata, *J. Phys. Chem.*, 1980, **84**, 751.
[26] A. Harriman, G. Porter and M-C. Richoux, *J. Chem. Soc., Faraday Trans. 2*, 1981, **77**, 1175.

J. Chem. Soc., Faraday Trans. 2, 1981, **77**, 2373–2383

Photo-oxidation of Water to Oxygen Sensitised by Tris(2,2'-bipyridyl)ruthenium(II)

BY ANTHONY HARRIMAN,* GEORGE PORTER AND PHILIP WALTERS

Davy Faraday Research Laboratory, The Royal Institution, 21 Albemarle Street, London W1X 4BS

Received 13*th May*, 1981

Irradiation of tris(2,2'-bipyridyl)ruthenium(II) in aqueous solution containing a sacrificial electron acceptor results in formation of $bipy_3Ru^{3+}$, a powerful oxidant which is able, in the presence of a redox catalyst, to oxidise water to O_2. Both colloidal RuO_2 and simple cobalt(II) ions are effective catalysts for O_2 production and, with $[Co(NH_3)_5Cl]^{2+}$ as sacrificial electron acceptor, the quantum efficiency for formation of O_2 was *ca.* 12%. However, when the sacrificial electron acceptor was replaced with Fe^{3+}, which is a reversible redox couple, no O_2 was observed upon prolonged irradiation. The absence of O_2 in this latter system can be explained by kinetic considerations.

Photoreduction of water to molecular hydrogen using sunlight as the excitation source offers a particularly interesting means of storing chemical potential, and in recent years considerable success has been achieved in this field.[1-7] Several different approaches towards this ideal have been advocated[8] but an inherent feature of the mechanism for any cyclic system is that oxidising equivalents must be produced. These oxidising equivalents have to be channelled into doing useful work or else they may result in destruction of the chromophore, as found with CdS.[9] In an ideal system, the oxidising equivalents would be used to oxidise water to molecular oxygen but, for small scale systems at least, this is not essential and more readily oxidised substrates can be substituted for water (*e.g.* EDTA, cysteine, triethanolamine, H_2S). For practical systems, these sacrificial electron donors would have to be waste products from industrial processes but, as such, their availability might be uncertain and for long-term purposes it is necessary that the ultimate electron donor is water.

The oxidation of water to O_2 requires the cooperation of four oxidising equivalents of sufficient redox power ($E_0 = +0.83$ V) and this has been difficult to achieve in the laboratory, except by band-gap illumination of semiconductors,[10,11] although green-plant photosynthesis solved the problems long ago. Green plants are believed[12] to use a manganese complex as the catalyst for water oxidation but *in vitro* studies with manganese compounds have not led to sustained photogeneration of O_2.[13-15] However, recent work by Kiwi and Grätzel[16] has established that noble-metal oxides such as PtO_2, IrO_2 and RuO_2 in powdered or colloidal form are capable of mimicking the functions of the manganese catalyst in photosynthesis. These redox catalysts facilitate coupling of the one-electron reduction of a suitable electron acceptor (D^+) to water oxidation according to

$$4D^+ + 2H_2O \xrightarrow{\text{CATALYST}} 4D + 4H^+ + O_2 \qquad (1)$$

and, so far, production of O_2 has been observed[16-19] when the electron acceptor has been Ce^{4+}, $bipy_3Ru^{3+}$, $bipy_3Fe^{3+}$, PbO_2 or $IrCl_6^{2-}$. In order for the production

PHOTO-OXIDATION OF H_2O TO O_2

of O_2 to be cyclic with respect to D^+ it is necessary to bring about the one-electron oxidation of D; if this can be achieved by a photochemical process then the overall photo-oxidation of water to O_2 can be realised. To date, this has been achieved only with the bipy$_3$Ru$^{3+/2+}$ couple since the photophysical/photochemical properties of the other oxidants mentioned above preclude their use as photosensitisers (*i.e.* the excited-state lifetimes of D are too short or D does not absorb in the visible region of the spectrum) and both Grätzel[20] and Lehn[18] and their coworkers have devised systems capable of continuous photogeneration of O_2 from water using bipy$_3$Ru^{2+}/RuO$_2$ systems. In addition, bipy$_3$Ru^{2+} has been used as the photosensitiser for photoelectrochemical generation of O_2 using Pt and RuO$_2$ electrodes;[21,22] these studies are worthy of special attention since they allow O_2 formation to occur at a site remote from the photolysis cell.

These systems have not been investigated in any detail and, as yet, quantum-yield and kinetic data are not available. However, these reactions are very important since they may provide a means of coupling together photoreduction of water to H_2 (which has reached an advanced stage) and photo-oxidation of water to O_2.[23] Consequently, we report here the results of our studies concerning the bipy$_3$Ru^{2+} photosensitised oxidation of water to O_2.

EXPERIMENTAL

MATERIALS

Samples of bipy$_3$Ru^{2+} (Strem Chemicals) were used as received whilst sodium persulphate, cobaltous sulphate and ferric ammonium alum were purchased from B.D.H. (AR grade). A literature method[24] was used for the preparation and purification of $[Co(NH_3)_5Cl]^{2+}$. Hydrated RuO$_2$ powder was obtained from Aldrich Chemicals (lot. no. 5117 CE) and colloidal RuO$_2$, supported on TiO$_2$ was a gift from Dr. A. Mackor (T.N.O., Utretcht). The concentration of Ru in each sample was determined by atomic absorption.

All solutions were prepared from doubly distilled deionised water and the pH was adjusted to 5.0 by addition of an acetate buffer (0.1 mol dm^{-3}). The solutions were deaerated by purging with N_2 and in most experiments the concentration of bipy$_3$Ru^{2+} was 5×10^{-5} mol dm^{-3}.

METHODS

Absorption spectra were recorded with a Perkin-Elmer 554 spectrophotometer and luminescence spectra were recorded with a Perkin-Elmer MPF 4 spectrofluorimeter. Steady-state irradiations were performed with a 100 W quartz/iodine projector lamp filtered through a glass filter to remove light of $\lambda < 420$ nm. Quantum-yield measurements were made with a 950 W Xe arc lamp filtered through a high radiance monochromator (Applied Photophysics) to isolate a 468 ± 2 nm wavelength region and the incident light intensity was calibrated with potassium ferrioxalate.

The concentration of evolved O_2 was measured with a Clark membrane oxygen electrode, purchased from Rank Bros.; the experimental arrangement, sensitivity and calibration of this instrument have been described previously.[25] For the sacrificial systems, the photolysate solutions (pH 5.0) contained bipy$_3$Ru^{2+} (5×10^{-5} mol dm^{-3}), $[Co(NH_3)_5Cl]^{2+}$ of persulphate (1×10^{-2} mol dm^{-3}) and a catalyst and were purged thoroughly with N_2 prior to irradiation. For the reversible systems, the sacrificial electron acceptor was replaced with ferric ammonium alum (4×10^{-4} mol dm^{-3}). In all cases, the pH was adjusted with acetate buffer and the solutions were stirred constantly throughout the experiment. The solutions were thermostatted at 29 °C and blank experiments were performed to ensure that the measured O_2 was not due to leakages. Formation of O_2 was confirmed by gas chromatography and "End-O-Mess" analysis techniques although all reported yields refer to the membrane polarography studies.

Flash-photolysis studies were made with an Applied Photophysics K2 (flash energy 200 J, flash duration 10 μs). The excitation lamps were filtered with aqueous potassium chromate to remove light of $\lambda < 400$ nm. Solutions contained $bipy_3Ru^{2+}$ (1×10^{-5} mol dm^{-3}) and ferric ammonium alum (4×10^{-4} mol dm^{-3}) and were deaerated by purging vigorously with N_2. All kinetic measurements were made at 452 nm where the return to ground-state $bipy_3Ru^{2+}$ could be monitored easily; the details of the flash-photolysis measurements have been reported previously.[26]

ABBREVIATIONS

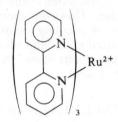

$bipy_3Ru^{2+}$

RESULTS AND DISCUSSION

Previous work has established that addition of RuO_2 (either as a powder, supported on alumina or TiO_2 or in colloidal form) to an aqueous solution of a strong oxidant such as Ce^{4+}, $bipy_3Ru^{3+}$ or $bipy_3Fe^{3+}$ accelerates the one-electron reduction of the oxidant and, within certain pH limits, results in production of O_2.[16-20] We have confirmed these original findings and, in particular, the $Ce^{4+/3+}$ couple provided a very clean system (a stoichiometric yield of O_2 has been claimed).[16] In the other cases, the yield of O_2 never reached the stoichiometric value expected from eqn (1) although there was very little overall decomposition of the oxidant couple. The RuO_2 catalysts were also very effective in catalysing O_2 evolution from PbO_2 (as a suspension in 0.05 mol dm^{-3} H_2SO_4) and H_2O_2 (as a dilute aqueous solution).

Of the above oxidants, only $bipy_3Ru^{3+}$ can be produced easily by photochemical methods, and consequently this oxidant has received extensive study. Creutz and Sutin[27] were the first to show that $bipy_3Ru^{3+}$ was capable of oxidising water to molecular oxygen although the mechanism of this four-electron process remains obscure. Later, Kiwi and Grätzel[16] found that this process was catalysed by RuO_2 but the pH at which the maximum yield of O_2 was obtained was found to depend upon the nature of the catalyst. Furthermore, these authors[16] together with Lehn et al.[18] showed that irradiation of $bipy_3Ru^{2+}$ in the presence of a sacrificial electron acceptor (such as $S_2O_8^{2-}$ or $[Co(NH_3)_5Cl]^{2+}$) and RuO_2 resulted in continuous O_2 formation. The overall reaction scheme may be expressed

$$*bipy_3Ru^{2+} + ox \rightarrow bipy_3Ru^{3+} + red \qquad (2)$$

$$4\,bipy_3Ru^{3+} + 2H_2O \xrightarrow{RuO_2} 4\,bipy_3Ru^{2+} + 4H^+ + O_2 \qquad (3)$$

where ox and red refer to the oxidised and reduced forms, respectively, of the sacrificial electron acceptor.

As regards the yield of $bipy_3Ru^{3+}$, the most effective sacrificial electron acceptor for these systems is the persulphate anion. At pH 5 (0.1 mol dm^{-3} acetate), persulphate ions quenched the triplet excited state of $bipy_3Ru^{2+}$ with a bimolecular quenching rate constant of 8×10^8 dm^3 mol^{-1} s^{-1}, as measured by luminescence techniques, and the quantum yield for formation of $bipy_3Ru^{3+}$ has been reported[28] to be 2.0. The high quantum yield arises from thermal oxidation of $bipy_3Ru^{2+}$ by a sulphate radical:

$$*bipy_3Ru^{2+} + S_2O_8^{2-} \rightarrow bipyRu^{3+} + SO_4^{2-} + SO_4^{\cdot -} \qquad (4)$$

$$bipy_3Ru^{2+} + SO_4^{\cdot -} \rightarrow bipy_3Ru^{3+} + SO_4^{2-}. \qquad (5)$$

In fact, the high quantum yield presents a serious experimental problem since high concentrations of $bipy_3Ru^{3+}$ are built-up rapidly and this species itself undergoes a photoreaction that leads to extensive loss of the chromophore. Furthermore, there exists the strong possibility that some of the evolved O_2 originates from the persulphate ions since it has been established that several transition-metal ions catalyse the decomposition of $S_2O_8^{2-}$ in aqueous solution.[29] Thus, we are wary of using persulphate as a sacrificial electron acceptor for O_2 evolving systems and, instead, we have found that cobalt(III) complexes provide less problems.

Lehn et al.[18] were the first to use cobalt(III) complexes as sacrificial electron acceptors for the $bipy_3Ru^{2+}$ photosensitised oxidation of water to O_2. Using $[Co(NH_3)_5Cl]^{2+}$, the triplet quenching rate constant for the process

$$*bipy_3Ru^{2+} + Co(NH_3)_5Cl^{2+} + 5H^+ \rightarrow bipy_3Ru^{3+} + Co^{2+} + Cl^- + 5 NH_4^+ \qquad (6)$$

was found to be 9.3×10^8 dm^3 mol^{-1} s^{-1}, which is very close to the diffusion-controlled limit for reaction between positively charged ions, and the quantum yield for formation of $bipy_3Ru^{3+}$ was found to be 0.31 for a $[Co(NH_3)_5Cl]^{2+}$ concentration of 1×10^{-2} mol dm^{-3}. As shown by eqn (6), electron transfer to the Co^{III} complex results in reduction to the corresponding Co^{II} complex which is much less stable than the parent Co^{III} complex and undergoes irreversible decomposition to form the simple $Co(H_2O)_6^{2+}$ cation. The redox potential of the aquo complex $[Co(H_2O)_6^{3+/2+} = +1.81$ V$]$ is considerably more positive than that of the $bipy_3Ru^{3+/2+}$ couple ($+1.27$ V) so that reverse electron transfer between $Co(H_2O)_6^{2+}$ and $bipy_3Ru^{3+}$ should be prohibited on thermodynamic grounds.[18] However, Shafirovich et al.[19] have reported that simple Co^{II} salts are effective catalysts for reduction of one-electron oxidants such as $bipy_3Ru^{3+}$, $bipy_3Fe^{3+}$ and $IrCl_6^{2-}$ and that these reduction reactions lead to oxidation of water to O_2. In fact, these authors have claimed[19] that irradiation of $bipy_3Ru^{2+}$ in aqueous solution at pH 7 containing $[Co(NH_3)_5Cl]^{2+}$ resulted in formation of O_2 with a quantum yield of ca. 0.025. These latter observations are particularly interesting since, if substantiated, they provide a route for O_2 evolution in homogeneous aqueous solution without the need to use powders or colloidal dispersions.

As shown in fig. 1, irradiation of $bipy_3Ru^{2+}$ (5×10^{-5} mol dm^{-3}) in aqueous solution at pH 5 containing $[Co(NH_3)_5Cl]^{2+}$ (1×10^{-2} mol dm^3) with light of $\lambda >$ 400 nm did not result in formation of O_2, at least over a 15 min irradiation period. Similar negative results were found for 15 min irradiations throughout the pH range 4–8 (previous work[16,17] has established that the thermal oxidation of water by $bipy_3Ru^{3+}$ is strongly dependent upon pH). However, when RuO_2 or $CoSO_4$ was

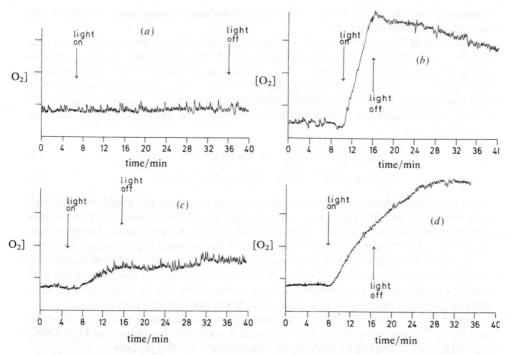

FIG. 1.—Typical traces for formation of O_2 from the irradiation of $bipy_3Ru^{2+}$ in aqueous solution at pH 5 containing $[Co(NH_3)_5Cl]^{2+}$ (10^{-2} mol dm^{-3}); (a) no added catalyst; (b) colloidal RuO_2 (3.6×10^{-6} mol dm^{-3}); (c) RuO_2 powder (5.5×10^{-4} mol dm^{-3}); (d) $CoSO_4$ (4.4×10^{-3} mol dm^{-3}).

present in the solution then irradiation resulted in formation of O_2 (fig. 1). With RuO_2 (powder or colloidal) as catalyst and using fixed experimental conditions, the rate of O_2 production increased with increasing concentration of RuO_2 until a limiting value was reached. Above this critical concentration there was no further increase in the rate of O_2 formation (R_{O_2}) upon increasing the concentration of RuO_2 (table 1). A similar saturation point has been observed[30] for the analogous photoreduction of water to H_2 using $bipy_3Ru^{2+}$ as sensitiser and colloidal Pt as catalyst and it reflects the concentration of catalyst at which the pseudo-first-order rate constant for diffusional encounter between $bipy_3Ru^{3+}$ and a RuO_2 particle is no longer rate limiting. Thus, with a high concentration of RuO_2 the rate-limiting step for production of O_2 is the photogeneration of $bipy_3Ru^{3+}$ whilst at low concentrations of catalyst the rate-limiting step is diffusion of $bipy_3Ru^{3+}$ to the surface of a RuO_2 particle.

As shown in table 1, the colloidal RuO_2 catalyst gave higher R_{O_2} values than found with RuO_2 powder. This finding might be expected on the grounds of the effective concentrations of the catalysts used in these experiments. Thus, for both catalysts RuO_2 is present in an aggregated form and, if it is assumed that the aggregates are spherical, the aggregation number (n) is related to the radius (r) of the particle by

$$n = 4\pi r^3 pN/3M \qquad (7)$$

PHOTO-OXIDATION OF H_2O TO O_2

TABLE 1.—YIELD AND RATE OF O_2 FORMATION FROM THE IRRADIATION OF $bipy_3Ru^{2+}$ IN AQUEOUS SOLUTION AT pH 5 CONTAINING $[Co(NH_3)_5Cl]^{2+}$ $(1 \times 10^{-2} \text{ mol dm}^{-3})$

catalyst	catalyst concentration/mol dm^{-3}	$R_{O_2}/10^{-6}$ mol dm^{-3} min^{-1}	$[O_2]/10^{-4}$ mol dm^{-3}
colloidal RuO_2	0.9×10^{-6}	2.85	1.06
	1.8×10^{-6}	5.61	1.23
	3.6×10^{-6}	6.86	1.17
	7.2×10^{-6}	7.08	1.17
RuO_2 powder	1.2×10^{-4}	0.27	
	2.0×10^{-4}	0.38	
	4.4×10^{-4}	0.42	
	6.0×10^{-4}	0.41	
$CoSO_4$	1.7×10^{-3}	1.01	1.10
	2.8×10^{-3}	2.16	1.01
	4.4×10^{-3}	2.57	0.41
	8.8×10^{-3}	1.88	0.14
	1.8×10^{-2}	1.24	0.07
	2.6×10^{-2}	0.77	0.05
	3.5×10^{-2}	0.56	0.03

where p is the density of RuO_2 (6.97) and M is the molecular weight (133.07). For the colloidal RuO_2 catalyst the maximum radius of the particles was *ca.* 50 nm, as determined from filtration studies, whilst the RuO_2 powder was of a very heterogeneous nature and we estimate that the average radius is *ca.* 1000 nm. Thus, from eqn (7), the aggregation numbers for the two catalysts were calculated to be 1.65×10^7 and 1.32×10^{11} for the colloid and powder, respectively. Consequently, for the optimum concentrations of RuO_2 catalyst found from table 1, the effective concentrations of particles

$$[RuO_2]^{particles} = [RuO_2]/n \qquad (8)$$

were 2.2×10^{-13} and 4.2×10^{-15} particle mol^{-1} for the colloid and powder, respectively, so that the colloidal RuO_2 catalyst was present at higher concentration. Now, the rate of diffusional encounter between $bipy_3Ru^{3+}$ and a RuO_2 particle (k_{diff}) can be expressed as

$$k_{diff} = 4\pi NrD/10^3 \qquad (9)$$

where the diffusion coefficient D is given a typical value of 1×10^{-5} cm^2 s^{-1} and has values of 3.8×10^{11} and 7.6×10^{12} particle mol^{-1} s^{-1} for colloidal and powder RuO_2, respectively. Therefore, the pseudo-first-order rate constant (k_{diff} $[RuO_2]^{particles}$) for encounter between $bipy_3Ru^{3+}$ and the effective catalyst was considerably higher for the colloidal catalyst. Presumably, at the optimum catalyst concentrations, the R_{O_2} values are related directly to these pseudo-first-order encounter rate constants.

The total yield of O_2 produced in each experiment, measured after prolonged standing in the dark (*ca.* 1 h), remained constant, within experimental limits, for the colloidal RuO_2 catalyst (table 1). Again, this finding might be expected since the same concentration of $bipy_3Ru^{3+}$ was formed in each experiment but it does infer that practically all of the oxidant produced in the photochemical process can be used for oxidation of water to O_2. The total yield of O_2 was dependent upon pH (the maximum yield was found at pH 5), which can be a problem since the overall reaction leads to a marked decrease in the concentration of protons, and it is essential that the irradiation time is kept to a minimum so as to maintain the original pH throughout the whole experiment:

$$2H_2O + 4[Co(NH_3)_5Cl]^{2+} + 16H^+ \rightarrow O_2 + 4Co^{2+} + 2ONH_4^+ + 4Cl^-. \quad (10)$$

With RuO_2 powder, the total yield of O_2 was irreproducible and all experiments performed with this catalyst are subject to considerable error. The powder adheres to the surface of the vessel and it is difficult to remove and, furthermore, there exists the strong possibility that O_2 adsorbs onto the surface of the powder. Overall, our experiences with RuO_2 powder have led us to the conclusions that quantitative measurements made with such materials are liable to gross irreproducibilities.

As shown in table 1, the presence of cobalt(II) ions also catalysed O_2 evolution although, in this case, the concentration of catalyst influenced both the yield and the rate of O_2 production. Rather surprisingly, it was found that the yield and rate of O_2 formation decreased with increasing concentration of cobalt(II) ions throughout the range from 4.4×10^{-3} to 3.5×10^{-2} mol dm^{-3}. It must be remembered that, according to eqn (10), the overall photoreaction results in formation of $Co(H_2O)_6^{2+}$ so that, even in the absence of added cobalt(II) ions, at the end of the irradiation period there will be $Co(H_2O)_6^{2+}$ present in solution. From the measured quantum yield (at an initial concentration of $[Co(NH_3)_5Cl]^{2+}$ of 1×10^{-2} mol dm^{-3}) the concentration of $Co(H_2O)_6^{2+}$ formed during a 15 min irradiation period corresponds to ca. 3×10^{-5} mol dm^{-3}. Shafirovich et al.,[19] during their detailed investigation of the cobalt(II) catalysed reduction of bipy$_3$Ru^{3+}, observed that the yield of O_2 was dependent upon the concentration of $Co(H_2O)_6^{2+}$ in that the yield of O_2 increased with increasing concentration of $Co(H_2O)_6^{2+}$ until a maximum value was reached ($[Co(H_2O)_6^{2+}] = 1 \times 10^{-4}$ mol dm^{-3}) and above this critical concentration the yield of O_2 began to decrease with increasing concentration of $Co(H_2O)_6^{2+}$. Our studies have been concerned with the higher concentration range and we have confirmed that the yield of O_2 decreases with increasing concentration of $Co(H_2O)_6^{2+}$, as shown in fig. 2. The rate of O_2 formation increases with increased concentration of $Co(H_2O)_6^{2+}$ until a maximum value was reached ($[Co(H_2O)_6^{2+}] =$

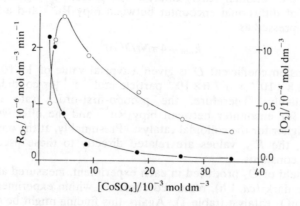

FIG. 2.—Effect of $CoSO_4$ concentration on the rate (○) and yield (●) of O_2 formation at pH 5.

4.4×10^{-3} mol dm^{-3}) and above this limiting concentration R_{O_2} decreases with increasing concentration of catalyst. These findings suggest that the intermediate species responsible for O_2 formation becomes involved in competitive processes at high concentration of $Co(H_2O)_6^{2+}$. Thus, based on electrochemical studies,[31] a plausible scheme for O_2 formation involves the following sequence:

$$Co^{2+} + bipy_3Ru^{3+} \rightarrow Co^{3+} + bipy_3Ru^{2+} \qquad (11)$$

$$Co^{3+} + bipy_3Ru^{3+} \rightarrow Co^{4+} + bipy_3Ru^{2+} \qquad (12)$$

$$Co^{4+} + 2H_2O \rightarrow Co^{2+} + H_2O_2 + 2H^+ \qquad (13)$$

$$H_2O_2 + 2bipy_3Ru^{3+} \rightarrow O_2 + 2H^+ + 2bipy_3Ru^{2+} \qquad (14)$$

$$Co^{3+} \rightarrow products. \qquad (15)$$

In this scheme we have deliberately avoided assigning the actual nature of the cobalt species with respect to the axial ligands although presumably these are water molecules in various stages of acid/base equilibria and the formation of cobaltous (hydrous) oxides is a distinct possibility. In fact, as shown in fig. 3 the rate and yield of O_2 formation from the cobalt(II) catalysed oxidation of water was dependent upon pH, the maxima being at pH 4.75 (as found for colloidal RuO_2) although Shafirovich[19] found the maximum yield to be at pH 7 for this process. Similarly, the course of reaction (15) remains unknown although cobalt(III) ions are believed to dimerise in aqueous solution.[32]

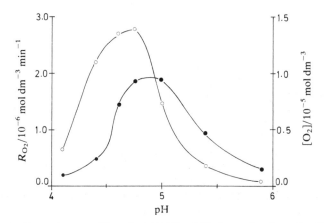

FIG. 3.—Effect of pH on the rate (○) and yield (●) of O_2 formation from the $CoSO_4$ catalysed oxidation of water by $bipy_3Ru^{3+}$ ($[Co(H_2O)_6^{2+}] = 4.4 \times 10^{-3}$ mol dm^{-3})

In general, the above scheme agrees well with the observed concentration dependence for the $Co(H_2O)_6^{2+}$ catalysed oxidation of water by $bipy_3Ru^{3+}$. At low concentrations of $Co(H_2O)_6^{2+}$, the ratio of $[bipy_3Ru^{3+}]/[Co^{3+}]$ will be high so that reaction (12) should be favoured over reaction (15) whilst at high concentrations of $Co(H_2O)_6^{2+}$ the ratio of $[bipy_3Ru^{3+}]/[Co^{3+}]$ will be low and reaction (15) becomes more favourable. Thus, with this system it is essential that the ratio of $[bipy_3Ru^{3+}]/[Co^{3+}]$ be kept high.

Quantum yields for formation of O_2 (Φ_{O_2}) from the $bipy_3Ru^{2+}/[Co(NH_3)_5Cl]^{2+}$ system at pH 5 were determined for each of the above catalysts using the optimum concentration of catalyst in each case. The observed values are collected in table

2 and it can be seen that colloidal RuO_2 and $Co(H_2O)_6^{2+}$ ions are quite effective catalysts, the final concentrations of evolved O_2 obtained from the two systems being comparable, although the rate of O_2 formation was considerably higher for colloidal RuO_2. With these catalysts, the ratio of quantum yields for formation of $bipy_3Ru^{3+}$ ($\Phi = 0.31$) and O_2 exceeds 10, rather than being 4 as expected from eqn (3), showing that, under these experimental conditions, there are important side reactions that lead to reduction of $bipy_3Ru^{3+}$ without formation of O_2 [i.e. reaction (15)] and the overall quantum efficiency for formation of O_2 is ca. 12% (where the concentration of $[Co(NH_3)_5Cl]^{2+}$ is 1×10^{-2} mol dm^{-3}). Table 2 suggests that colloidal RuO_2 and $Co(H_2O)_6^{2+}$ ions are more efficient catalysts than RuO_2 powder but this may be due to adsorption of O_2 onto the surface of the powder (TiO_2 may also adsorb some O_2). In this respect, it should be noted that previous workers have experienced problems with commercial samples of RuO_2 powder and the efficiency of these powders can vary enormously from batch to batch.

TABLE 2.—QUANTUM YIELDS FOR FORMATION OF O_2 FROM THE IRRADIATION OF $bipy_3Ru^{2+}$ IN AQUEOUS SOLUTION AT pH 5 CONTAINING $[Co(NH_3)_5Cl]^{2+}$ $(1 \times 10^{-2}$ mol dm$^{-3})$

catalyst	catalyst concentration/mol dm^{-3}	Φ_{O_2}
colloidal RuO_2	3.6×10^{-6}	0.030
RuO_2 powder	5.5×10^{-4}	0.003
$CoSO_4$	1.7×10^{-3}	0.020

The above work has been concerned with O_2 production from sacrificial systems and, whilst such systems are very useful for optimising the reaction conditions, they have little application for construction of practical solar energy storage devices. Instead, it is necessary that the sacrificial electron acceptor is replaced with a reversible redox couple. This introduces severe kinetic problems since in most cases where a reversible couple is used in a photoinduced electron-transfer process the rate of reverse electron transfer is comparable with that of the forward reaction. However, we have reported[26] previously that, whilst Fe^{3+} ions quenched the triplet excited state of $bipy_3Ru^{2+}$ at the diffusion-controlled rate limit, the reverse reaction between $bipy_3Ru^{3+}$ and Fe^{2+} ions was relatively slow:

$$*bipy_3Ru^{2+} + Fe^{3+} \xrightarrow{k_F} bipy_3Ru^{3+} + Fe^{2+} \tag{16}$$

$$k_F = 3 \times 10^9 \text{ dm}^3 \text{ mol}^{-1} \text{ s}^{-1} \quad (\text{pH 1})$$

$$bipy_3Ru^{3+} + Fe^{2+} \xrightarrow{k_R} bipy_3Ru^{2+} + Fe^{3+} \tag{17}$$

$$k_R = 1 \times 10^4 \text{ dm}^3 \text{ mol}^{-1} \text{ s}^{-1} \quad (\text{pH 1}).$$

Unfortunately, k_R was found to be highly dependent upon pH and increased sharply with increasing pH due to the acid-dependent self-exchange reaction for the $Fe^{3+/2+}$ redox couple.[26] For O_2 formation from the $bipy_3Ru^{3+/2+}$ couple,[16,17] the optimum pH seems to be ca. 5 and at this pH luminescence quenching measurements showed that k_F was 5.5×10^8 dm^3 mol^{-1} s^{-1} whilst, from flash-photolysis studies, k_R was

PHOTO-OXIDATION OF H_2O TO O_2

found to be $3.4 \times 10^8 \, dm^3 \, mol^{-1} \, s^{-1}$. Thus, at pH 5 the reverse electron-transfer step ensures that the lifetime of bipy$_3$Ru^{3+} will be very low relative to comparable sacrificial systems.

Steady-state irradiation of bipy$_3$Ru^{2+} in aqueous solution at pH 5 containing Fe^{3+} ($4 \times 10^{-4} \, mol \, dm^{-3}$) gave no detectable yield of O_2. Similarly, no O_2 was observed when the irradiation was carried out in the presence of Co(H$_2$O)$_6^{2+}$ ($4.4 \times 10^{-3} \, mol \, dm^{-3}$), RuO$_2$ powder ($5.5 \times 10^{-4} \, mol \, dm^{-3}$) or colloidal RuO$_2$ ($3.6 \times 10^{-6} \, mol \, dm^{-3}$). Even at pH 1, where k_R is much slower than that at pH 5, O_2 could not be observed as a reaction product. (In separate experiments, it was shown that the catalysts were effective for O_2 formation at pH 1 for the sacrificial systems.) The absence of O_2 from these systems (we estimate that $\Phi_{O_2} < 10^{-4}$) suggests either that the catalysts cannot intercept reverse electron transfer due to unfavourable kinetics or that the catalysts serve to enhance reverse electron transfer (*i.e.* the system is short circuited).

As described above, the pseudo-first-order rate constant for diffusional encounter between bipy$_3$Ru^{3+} and a catalyst particle depends upon the radius of the particle. Using our estimates for the radii of the RuO$_2$ catalysts we have calculated that, under the conditions used for the above experiments with iron(III) as reversible redox couple, k_{diff}[RuO$_2$]^{particles} was approximately 3.2×10^{-2} and $8.4 \times 10^{-2} \, s^{-1}$ for RuO$_2$ powder and colloid, respectively. Under the conditions of the flash-photolysis experiments, the half-life of bipy$_3$Ru^{3+} in the absence of an added catalyst was *ca.* 12 ms so that we would expect only a very small fraction of the bipy$_3$Ru^{3+} molecules to decay *via* encounter with a catalyst surface.

The above calculation cannot be made for Co(H$_2$O)$_6^{2+}$ since we do not know the identity nor concentration of the active catalyst. However, since this system is completely homogeneous it can be studied by flash-photolysis techniques and a set of experiments were performed to investigate the effect of cobalt(II) ions on the rate of return to ground-state bipy$_3$Ru^{2+} following flash excitation of bipy$_3$Ru^{2+} in the presence of iron(III) ($4 \times 10^{-4} \, mol \, dm^{-3}$). As shown in table 3, the bimolecular rate constant for reverse electron transfer between bipy$_3$Ru^{3+} and Fe^{2+} (k_R), at fixed ionic strength, increased with increasing concentration of Co(H$_2$O)$_6^{2+}$,

TABLE 3.—BIMOLECULAR RATE CONSTANTS FOR REVERSE ELECTRON TRANSFER BETWEEN bipy$_3$Ru^{3+} AND Fe^{2+} IN AQUEOUS SOLUTION AT pH 5

[Co(H$_2$O)$_6^{2+}$]/mol dm^{-3}	k_R/10^8 dm^3 mol^{-1} s^{-1}
—	3.4
3.5×10^{-3}	4.9
1.4×10^{-2}	6.8

within the very narrow range studied. This finding suggests that cobalt(II) ions perturb the self-exchange rate constant for the Fe$^{3+/2+}$ couple, in much the same manner as do protons.[26] It is possible that RuO$_2$ also catalyses the Fe$^{3+/2+}$ self-exchange rate so that iron(III) does not appear to be a suitable reversible redox couple for non-sacrificial O_2-evolving systems.

The work described above has confirmed earlier findings that bipy$_3$Ru^{2+} functions as a photosensitiser for O_2 evolution from sacrificial systems when RuO$_2$ or Co(H$_2$O)$_6^{2+}$ are incorporated as redox catalysts. However, the Co(H$_2$O)$_6^{2+}$ catalyst has not been well characterised and neither the identity of the active catalyst species nor the reason for the observed concentration dependence have been explained, as yet. Colloidal RuO_2, supported on TiO_2 particles, is a particularly effective catalyst for O_2 evolution from water but the large particle size imposes severe problems when diffusion to the catalyst surface has to compete with reverse electron transfer. Thus, there is an urgent need to produce colloidal RuO_2 particles having a very small diameter and work is now in progress aimed at producing such particles.

We thank the S.R.C., the E.E.C. and G.E. (Schenectady) for financial support. We also thank Dr. A. Mackor (T.N.O., Utrecht) for providing a sample of colloidal RuO_2 and A. Mills for many helpful discussions.

[1] B. V. Koryakin, T. S. Dzhabiev and A. E. Shilov, *Dokl. Akad. Nauk SSSR*, 1977, **238**, 620.
[2] J. M. Lehn and J. P. Sauvage, *Nouv. J. Chim.*, 1977, **1**, 449.
[3] K. Kalyanasundaram, J. Kiwi and M. Grätzel, *Helv. Chim. Acta*, 1978, **61**, 2720.
[4] A. Moradpour, E. Amouyal, P. Keller and H. Kagan, *Nouv. J. Chim.*, 1978, **2**, 547.
[5] A. I. Krasna, *Photochem. Photobiol.*, 1979, **29**, 267.
[6] M. Kirsch, J. M. Lehn and J. P. Sauvage, *Helv. Chim. Acta*, 1979, **62**, 1345.
[7] A. Harriman, G. Porter and M. C. Richoux, *J. Chem. Soc., Faraday Trans. 2*, 1981, **77**, 833.
[8] J. R. Darwent, P. Douglas, A. Harriman, G. Porter and M. C. Richoux, *Coord. Chem. Rev.*, submitted for publication.
[9] H. Gerischer and J. Gobrecht, *Ber. Bunsenges. Phys. Chem.*, 1978, **82**, 520.
[10] E. Borgarello, J. Kiwi, E. Pelizzetti, M. Visca and M. Grätzel, *Nature (London)*, 1981, **289**, 158.
[11] J. M. Lehn, J. P. Sauvage and R. Ziessel, *Nouv. J. Chim.*, 1980, **4**, 623.
[12] A. Harriman and J. Barber, *Topics in Photosynthesis*, ed. J. Barber (Elsevier, Amsterdam, 1979), vol. 3, chap. 8.
[13] R. G. Brown, A. Harriman and G. Porter, *J. Chem. Soc., Faraday Trans. 2*, 1977, **73**, 103.
[14] I. A. Duncan, A. Harriman and G. Porter, *J. Chem. Soc., Faraday Trans. 2*, 1978, **74**, 1920.
[15] A. Harriman and G. Porter, *J. Chem. Soc., Faraday Trans. 2*, 1979, **75**, 1532; 1979, **75**, 1543; 1980, **76**, 1415; 1980, **76**, 1429.
[16] J. Kiwi and M. Grätzel, *Angew. Chem. Int. Ed. Engl.*, 1978, **17**, 860; 1979, **18**, 624; *Chimia*, 1979, **33**, 289.
[17] K. Kalyanasundaram, O. Micic, E. Promauro and M. Grätzel, *Helv. Chim. Acta*, 1979, **62**, 2432.
[18] J. M. Lehn, J. P. Sauvage and R. Ziessel, *Nouv. J. Chim.*, 1979, **3**, 423.
[19] V. Ya. Shafirovich, N. K. Khannov and V. V. Strelets, *Nouv. J. Chim.*, 1980, **4**, 81.
[20] M. Grätzel, *Ber. Bunsenges. Phys. Chem.*, 1980, **84**, 981.
[21] D. P. Rillema, W. J. Dressick and T. J. Meyer, *J. Chem. Soc., Chem. Commun.*, 1980, 247.
[22] M. Neumann-Spallart, K. Kalyanasundaram, C. Grätzel and M. Grätzel, *Helv. Chim. Acta*, 1980, **63**, 1111.
[23] K. Kalyanasundaram and M. Grätzel, *Angew. Chem. Int. Ed. Engl.*, 1979, **18**, 701.
[24] W. A. Hynes, L. K. Yanowski and M. Schiller, *J. Am. Chem. Soc.*, 1938, **60**, 3053.
[25] A. Mills, A. Harriman and G. Porter, *Anal. Chem.*, 1981, **53**, 1254.
[26] M. I. C. Ferreira and A. Harriman, *J. Chem. Soc., Faraday Trans. 2*, 1979, **75**, 874.
[27] C. Creutz and N. Sutin, *Proc. Natl. Acad. Sci. USA*, 1975, **72**, 2858.
[28] F. Boletta, A. Juris, M. Maestri and D. Sandrini, *Inorg. Chim. Acta*, 1980, **44**, L175.
[29] D. A. House, *Chem. Rev.*, 1962, **62**, 185.
[30] A. Harriman and A. Mills, *J. Chem. Soc., Faraday Trans. 2*, 1981, **77**, 1939.
[31] V. Ya. Shafirovich and V. V. Strelets, *Nouv. J. Chim.*, 1978, **2**, 199.
[32] M. Anbar and I. Precht, *J. Am. Chem. Soc.*, 1967, **89**, 2553.

Further publications

IN VITRO PHOTOSYNTHESIS.
G. Porter and Mary D. Archer, *Interdisciplinary Science Reviews*, 1976, **1** (No. 2), 119.

IN VITRO MODELS FOR PHOTOSYNTHESIS
(THE BAKERIAN LECTURE, 1977)
G. Porter, *Proc. Roy. Soc.*, 1978, **A362**, 281.

MODEL SYSTEMS FOR PHOTOSYNTHESIS. III. PRIMARY
PHOTO-PROCESSES OF CHLOROPLAST PIGMENTS IN
MONOMOLECULAR ARRAYS ON SOLID SURFACES.
S. M. de B. Costa, J. R. Froines, J. M. Harris, R. M. Leblanc, B. H. Orger
and G. Porter, *Proc. Roy. Soc.*, 1972, **A326**, 503.

SOLVATE AND DIMER EQUILIBRIA IN SOLUTIONS OF CHLOROPHYLL.
R. L. Amster and G. Porter, *Proc. Roy. Soc.*, 1966, **A296**, 38.

MODEL SYSTEMS FOR PHOTOSYNTHESIS. IV. PHOTOSENSITIZATION
BY CHLOROPHYLL a MONOLAYERS AT A LIPID/WATER INTERFACE.
S. M. de B. Costa and G. Porter, *Proc. Roy. Soc.*, 1974, **A341**, 167.

DECAY OF HIGH-VALENT MANGANESE PORPHYRINS IN
AQUEOUS SOLUTION AND CATALYSED FORMATION OF OXYGEN.
A. Harriman, P. A. Christensen, G. Porter, K. Moorhouse, P. Neta and M.-C. Richoux,
J. Chem. Soc., Faraday Trans. I, 1986, **82**, 3215.

PORPHYRINS IN AQUEOUS SOLUTION.
N. Carnieri, A. Harriman and G. Porter, *J.C.S. Dalton Trans.*, 1982, 931.

REACTIONS OF PHOTOEXCITED CHLOROPHYLL-a WITH
MANGANESE COMPLEXES IN SOLUTION.
R. G. Brown, A. Harriman and G. Porter, *J.C.S. Faraday Trans. II*, 1977, **73**.

PHOTOREDOX PROCESSES IN METALLOPORPHYRIN-CROWN
ETHER SYSTEMS.
G. Bondeel, A. Harriman, G. Porter and A. Wilowska,
J. Chem. Soc., Faraday Trans. II, 1984, **80**, 867.

PHOTOREDOX PROPERTIES OF ZINC PORPHYRIN/VIOLOGEN COMPLEXES.
A. Harriman, G. Porter and A. Wilowska, *J.C.S. Faraday Trans. II*, 1984, **80**, 191.

LUMINESCENCE OF PORPHYRINS AND METALLOPORPHYRINS.
Part 11. — ENERGY TRANSFER IN ZINC-METAL-FREE PORPHYRIN DIMERS.
R. L. Brookfield, H. Ellul, A. Harriman and G. Porter,
J. Chem. Soc., Faraday Trans. II, 1986, **82**, 219.

PHOTOREDUCTION OF METHYL VIOLOGEN SENSITIZED BY
SULPHONATED PHTHALOCYANINES IN MICELLAR SOLUTIONS.
J. R. Darwent, I. McCubbin and G. Porter, *J.C.S. Faraday Trans. 2*, 1982, 903.

DESIGN, PREPARATION, AND CHARACTERIZATION OF
RuO_2/TiO_2 COLLOIDAL CATALYTIC SURFACES ACTIVE IN
PHOTOOXIDATION OF WATER.
G. Blondeel, A. Harriman, G. Porter, D. Unwin and J. Kiwi, *J. Phys. Chem.*, 1983, **87**, 2629.

DETECTION OF SMALL QUANTITIES OF PHOTOCHEMICALLY
PRODUCED OXYGEN BY REACTION WITH ALKALINE PYROGALLOL.
I. A. Duncan, A. Harriman and G. Porter, *Anal. Chem.*, 1979, **51**, No. 13, 2206.

MODEL SYSTEMS FOR PHOTOSYNTHESIS. VII.
CHLOROPHYLL-a PHOTOSENSITIZED REDUCTION OF METHYL
VIOLOGEN BY HYDRO QUINONES.
J. R. Darwent, K. Kalyanasundaram and G. Porter, *Proc. R. Soc.*, 1980, **A373**, 179.

CONCENTRATION QUENCHING IN CHLOROPHYLL.
G. S. Beddard and G. Porter, *Nature*, 1976, **260**, 366.

PHOTOCHEMICAL HYDROGEN PRODUCTION USING
CADMIUM SULPHIDE SUSPENSIONS IN AERATED WATER.
J. R. Darwent and G. Porter, *Chem. Comm.*, 1981, 145.

PHOTOSENSITISED REDUCTION OF WATER TO HYDROGEN
USING WATER-SOLUBLE ZINC PORPHYRINS.
A. Harriman, G. Porter and M.-C. Richoux, *J.C.S. Faraday*, 1981, **77**, 833.

PHOTOREDUCTION OF METHYL VIOLOGEN SENSITISED BY
THE EXCITED SINGLET STATE OF A MAGNESIUM PHTHALOCYANINE.
A. Harriman, G. Porter and M.-C. Richoux, *J.C.S. Faraday Trans. II*, 1981, **77**, 1187.

MEMBRANE POLAROGRAPHIC DETECTORS FOR DETERMINATION OF
HYDROGEN AND OXYGEN PRODUCED BY THE PHOTODISSOCIATION
OF WATER.
A. Mills, A. Harriman and G. Porter, *Anal. Chem.*, 1981, **53**, 1254.

ATTEMPTED PHOTO-OXIDATION OF WATER TO OXYGEN
USING ZINC (II) PORPHYRINS.
A. Harriman, G. Porter and P. Walters, *J. Photochem.*, 1982, **19**, 183.

METAL PHTHALOCYANINES AND PORPHYRINS AS PHOTOSENSITIZERS
FOR REDUCTION OF WATER TO HYDROGEN.
J. R. Darwent, P. Douglas, A. Harriman, G. Porter and M.-C. Richoux,
Coord. Chem. Revs., 1982, **44**, 83.

VIOLOGEN/PLATINUM SYSTEMS FOR HYDROGEN GENERATION.
A. Harriman and G. Porter, *J.C.S. Faraday Trans. 2*, 1982, **78**, 1937.

COLLOIDAL PLATINUM CATALYSTS FOR REVERSIBLE
PHOTOREDOX PROCESSES.
A. Harriman, G. Porter and M. C. Richoux, *J.C.S. Faraday Trans. 2*, 1982, **78**, 1955.

A THEORETICAL INVESTIGATION OF THE DYNAMICS OF
ENERGY TRAPPING IN A TWO-DIMENSIONAL MODEL
OF THE PHOTOSYNTHETIC UNIT.
J. A. Altmann, G. S. Beddard and G. Porter, *Chem. Phys. Letters*, 1978, **58**, 54.

MODEL SYSTEMS FOR PHOTOSYNTHESIS. VI. CHLOROPHYLL-a
SENSITIZED REDUCTION OF METHYL VIOLOGEN IN NON-IONIC MICELLES.
K. Kalyanasunaram and G. Porter, *Proc. Roy. Soc.*, 1978, **A364**, 29.

QUENCHING OF CHLOROPHYLL FLUORESCENCE BY NITROBENZENE.
G. S. Beddard, Sheena Carlin, L. Harris, G. Porter and C. J. Tredwell,
Photochem. & Photobiol., 1978, **27**, 433.

THIONINE SENSITISED PHOTOCHEMISTRY OF MANGANESE GLUCONATE
I. A. Duncan, A. Harriman and G. Porter, *J.C.S. Faraday Trans. II*, 1978, **74**, 1920.

REVERSIBLE PHOTO-OXIDATION OF ZINC TETRAPHENYLPORPHINE
BY BENZO-1, 4-QUINONE.
A. Harriman, G. Porter and N. Searle, *J.C.S. Faraday Trans. II*, 1979, **75**, 1515.

PHOTOCHEMISTRY OF 5-METHYLPHENZINIUM SALTS IN
AQUEOUS SOLUTION. 2. OPTICAL FLASH PHOTOLYSIS AND
FLUORESCENCE RESULTS AND A PROPOSED MECHANISM.
V. S. F. Chew, J. R. Bolton, R. G. Brown and G. Porter, *J. Phys. Chem.*, 1980, **84**, 1909.

TIME-RESOLVED FLUORESCENCE FROM BIOLOGICAL SYSTEMS:
TRYPTOPHAN AND SIMPLE PEPTIDES.
G. S. Beddard, G. R. Fleming, G. Porter and R. J. Robbins,
Phil. Trans. R. Soc., 1980, **A298**, 321.

PHOTOSENSITISED DISSOCIATION OF WATER USING
DISPERSED SUSPENSION OF n-TYPE SEMICONDUCTORS.
A. Mills and G. Porter, *J.C.S. Faraday Trans. 1*, 1982, **78**, 3659.

A PULSE-RADIOLYTIC AND PHOTOCHEMICAL STUDY OF THE
OXIDATION OF WATER BY ZINC PORPHYRIN p-RADICAL CATIONS.
P. A. Christensen, A. Harriman, G. Porter and P. Neta,
J.C.S. Faraday Trans. 2, 1984, **80**, 1451.

PHOTOCHEMISTRY OF MANGANESE PORPHYRINS.

PART I. CHARACTERISATION OF SOME WATER SOLUBLE COMPLEXES.
A. Harriman and G. Porter, *J.C.S. Faraday Trans. II*, 1979, **75**, 1532.
PART II. PHOTOREDUCTION.
A. Harriman and G. Porter, *J.C.S. Faraday Trans. II*, 1979, **75**, 1543.
PART III. INTER-CONVERSION OF Mn^{II}/Mn^{III}.
I. A. Duncan, A. Harriman and G. Porter, *J.C.S. Faraday II*, 1980, **76**, 1415.
PART IV. PHOTO-SENSITISED REDUCTION OF QUINONES.
A. Harriman and G. Porter, *J.C.S. Faraday II*, 1980, **76**, 1429.
PART VI. OXIDATION-REDUCTION OF MANGANESE (III) PORPHYRINS.
N. Carnieri, A. Harriman and G. Porter, *J.C.S. Dalton Trans.*, 1982, 931.
PART VII. CHARACTERISATION OF MANGANESE PORPHYRINS IN
ORGANIC AND AQUEOUS/ORGANIC MICROHETEROGENEOUS SYSTEMS.
N. Carnieri, A. Harriman and G. Porter, *J.C.S. Dalton Trans.*, 1982, 1231.

Chapter 10

PHOTOSYNTHESIS *IN VIVO*

Photosynthesis is by far the most important photochemical, and even chemical, process. It is the source of our food and nearly all our fuels, the driving force behind evolution and life itself. It uses mechanisms of energy storage and energy transduction that are more sophisticated than anything that the chemist has achieved in the laboratory. For these reasons alone it engages the interest of research scientists of many disciplines.

Photosynthesis is also the base of the largest of all industrial processes — agriculture. The production of energy from biomass is of vital importance in the developing countries but is very inefficient and uncompetitive with other sources of energy in developed countries. Rapid advances in genetic engineering and breeding of plants indicate that it may be possible greatly to improve the efficiency of photosynthesis, and this provides a second incentive to research into the basic mechanisms of plant phosynthesis.

The work described below is concerned with the first steps that follow absorption of light by the photosynthetic unit. These steps are very fast indeed and it has only been possible to time-resolve the processes after the development of flash photoysis and other fast reaction techniques.

Early measurements in the picosecond region used flashes of high energy with the result that the kinetics were complicated by interactions between two excited states or excitons, leading to anomalously short observed lifetimes of fluorescence and absorption. Fluorescence, where the base line is zero intensity, could sometimes be measured at low enough intensity to tolerate these distortions (see for example the streak-camera measurements of the picosecond kinetics of energy transfer in phycobilisomes) but in absorption, where small relative intensity changes have to be measured, single flash experiments were quite inadequate. To use low intensity flashes but to retain the precision of measurements, high repetition rates and integration of the results from many flashes, by single photon counting (fluorescence) or multichannel detectors (absorption) was necessary, and this is routine practice today.

Reprinted from

Biochimica et Biophysica Acta, 501 (1978) 232—245
© Elsevier/North-Holland Biomedical Press

BBA 47446

PICOSECOND TIME-RESOLVED ENERGY TRANSFER IN *PORPHYRIDIUM CRUENTUM*

PART I. IN THE INTACT ALGA

G. PORTER [a], C.J. TREDWELL [a], G.F.W. SEARLE [b] and J. BARBER [b]

[a] *Davy Faraday Research Laboratory of the Royal Institution, 21 Albemarle Street, London W1X 4BS, and* [b] *Department of Botany, Imperial College, London S.W.7. (U.K.)*

(Received June 28th, 1977)

Summary

The wavelength-resolved fluorescence emission kinetics of the accessory pigments and chlorophyll *a* in *Porphyridium cruentum* have been studied by picosecond laser spectroscopy. Direct excitation of the pigment B-phycoerythrin with a 530 nm, 6 ps pulse produced fluorescence emission from all of the pigments as a result of energy transfer between the pigments to the reaction centre of Photosystem II. The emission from B-phycoerythrin at 576 nm follows a nonexponential decay law with a mean fluorescence lifetime of 70 ps, whereas the fluorescence from R-phycocyanin (640 nm), allophycocyanin (660 nm) and chlorophyll *a* (685 nm) all appeared to follow an exponential decay law with lifetimes of 90 ps, 118 ps and 175 ps respectively. Upon closure of the Photosystem II reaction centres with 3-(3,4-dichlorophenyl)-1,1-dimethylurea and preillumination the chlorophyll *a* decay became non-exponential, having a long component with an apparent lifetime of 840 ps. The fluorescence from the latter three pigments all showed finite risetimes to the maximum emission intensity of 12 ps for R-phycocyanin, 24 ps for allophycocyanin and 50 ps for chlorophyll *a*.

A kinetic analysis of these results indicates that energy transfer between the pigments is at least 99% efficient and is governed by an exp $-At^{1/2}$ transfer function. The apparent exponential behaviour of the fluorescence decay functions of the latter three pigments is shown to be a direct result of the energy transfer kinetics, as are the observed risetimes in the fluorescence emissions.

Abbreviation: DCMU, 3-(3,4-dichlorophenyl)-1,1-dimethyl urea.

Introduction

The red alga, *Porphyridium cruentum*, is a unicellular member of the lower rhodophyceae. It possesses water-soluble accessory light harvesting pigments that are contained within structures known as phycobilisomes attached to the thylakoid membrane [1]. Phycobilisomes contain three main pigments, namely B-phycoerythrin, R-phycocyanin and allophycocyanin; the presence of B-phycoerythrin [2], and allophycocyanin B [3] or an aggregated form of allophycocyanin [4] is still uncertain (see Part II). Each of these pigments fluoresce in a well-defined spectral region, and the emissions may be wavelength-resolved without difficulty [5]. Chlorophyll *a* in *P cruentum* is contained within the thylakoid membrane, as are the carotenoids, the most important of which is β-carotene [6].

Early steady-state fluorescence studies of energy transfer in *P. cruentum* [7] indicated that the phycobilisomes preferentially serve Photosystem II and subsequently the energy transfer sequence has been proposed as:

B-phycoerythrin → R-phycocyanin → allophycocyanin → chlorophyll *a*

(an excellent review on the phycobilins can be found in ref. 5). Probably as a result of this highly efficient light harvesting system, Photosystem II has less chlorophyll *a* associated with it than Photosystem I [8]. The fluorescence lifetimes of chlorophyll *a* and the accessory pigments in *P. cruentum* were investigated some time ago by Rabinowitch and co-authors, who used a nanosecond flashlamp technique which entailed deconvolution of the fluorescence decay from the flashlamp profile [9,10,11]. B-phycoerythrin was found to transfer energy to R-phycocyanin with a transfer time of 300 ± 200 ps, and the transfer times of R-phycocyanin to chlorophyll *a* via allophycocyanin was reported to be 500 ± 200 ps [10]. Considering the efficiency of energy transfer [7] and the square root of time dependence of the transfer kinetics [12,13], it seems probable that the transfer times are much shorter than those reported. Similarly, the kinetic analysis presented [10] needs to be modified to include the non-exponential behaviour of the transfer kinetics. Previous studies of the chlorophyll *a* fluorescence kinetics have used excitation with blue light to reduce the interference from the overlapping phycobilin emission [14], and chlorophyll lifetimes close to 0.5 ns have been reported for the dark adapted state [11,15]. Mar et al. [16] have measured a chlorophyll fluorescence lifetime of approximately 1 ns for *P. cruentum* with the Photosystem II reaction centres fully closed, from which they deduced that the lifetime for the dark adapted state should be in the region of 350 ps.

In this communication we report our investigation of energy transfer between the accessory pigments and chlorophyll *a* in *P. cruentum* using picosecond time-resolved fluorescence spectroscopy.

Materials and Methods

Cultures of *P. cruentum* were grown at room temperature (20—50°C) under continuous illumination from an incandescent lamp (1 mW/cm²) supplemented with daylight. The culture medium, a sterile artificial sea water medium [17], was agitated by bubbling with filtered air. Cells were normally harvested 10—12 days after innoculation by centrifugation at $3000 \times g$ for 2 min. When young cells were required, the culture was innoculated with the minimum amount of a previous culture and the cells were harvested after 2—3 days growth when the cell density was still low. Gantt and Lipschultz [18] have shown that these young cells have a higher ratio of chlorophyll a : phycobilins in comparison with 5—12 day old cultures. The harvested cells were resuspended in growth medium to give a transmission of approximately 50% at 530 nm in a 1 mm cuvette. Measurements were performed with dark adapted *P. cruentum*, except in the case of chlorophyl a where the emission was also studied with the Photosystem II reaction centres fully closed by the addition of 10^{-4} M DCMU and continuous irradiation with 633 nm light (1.25 mW/cm^2) from a CW helium: neon laser. Young cells were used only when expressly stated.

The picosecond laser and streak camera system have been described in detail elsewhere [12,13,19]. A train of 6 ps (full width at half maximum height), 530 nm light pulses is generated by a frequency-doubled, mode-locked neodymium: glass laser oscillator. A single pulse is extracted from the centre of the train by a Pockels cell electro-optic shutter. The intensity of the 530 nm excitation pulse was controlled by a non-saturable neutral optical density filter situated before the sample. An area of 0.28 cm^2 of the sample was irradiated which limited the maximum unattenuated intensity to 10^{15} photons/cm^2; in general the intensity was kept below 10^{14} photons/cm^2 for these measurements. Fluorescence from the sample was passed through a wavelength selection filter and focussed onto the slit of an S20 photocathode Imacon 600 streak camera (John Hadland (P.I.) Ltd.). A vidicon optical multichannel analyser (OMA 1205 A and B, Princeton Applied Research) stored the resulting streak trace in digital form which could then be displayed on an oscilloscope or transferred to punch tape for analysis. Streak speeds were varied between 120 ps/50 OMA channels (120 ps per major division on the oscilloscope traces) and 540 ps/50 OMA channels depending upon the duration of the fluorescence decay. The linearity of the detection system is better than ±3% between 30 and 3000 counts in any channel of the OMA memory. Fluorescence emission components from *P. cruentum* were resolved with the following filters:

(i) B-phycoerythrin — 576 nm, 9 nm bandwidth Balzer B-40 Filtraflex interference filter.

(ii) R-phycocyanin — 640 nm, 13 nm bandwidth Balzer B-40 Filtraflex interference filter.

(iii) allophycocyanin — 661 nm, 14 nm bandwidth MTO Intervex A interference filter.

(iv) chlorophyll a — 685 nm, 11 nm bandwidth Balzer B-40 Filtraflex interference filter.

Fluorescence emission and excitation spectra were recorded on a Perkin-Elmer MPF-3 spectrofluorimeter, and were not corrected for the spectral response of the photomultiplier (Hamamatsu R446S) or the monochromator, or for the emission spectrum of the xenon arc lamp.

Results

Spectral characteristics

The absorption spectra of the *P cruentum* cultures grown for these experiments (see Fig. 1) agreed with previously published spectra of this alga. Steady-state fluorescence emission and excitation spectra of the intact algal cells, under the conditions used in the kinetic measurements, are shown in Fig. 2. The 10—12 day old cultures gave a relatively high intensity B-phycoerythrin emission band at 578 nm (Fig. 2A) upon excitation at 530 nm. Only a small change was observed in the chlorophyll *a* emission at 685 nm when 3-(3,4-dichlorophenyl)-1,1-dimethyl urea (DCMU) was added to these older cultures, as shown in Fig. 2A. In contrast, the younger cultures (2—3 days old) gave a much larger change in the emission intensity at 685 nm on DCMU addition (Fig. 2B). The emission intensities of B-phycoerythrin (578 nm), R-phycocyanin (640 nm) and allophycocyanin (660 nm) from the young cells also showed slight differences from those obtained from the older cells. It is apparent that excitation of B-phycoerythrin with 530 nm light results in energy transfer to the other pigments, as these do not have a significant absorption at this wavelength. The fluorescence from these pigments can be wavelength-resolved with suitable interference filters, although a certain amount of overlap is unavoidable. Fluorescence emission spectra of the isolated phycobiliproteins are well documented [2] and the reported fluorescence maxima are close to those observed in the spectra of these intact cells.

Excitation spectra of *P. cruentum* at various monitoring wavelengths are shown in Fig. 2C-F; at each wavelength the B-phycoerythrin band is predominant and the large amount of chlorophyll *a* observed in the absorption spectrum does not contribute significantly to the fluorescence emission in agreement with previous studies [7]. However, a weak chlorophyll *a* contribution

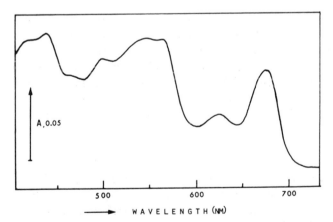

A, 0.05

500   600   700

WAVELENGTH (NM)

Fig. 1. The absorption spectrum of *Porphyridium cruentum*. A dilute suspension of 10—12 day old cells, pathlength 10 mm, was measured in an Aminco-Chance DW2 spectrophotometer with opal glass to reduce scattering artefacts. The peak at 678 nm is due to chlorophyll *a* only, that at 625 nm is due both to chlorophyll *a* and R-phycocyanin. B-phycoerythrin contributes the intense twin peak at 565—550 nm, and carotenoid absorption bands are apparent at 500 and 470 nm. The major band in the blue at 436 nm is the Soret band of chlorophyll *a*.

236

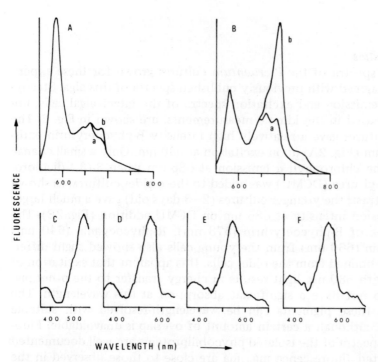

Fig. 2. The fluorescence emission and excitation spectra of *P. cruentum* measured on a Perkin Elmer MPF 3 fluorescence spectrophotometer using wavelengths employed in fluorescence kinetics experiments. In (A) and (B) the emission spectrum on excitation at 530 nm is shown for normal and young cultures respectively both without addition (a), and also on addition of 10^{-4} M DCMU (b). In (C)—(F) the excitation spectra for emission at 576 nm (C), 661 nm (D) and 730 nm (E and F) are shown. The excitation spectrum for 640 nm emission was similar to (D). Chlorophyll emission was monitored in young cells at 730 nm rather than 685 nm, both with out addition (E), and on addition of 10^{-4} M DCMU (F). No corrections were made for photomultiplier response, monochromator sensitivity or xenon arc lamp emission. Fluorescence intensity is expressed on a linear scale.

was observed for emission at 730 nm as indicated by the small peak at 436 nm which corresponds to the chlorophyll Soret band (Fig. 2E and 2F). When DCMU was added, the chlorophyll *a* and carotenoid contribution to the 730 nm fluorescence (light absorbed in the region of 350—450 nm) increased. Since these pigments are primarily associated with Photosystem I [20,21], this may indicate an increased contribution from Photosystem I to the emission at 730 nm. Minor peaks observed at 620 nm and 650 nm can be ascribed to R-phycocyanin and allophycocyanin respectively.

Fluorescence kinetics

Fluorescence decay curves for the three accessory pigments, B-phycoerythrin, R-phycocyanin and allophycocyanin, are shown in Fig. 3; all of these traces were obtained by single pulse excitation at 530 nm with an intensity of less than 10^{14} photons/cm^2. The fluorescence kinetics of B-phycoerythrin were found to follow an exp $-At^{1/2}$ decay law, similar to that reported for the fluorescence decay of chlorophyll *a* in *Chlorella* and spinach chloroplasts [12,13]. However, the emissions from R-phycocyanin and allophycocyanin both appeared to be governed by an exp $-kt$ decay law. The measured fluorescence

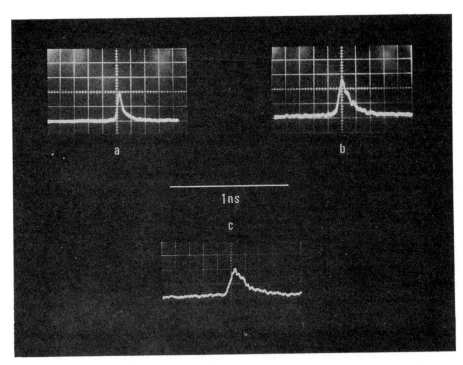

Fig. 3. The wavelength-resolved fluorescence kinetics of phycobiliproteins in *P. cruentum* measured on the picosecond laser apparatus. The monitoring wavelengths were a, 576 nm for B-phycoeryhthrin; b, 640 nm for R-phycocyanin; and c, 661 nm for allophycocyanin. An upward deflection represents an increase in fluorescence intensity on a linear scale. For further details see text.

lifetimes were 70 ps (mean lifetime) for B-phycoerythrin, 90 ps for R-phyco-cyanin and 118 ps for allophycocyanin. Since exciton-exciton annihilation could cause severe distortion of the decay traces [12,13], the measurements were repeated with a cut-off filter that transmitted above 600 nm (RG 600,

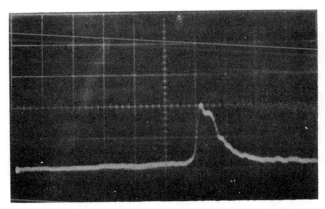

Fig. 4. The fluorescence kinetics of the photosynthetic pigments of *P. cruentum* at a low laser pulse intensity. A red cut-off wavelength selection filter was used (50%T at 600 nm), and the photon density was 10^{13} photon/cm^2. An upward deflection represents an increase in fluorescence intensity on a linear scale. Each major division on the x-axis represents 188 ps.

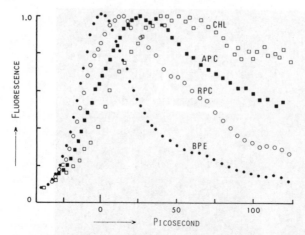

Fig. 5. The risetimes of wavelength-resolved pigment fluorescence in *P. cruentum*. Each curve was measured separately with interference filters as wavelength selection filters, and aligned on the time axis as described in the text. BPE is B-phycoerythrin, RPC is R-phycocyanin, APC is allophycocyanin and CHL is chlorophyll *a*. Fluorescence intensity is on a normalised linear scale.

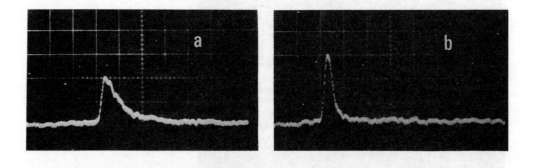

Fig. 6. The fluorescence kinetics of chlorophyll *a* in dark-adapted young cells of *P. cruentum*. In (a) on a short timescale (188 ps per major division), and on a longer timescale in (b) (543 ps per major division) to show the absence of a long component. A Balzer B40-685 nm interference filter was used as wavelength selection filter, and an upward deflection represents an increase in fluorescence.

Schott) which enabled the excitation intensity to be reduced to 10^{13} photons/ cm^2 (see Fig. 4). Although the fluorescence components were not wavelength-resolved, there was no indication that the decay rates were slower than those observed at 10^{14} photons/cm^2. The reproducible decay curve shown in Fig. 4 indicated that at least one of the fluorescence components had a significant risetime. The initial part of the wavelength-resolved decay of each of the pigments is shown in Fig. 5; the risetime of the B-phycoerythrin emission was consistent with the time resolution of the detection system and the width of the excitation pulse, whereas those of R-phycocyanin and allophycocyanin were significantly longer. After matching the initial rise of the fluorescence emissions with that of B-phycoerythrin, the times taken to reach maximum emission intensity, relative to the maximum of B-phycoerythrin, were approximately 12 ps and 24 ps for R-phycocyanin and allophycocyanin respectively.

The fluorescence emission from chlorophyll a in the 10—12 day old cultures of $P. cruentum$ was relatively weak and was obscured by the strong allophyco-cyanin emission. Consequently, young cells (2—3 days old) were used to study the kinetics of the chlorophyll a emission. The chlorophyll fluorescence decay traces obtained from these cultures are shown in Figs. 6 and 7; Fig. 6 shows the decay from the dark adapted alga and Fig. 7 the decay after treatment with DCMU and preillumination. The fluorescence decay kinetics of the 685 nm component (predominantly chlorophyll a) appeared to be adequately described by an exp $-kt$ decay law with a lifetime of 175 ps. A risetime to maximum emission of 50 ps was measured for chlorophyll a and is shown with the other pigments in Fig. 5. Upon the addition of DCMU and preillumination the decay curves became nonexponential as shown in Fig. 7. The initial phase of this decay had a 1/e lifetime of 110 ps, and the longer component had a lifetime of 840 ps.

A summary of the observed fluorescence decay rates lifetimes and risetimes is given in Table I.

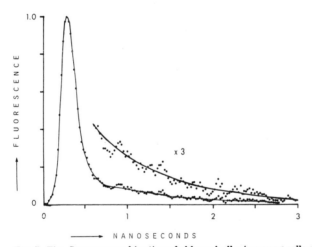

Fig. 7. The fluorescence kinetics of chlorophyll a in young cells of $P. cruentum$ with Photosystem II reaction centres closed by addition of DCMU and preilluminated as described in the text. Other conditions as for Fig. 6.

240

TABLE I

THE FLUORESECENCE CHARACTERISTICS OF THE PIGMENTS OF *PORPHYRIDIUM CRUENTUM*

a, mean lifetime; b, calculated rate constants for fluorescence decay by direct excitation; c, apparent rate constants caused by energy transfer; *, calculated values obtained from the kinetic equations using the A values (see text); (i), dark adapted algae; (ii), algae treated with DCMU and preillumination; A and k are fluorescence decay constants; $\tau_{1/e}$ is the 1/e fluorescence lifetime; ϕ_{calc} is the quantum of fluorescence calculated from the measured fluorescence decay constants; τ_{rise} is the risetime of fluorescence.

Pigment	Emission wavelength (nm)	$\tau_{1/2}$ (ps)	A $(ps^{-1/2})$	k (ps^{-1})	τ_{rise} (ps)	$\tau^{*}_{1/e}$ (ps)	τ^{*}_{rise} (ps)	ϕ_{calc}
B-phycoerythrin	578	70 ± 5 [a]	0.26	—	0	—	0	0.0036
R-phycocyanin	640	90 ± 10	0.48 [b]	0.0110 [c]	12	85 ± 5	12	0.0029
Allophycocyanin	660	118 ± 8	0.52 [b]	0.0085 [c]	24	115 ± 8	22	0.0018
Chlorophyll a (i)	685	175 ± 10	0.40 [b]	0.0057 [c]	50	176 ± 8	52	0.0007
Chlorophyll a (ii)	685	110 ± 5	—	0.0091	—	—	—	
	685	840 ± 10	—	0.0012	—	—	—	0.0021

Discussion

We have previously shown that energy transfer in the chlorophyll a antenna system of Photosystem II results in a fluorescence decay law of the form [12, 13,22];

$$I(t) = I_0 \exp - 2At^{1/2} \qquad (1)$$

where I_0 is the initial fluorescence intensity and $I(t)$ is the fluorescence intensity at time t. The rate constant, $2A$, is a combination of the rate constant controlling donor:donor energy transfer and that controlling donor:acceptor energy transfer. Since B-phycoerythrin also has a fluorescence decay law of this form, it may not be unreasonable to assume that energy transfer among the other light harvesting pigments of *P. cruentum* is kinetically similar. As yet it is not possible to draw any conclusions concerning the validity of the Förster equation for energy transfer [23] in these systems. However, the general technique described by Birks [24] may be used to extend the kinetics implied by Eqn. 1 to a multiple donor:acceptor system.

If we consider a four component energy transfer system, where the pigments J, K, L and M transfer their energy in the following sequence:

$$J \rightarrow K \rightarrow L \rightarrow M \rightarrow X$$

where X is a non-fluorescent quencher such as the reaction centre, then direct excitation of any pigment will produce fluorescence emission with a decay law given by Eqn. 1; it is assumed that fluorescence emission and intersystem crossing only make a very minor contribution to the depletion of the excited state population. For direct excitation the rate constants for the decay of the excited state population of J, K, L and M are A_1, A_2, A_3 and A_4 respectively. Excitation of pigment J with a light pulse described by the function $P(t)$ will produce an excited state population J^* whose decay is described by the rate equation:

$$\frac{dJ^*}{dt} = P(t) - J^* \cdot A_1 \cdot t^{-1/2}$$

If $P(t)$ is a δ-pulse, then this function may be neglected for $t > 0$. Since the energy lost by J is gained by pigment K, the rate equation for the excited state population of K (i.e. K^*) is:

$$\frac{dK^*}{dt} = J^* \cdot A_1 \cdot t^{-1/2} - K^* \cdot A_2 \cdot t^{-1/2}$$

Substituting for J^* using eqn. 1, the equation becomes:

$$\frac{dK^*}{dt} + K^* \cdot A_2 \cdot t^{-1/2} = J_0^* \cdot A_1 \cdot t^{-1/2} \cdot \exp - 2A_1 t^{1/2}$$

where J_0^* is the excited state population of J at time $t = 0$. Solving the differential equation the time course of K^* is given by:

$$K^*(t) = J_0^* \cdot A_1 \cdot \exp(2A_2 t^{1/2}) \cdot \int_0^t t^{-1/2} \cdot \exp(2A_2 - 2A_1)t^{1/2} \, dt$$

which gives

$$K^*(t) = \frac{J_0^* \cdot A_1}{(A_2 - A_1)} \left[\exp(-2A_1 t^{1/2}) - \exp(-2A_2 t^{1/2})\right] \tag{2}$$

Similarly the rate equation for the excited state population of L is:

$$\frac{dL^*}{dt} = K^* \cdot A_2 \cdot t^{-1/2} - L^* \cdot A_3 \cdot t^{-1/2}$$

which, after substitution for K^* from eqn. 2 becomes:

$$\frac{dL^*}{dt} + L^* \cdot A_3 \cdot t^{-1/2} = \frac{J_0^* \cdot A_1 \cdot A_2}{(A_2 - A_1)} \cdot t^{-1/2} \left[\exp(-2A_1 t^{1/2}) - \exp(-2A_2 t^{1/2})\right]$$

Again solving the differential equation the function $L^*(t)$ has the form:

$$L^*(t) = \frac{J_0^* \cdot A_1 \cdot A_2}{(A_2 - A_1)} \left[\frac{\exp(-2A_1 t^{1/2}) - \exp(-2A_3 t^{1/2})}{(A_3 - A_1)} \right.$$

$$\left. - \frac{\exp(-2A_2 t^{1/2}) + \exp(-2A_3 t^{1/2})}{(A_3 - A_2)} \right] \tag{3}$$

and by the same method the function $M^*(t)$ is:

$$M^*(t) = \frac{J_0^* \cdot A_1 \cdot A_2 \cdot A_3}{(A_2 - A_1)} \left[\frac{\exp(-2A_1 t^{1/2}) - \exp(-2A_4 t^{1/2})}{(A_4 - A_1)(A_3 - A_1)} \right.$$

$$- \frac{\exp(-2A_4 t^{1/2}) + \exp(-2A_3 t^{1/2})}{(A_3 - A_4)(A_3 - A_1)}$$

$$\left. - \frac{\exp(-2A_4 t^{1/2}) + \exp(-2A_2 t^{1/2})}{(A_2 - A_4)(A_3 - A_2)} + \frac{\exp(-2A_4 t^{1/2}) - \exp(-2A_3 t^{1/2})}{(A_3 - A_4)(A_3 - A_2)} \right] \tag{4}$$

242

The final form of the Eqns. 2—4 are similar to those derived by Tomita and Rabinowitch [10], except that these are all dependent upon the square root of time. It should be noted that these equations are probably not valid for long time intervals after excitation since the normal fluorescence decay of the pigment will eventually become significant. In fact, the streak camera's sensitivity to incident light precludes measurements in this region and the approximation should describe the observable initial fluorescence decay. Since A_1 is given by the fluorescence decay rate of B-phycoerythrin, the values of A_2, A_3 and A_4 may be evaluated sequentially so that each gives the experimentally recorded fluorescence decay rate and risetime when inserted into the relevant equation. The rate constants that gave the best fit to the experimental data were as follows, $2A_1 = 0.26$ ps$^{-1/2}$, $2A_2 = 0.48$ ps$^{-1/2}$, $2A_3 = 0.52$ ps$^{-1/2}$ and $2A_4 = 0.40$ ps$^{-1/2}$. Apart from Eqn. 1, all of the equations gave decay curves which were very close to a normal exponential decay of the form exp $-kt$; the slight deviation from exponentiality would not have been discernable on the experimental decay curves. In order to demonstrate the agreement between the calculated and experimental data, the calculated curves were convoluted with the resolution function of the streak camera. This function is a Gaussian curve with a full width, at half maximum height, of 28 ps on this time scale and was determined from the apparent profile of a 6 ps laser pulse recorded at the same streak speed. Fig. 8 shows the convoluted theoretical decay curves of the four pigments obtained from the numerical evaluation of the expression:

$$F(t) = \int\limits_{0}^{\infty} f(\propto) \cdot R(t - \propto) \mathrm{d}\propto$$

where $F(t)$ is the convoluted fluorescence function, $f(\propto)$ is the true fluorescence function, $R(t)$ is the camera resolution function and $\propto$ is the time spread

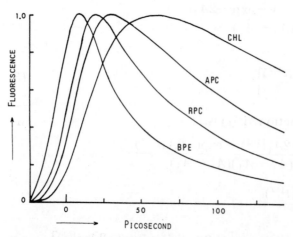

Fig. 8. Theoretical curves for the rise and decay of fluorescence of the pigments of *P. cruentum* on direct excitation of B-phycoerythrin and transfer of excitation to the other pigments according to the kinetic treatment derived in the text. Other conditions as for Fig. 5.

of the resolution function. This procedure only affects the fluorescence function of B-phycoerythrin to any significant extent, but does distort the initial rise in the fluorescence of the remaining curves. However, the time intervals between the maxima of the curves were not altered significantly. A summary of the calculated rate constants, apparent fluorescence lifetimes and risetimes is given in Table I for comparison. Within the limits of experimental error the calculated curves are in reasonable agreement with those observed experimentally.

The fluorescence quantum yield of each of the pigments when subjected to direct excitation, may therefore be calculated from the expression:

$$\phi_{\text{calc}} = \frac{\phi_0}{\tau_0} \int\limits_0^\infty \exp(-2At^{1/2})\mathrm{d}t = \frac{2\phi_0}{\tau_0 \cdot (2A)^2}$$

where ϕ_0 and τ_0 are the fluorescence quantum yield and lifetime respectively of the pigment in vitro. Brody and Rabinowitch [9] have reported the fluorescence lifetimes of B-phycoerythrin and R-phycocyanin in vitro to be 7.1 ns (ϕ_0 = 0.85) and 1.8 ns (ϕ_0 = 0.53) respectively; the lifetime of chlorophyll a is 5.7 ns (ϕ_0 = 0.33) [12]; the radiative lifetime of allophycocyanin was estimated to be 4.2 ns from the results obtained in Part II of this work. The values of ϕ_{calc} obtained from these results and the calculated decay rates for each pigment are summarised in Table I. Since the calculated fluorescence quantum yields are all less than 0.01, the assumption that fluorescence emission only makes a minor contribution to the depletion of the excited state population appears to be valid over a major part of the excited state decay profile. The calculated quantum yields also indicate that energy transfer between any two pigments in the sequence is approximately 99% efficient, in contrast to the transfer efficiencies reported by Tomita and Rabinowitch [10], (phycoerythrin:phycocyanin 96 ± 3%, phycocyanin:allophycocyanin:chlorophyll a 78 ± 8%).

The pigment b-phycoerythrin [2] has been omitted from the discussion since its spectral characteristics are so similar to those of B-phycoerythrin that the two pigments may be considered to operate in parallel. Hence the rate constant A_1 includes the rate constants for both pigments acting in parallel. The existence of allophycocyanin B [3] or an aggregated form of allophycocyanin [4] in intact *P. cruentum* remains to be confirmed. However, this pigment would probably have a very high transfer efficiency to chlorophyll a, and would not affect the observed kinetics to any great extent. Reabsorption of the fluorescence emission by the pigments should not result in any distortion of the fluorescence decay kinetics since excitation by this process can only produce less than 1% of the total excited state population. Similarly, direct energy transfer from B-phycoerythrin to allophycocyanin or chlorophyll a and from R-phycocyanin to chlorophyll a are minor processes owing to the very small overlap between the emission and absorption spectra of these pigments [2].

The absence of exciton-exciton annihilation at the intensities used in these measurements is probably a direct result of the transfer kinetics and the geometry of the phycobilisome. Gantt et al. [4] suggested that the allophycocyanin is in contact with the thylakoid membrane and surrounded by a layer of R-phycocyanin, which in turn is surrounded by a layer of B-phycoerythrin. In this structured sphere energy migration between layers is more probable than migration within the layer. For an initial random distribution of excitons in the B-phycoerythrin layer, the probability of exciton-exciton interactions is much lower than for a comparable situation in a chlorophyll a-containing antenna system. The relative magnitudes of the rate constants for energy transfer also preclude the formation of a large exciton population in R-phycocyanin, allophycocyanin and chlorophyll a in the dark-adapted state. Clearly exciton-exciton annihilation will only be observed at very high excitation intensities ($>10^{15}$ photons/cm^2) or when energy transfer between the pigments is prevented, as in the case of isolated phycobilisomes described in part II.

The energy transfer rate constant found for the chlorophyll a of Photosystem II in the dark adapted state, 0.40 ps$^{-1/2}$, is much larger than the value of 0.047 ps$^{-1/2}$ observed in both *Chlorella* [12] and spinach subchloroplast fragments [13]. This could be explained by the smaller amount of chlorophyll associated with Photosystem II in *P. cruentum*, which would reduce the migration distance to an active reaction centre and consequently increase the trapping rate. Ley and Butler [8] have reported that the quantum yield of energy transfer from Photosystem II to I is high in dark-adapted *P. cruentum*, which might also cause a faster Photosystem II fluorescence decay compared to that in *Chlorella*.

Closure of the Photosystem II reaction centres with DCMU and preillumination resulted in complex decay kinetics for chlorophyll fluorescence (Fig. 7). It is possible that the shorter component is due to energy transfer from Photosystem II to I, for it has been reported that the quantum yield of this process increases to 0.90—0.95 with the Photosystem II reaction centres closed [8]. Thus the initial component ($k = 0.0091$ ps^{-1}, $\tau_{1/e} = 110$ ps) might reflect this Photosystem II to I energy transfer, whereas the second component ($k = 0.0012$ ps^{-1}, $\tau_{1/e} = 840$ ps) would represent the chlorophyll fluorescence lifetime of Photosystem II with reaction centres closed. Alternatively it is possible that exciton-exciton annihilation might become more important under these conditions owing to the small size of the Photosystem II chlorophyll antenna system in *P. cruentum* and the longer lifetime of excitons on addition of DCMU and preillumination. However, when the laser pulse intensity was varied, although the relative intensities of the two components changed slightly the decay rate of the initial component altered only slightly and not to the appreciable extent expected from previous observations of exciton-exciton annihilation [12,13,22]. Therefore the contribution of exciton-exciton annihilation to the observed decay law would not appear to be a major factor at these excitation intensities (10^{14} photons/cm^2).

A study of the kinetics of energy transfer within the isolated phycobilisome is reported in the accompanying paper (part II).

Acknowledgements

This work was supported by the EEC Solar Energy Research and Development Programme, the Science Research Council, and by the award of a Ministry of Defence Post-doctoral Fellowship to C.J.T. We are grateful to Elizabeth Dibb for technical assistance and to John Ferreira for discussions on energy transfer.

References

1 Gantt, E. and Conti, S.F. (1966) J. Cell Biol. 29, 423—434
2 Gantt, E. and Lipschultz, C.A. (1974) Biochemistry 13, 2960—2966
3 Glazer, A.N. and Bryant, D.A. (1975) Arch. Microbiol. 104, 15—22
4 Gantt, E., Lipschultz, C.A. and Zilinskas, B.A. (1977) Brookhaven Symp. Biol. 28, 347—357
5 Glazer, A.N. (1976) in Photochem. Photobiol. Revs. (Smith, K.C., ed.), Vol. 1, pp. 71—115, Plenum Press, New York
6 Goodwin, T.W. (1965) in Chemistry and Biochemistry of Plant Pigments (Goodwin, T.W., ed.), pp. 127—142, Academic Press, New York
7 Duysens, L.N.M. (1952) Transfer of Excitation Energy in Photosynthesis, Ph.D. thesis, University of Utrecht, The Netherlands
8 Ley, A.C. and Butler, W.L. (1976) Proc. Natl. Acad. Sci. U.S. 73, 3957—3960
9 Brody, S.S. and Rabinowitch, E. (1957) Science 125, 555
10 Tomita, G. and Rabinowitch, E. (1962) Biophys. J. 2, 483—499
11 Singhal, G.S. and Rabinowitch, E. (1969) Biophys. J. 9, 586—591
12 Porter, G., Synowiec, J.A. and Tredwell, C.J. (1977) Biochim. Biophys. Acta 459, 329—336
13 Searle, G.F.W., Barber, J., Harris, L., Porter, G. and Tredwell, C.J. (1977) Biochim. Biophys. Acta 459, 390—401
14 French, C.S. and Young, V.K. (1952) J. Gen. Physiology 35, 873—890
15 Nicholson, W.J. and Fortoul, J.I. (1967) Biochim. Biophys. Acta 143, 577—582
16 Mar, T., Govindjee, Singhal, G.S. and Merkelo, H. (1972) Biophys. J. 12, 797—808
17 Jones, R.F., Speer, H.L. and Kury, W. (1963) Physiol. Plant. 16, 636—645
18 Gantt, E. and Lipschultz, C.A. (1972), J. Cell Biol. 54, 313—324
19 Archer, M.D., Ferreira, M.I.C., Porter, G. and Tredwell, C.J. (1977) Nouveau J. Chim. 1, 9—12
20 Goedheer, J.C. (1969) Biochim. Biophys. Acta 172, 252—265
21 Butler, W.L. (1977) Brookhaven Symp. Biol. 28, 338—346
22 Harris, L., Porter, G., Synowiec, J.A., Tredwell, C.J. and Barber, J. (1976) Biochim. Biophys. Acta 449, 329—339
23 Förster, Th. (1949) Z. Naturforschg 4a, 321—327
24 Birks, J.B. (1968) J. Phys. B. (Proc. Phys. Soc.), Ser 2, 1, 946—957

Biochimica et Biophysica Acta, 545 (1979) 165—174
© Elsevier/North-Holland Biomedical Press

BBA 47592

THE FLUORESCENCE DECAY KINETICS OF IN VIVO CHLOROPHYLL MEASURED USING LOW INTENSITY EXCITATION

G.S. BEDDARD [a], G.R. FLEMING [a], G. PORTER [a], G.F.W. SEARLE [b] and J.A. SYNOWIEC [a]

[a] *The Davy Faraday Research Laboratory of the Royal Institution, 21 Albemarle Street, London W1X 4BS and* [b] *Department of Botany, Imperial College, London SW7 2BB (U.K.)*

(Received May 5th, 1978)

Key words: Chlorophyll; Fluorescence decay kinetics; Photosystem I

Summary

We report fluorescence lifetimes for in vivo chlorophyll *a* using a time-correlated single-photon counting technique with tunable dye laser excitation. The fluorescence decay of dark-adapted chlorella is almost exponential with a lifetime of 490 ps, which is independent of excitation from 570 nm to 640 nm.

Chloroplasts show a two-component decay of 410 ps and approximately 1.4 ns, the proportion of long component depending upon the fluorescence state of the chloroplasts. The fluorescence lifetime of Photosystem I was determined to be 110 ps from measurements on fragments enriched in Photosystem I prepared from chloroplasts with digitonin.

Introduction

An accurate determination of the kinetic law governing the excited state decay of in vivo chlorophyll is of fundamental importance to the understanding of the excitation energy transfer process in photosynthesis [1]. Conventional single photon counting techniques have been used, but these lacked sufficient temporal resolution for accurate in vivo lifetime determinations [2—4]. The advent of mode-locked lasers and streak cameras renewed interest in these measurements [5]. However, it soon became apparent that the high power of the laser pulses could give rise to anomalies as a result of exciton annihilation [6,7]. Once these effects were recognised and the laser pulse intensities controlled, it was possible to obtain results which correlated well with those predicted by steady-state fluorescence yield measurements [8,9]. Although there

---

Abbreviations: PS I, Photosystem I; PS II, Photosystem II; DCMU, 3-(3,4-dichlorophenyl)-1,1-dimethylurea.

is general agreement between ourselves and other investigators on the gross values of the fluorescence lifetimes under various conditions [5], there is still much controversy over the details of the form of the decay kinetics, i.e. whether it is a single or a sum of exponentials or a time-dependent function. In this paper we describe measurements of the fluorescence decay of in vivo chlorophyll a by a single-photon counting technique using low power excitation from a tunable dye laser pumped by an argon ion laser.

Experimental

Picosecond-tunable pulses were obtained from a Rhodamine 6G dye laser synchronously pumped by a mode-locked argon ion laser (CR 12 Coherent Radiation Ltd.). The dye laser output pulses were determined to be less than 10 ps full width at half maximum by a zero background second harmonic generation auto-correlation technique over the wavelength range of 580—640 nm. For the photon counting measurements the pulse repetition rate was reduced from 75 MHz to 33 kHz using a Pockels cell between crossed polarisers. A contrast ratio of better than 500 : 1 between the transmitted and rejected pulses was achieved. The subsequent laser output was divided along two paths by a beamsplitter. One was attenuated and used to excite the sample while the other was incident upon a Texas Instruments TI XL 56 silicon avalanche photo-diode which provided the start signal for the time-to-amplitude converter. Fluorescence emitted at right angles to the excitation beam was detected through appropriate filters by a Mullard 56 TUVP photomultiplier tube. Temporal linearity of the photomultiplier tube was obtained by reducing the light-sensitive area of the photocathode to 3 mm diameter. Time calibrations were carried out by monitoring the excitation pulses through suitable optical delays. The laser power at the sample cell was measured using an Alphametrics photometer. Experiments were performed with incident laser intensities within the range 10^9—10^{11} photons/cm^2 per pulse.

The green alga *Chlorella pyrenoidosa* was cultured as described previously [10]. Pea (*Pisum cativum*) chloroplasts were isolated with the outer envelope intact and hypotonically shocked immediately before additions were made and the fluorescence measured [11]. Details of the media are given in the text. Sample suspensions were flowed at a rate of 1 l/min for dark-adapted samples through a 1 cm pathlength cell from a reservoir and had a concentration of approx. 5—8 μg chlorophyll/ml or $A_{680nm} = 0.3$—0.5, as measured using an integrating sphere. Photosystem II reaction centres of chloroplasts were closed by addition of 3-(3,4-dichlorophenyl)-1,1-dimethylurea (DCMU) to a final concentration of 10 μM and pre-illumination with 633 nm light from a 0.5 mW CW HeNe laser; the sample flow rate of the chloroplast suspension was also decreased. A purified Photosystem I preparation was obtained from pea chloroplasts by isolation of a stroma lamellae vesicle fraction using 0.2% digitonin, as previously described [12]. Stroma lamellae vesicle samples were not flowed. All measurements were carried out at room temperature, and the fluorescence emission was observed at wavelengths greater than 665 nm using a Schott RG 665 filter.

Results

The data were analysed over 3 orders of magnitude of decay with single and two exponential decay characterisitics by an iterative convolution technique using a gradient expansion algorithm. As a check on the instrument's performance the dye molecule Rose Bengal was measured using 580 nm excitation and the same emission filters as used with the photosynthetic systems. The fluorescence had an exponetial decay over three decades decrease in fluorescence intensity of 597 ps in methanol and of 122 ps in water, which compares well with previous measurements of 543 ps [13], 655 ps [14], 118 ps [13] and 95 ps [14], respectively. Table I summarises the results obtained for chlorophyll in vivo.

Chlorella. Dark-adapted *Chlorella* analysed with the assumption of a single exponential decay gave reasonably good fits, as judged by a chi-square criterion but the calculated best fit data revealed small systematic deviations from the actual data which indicated that the decay was probably non-exponential. This may be seen in Fig. 1, where the fluorescence decay is close to, but not quite exponential over a 1000-fold decrease in intensity. The fit to the data could be considerably improved using two exponential terms, although the lifetimes varied slightly between different experiments. The two lifetimes obtained for dark-adapted *Chlorella* were found to be in the ranges 270—350 ps and 530—650 ps with the long component accounting for between 38 and 27% of the initial intensity.

No effect upon the lifetimes was discerned when the excitation wavelength was varied within the range 580—640 nm. Similarly, variation of the incident laser intensity from 10^9 to 10^{11} photons/cm^2 per pulse did not affect the fluo-

TABLE I

CHARACTERISTICS OF THE FLUORESCENCE DECAY OF IN VIVO CHLOROPHYLL *a*

The fluorescence yields (ϕ_{calc}) were calculated from the expression

$$\phi_{calc} = \frac{1}{\tau_0 I_0} \int_0^\infty I(t)dt$$

where τ_0 is the natural lifetime (19.5 ns)[30] of in vitro chlorophyll *a*. The mean lifetimes (τ_{mean}) were calculated from the expression $\tau_{mean} = (\alpha_1\tau_1 + \alpha_2\tau_2)/(\alpha_1 + \alpha_2)$ where α_1 and α_2 are the ratios of the lifetimes of τ_1 and τ_2, respectively, and $(\alpha_1 + \alpha_2) = 1$

Sample	τ_1 (ps) *	τ_2 (ps) *	% $\frac{\tau_2}{\tau_1}$	ϕ_{calc}	τ_{mean} (ps)
Chlorella					
Dark adapted	492	—	—	0.025	—
Chloroplasts					
Dark adapted	413	1463	3.8	0.023	453
Light + DCMU	453	1328	9.9	0.028	540
Light + DCMU + Mg^{2+}	462	1342	36.6	0.040	784
Stroma lamellae vesicle fraction	113	1192	3—9	0.0058	—

* Estimated error, 5%.

168

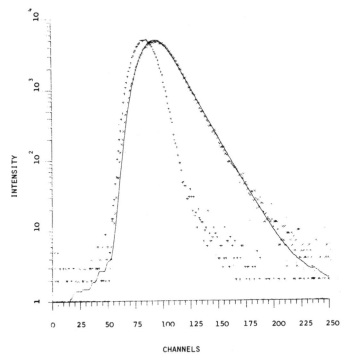

CHANNELS

Fig. 1. Time-resolved fluorescence emission of dark-adapted chlorella. Time scale = 32.9 ps/channel.

rescence decay. This gives us confidence that no excitation annihilation processes were occurring.

Chloroplasts. The fluorescence decay for dark-adapted chloroplasts was fitted by a double exponential consisting of a major short component (413 ps) and a long component (1463 ps) comprising about 3.8% of the initial intensity (Table I). In low salt buffer upon pre-illumination and the addition of DCMU, the proportion of the long component increased to about 10%; the lifetime of the short component increased slightly. The addition of 5 mM Mg^{2+} caused a further increase in the proportion of the long component to about 37% of the total. The lifetime of the longer component under the three different conditions is the same within experimental error. The decay kinetics of dark-adapted chloroplasts are presented in Fig. 2.

Stroma lamellae vesicle fraction. Fig. 3 shows the time-resolved room temperature emission of the stroma lamellae vesicle Photosystem I (PS I) fraction. The fluorescence decay was again fitted by a double exponential, the major short component (113 ps) being attributed to the PS I emission. The proportion of long component (approximately 1.2 ns) varied for different preparations of the stroma lamellae vesicle fractions, being less the smaller the amount of Photosystem II (PS II) left in the preparation, determined from the 77 K emission spectrum. The fit to the tail of the emission curve is not the best possible since the complete decay curve was fitted by a double exponential, whereas it should have been fitted to a triple exponential decay (which unfortunately is not possible on our convolution programme). The shortest life-

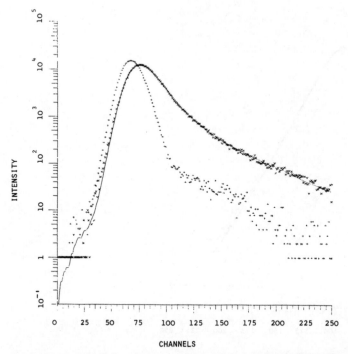

Fig. 2. Time-resolved fluorescence emission of dark-adapted chloroplasts. Time scale = 32.9 ps/channel. For the lifetime measurements the chloroplasts were diluted in water and then double strength low salt buffer was added. (Low salt buffer: 0.33 M Sorbitol/10 mM N-2-hydroxyethylpiperazine-N'-2-ethane-sulphonic acid, adjusted to pH 7.6 with tris(hydroxymethyl)aminomethane).

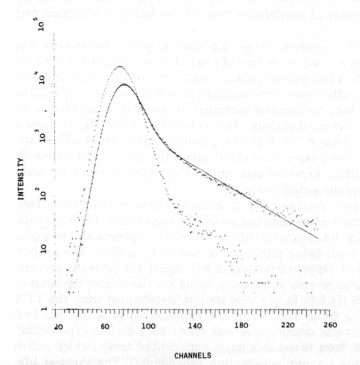

Fig. 3. Time-resolved fluorescence emission of the stroma lamellae vesicle fraction. Time scale = 31.9 ps/channel. The stroma lamellae vesicle fraction was resuspended in 50 mM Tris-HCl (pH 7.8)/2% (w/v) NaCl.

time component would then be due to PS I and the two residual longer components due to PS II. (see Discussion)

Discussion

Non-exponential decays of in vivo chlorophyll fluorescence. If the observed fluorescence decay is non-exponential, then its precise form contains information not accessible in a quantum yield determination. We wish to distinguish clearly between the different causes of non-exponentiality and suggest what kind of information may be obtained in the different cases.

Deviations from exponentiality can arise from two distinct causes: (a) an intrinsic time dependence in the emission probability. This form of non-exponentiality can give information on the mechanism of the energy transfer processes, (b) an inhomogeneity in the emitting species. Inhomogeneity may arise from trivial causes such as differences between individual chloroplasts or cells or may reflect different quenching probabilities in different physical regions of the same chloroplast or photosynthetic unit, and thus contain structural information. Additionally, non-exponential decays can be experimentally induced, for example from high intensity effects such as exciton-exciton annihilation [8,9] or stimulated emission [15], or from reabsorption [16]. These effects may also contain structural and dynamic information if properly understood [17].

The physical basis for an intrinsic time dependence for the quenching of excited chlorophylls is that some of the initially excited molecules may be very close to a trap and the excitation will reach the trap much more quickly than from initially excited chlorophylls far from the trap. This 'transient term' may be reasonably expected to affect only the initial portion of the fluorescence decay; at later times the probability of quenching will be time independent. This expectation is confirmed by recent calculations of Altmann et al. [18]. A transient term of the form predicted for this effect $[I(t) \alpha \exp - (at + bt^{0.5})]$ was not observed in the present experiments. Our data were always much better fitted by a single exponential (chlorella) or double exponential (chloroplasts) than by an equation of the above form. Such a transient term was reported with a Nd : glass/streak camera system at excitation intensities of $10^{13}-10^{14}$ photons/cm^2 per pulse [9,12]. At present, the time resolution of our apparatus is insufficient to definitely resolve this difference and we are working to improve our time resolution capability.

Non-exponential decays arising from structural heterogeneity may have several causes. (1) Contributions to the total emission from PS I as well as PS II. Although steady-state measurements of fluorescence have indicated that the 685 nm and 730 nm emissions be assigned to PS II and PS I respectively [19], attempts to separate spectrally the time-resolved emission have not been satisfactory. Sub-chloroplast particles enriched in PS I have been shown to emit mainly at 685 nm at room temperature [12]. Thus, spectrally resolved fluorescence decays may still contain contributions from both fluorescing photosystems. (2) Contributions from excitations near closed or open traps, i.e on the state of the primary donors and acceptors. Since the fluorescence yield of PS II is dependent upon the state of its reaction centre [20], further inhomogeneity may be introduced depending upon the ratio of open to closed traps which have different fluorescence lifetimes. (3) Heterogeneity in the light collection protein pigment complexes. The time scale of the experiment, i.e. the fluorescence decay time, is important in these cases. If there are a number of distinct sites (e.g. open and closed traps) and the excitation is able to visit them all during its lifetime, then the decay will be exponential; the sample will be homogeneous on this time scale.

Although it should be possible to distinguish the various cases, precise fluorescence decay curves are required, free, as far as possible, from any experimentally induced non-exponentiality. We have avoided annihilation or stimulated emission processes by using low incident light levels, typically 10^9 photons/cm^2 per pulse which is at least four orders of magnitude less than the threshold for exciton annihilation processes [6]. Quenching by species built up by previous pulses was avoided by flowing the sample and by reducing the pulse repetition rate to 33 kHz. Normally, the contrast ratio of our electro-optic modulator was better than 500 : 1; this could be reduced to 200 : 1 without any discernible effect on the fluorescence decays. Finally, dilute samples were used to avoid reabsorption of fluorescence.

Chlorella and chloroplasts. Most of the early measurements of in vivo chlorophyll fluorescence decays utilising high power mode-locked solid-state lasers and optical shutter or streak camera detection gave anomalous decay times ($\lesssim$200 ps) for both *Chlorella* and chloroplasts, with various degrees of non-exponentiality in the decays [5]. The high incident laser intensities produced multiple excitations which then produced annihilation effects within the photosynthetic unit. When moderate (approx. 10^{13} photons/cm^2 per pulse) single-pulse excitation intensities were employed, lifetimes in the region of 450—650 ps [8,9] were observed, which compare well with the range of 350—800 ps obtained by phase fluorimetric and standard photon-counting methods, (refs. 21—24 and also references cited in 25) and with values predicted from steady-state fluorescence yield determinations [26]. The gross values of the fluorescence lifetimes seem now to be accepted and the main point of conjecture is the form of the fluorescence decay.

Sauer and Brewington [3] have applied a similar technique to ours, but using a longer excitation pulse from a spark lamp. For dark-adapted *Chlorella* they obtained a fluorescence lifetime of 0.4 ns, in reasonable agreement with our result (Table I). At present, we cannot account for the slight non-exponentiality we observed in the fluorescence decay of *Chlorella*. It seems unlikely that the longer component of 530—650 ps is due to a proportion of the reaction centres being closed, since a lifetime of about 1.5 ns [9] would then be expected, and attempts to fit the data with decays longer than 1 ns proved unsuccessful. For dark-adapted spinach chloroplasts, Sauer and Brewington [3] obtained single exponential decays of 0.2—0.32 ns (depending upon the ionic medium) in contrast to our value of 413 ps (with a 3.8% component of 1463 ps) for pea chloroplasts. We do not know the reason for this discrepancy. For pre-illuminated chloroplasts with DCMU and NH$_2$OH they obtained [3] a biphasic decay of 0.48 ns (93%) and 2.0 ns (6.5%) which is in reasonable agreement with our result (Table I). We believe that our measurements of the dark-adapted state of chloroplasts indicate that some PS II reaction centres are non-quenchers, i.e. effectively closed, on a statistical basis at any given time. This could possibly be due to some pre-illumination by scattered laser light (which seems unlikely since the effect was not observed in chlorella) or a result of damage to some of the chloroplasts during preparation. This explains the 3.8% of approx. 1.4 ns decay in the dark-adapted samples. Even upon addition of DCMU many PS II reaction centres remain open, presumably because of the low level of illumination, but the proportion of the long lifetime is now increased to approx. 10% (Table I) and addition of Mg^{2+} further increases the proportion of long component present to approx. 37%. The fluorescence yields and mean lifetimes are also consistent with this viewpoint. The weighted mean lifetimes calculated from our data can be compared to the lifetimes obtained by Moya et al. [27] who, using a phase fluorimetric technique and assuming a simple exponential decay, obtained a 4-fold increase in lifetime upon illumination, whereas our calculated mean lifetime increases by a factor of 1.7. The change in fluorescence yield also reflects this incomplete pre-illumination. Unfortunately, it was not possible to increase the level of light used for pre-illuminating the samples. The main effect of pre-illumination appears to be to increase the ratio of the long to the short components. We propose that the shorter lifetime is due to quenching of the exciton within the photosynthetic unit by an open reaction centre and the longer component is the result of excitation in a photosynthetic unit with a closed trap. We believe that our data are consistent with an isolated unit model, irrespective of whether migration or trapping is the rate determining step. If an excitation was able to visit a number of traps, then an exponential decay varying continuously from approx. 400 ps to approx. 1400 ps is expected as more traps are closed. The model we envisage consists of light-harvesting pigments which can transfer their energy to many photosynthetic units. However, the photosynthetic units are located in potential energy wells with the traps at the minima *. Thus, once an exciton moves within the vicinity of a trap it cannot return to the light-harvesting pigments, and thus to another photosynthetic unit. The proportion of closed traps would be reflected in the initial intensity ratio of long to short components, which, as can be seen from Table I, increases in proportion to the calculated fluorescence yield.

The effect of Mg^{2+} is also to cause an increase in the ratio of long to short components. This supports the idea that cation-induced changes in fluorescence yield reflect changes in the partitioning of absorbed light between the two Photosystems in agreement with the conclusions of Butler and Kitajima [28] since, if spillover was predominant, then one would expect a lengthening of the value of the lifetime of the short component on addition of Mg^{2+}, which we have not observed, because quenching of PS II fluorescence by PS I would be absent.

Stroma lamellae vesicle fraction. The PS I lifetime of 113 ps (±10%) is longer than most previous determinations (as cited by Searle et al. [12]) but is in agreement with the value of about 100 ps reported by Searle et al. [12]. We have assigned the minor long component (approx. 1.2 ns) to a residual amount of

---

* Very recently Hipkins [31] independently reached similar conclusions from fluorescence-induction studies.

PS II still present in the preparation. Since the sample was not flowed, it was expected that the PS II reaction centres would be closed due to illumination by the laser, which resulted in this fluorescence lifetime being comparable to the long component seen in chloroplasts (Table I). On the basis of this lifetime (113 ps), and with the reasonable assumption that the natural radiative lifetime of chlorophyll is unchanged in vivo, we calculate a fluorescence yield of 0.0058. This is higher than the value of 0.003 measured by Boardman et al. [26], however, who also report that the fluorescence yield of PS II is a factor of 5 greater than that of PS I. The fluorescence yields of PS I and PS II reported by Brown [29] for different organisms also show a difference of a factor of 4—5 between the two Photosystems. The ratio of our calculated fluorescence yields for PS II and PS I is 3.9, which is in good agreement with the steady-state determinations.

Acknowledgements

We would like to thank the Science Research Council for the support of this work and for the award of a studentship to J.A.S. G.S.B. thanks the Royal Society for a Mr. and Mrs. John Jaffé Donation Research Fellowhip and G.R.F. the Leverhulme Trust Fund for a Fellowship. We are grateful to Dr. C.J. Tredwell for helful discussions, to S.B. Morris and P.T. Williams for technical support, to Queen Mary College for computing facilities and to Dr. J. Barber for laboratory facilities used in the preparation of the stroma lamellae vesicle fraction.

References

1 Knox, R.S. (1977) in Primary Processes of Photosynthesis (Barber, J., ed.), pp. 55—97, Elsevier, Amsterdam
2 Ware, W.R. (1971) in Creation and Detection of the Excited State (Lamola, A., ed.), Vol. 1, Part A, pp. 213—302, Marcel Dekker, New York
3 Sauer, K. and Brewington, G.T. (1978) Proceedings of the Fourth International Congress on Photosynthesis (Hall, D.O., Coombs, J. and Goodwin, T.W., eds.), pp. 409—421, The Biochemical Society
4 Hervo, G., Paillotin, G. and Thiery, J. (1975) J. Chim. Phys. 72, 761—766
5 Govindjee (1978) Photochem. Photobiol., in press
6 Campillo, A.J., Shapiro, S.L., Kollman, V.H., Winn, K.R. and Hyer, R.C. (1976) Biophys. J. 16, 93—97
7 Mauzerall, D. (1976) Biophys. J. 16, 87—91
8 Campillo, A.J., Kollman, V.H. and Shapiro, S.L. (1976) Science 193, 227—229
9 Porter, G., Synowiec, J.A. and Tredwell, C.J. (1977) Biochim. Biophys. Acta 459, 329—336
10 Barber, J. (1968) Biochim. Biophys. Acta 150, 618—625
11 Barber, J., Searle, G.F.W. and Tredwell, C.J. (1978) Biochim. Biophys. Acta 501, 174—182
12 Searle, G.F.W., Barber, J., Harris, L., Porter, G. and Tredwell, C.J. (1977) Biochim. Biophys. Acta 459, 390—401
13 Cramer. L.E. and Spears, K.G. (1978) J. Am. Chem. Soc. 100, 221—227
14 Fleming, G.R., Knight, A.E.W., Morris, J.M., Morrison, R.J.S. and Robinson, G.W. (1977) J. Am. Chem. Soc. 99, 4306—4311
15 Hindman, J.C., Kugel, R., Svirmickas, A. and Katz, J.J. (1978) Chem. Phys. Lett. 53, 197—200
16 Birks, J.B. (1970) Photosynthetics of Aromatic Molecules, pp. 92—93, Wiley-Interscience, New York
17 Geacintov, N.E., Breton, J., Swenberg, C.E. and Paillotin, G. (1977) Photochem. Photobiol. 26, 629—638
18 Altmann, J.A., Beddard, G.S. and Porter, G. (1978) Chem. Phys. Lett., in the press
19 Barber, J. (1976) in The Intact Chloroplast (Barber, J., ed.), pp. 89—134, Elsevier, Amsterdam

174

20 Govindjee and Papageogiou, G. (1971) in Photophysiology (Giese, A.C., ed.), Vol. VI, pp. 1—46, Academic Press, New York
21 Butler, W.L. and Norris, K.H. (1963) Biochim. Biophys. Acta 66, 72—77
22 Muller, A., Lumry, R. and Walker, M.S. (1969) Photochem. Photobiol 9, 113—126
23 Mar, T., Govindjee, Singhal, G.S. and Merkelo, M. (1972) Biophys. J. 12, 797—808
24 Hervo, G., Paillotin, G. and Thiery, J. (1975) J. Chim. Phys. 72, 761—766
25 Harris, L., Porter, G., Synowiec, J.A., Tredwell, C.J. and Barber, J. (1976) Biochim. Biophys. Acta 449, 329—339
26 Boardman, N.K., Thome, S.W. and Anderson, J.M. (1966) Proc. Natl. Acad. Sci. U.S. 56, 586—593
27 Moya, I., Govindjee, Vernotte, C. and Briantais, J.-M. (1977) FEBS Lett. 75, 13—18
28 Butler, W.L. and Kitajima, M. (1975) Biochim. Biophys. Acta 396, 72—85
29 Brown, J.S. (1969) Biophys. J. 9, 1542—1552
30 Beddard, G.S., Porter, G. and Weese, M. (1975) Proc. Roy. Soc. Lond. A342, 317—325
31 Hipkins, M.F. (1978) Biochim. Biophys. Acta 502, 514—523

518

Further publications

PICOSECOND TIME-RESOLVED ENERGY TRANSFER IN
PORPHYRIDIUM CRUENTUM. Part II. IN THE ISOLATED LIGHT
HARVESTING COMPLEX (PHYCOBILISOMES).
G. Porter, C. J. Tredwell, G. F. W. Searle and J. Barber,
Biochim. Biophys. Acta, 1978, **501**, 246.

PICOSECOND FLUORESCENCE STUDIES OF PHOTOSYSTEM II.
G. S. Beddard, G. R. Fleming, G. Porter and J. A. Synowiec,
Biochem. Soc. Trans., 1979, **6**, 1385.

PICOSECOND TIME RESOLVED FLUORESCENCE OF
CHLOROPHYLL IN VIVO.
C. J. Tredwell, J. A. Synowiec, G. F. W. Searle, G. Porter and J. Barber,
Photochem. and Photobiol., 1978, **28**, 1013.

FLUORESCENCE LIFETIMES IN THE PHOTOSYNTHETIC UNIT.
G. S. Beddard, G. Porter and C. J. Tredwell, *Nature*, 1975, **258**, 166.

FLUORESCENCE LIFETIMES OF CHLORELLA PYRENOIDOSA.
L. Harris, G. Porter, J. A. Synowiec, C. J. Tredwell and J. Barber,
Biochim. Biophys. Acta, 1976, **449**, 329.

INTENSITY EFFECTS ON THE FLUORESCENCE OF *IN VIVO* CHLOROPHYLL.
G. Porter, J. A. Synowiec and C. J. Tredwell, *Biochim. Biophys. Acta*, 1977, **459**, 329.

PICOSECOND LASER STUDY OF FLUORESCENCE LIFETIMES
IN SPINACH CHLOROPLAST PHOTOSYSTEM I AND
PHOTOSYSTEM II PREPARATIONS.
G. F. W. Searle, J. Barber, L. Harris, G. Porter and C. J. Tredwell,
Biochim. Biophys. Acta, 1977, **459**, 390.

EXCITED STATE ANNIHILATION IN THE PHOTOSYNTHETIC UNIT.
G. S. Beddard and G. Porter, *Biochim. Biophys. Acta*, 1977, **462**, 63.

FLUORESCENCE AND ENERGY TRANSFER IN PHOTOSYNTHESIS.
G. S. Beddard, G. R. Fleming, G. Porter and C. J. Tredwell,
"Picosecond Phenomena", *Chem. Phys.*, 1978, **4**, 149.

CHLOROPHYLL ORGANIZATION AND ENERGY TRANSFER
IN PHOTOSYNTHESIS.
J. A. Altmann, G. S. Beddard and G. Porter,
Ciba Foundation Symposium 61, pub. Excerpta Medica, 1979, pp. 191–199.

PICOSECOND TIME-RESOLVED FLUORESCENCE STUDY OF
CHLOROPHYLL ORGANISATION AND EXCITATION ENERGY
DISTRIBUTION IN CHLOROPLASTS FROM WILD-TYPE BARLEY
AND A MUTANT LACKING CHLOROPHYLL-b.
G. F. W. Searle, C. J. Tredwell, J. Barber and G. Porter,
Biochim. Biophys. Acta, 1979, **545**, 496.

PICOSECOND ENERGY TRANSFER IN ANACYSTIS NIDULANS.
S. S. Brody, G. Porter, C. J. Tredwell and J. Barber,
Photobiochem. Photobiophys., 1981, **2**, 11.

PICOSECOND FLUORESCENCE AND ABSORPTION SPECTROSCOPY
OF LIGHT-HARVESTING CHLOROPHYLL-PROTEIN COMPLEX
FROM PEA CHLOROPLASTS.
J. P. Ide, D. R. Klug, W. Kuhlbrandt, L. B. Giorgi, G. Porter, B. Gore,
T. Doust and J. Barber,
Biochem. Soc. Trans., 1986, **14**, 34.

PICOSECOND TRANSIENT ABSORPTION SPECTROSCOPY
OF PHOTOSYSTEM I REACTION CENTRES FROM HIGHER PLANTS.
L. B. Giorgi, T. Doust, B. L. Gore, D. R. Klug, G. Porter and J. Barber,
Biochem. Soc. Trans., 1986, **14**, 47.

DETERGENT EFFECTS UPON THE PICOSECOND DYNAMICS
OF HIGHER-PLANT LIGHT HARVESTING CHLOROPHYLL (LHC2).
J. P. Ide, D. R. Klug, B. Crystall, B. L. Gore, L. B. Giorgi and G. Porter,
J. Chem. Soc., Faraday Trans. 2, 1986, **82**, 2263.

PICOSECOND TRANSIENT ABSORPTION SPECTROSCOPY
OF GREEN PLANT PHOTOSYSTEM I REACTION CENTRES (III).
B. L. Gore, L. B. Giorgi and G. Porter, in "Progress in Photosynthesis
Research", Ed. J. Biggins, Martinus–Nijhoff Publishers, 1987, Vol. I, pp. 257–260.

SPECTRAL SHIFTS IN PICOSECOND TRANSIENT ABSORPTION
SPECTRA DUE TO STIMULATED EMISSION FROM
CHLORPHYLL *IN VITRO* AND IN PROTEIN COMPLEXES (IV).
D. R. Klug, B. L. Gore, L. B. Giorgi and G. Porter, in "Progress in Photosynthesis
Research", Ed. J. Biggins, Martinus–Nijhoff Publishers, 1987, Vol. I, pp. 95–98.

THE DEPENDENCE OF THE ENERGY TRANSFER KINETICS
OF THE HIGHER PLANT LIGHT HARVESTING CHLOROPHYLL
PROTEIN COMPLEX ON CHLORPHYLL/DETERGENT RATIO (V).
J. P. Ide, D. R. Klug, B. Crystall, B. L. Gore, L. B. Giorgi, W. Kühlbrandt,
J. Barber and G. Porter, in "Progress in Photosynthesis Research",
Ed. J. Biggins, Martinus–Nijhoff Publishers, 1987, Vol. I, pp. 131–134.

PICOSECOND TRANSIENT ABSORPTION SPECTROSCOPY
OF PHOTOSYSTEM I REACTION CENTRES FROM HIGHER PLANTS (II).
L. B. Giorgi, B. L. Gore, D. R. Klug, J. P. Ide, J. Barber and G. Porter, in
"Progress in Photosynthesis Research", Ed. J. Biggins, Martinus–Nijhoff Publishers,
1987, Vol. I, pp. 257–260 (VI).

THE STATE OF DETERGENT SOLUBILISED LIGHT HARVESTING
CHLOROPHYLL-a/b PROTEIN COMPLEX AS MONITORED BY
PICOSECOND TIME-RESOLVED FLUORESCENCE AND CIRCULAR
DICHROISM (VII).
J. P. Ide, D. R. Klug, W. Kühlbrandt, L. B. Giorgi and G. Porter,
Biochim. Biophys. Acta, 1987, **893**, 349.

RESOLUTION OF A LONG-LIVED FLUORESCENCE COMPONENT
FROM D1/D2/CYTOCHROME B-559 REACTION CENTRES (IX).
B. Crystal, B. J. Booth, D. R. Klug, J. Barber and G. Porter,
FEBS Lett., 1989, **249**(1), 75.

ENERGY TRANSFER TO LOW ENERGY CHLOROPHYLL SPECIES PRIOR
TO TRAPPING BY P700 AND SUBSEQUENT ELECTRON TRANSFER (X).
D. R. Klug, L. B. Giorgi, B. Crystal, J. Barber and G. Porter, *Photosynth. Res.*, 1989,
22, 277.

MICROSECOND AND NANOSECOND KINETICS OF ISOLATED
PHOTOSYSTEM 2 REACTION CENTRES STUDIED BY SINGLE-PHOTON
COUNTING AND TRANSIENT ABSORPTION.
L. B. Giorgi, B. Crystall, P. J. Booth, J. R. Durrant, D. R. Klug, J. Barber and G. Porter,
Proceedings of 3rd EEC Workshop on Photochemical, Photoelectrochemical and
Photobiological Research and Development, 1989.

Chapter 11

FEMTOSECOND KINETIC STUDIES OF PHOTOSYSTEM 2

The fastest molecular processes which can be followed in real time occupy a few femtoseconds; at faster times than this the uncertainty principle introduces serious limitations to the precision of measurements in the energy range of chemical and biological processes. As a result of striking advances in the production of very short laser pulses, measurements every few femtoseconds became possible about a decade ago and it was soon apparent that the first steps of energy and electron transfer in photosynthesis did indeed occur over these very short times.

The systems which command most attention at present are the light harvesting units and the reaction centres of photosynthetic bacteria and of green plants. These are being isolated, crystallised and their structures determined, simultaneously with time resolved flash photolysis studies of their kinetics in the picosecond and femtosecond regions of time. The precision of these spectroscopic and kinetic measurements results not only from the short laser pulses (time), but also from the wide spectral range available in both pulse and probe pulses (wavelength) and the analysis of the large numbers of experiments made possible by the high repetition rate (∇ optical density).

In green plants there are two photosystems working in series to transport charge across the membrane. The photosystem 2 is on the oxidising side and is ultimately responsible for the oxidation of water. The isolation of the reaction centre from photosystem 2 has greatly simplified spectroscopic measurements of the electron transfer chain, though the system is complex enough that unequivocal interpretations and distinctions between the processes of energy and electron transfer are still somewhat controversial. The latest papers below represent the state of knowledge in 1996.

Reprinted from "Femtosecond Chemistry", Vol. I, pp. 625–632.
Eds. Jörn Manz and Ludger Wöste, VCH Publishers Inc.

21 Femtosecond Processes in Photosynthesis

George Porter

21.1 Introduction

There are two basic photomolecular processes that determine the course of photosynthesis. First there is energy transfer by which the molecules of chlorophyll, and its ancillary pigments which form the light-harvesting antennae, collect and transfer the absorbed solar energy to the reaction center and to the primary trap. Second are the electron transfer processes, by which the excited pigment in the trap transfers its electrons, via intermediate pigments, across the membrane to create an oxidizing surface on the donor side which liberates molecular oxygen from water, and a reducing surface on the other side which ultimately reduces carbon dioxide to carbohydrate.

Since the reactions begin from chlorophyll in its excited singlet state, the primary events have to compete with fluorescence and, for high efficiency, this means that lifetimes must be less than one nanosecond.

21.2 Energy Transfer

One example of energy transfer in the light-harvesting antenna is given by Chapter 22 and was seen to occur in a few picoseconds. The chlorophyll and other pigment molecules in a photosynthetic antenna are separated from their nearest neighbours by distances (center to center) of the order of 10 Å which, if transfer occurs by the Förster resonance transfer mechanism, leads to predicted transfer times between chlorophyll molecules of 150 fs at 10 Å and 1.15 ps at 14 Å separation. In an array, or cluster, of 300 chlorophyll molecules, one of which is trap, the corresponding times (calculated for a regular cubic array) for random walk diffusion to the trap are 69 ps at 10 Å and 528 ps at 14 Å.

The structures of the light-harvesting units have not yet been determined in detail. The pigment molecules are held in position by proteins and the unit is by no means homogeneous but is composed of smaller units which are often separated from each other by greater distances than the average separation of the molecules. Suffice it to say that the measured lifetimes of transfer are not inconstistent with those calculated on the basis of what is known of the structures involved and there seem to be no problems in accounting for the fast rates in terms of established theories.

One outstanding problem which does remain is the absence of concentration quenching at the high concentrations in the photosynthetic unit, whilst such quenching occurs at much lower concentrations in model systems prepared *in vitro*. This is presumably avoided by special arrangement of pigments within the protein scaffolding but further progress here is impeded by lack of structural data rather than of kinetic data.

21.3 Electron Transfer in the Reaction Centre

This is one of the most intriguing processes of nature and its study is at present one of the most rapidly advancing fields of science. The skills of the biochemist have made possible the isolation and often the crystallization of various pigment–protein complexes including the reaction centre or its separate parts. In a very few cases it has been possible, by X-ray diffraction methods, to determine the structure down to a resolution of nearly 2 Å and also, in other cases, by image-analysis electron microscopy, which gives lower resolution but is available for less complete (two-dimensional) crystals. Finally, whilst the time resolution of the first flash photolysis studies of reaction centers (a few picosecond) was rather too long for reliable kinetic measurements, the present femtosecond techniques are able to give precise data at times of 100 fs or less.

21.4 Photosynthetic Bacteria

The first reaction centers to be isolated were those of the purple bacteria [1]. Flash photolysis studies in the picosecond region were carried out in 1975 before the structure was known [2], although the composition of the pigments in the electron trasport chain was fairly well established. The sequence of transfers across the reaction center and hence across the membrane, was shown to be as follows, with the corresponding rates:

$$(BCl_2)^* + BPh \xrightarrow{(4\ ps)^{-1}} (BCl_2)^+ + BPh^-$$

$$BPh^- + UQ \xrightarrow{(200\ ps)^{-1}} BPh + UQ^-$$

where BCl = bacteriochlorophyll; (BCl_2) = bacteriochlorophyll special pair; BPh = bacteriopheophytin; UQ = ubiquinone.

The structure of this reaction centre determined in 1984 at high resolution [3] was in excellent accord with the sequence of transfers given above, both in the order of pigment arrangements across the membrane and in the rates of transfer predicted from the separation distances. In fact this system has become one of the best test-beds for theories of electron transfer. The separation between the (BCl_2) and BPh was found to be 17 Å, which is rather large for the measured rate of $(4\,ps)^{-1}$ and it now seems probable that the intermediate chlorophyll molecule, which is found between (BCl_2) and BPh, acts as an intermediate in the transfer.

21.5 Higher-Plant Reaction Centres

There are two such reaction centres, corresponding to the two photosystems. It has not yet been possible to prepare isolated reaction centres of photosystem 1 (PS1), and it is almost impossible to study the multifarious processes of electron transfer in the presence of an excess of light-harvesting chlorophyll. Most studies on PS1 have therefore been concerned with energy transfer.

The preparation of isolated reaction centres of PS2 by Nanba and Satoh and by Barber and his colleagues at Imperial College has been more successful. Photosystem 2, which is able to oxidize water to oxygen, is of special interest, and recent work on events in the femtosecond range, carried out in our laboratory by David Klug, James Durant and others listed in the references, will now be described.

Although there are many similarities with the bacterial reaction centre, there are also some surprises.

21.5.1 Femtosecond Kinetics in Photosystem 2

The reaction centres were isolated from pea chloroplasts. The protein sequences of the D_1 and D_2 polypeptides show close homologies with those of the *L* and *M* polypeptides in the reaction centres of purple bacteria.

Studies at lower time resolution, including studies of larger particles than the isolated reaction centre, have shown the sequence of transfers to be that shown in Fig. 21.1.

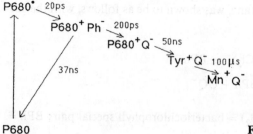

Figure 21.1

The quinones Q_A and Q_B and the manganese complex are absent in the reaction centres used by us, so the electron transfers are limited to primary charge separation from the special pair (which absorbs at 680 nm) to pheophytin, with the possible intervention of a single chlorophyll molecule as was suggested in the studies of bacteria.

Our isolated reaction centres contain six chlorophylls, two pheophytins (Ph), two β-carotenes, and one cytochrome b_{559}. Two of the chlorophylls are assigned to the special pair P680, two others may be intermediate between P680 and Ph, one next to each Ph, and the function of the other two is unknown except that, as we shall see, they take part in light harvesting and energy transfer.

It is a much more difficult problem to identify the separate transient spectra in PS2 than in the bacteria because of the small energy differences between them. Whilst in bacteria the first excited state of the special pair lies at least 150 meV below the corresponding excited state of all other pigments, in PS2 the plant chlorins are all within 30 meV of each other. It is therefore necessary to be able to vary the excitation wavelength over small intervals so that, in our apparatus, white light is used to generate the excitation pulse as well as the probe pulse.

The instrumental arrangement used in this work is shown in Chapter 22. For the excitation pulse an interference filter or monochromator selects a small-wavelength band from the white light and this pulse is once more amplified to an energy above 0.1 μJ. Low-temperature studies have shown that the plant chlorins occur in two pools absorbing at 680 and 670 nm in the main band. To distinguish these we used 665 and 690 nm pulses for femtosecond studies. In the picosecond work we used 612 or 694 nm and monitored at 612 or 695 nm in both cases.

We will refer first to Fig. 21.2, which describes all the kinetic processes identified in the PS2 core particle. The upper half of the diagram describes the slower processes that occur after the formation of $P680^+Pheo^-$ in the core particle when quinone and the manganese complex are present. These components are not present in the isolated reaction centres that we have studied and all the events shown in the lower part of Fig. 21.2 were measured in these reaction centres using femtosecond time resolution.

At least four components were found via global analysis of all our data, and are assigned to the various processes as follows.

(1) A 200 $\pm$ 100 ps component appearing only with lower-wavelength (612 nm) excitation. This was assigned to slow energy transfer from other pigments and will not be considered further here.

(2) A 100 fs transient which decayed at this rate at 670 nm and showed a corresponding grow-in at 680 nm.

(3) A 21.4 $\pm$ 1.5 ps transient which is clearly shown to be the grow-in of the charge-transfer product $P680^+Pheo^-$.

(4) A 1.6 ps component which is less well characterized and, under some conditions, itself appears to have two components of 0.4–0.6 and 2.6–3.7 ps. It is tentatively assigned to a process of energy localization in the $P680^*$.

Processes (2) and (3) are particularly interesting and will be discussed a little further.

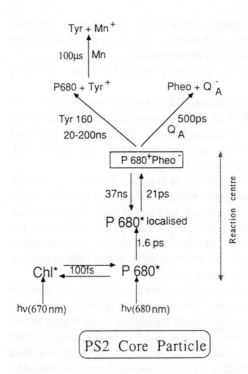

Figure 21.2 PS2 core particle kinetics. The isolated reaction centre undergoes the processes in the lower part only.

21.5.2 Energy Transfer in PS2 [Process (2)]

It was possible [4] to obtain 75% selective excitation of either the (component) C680 or C670 pigment pools using 694 or 665 nm excitation. The absorption changes observed are shown in Fig. 21.3. They were obtained with pulses of 100–140 fs FWHM (full width at half-maximum) (0.1 μJ). The records are on a 0–5.5 ps timescale with time intervals of 13.2 fs between points.

The 100 fs transients were quite different from other transients in that excitation of 680 and 670 nm pigments gave quite opposite results. This is shown in Fig. 21.4, which gives the amplitudes of the transient absorption components which are attributed, when excitation is at 665 nm, to energy transfer from C670* to C680 and, when excitation is at 694 nm, to energy transfer from C680* to C670.

We conclude that energy equilibration between these pigments occurs at a rate of $(100 \pm 50 \text{ fs})^{-1}$. This rate is similar to that referred to above for energy transfer between two chlorophyll molecules separated, centre to centre, by 10 Å.

The relative amplitudes of the 100 fs components following excitation at 665 and 694 nm lead to an apparent equilibrium constant of 1 ± 0.5 between the 670 and 680 nm pigment pools. This is consistent with (a) one pigment for a localised P680; (b) two to four pigments for the 670 nm pool; (c) an energy separation of ≈ 30 meV ($kT = 25$ meV) giving a Boltzmann factor of $1/e$ or 0.37 in favor of the P680.

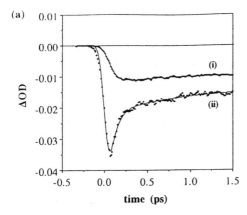

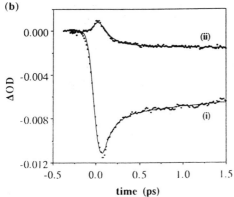

Figure 21.3 Absorption changes, recorded at intervals of 13 fs, in the PS2 reaction center: (a) probe at 685 nm, pulse at 665 nm (i) and 694 nm (ii); (b) (i) probe at 670 nm, pulse at 665 nm, and (ii) probe at 667 nm, pulse at 694 nm.

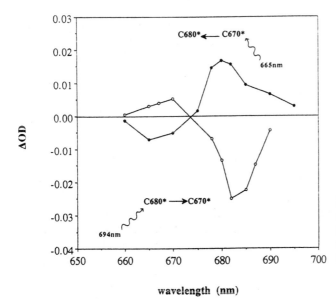

Figure 21.4 Spectra of amplitudes of 100 fs transient absorption components observed after excitation at 665 nm (•) and 684 nm (○).

21.5.3 Electron Transfer from P680 to Pheo (21 ps)

This corresponds to the process in photosynthetic bacteria which occurs with a rate of $(3.5 \text{ ps})^{-1}$. This is a large difference for such similar systems, and has been questioned by other workers who find a shorter lifetime in PS2. There is no question about the existence of our 21 ps transient, but one must be able to identify it with the electron-transfer step proposed. This has now been done convincingly by observing both the bleaching of the Pheo band at 545 nm and the grow-in of the Pheo^- band at 460 nm, with a rate of $(21 \pm 3 \text{ ps})^{-1}$. The spectrum after 100 ps is identical with the well-characterized spectrum at 9 ns, which is unchanged up to 37 ns and is that of the radical pair $\text{P680}^+\text{Ph}^-$.

One may now enquire why electron transfer is slower $(21 \pm 3 \text{ ps})$ in PS2 than in bacteria where, in all three studied – *Rhodobacter (RB.) sphaeroides, Rhodopseudomonas (Rhps.) viridis* and *Rb. capsulatus* – it is 3.5 ps. In spite of close homologies in the amino acid sequences there is a difference in the amino acids in the key positions between the special pair and the BPheo or Pheo, as shown in the following comparison: The amino acid leucine in position D2 206 of the higher plant is replaced by tyrosine in the equivalent position M 210 of native bacteria with a concomitant increase in electron-transfer rate from $(21 \text{ ps})^{-1}$ to $(3.5 \text{ ps})^{-1}$:

Higher plant PS2 (pea)

D2 206	D1 206	
leucine	phenylalanine	Pheo reduction 21 ± 3 ps

Bacteria (native)

M 210	L 181	
Tyrosine	Phenylalanine	Pheo reduction 3.5 ps

There is, however, a mutant strain of bacteria in which the M 210 amino acid has been changed from tyrosine to leucine, as in the corresponding D2 206 position of PS2 in higher plants. In this mutant the rate constant for pheophytin reduction is increased to $(22 \pm 8 \text{ ps})^{-1}$.

Bacteria (mutant)

M 210	L 181	
Leucine	Phenylalanine	Pheo reduction 22 ± 8 ps

21.5.4 Conclusions

The significance of this interesting observation will become clearer as more mutants become available for study. Mutants especially designed for comparative studies of energy and electron transfer in photosynthetic systems are likely to play an important role in understanding these processes and perhaps in their application to purposes.

Acknowledgement

Most of the work described in this paper was carried out by my colleagues at Imperial College listed in [4] and [5]. We are grateful to the Science and Engineering Research Council, the Agriculture and Food Research Council, The Royal Society and RITE (NEDO, Japan) for financial support of the work.

Final version received: 30th January 1994.

References

[1] R. K. Clayton, B. J. Clayton, *Biochim. Biophys. Acta* **1978**, *501*, 478–487.

[2] M. W. Windsor, *J. Chem. Soc. Faraday Trans.* **1986**, *82*, 2237–2243.

[3] J. Deisenhöfer, O. Epp, K. Miki, R. Hüber, M. Michel, *J. Mol. Biol.* 1984, *80*, 365.

[4] J. R. Durrant, G. Hastings, D. M. Joseph, J. Barber, G. Porter, D. R. Klug, *Proc. Natl. Acad. Sci. USA (Biophys.)* **1992**, *89*, 11632–11636.

[5] G. Hastings, J. R. Durrant, J. Barber, G. Porter, D. R. Klug, *Biochemistry* **1992**, *31*, 7638–7647.

[6] J. Barber (Ed.), *Topics in Photosynthesis,* Vol. 8. The Light Reactions, Elsevier Science, Amsterdam, **1987**.

Proc. Natl. Acad. Sci. USA
Vol. 89, pp. 11632–11636, December 1992
Biophysics

Subpicosecond equilibration of excitation energy in isolated photosystem II reaction centers

(energy transfer/electron transfer/P680)

James R. Durrant[†‡], Gary Hastings[†], D. Melissa Joseph[†‡], James Barber[‡], George Porter[†], and David R. Klug[†]

[†]Photochemistry Research Group, Department of Biology and [‡]Agriculture and Food Research Council Photosynthesis Research Group, Department of Biochemistry, Centre for Photomolecular Sciences, Imperial College, London SW7 2BB, United Kingdom

Contributed by George Porter, August 31, 1992

ABSTRACT   Photosystem II reaction centers have been studied by femtosecond transient absorption spectroscopy. We demonstrate that it is possible to achieve good photoselectivity between the primary electron donor P680 and the majority of the accessory chlorins. Energy transfer can be observed in both directions between P680 and these accessory chlorins depending on which is initially excited. After excitation of either P680 or the other chlorins, the excitation energy is observed to equilibrate between the majority of these pigments at a rate of 100 ± 50 fs^{-1}. This energy-transfer equilibration takes place before any electron-transfer reactions and must therefore be taken into account in studies of primary electron-transfer reactions in photosystem II. We also show further evidence that the initially excited P680 excited singlet state is delocalized over at least two chlorins and that this delocalization lasts for at least 200 fs.

In photosynthetic organisms, solar energy is absorbed by pigments in light-harvesting complexes and then transferred in a series of ultrafast energy-transfer steps to a primary electron donor within a photosynthetic reaction center (for reviews, see refs. 1 and 2). This initiates a sequence of electron-transfer reactions within the reaction center, which results in the excitation energy being trapped in progressively longer-lived radical pair states. In the case of the photosystem II (PSII) reaction center of higher plants, these electron-transfer reactions produce an unusually high oxidizing potential of ≈ 1 V, approximately twice that of purple bacteria. This oxidizing potential is used to drive water splitting, which gives rise to oxygen evolution.

The primary electron donor of PSII is thought to correspond to a spectral feature at 680 nm and is referred to as P680 (3), while a pheophytin (Ph) molecule functions as an electron acceptor (4–6). Studies of PSII core complexes binding 60 and 80 antenna chlorophylls have suggested that the primary radical pair P680$^+$Ph$^-$ is formed at a rate of ≈ 100 ps^{-1} following the absorption of a photon by antenna pigments (7). A similar conclusion was reached from photovoltage studies of larger PSII complexes (8). A kinetic model, in which there is a rapid (≈ 1 ps) equilibration of excitation energy between the antenna pigments and P680, followed by the observed trapping of the excitation energy by radical pair formation in ≈ 100 ps, has been proposed for this process (refs. 9 and 10; reviewed in ref. 2). This so-called trapping limited model (11) is valid when the rate of electron transfer from the primary electron donor is slower than energy transfer back to the antenna pigments. This model has also been applied to other photosynthetic antenna/reaction center complexes (12–14). However, previous studies have not been able to time resolve

the energy-transfer processes that are predicted to cause the equilibration of excitation energy between the antenna and primary electron donor pigments prior to radical pair formation.

We report here a study of excitation energy equilibration in the isolated D1/D2/cytochrome b_{559} complex, which is the reaction center of PSII. This complex is much smaller than the isolated PSII core complexes discussed above, binding only six chlorophyll a and two Ph a pigments (15, 16). While several of these pigments are presumably involved in electron-transfer processes, these pigments will also function in an energy-transferring capacity before charge separation. Time-resolved fluorescence studies have determined that at least 94% of the chlorin pigments in our PSII reaction center preparation are able to transfer excitation energy to P680, resulting in a near unity quantum yield of the primary radical pair (17, 18). In a previous study, we determined that when P680 is directly excited, Ph reduction occurs primarily at a rate of 21 ps^{-1} in isolated PSII reaction centers at room temperature (4).

There have been several discussions of the similarities between the PSII reaction center of higher plants and the reaction center of purple bacteria (19, 20). However, these two complexes are likely to be very different in terms of their energy-transfer kinetics when isolated from their antenna systems. The lowest S$_0$–S$_1$ optical transition for the primary electron donor (P) of purple bacteria is at least 150 meV below those of the other accessory pigments bound to the isolated reaction center. The special pair is therefore an energetic trap for excited singlet states ($k_BT \approx 25$ meV at room temperature), and there is little thermally activated back energy transfer from P to the other pigments. It has been reported that excitation of the Q$_y$-absorption bands of any of the chlorins in reaction centers of purple bacteria results in localization of the excitation energy on P within 100 fs (21, 22), followed by Ph reduction at a rate of ≈ 3 ps^{-1} at room temperature (23, 24). In contrast, the S$_0$–S$_1$ transitions (Q$_y$-absorption bands) for chlorins bound to the isolated PSII reaction centers overlap, in such a way that all eight pigments have their lowest excited singlet states separated by no more than ≈ 30 meV (25, 26). Therefore, the P680 excited singlet state is not likely to be a deep trap for excitation energy. Extensive forward and reverse energy transfer between chlorins is therefore expected, and distinguishing these from electron-transfer reactions is a prerequisite for a meaningful study of the mechanism of primary charge separation in PSII.

Low-temperature and gaussian deconvolution studies (25, 26) have indicated that the chlorins in the isolated PSII reaction center can be approximately grouped into two pools according to their Q$_y$-absorption maxima: those with absorption maxima near 680 nm (referred to hereafter as C680), and the remaining chlorins with maxima near 670 nm (C670). The

Abbreviations: PSII, photosystem two; Ph, pheophytin.

531

Biophysics: Durrant et al.   Proc. Natl. Acad. Sci. USA 89 (1992)   11633

C680 absorption band is dominated by P680. The location of the Q_y-absorption maxima of the Ph are uncertain (25–29). As the 670- and 680-nm absorption bands are separated by approximately $k_B T$ at room temperature, equilibration of excitation energy following selective excitation of either chlorin pool will result in energy transfer to the other pool.

There have been several reports of energy-transfer processes in the PSII reaction center with lifetimes ranging from several tens to hundreds of picoseconds (4, 18, 28, 30–32). In particular, Small et al. (28, 32) have reported that energy transfers from the C670 chlorophylls and the photoreducible Ph to P680 have lifetimes of 12 and 50 ps, respectively, at 4.2 K on the basis of spectral hole burning studies. This would imply that the trapping limited kinetic model discussed above may not be appropriate for PSII. Our own previous transient absorption studies resolved a 200-ps energy-transfer/trapping process, which we assigned to a small minority of the reaction center chlorins or to partially damaged reaction centers (4). This 200-ps component was not observed, however, when P680 was directly excited (4). We also reported the observation of two faster kinetic components with lifetimes of 400 fs and 3.5 ps after direct excitation of P680 (33). These components were both assigned to decay of an initially delocalized P680 singlet excited state, although it was not possible to determine whether these decays resulted from energy- or electron-transfer processes.

The results presented here are a continuation of our previous studies (4, 33) of isolated PSII reaction centers using femtosecond transient absorption spectroscopy. In this paper, we report a femtosecond equilibration of excitation energy between reaction center pigments, which occurs prior to those processes we have resolved previously; we also discuss further evidence for delocalization of the P680 singlet excited state.

MATERIALS AND METHODS

Experimental details were as described (4, 33). PSII reaction centers were isolated from pea thylakoid membranes (18, 34) and studied under anaerobic conditions at room temperature (295 K). PSII reaction centers were excited with 0.1-μJ pulses centered at either 665 or 694 nm. The excitation and probe pulses were parallel polarized to better than 99%. Between 5% and 10% of reaction centers in the pumped volume were excited by each excitation pulse. The instrument response of the spectrometer was estimated from the rise time of absorption changes in dye standards for each combination of excitation and probe wavelengths used in this paper. These instrument responses were assumed to have gaussian temporal profiles, an approximation that resulted in adequate, but not perfect, fits to data obtained with both the laser dyes and PSII reaction centers (see Fig. 1). The instrument responses had full width at half maximum values of 100–140 fs. Zero time delay was also determined from dye standards. The white light probe pulses were spectrally dispersed by <2 fs/nm between 660 and 695 nm.

Absorption changes were monitored as a function of time at single wavelengths between 660 and 695 nm (detection bandwidth, 2 nm). Data presented in this paper were collected over a 0- to 2-ps time scale, with a time delay of 13.2 fs between points. Data were analyzed assuming multiexponential kinetics as described by Hastings et al. (4), with decays being analyzed both individually and globally. These analyses included information previously obtained on relatively slow transient absorption components from data collected on longer time scales (4, 33). Data were globally analyzed either without deconvolution for time delays >100 fs or using iterative reconvolution with the instrument response. Similar lifetimes and spectra were obtained from

both types of analyses, which strongly supports the validity of our fitting procedures.

Transient absorption data collected by using 665-nm excitation have been scaled to take account of differences in the proportion of reaction centers excited by the 665- and 694-nm pulses. The scale factor was calculated to give the same final radical pair spectrum, as we have discussed (4, 33).

For each combination of excitation and probe wavelengths, transient absorption data were collected by using sample cuvettes containing the PSII reaction center suspension, the suspension buffer only, and dye standards in methanol to assess possible artifactual distortions of the data. In all the data reported in this paper, appropriate conditions were chosen to ensure that oscillations resulting from impulsive stimulated Raman scattering in the glass windows of the sample cuvette (Q. Hong, D.R.K., J.R.D., G.H., and G.P., unpublished data) were at least 10 times smaller than the transient absorption components of interest.

RESULTS

Excitation wavelengths were chosen with the intention of selectively exciting either the C680 (using 694-nm excitation) or C670 (using 665-nm excitation) chlorin pools. By comparing the spectra of the excitation pulses and the two pigment pools, it is possible to estimate that 694- and 665-nm excitation pulses should achieve an ≈75% selective excitation of the C680 and C670 pigment pools. In fact, our data indicate that the photoselection was actually better than this.

Transient absorption data presented in this paper were collected on a 0- to 2-ps time scale at probe wavelengths between 660 and 695 nm. Typical data are shown in Fig. 1. We have reported previously from global analyses of data collected on 0- to 12-ps and 0- to 80-ps time scales that a

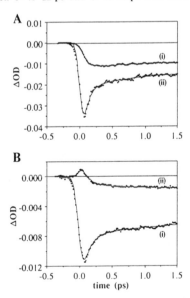

FIG. 1. (A) Kinetics of transient absorption changes observed at 685 nm after excitation of PSII reaction centers with 665-nm (i) and 694-nm (ii) pulses. (B) Kinetics of transient absorption changes observed at 670 nm using 665-nm excitation (i) and 667 nm using 694-nm excitation (ii). Circles are data points and solid lines are best fits to the data obtained using two exponential components with lifetimes of 100 and 400 fs for 665-nm excitation, and 100 and 600 fs for 694-nm excitation, and a nondecaying component. The instrument response and zero time delay were determined with laser dye standards.

11634   Biophysics: Durrant *et al.*

Proc. Natl. Acad. Sci. USA 89 (1992)

minimum of three exponential components with lifetimes of 400 fs, 3.5 ps, and 21 ps were required to fit the data in addition to a nondecaying component (4, 33). When data collected on the 0- to 2-ps time scale were included in the global analyses, an additional exponential component was required to fit the data. This resulted in the 400-fs component splitting into two components with lifetimes of 100 and 600 fs; the lifetimes of the slower 3.5- and 21-ps components were unaffected.

When the excitation wavelength was changed to 665 nm, a clear qualitative difference became apparent between the fastest component (having a lifetime of 100 ± 50 fs) and the slower components (lifetimes of ≈600 fs and ≈3.5 ps). The 100-fs component changed sign at all probed wavelengths when the excitation wavelength was changed (see below), while the slower components did not. The lifetime of the fastest component after excitation at either 665 or 694 nm was 100 ± 50 fs as the lifetimes were indistinguishable. Differences in the 100-fs components are clearly visible in the kinetic data shown in Fig. 1. For example, there is an ≈50-fs lag in the appearance of the transient signal at 685 nm after excitation at 665 rather than 694 nm (Fig. 1*A*), while Fig. 1*B* shows particularly clearly the change in sign of the 100-fs component, which occurs when the excitation wavelength is changed.

Fig. 2 shows spectra of the amplitudes of the 100-fs components after excitation at 665 and 694 nm [i.e., kinetic spectra (4)]. These spectra are dominated by the transient bleaching of ground state Q_y-absorption bands and stimulated emission from chlorin excited singlet states. Both spectra show features characteristic of energy transfer between pigments having different band maxima and oscillator strengths (see below). The spectrum of the ≈100-fs component after 694-nm excitation is inverted compared to that obtained after 665-nm excitation. This is what would be expected from components that are due to energy transfer in opposite directions between the two pigment pools.

The transient absorption changes resulting from the 100-fs components are further illustrated in Fig. 3. This figure shows both the transient absorption spectra at 0 ± 20 fs time delay (obtained from extrapolation of the fitted functions back to *t* = 0 fs) and the spectra that would result from the decay of the amplitude of the 100-fs component to zero. The transient

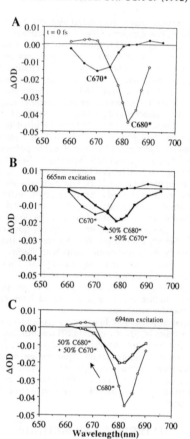

FIG. 3. Transient absorption spectra before (○, ●) and after (□, ■) the 100-fs components. Excitation was at 665 (●, ■) or 694 (○, □) nm. These spectra were calculated from the sum of amplitudes of all the kinetic components resolved in the global analyses, either including (○, ●) or excluding (□, ■) contributions from the 100-fs components. The spectra before the 100-fs components correspond to the spectra at *t* = 0 ± 20 fs and are shown together in *A*. These data do not take account of any components with lifetimes of ≪100 fs.

spectra at *t* = 0 (Fig. 3*A*) clearly demonstrate the effectiveness of the photoselection achieved with both the 665- and 694-nm excitation pulses. This also confirms the validity of grouping the PSII reaction center chlorins into these two pools. Fig. 3 *B* and *C* is discussed in more detail below.

DISCUSSION

We have shown that it is possible to excite selectively the C670 or C680 chlorin pools in the isolated PSII reaction center. After excitation of either pigment pool, complex multiexponential transient absorption kinetics are observed. The fastest component has a lifetime of 100 ± 50 fs after excitation of either pigment pool and can be qualitatively distinguished from the slower components (lifetimes of ≈600 fs and ≈3.5 ps) as follows. The spectrum of the 100-fs components between 660 and 690 nm completely inverts when the excitation wavelength is changed from 694 to 665 nm, while the spectra of the slower components do not. This 100-fs component is assigned to energy transfer between the C670 and C680 chlorin pools, presumably resulting in excitation energy equilibration between the C670 and C680 pigments. The slower 600-fs and 3.5-ps components are as-

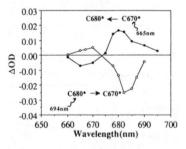

FIG. 2. Spectra of amplitudes of the ≈100-fs transient absorption components observed between 660 and 695 nm after excitation at 665 (●) and 694 (○) nm. These kinetic spectra (sometimes called decay-associated spectra) were determined from global analyses of kinetic data such as those shown in Fig. 1. The spectrum observed using 665-nm excitation has a negative maximum at 670 nm corresponding to a recovery of the initial C670 bleach/stimulated emission and a positive maximum at 680 nm corresponding to production of C680 bleach/stimulated emission. The positive 680-nm maximum is larger than the negative 670-nm maximum due to the larger mean oscillator strength of the C680 chlorins relative to the C670 chlorins (see *Discussion*). The spectrum obtained using 694-nm excitation is inverted with respect to the spectrum obtained using 665-nm excitation. C670* and C680* are lowest excited singlet states of C670 and C680 chlorins, respectively.

signed, as previously (33), to decay of a delocalized P680 excited singlet state.

Our observation that the 100-fs energy-transfer processes can be completely reversed by changing the excitation wavelength indicates that all of the states involved are optically accessible singlet excited states. Moreover, neither chlorin cation nor anion states are present at these early times (unpublished data). The observation that the spectra of these two components are opposite (Fig. 2) indicates that neither polarization effects nor complications of excited state absorption significantly affect the observation of this equilibration.

Composition of the C680 and C670 Pigment Pools and Delocalization of P680. The structural identity of P680, the primary electron donor of PSII, is currently the subject of some controversy (25, 26, 35–38), and the extent of the analogy between P680 and the special pair of bacteriochlorophylls found in reaction centers of purple bacteria is unclear. The results presented here, however, do allow us to determine the degree of delocalization of the P680 excited singlet state. This can be achieved by comparing the absorption changes resulting from energy transfer between the C670 and C680 pigment pools (Fig. 3). Energy transfer from C670 pigments to C680 pigments results in a 60% increase in the area of the transient spectrum (bleached oscillator strengths), while energy transfer in the reverse direction results in a 40% reduction in the area.

These large changes cannot be attributed to differences in excited state absorption. The excited state absorption (S_1–S_n) of chlorophyll a is known (39). The extreme case of C680* exhibiting no excited state absorption can only account for a quarter of the observed changes.

From the data described above, it can therefore be concluded that the C680 pigments exhibit a significantly higher Q_y-band oscillator strength than the C670 pigments. The relative magnitudes of the C670 and C680 oscillator strengths can be quantified by using the 0-fs spectra shown in Fig. 3A to simulate the kinetic spectra of the 100-fs components shown in Fig. 2. Alternatively the areas of these 0-fs spectra following 665- and 694-nm excitation can be compared directly, as these spectra have been normalized to take account of differences in excited reaction center populations (see *Materials and Methods*). Both estimates indicate that C680 excited singlet states result in an absorption change 2.5 ± 0.6 times larger than those of C670 pigments. The 670-nm-absorbing chlorins are primarily monomeric chlorophylls (25, 26), and we therefore conclude that the C680 pigments exhibit a Q_y-band oscillator strength at least twice that of monomeric chlorophyll species. This observation presumably results from P680 dominating the C680 absorption changes observed here, with the excited singlet state(s) of P680 being delocalized over at least two chlorin molecules.

Our conclusion that P680 comprises at least two coupled chlorin molecules is in agreement with a previous study by us (33) that was based on a comparison of data collected using 612- and 694-nm excitation. This conclusion is supported by absorption spectroscopy of the P680 triplet state (26, 37, 38), circular dichroism (25, 40), fluorescence anisotropy (41), and gaussian deconvolution (25, 26) studies of PSII reaction centers. Kwa *et al.* (41) interpreted their fluorescence anisotropy measurements as indicating that the exciton coupling between the P680 chlorins is broken sometime after the initial excitation. Our results indicate that the P680 singlet state remains delocalized for at least a few hundred femtoseconds. It is possible, however, that the initially excited delocalized P680 singlet state may localize on a single pigment on a slower time scale. Our conclusion that the initially excited P680 singlet state is delocalized over at least two pigments, however, does not necessarily imply that the pigments are structurally analogous to the special pair of bacterial reaction centers.

We have concluded that the C680 pigment pool is dominated by P680, in agreement with previous gaussian deconvolution studies (25, 26). The large amplitude of the 100-fs components observed here after excitation of either the C670 or C680 pigments indicates that the majority of the chlorins in the PSII reaction center participate in these femtosecond energy-transfer processes. Several other studies, including our own, have observed slower energy-transfer processes in PSII reaction centers with lifetimes ranging from several tens to hundreds of picoseconds (4, 28, 30, 31). It must be concluded that, at least at room temperature, these slower energy-transfer processes are associated with only a minority of chlorins in the reaction center, in agreement with our previous studies (4). We have not yet determined whether the two Ph in the PSII reaction center are included in the femtosecond equilibration of excitation energy we have observed.

Establishment of a Boltzmann Equilibrium? The spectra of the excitation pulses indicate that 694-nm excitation will produce vibrationally cool C680 pigments, while 665-nm excitation will excite C670 pigments with excess vibrational energy. However, the same rate of equilibration of the excitation energy is observed after excitation of either pigment pool. Moreover, we have observed an endothermic energy transfer from C680 chlorins to C670 chlorins. These observations imply that some thermalization of vibrational energy occurs within 100 fs, which suggests that the equilibrium produced by the 100-fs energy-transfer processes may be a Boltzmann equilibrium.

If there is indeed a Boltzmann equilibrium between the C670 and C680 pigment pools, the equilibrium constant can be determined from the relative amplitudes of the 100-fs components after excitation at 665 and 694 nm (the amplitudes of these components reflect the difference between the initial excitation energy distributions and those at equilibrium). This yields an equilibrium constant of 1.0 ± 0.5 between the C670 and C680 pigment pools. This equilibrium constant is consistent with that expected for such a Boltzmann equilibrium, given the 10-nm separation of the pools' Q_y-absorption bands and assuming that a single species (P680) dominates the C680 pool and that the C670 pool comprises two to four chlorins.

Fig. 3 *B* and *C* shows the transient spectra before and after the 100-fs energy-transfer equilibration is complete. Similar transient spectra are obtained after equilibration with either 665- or 694-nm excitation (Fig. 3 *B* and *C*; square symbols). The observation of similar spectra supports our conclusion described above that excess excitation energy is largely thermalized on this time scale. The transient spectra at equilibrium are dominated by P680 bleach/stimulated emission due to the higher oscillator strength of P680 compared with those of the C670 chlorins. Therefore, energy transfer from C670 chlorins to P680, thereby forming the equilibrium state, results in a large shift in the peak of the transient spectrum from ≈670 to ≈678 nm (Fig. 3B), whereas energy transfer from P680 to the C670 chlorins to form an equilibrium results in only an ≈2-nm shift in the peak of the transient spectrum in the opposite direction (Fig. 3C). Nevertheless, there is a large change in the amplitude of the transient spectrum at 680 nm and a change in sign at 670 nm (see Fig. 3C).

Close inspection of Fig. 3 *B* and *C* indicates a spectrum after the 100-fs component using excitation at 665 nm is blue shifted ≈2 nm relative to the spectrum observed using 694-nm excitation. This small difference may result from the exclusive use of parallel polarized excitation and probe pulses, as this will have the effect of underestimating the absorption changes of pigments different in orientation to those directly excited (this may also explain the small differences between the two spectra shown in Fig. 2). Alternatively, the 2-nm blue shift could result from an incomplete thermalization of excess vibrational energy or from the

Proc. Natl. Acad. Sci. USA 89 (1992)

FIG. 4. Simple kinetic model of isolated PSII reaction centers including energy transfer between P680 and the C670 pigment pool and charge separation resulting in the formation of $P680^+Ph^-$. k_1 and k_{-1} are energy-transfer rate constants, and k_2 is an electron-transfer rate constant. This model is correct only if there are no intermediate states between $P680^*$ and $P680^+Ph^-$ (see text).

presence of some 670-nm-absorbing chlorins, which are unable to transfer excitation energy to the other pigments on this time scale.

One of the decay pathways for the primary radical pair state is via charge recombination and radiative decay to the ground state. It follows from our observation of an ≈1:1 equilibrium between P680 and the C670 chlorins that the charge recombination or delayed fluorescence produced by this decay pathway will be emitted by both P680 and C670 chlorins. This conclusion provides an explanation for the broad width of the time-resolved emission spectrum for the charge recombination fluorescence measured by Booth *et al.* (18).

Tentative Kinetic Model for Trapping Excitation Energy in the PSII Reaction Center. We have reported elsewhere that Ph reduction is observed with a lifetime of 21 ± 3 ps following excitation of PSII reaction centers at 612 or 694 nm (4). It can therefore be concluded that equilibration of the excitation energy between the majority of pigments precedes Ph reduction. Therefore, it is appropriate to describe the kinetics within the PSII reaction center in terms of the trapping limited model (7, 9, 11), as illustrated in Fig. 4. Using the equilibrium constant (k_1/k_{-1}) of 1.0 ± 0.5 for distribution of excitation energy before radical pair formation and an observed rate of radical pair formation of 21 ps^{-1}, we obtain a value of 10.5 ± 3 ps^{-1} for the intrinsic rate of primary charge separation (k_2) in PSII. However, it must be appreciated that this model assumes that the only trapping process is the 21-ps Ph reduction. As other kinetic components are observed with lifetimes of a few picoseconds, this assumption may not be valid. Identification of these unassigned components will allow development of a more complete kinetic model for the electron- and energy-transfer pathways within the isolated PSII reaction center.

Solution of the kinetic model shown in Fig. 4 determines that the observed rate of equilibration of the excitation energy is approximately the sum of the two energy-transfer rates k_1 and k_{-1}. As we have determined the equilibrium constant k_1/k_{-1} to be 1 ± 0.5, it follows that $k_1 \approx k_{-1} \approx 200 \pm 100$ fs^{-1}. It should be noted that k_1 and k_{-1} correspond to the mean rates of energy transfer between the two pigment pools rather than to rates of energy transfer between two specific chromophores.

The C680 pool appears to be dominated by P680 alone. We can therefore conclude that the average energy-transfer rate from individual C670 chlorins to P680 is ≈200 fs^{-1}. The separation in energy between the C670 and C680 S_1 levels yields a rate of 600 fs^{-1} for energy transfer from P680 to each C670 chlorin. These energy-transfer rates are of the same order of magnitude as those previously estimated for antenna/reaction center complexes of PSI (12, 13) and PSII (9), and within LHC2 (42), and are also consistent with Förster energy transfer (42).

We would like to thank Niall Walsh and Caroline Woollin for preparing the reaction center samples, Qiang Hong for help with the femtosecond spectrometer, and Chris Barnett for excellent technical assistance. We also acknowledge financial support from the Science

and Engineering Research Council, the Agriculture and Food Research Council, and The Royal Society.

1. van Grondelle, R. (1985) *Biochim. Biophys. Acta* **811**, 147–195.
2. Renger, G. (1991) in *Topics in Photosynthesis—The Photosystems: Structure, Function and Molecular Biology*, ed. Barber, J. (Elsevier, Amsterdam), Vol. 11, pp. 45–100.
3. van Gorkom, H. J., Pulles, M. P. J. & Wessels, J. S. C. (1975) *Biochim. Biophys. Acta* **408**, 331–339.
4. Hastings, G., Durrant, J. R., Hong, Q., Barber, J., Porter, G. & Klug, D. R. (1992) *Biochemistry* **31**, 7638–7647.
5. Klimov, V. V., Klevanik, A. V., Shuvalov, V. A. & Krasnovsky, A. A. (1977) *FEBS Lett.* **82**, 183–186.
6. Danielius, R. V., Satoh, K., van Kan, P. J. M., Plijter, J. J., Nuijs, A. M. & van Gorkom, H. J. (1987) *FEBS Lett.* **213**, 241–244.
7. Schatz, G. H., Brock, H. & Holzwarth, A. R. (1987) *Proc. Natl. Acad. Sci. USA* **84**, 8414–8418.
8. Liebl, W., Breton, J., Deprez, J. & Trissl, H.-W. (1989) *Photosynth. Res.* **22**, 257.
9. Schatz, G. H., Brock, H. & Holzwarth, A. R. (1988) *Biophys. J.* **54**, 397–405.
10. Beauregard, M., Martin, I. & Holzwarth, A. R. (1991) *Biochim. Biophys. Acta* **1060**, 271–283.
11. Pearlstein, R. M. (1982) in *Photosynthesis: Energy Conversion by Plants and Bacteria*, ed. Govindjee (Academic, New York), Vol. 1, pp. 293–330.
12. Owens, T. G., Webb, S. P., Mets, R. S., Alberte, R. S. & Fleming, G. R. (1987) *Proc. Natl. Acad. Sci. USA* **84**, 1532–1536.
13. Jean, J. M., Chan, C.-K., Fleming, G. & Owens, T. G. (1989) *Biophys. J.* **56**, 1203–1215.
14. Beekman, L. M. P., Visschers, R. W., Visscher, K. J., Althuis, B., Barz, W., Oesterhelt, D., Sundström, V. & van Grondelle, R. (1992) in *Ultrafast Phenomena VIII* (Springer, Berlin), in press.
15. Gounaris, K., Chapman, D. J., Booth, P., Crystall, B., Giorgi, L. B., Klug, D. R., Porter, G. & Barber, J. (1990) *FEBS Lett.* **265**, 88–92.
16. Kobayashi, M., Maeda, H., Watanabe, T., Nakane, H. & Satoh, K. (1990) *FEBS Lett.* **260**, 138–140.
17. Booth, P. J., Crystall, B., Giorgi, L., Barber, J., Klug, D. R. & Porter, G. (1990) *Biochim. Biophys. Acta* **1016**, 141–152.
18. Booth, P. J., Crystall, B., Ahmad, I., Barber, J., Porter, G. & Klug, D. R. (1991) *Biochemistry* **30**, 7573–7586.
19. Barber, J. (1987) *Trends Biochem. Sci.* **12**, 321–326.
20. Michel, H. & Deisenhofer, J. (1988) *Biochemistry* **27**, 1–7.
21. Breton, J., Martin, J.-L., Migus, A., Antonetti, A. & Orszag, A. (1986) *Proc. Natl. Acad. Sci. USA* **83**, 5121–5125.
22. Johnson, S. G., Tang, D., Jankowiak, R., Hayes, J. M., Small, G. J. & Tiede, D. M. (1990) *J. Phys. Chem.* **94**, 5849–5855.
23. Martin, J. L., Breton, J., Hoff, A. J., Migus, A. & Antonetti, A. (1986) *Proc. Natl. Acad. Sci. USA* **83**, 957–961.
24. Woodbury, N. W., Becker, M., Middendorf, D. & Parson, W. W. (1985) *Biochemistry* **24**, 7516–7521.
25. Braun, P., Greenberg, B. M. & Scherz, A. (1990) *Biochemistry* **29**, 10376–10387.
26. van Kan, P. J. M., Otte, S. C. M., Kleinherenbrink, F. A. M., Nieveen, M. C., Aartsma, T. J. & van Gorkom, H. J. (1990) *Biochim. Biophys. Acta* **1020**, 146–152.
27. Ganago, I. B., Klimov, V. V., Ganago, A. O., Shuvalov, V. A. & Erokhin, Y. E. (1982) *FEBS Lett.* **140**, 127–130.
28. Tang, D., Jankowiak, R., Seibert, M. & Small, G. J. (1991) *Photosynth. Res.* **27**, 19–29.
29. Breton, J. (1990) in *Perspectives in Photosynthesis*, eds. Jortner, J. & Pullman, B. (Kluwer, Dordrecht, The Netherlands), pp. 23–38.
30. Roelofs, T. A., Gilbert, M., Shuvalov, V. A. & Holzwarth, A. R. (1991) *Biochim. Biophys. Acta* **1060**, 237–244.
31. Wasielewski, M. R., Johnson, D. G., Govindjee, Preston, C. & Seibert, M. (1989) *Photosynth. Res.* **22**, 89–99.
32. Reddy, N. R. S., Lyle, P. A. & Small, G. J. (1992) *Photosynth. Res.* **31**, 167–194.
33. Durrant, J. R., Hastings, G., Hong, Q., Barber, J., Porter, G. & Klug, D. R. (1992) *Chem. Phys. Lett.* **188**, 54–60.
34. Chapman, D. J., Gounaris, K. & Barber, J. (1991) in *Methods in Plant Biochemistry*, ed. Rogers, L. (Academic, London), pp. 171–193.
35. van Mieghem, F. J. E., Nitschke, W., Mathis, P. & Rutherford, A. W. (1989) *Biochim. Biophys. Acta* **977**, 207–214.
36. Rutherford, A. W. (1986) *Biochem. Soc. Trans.* **14**, 15–17.
37. Durrant, J. R., Giorgi, L. B., Barber, J., Klug, D. R. & Porter, G. (1990) *Biochim. Biophys. Acta* **1017**, 167–175.
38. den Blanken, H. J., Hoff, A. J., Jongenelis, A. P. J. M. & Diner, B. A. (1983) *FEBS Lett.* **157**, 21–27.
39. Shepanski, J. F. & Anderson, R. W. (1981) *Chem. Phys. Lett.* **78**, 165–173.
40. He, W. Z., Telfer, A., Drake, A., Hoadley, J. & Barber, J. (1990) in *Current Research in Photosynthesis*, ed. Baltscheffsky, M. (Kluwer, Dordrecht, The Netherlands), Vol. 1, pp. 431–434.
41. Kwa, S. L. S., Newell, W. R., van Grondelle, R. & Dekker, J. P. (1992) *Biochim. Biophys. Acta* **1099**, 193–202.
42. Eads, D. D., Castner, W. C., Alberte, R. S., Mets, L. & Fleming, G. R. (1989) *J. Phys. Chem.* **93**, 8271–8275.

Proc. Natl. Acad. Sci. USA
Vol. 92, pp. 4798–4802, May 1995
Biophysics

A multimer model for P680, the primary electron donor of photosystem II

(excitons/disorder/reaction center)

James R. Durrant*, David R. Klug*, Stefan L. S. Kwa†, Rienk van Grondelle†, George Porter*, and Jan P. Dekker†

*Centre for Photomolecular Sciences, Department of Biochemistry, Imperial College, London, SW7 2AY, United Kingdom; and †Department of Physics and Astronomy and Institute for Molecular Biological Sciences, Vrije Universiteit, De Boelelaan 1081, 1081 HV Amsterdam, The Netherlands

Contributed by George Porter, December 19, 1994

ABSTRACT   We consider a model of the photosystem II (PS II) reaction center in which its spectral properties result from weak (≈ 100 cm^{-1}) excitonic interactions between the majority of reaction center chlorins. Such a model is consistent with a structure similar to that of the reaction center of purple bacteria but with a reduced coupling of the chlorophyll special pair. We find that this model is consistent with many experimental studies of PS II. The similarity in magnitude of the exciton coupling and energetic disorder in PS II results in the exciton states being structurally highly heterogeneous. This model suggests that P680, the primary electron donor of PS II, should not be considered a dimer but a multimer of several weakly coupled pigments, including the pheophytin electron acceptor. We thus conclude that even if the reaction center of PS II is structurally similar to that of purple bacteria, its spectroscopy and primary photochemistry may be very different.

The primary processes of photosynthesis involve the absorption of solar energy by an array of light-harvesting pigments, typically chlorophyll, embedded in pigment–protein complexes within a lipid membrane. The resulting chlorophyll excited state is rapidly transferred to a primary electron donor species within a photosynthetic reaction center, where the energy is trapped by a sequence of electron transfer reactions (1). The close proximity of the chlorophylls within the pigment–protein complexes gives rise to dipole–dipole coupling between the pigments. This coupling is responsible for Förster energy transfer between the chlorophylls and may also result in energetic shifts and delocalization of the excited states (exciton interactions) (1). Exciton interactions are important for photosynthetic function in, for example, defining the precursor state to the initial charge separation reaction (2) and are also important in many nonbiological supramolecular systems (3). Moreover, as exciton interactions can strongly influence the properties of optical transitions monitored in many studies of photosynthetic complexes, their consideration can be essential in the interpretation of experimental results.

In this paper we consider the importance of exciton interactions within the photosystem two (PS II) reaction center. The initial charge separation reaction in PS II results in oxidation of the primary electron donor, a chlorophyll species referred to as P680 (due to a characteristic bleaching observed at 680 nm upon oxidation of this species) and reduction of a pheophytin molecule. This electron transfer reaction is of particular interest as the resulting species P680$^+$ is thought to be the most oxidizing species found in living organisms, with a potential of +1.1 eV (compare with 0.4–0.5 eV for the primary electron donors of other photosynthetic reaction centers). This high oxidizing potential is essential to PS II's ability to extract electrons from water, thereby releasing molecular oxygen and generating our oxygenic atmosphere.

There appear to be extensive similarities between the PS II reaction center and the reaction center of the photosynthetic purple bacteria. Therefore the structure of the purple bacterial reaction center, determined by x-ray crystallography (4) (see Fig. 1), has been widely used as a model for PS II (5–7). Some aspects of this structural model of PS II have been experimentally confirmed, such as the residues that bind the pheophytin electron acceptor (ref. 8 and references therein). Other aspects of the structural model remain poorly defined, and indeed there must be some differences in order to generate the high oxidizing potential of P680$^+$ (9).

The PS II reaction center exhibits a much greater degree of spectral overlap than the bacterial reaction center, particularly for the functionally important S_0–S_1 (Q_y) transitions. In purple bacteria, about half of the splitting of these transitions is attributed to excitonic interactions between the special pair bacteriochlorophylls (P_L and P_M) (2), which constitute the bacterial primary electron donor (P870/P960). This coupling ($V \approx 550$ cm^{-1} for P870 and 950 cm^{-1} for P960) results in a red-shifted special pair excited state, which is an energetic trap for excited states within the isolated reaction center. In contrast, in PS II, P680 exhibits only a small red shift relative to the other reaction center pigments and is therefore only a weak trap for excitation energy. Indeed after optical excitation of P680 in isolated PS II reaction centers, excitation energy rapidly equilibrates between the majority of reaction center singlet excited states, and primary charge separation proceeds from this equilibrated state (1, 10).

Studies of the PS II reaction center have resolved some, albeit relatively weak, exciton coupling between pigments. Evidence for such interactions has been obtained from the circular dichroism spectrum (11–13) and from absorption changes caused by formation of the P680 triplet state (14–16) or the P680 excited singlet state (10). Exciton transitions have been resolved at ≈ 680 nm and ≈ 667 nm, with the 680-nm band carrying a much greater oscillator strength (10, 12, 15, 16). The splitting of these transitions indicates a maximum coupling strength of $V \approx 140$ cm^{-1}, which is 3–4 times less than the coupling of the bacterial special pair. These observations have been interpreted in terms of P680 being a weakly coupled special pair of chlorophylls (12–14). However, such special pair models of P680 are complicated by the observation that the triplet state of P680 appears to be localized upon a single chlorophyll with an orientation similar to one of the monomeric bacteriochlorophylls (B_L and B_M) in the bacterial reaction center (17, 18). This has led to suggestions that P680 should be considered structurally analogous to B_L, with weak exciton coupling to its structural neighbors (18) or that P680

Abbreviations: PS II, photosystem II; P680, primary electron donor of PS II; Chn, chlorophyll n; Phn, pheophytin n.

Proc. Natl. Acad. Sci. USA 92 (1995)   4799

is an asymmetric special pair with one chlorophyll of the dimer orientated at a similar angle to B_L (9, 19).

The models of P680 discussed above have principally considered exciton coupling within a chlorophyll special pair but have not considered the effect of exciton interactions between the other reaction center chlorins. The possibility that these interactions may be important has, however, previously been suggested by Tetenkin *et al.* (11). Moreover, it has recently been pointed out by one of us (20) that if the structural arrangement of the chlorins is maintained between PS II and purple bacteria, then the dipole–dipole coupling between "monomeric" accessory chlorophylls (e.g., B_L/Ch1) and the adjacent pheophytin (H_L/Ph1) and special pair pigments (P_M/Ch2) are of the order of 100 cm^{-1} (see Fig. 1, discussed in detail below; Ch1 and Ch2 are chlorophylls 1 and 2; Ph1 is pheophytin 1). These couplings are of similar magnitude to the coupling strengths experimentally observed for P680. Therefore it may be essential to include them in models of the exciton interactions within the PS II reaction center (20).

In this paper, we develop a model of the excited singlet states of the PS II reaction center in which we take into account all of the dipole couplings between the pigments and the effect of transition energy disorder (inhomogeneous disorder). The inclusion of disorder in our calculations is important since the coupling strength and disorder are of similar magnitude. We conclude that P680 should be considered a "multimer" of several weakly coupled pigments whose excited states may be rather heterogeneous. We present a theoretical description of this multimer model of P680 and consider the extent to which such a model is in agreement with experimental observations.

THEORETICAL METHODS

The theoretical analysis used in this paper is based upon a description of Frenkel excitons as applied by Fidder *et al.* (21) to calculate the properties of molecular (J-) aggregates. Each reaction center chlorin is treated as a point dipole, and only one excited state is considered. Charge transfer states, simultaneous excitation of more than one reaction center chlorin, and electron–phonon coupling are neglected. Within these assumptions, the electronic exciton states of the reaction center can be described by a Hamiltonian:

$$H = \sum_n (\langle \varepsilon_n \rangle + d_n) |n\rangle\langle n| + \sum_{n,m} \sum_{n \neq m} V_{nm} |m\rangle\langle n|. \quad [1]$$

Here $|n\rangle$ denotes the state in which chlorin n is excited, and the summations are over all reaction center chlorins. $\langle \varepsilon_n \rangle$ is the average monomer transition energy of the chlorin n, and d_n is the (static) inhomogeneous offset energy of this chlorin, reflecting the effect of disorder imposed by the surrounding protein environment (diagonal disorder). In practice, the d_n may be considered as random variables taken from a Gaussian probability distribution with full width at half-maximum Γ_{inh}. V_{nm} is the dipole–dipole coupling between chlorins determined using the point dipole approximation (1), following Knapp *et al.* (2), who found the point dipole and extended dipole models yielded similar interaction energies for the *Rhodopseudomonas viridis* reaction center.

For a particular realization of the disorder, the exciton eigenstates are found by diagonalizing the matrix H_{nm}. Then the *i*th eigenvalue E_i gives the energy of the *i*th exciton state $|\psi_i\rangle$, whereas the normalized *i*th eigenvector $\mathbf{a}_i = (a_{i1}, ..., a_{i6})$ specifies the amplitude of each pigment's contribution to each exciton state:

$$|\psi_i\rangle = \sum_n a_{in} |n\rangle. \quad [2]$$

The optical properties of a particular reaction center can be readily calculated from these eigenvalues and eigenvectors (2, 21). Comparison of this model with experiment requires ensemble averaging to calculate the optical properties that a sample might exhibit. Further details of this numerical approach, in which the disorder is randomly generated according to the Gaussian probability distribution, are given by Fidder *et al.* (21).

Following Fidder *et al.* (21), we define the spatial extent (delocalization) of these exciton states $N_{del} = 1/\Sigma_n (a_{in})^4$. Using this definition, $N_{del} = 1$ for monomer, 2 for a undisordered, symmetric dimer, 3 for a trimer, etc.

RESULTS AND DISCUSSION

Exciton Calculations. The exciton calculations presented in this paper are based upon the structure of the reaction center of the purple bacterium *R. viridis* (4) but using chlorophyll a/pheophytin a dipole strengths (see Fig. 1). The coupling strengths shown in Fig. 1 are of similar magnitude to those observed experimentally for P680 (≤ 140 cm^{-1}) (11–13, 15), with the exception of the 418-cm^{-1} coupling between Ch2 and Ch3, corresponding to the bacterial special pair. Several structural modifications have been proposed previously to reduce the coupling between Ch2 and Ch3, including increasing their separation (12), rotation of either Ch2 or Ch3 (19), or deletion of Ch3 (9, 20). Because the main conclusions presented in this paper were found to be largely independent of which modification was used, results are presented here for only one modification, an increased separation of Ch2 and Ch3.

The reaction center model shown in Fig. 1 includes only four chlorophyll molecules, whereas the most widely studied isolated PS II reaction center preparations bind at least six chlorophyll molecules per two pheophytin molecules. However, it appears that these two additional chlorophyll molecules are only very weakly coupled ($V \approx 5$–10 cm^{-1}) to the other reaction center pigments, as demonstrated by slow (10–50 ps) energy transfer from these pigments to P680 (ref. 22 and references therein), and they are most probably bound on the exterior of the PS II reaction center (6). These "peripheral" pigments, which have absorption maxima near 670 nm (ref. 22

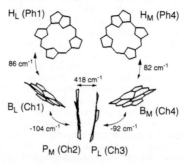

FIG. 1. Structural arrangement of the pigments in the reaction center of the *R. viridis*. This figure also shows the magnitudes of the strongest dipolar couplings between the pigments after substitution of chlorophyll a and pheophytin a for bacteriochlorophyll b and bacteriopheophytin b (the maximum coupling experimentally observed for P680 is ≈ 140 cm^{-1}). Dipole strengths of 23 and 14 debye^2 were used for chlorophyll a and pheophytin a, respectively (5) (compare with 51 debye^2 for bacteriochlorophyll b). This model is used in the exciton calculations for the PS II reaction center presented in the text, with the addition of a small structural modification to reduce the coupling between Ch2 and Ch3. However, the key conclusions of these calculations are found to be insensitive to the details of the structural model used.

Table 1.   Exciton interaction energies (cm^{-1}) V_{nm} for a model of the PS II reaction center based on the *R. viridis* structure but with the separation of Ch2 and Ch3 increased by 2.8 Å along their connecting axis

	Ph1	Ch1	Ch2	Ch3	Ch4	Ph4
Ph1	$\langle\varepsilon_1\rangle + d_1$	86.3	17.3	−1.2	−5.7	2.7
Ch1		$\langle\varepsilon_2\rangle + d_2$	−101	−42.7	15.8	−5.5
Ch2			$\langle\varepsilon_3\rangle + d_3$	120	−37.9	−1.7
Ch3				$\langle\varepsilon_4\rangle + d_4$	−90.2	17.2
Ch4					$\langle\varepsilon_5\rangle + d_5$	82.3
Ph4						$\langle\varepsilon_6\rangle + d_6$

Diagonal elements are the monomer transition energies, where $\langle\varepsilon_n\rangle = 14{,}860$ cm^{-1}.

and references therein), are therefore not included in the exciton calculations presented in this paper.

Table 1 gives the calculated dipole coupling strengths for the structural model of the PS II reaction center where the separation of Ch2 and Ch3 has been increased by 2.8 Å. Results of exciton calculations for this model are shown in Figs. 2 and 3. In these calculations, the transition energies $\langle\varepsilon_n\rangle$ of all six pigments were assumed to be at 14,860 cm^{-1} (673 nm), although we show below that this assumption is not critical. The calculations used $\Gamma_{inh} = 210$ cm^{-1}, based upon experimental observations of inhomogeneous broadening (full width

at half-maximum ≈ 120 cm^{-1}) at low temperature and assuming an exciton exchange narrowing of $\sqrt{N_{del}}$ (see below) (23). However we show below that our results are also rather insensitive to the value of Γ_{inh} used.

Fig. 2*A* shows the results of ensemble averaging over 2000 reaction centers. It is apparent that while the density of states is relatively broad and symmetrically distributed around 673 nm (the monomer transition wavelength), the reaction center absorption spectrum is dominated by the absorption from high oscillator strength states near 680 nm. Also shown are the results from the three individual reaction centers with different inhomogeneous shifts, d_n.

Fig. 2*B* shows the predicted change in reaction center absorption resulting from formation of a localized excited state (e.g., triplet, cation, or anion) upon either Ph1, Ch1, or Ch2 (this simulation neglects excited state absorption and electrochromic shifts). In all three cases, the absorption difference spectrum is dominated by a bleaching at 680 nm, consistent with experimental observations that formation of *any* PS II reaction center chlorin cation, triplet, or anion state (or singlet oxygen-induced photobleaching) results in an absorption bleach at 680 nm (except for bleaching of the peripheral 670 nm absorbing chlorophylls).

In Fig. 2*A* it can be seen that the lowest energy exciton state lies near 680 nm and carries the oscillator strength of two to three chlorophyll molecules. Fig. 3 shows how for the three individual particles of Fig. 2*A*, these three lowest energy states are distributed over the reaction center chlorins. Several chlorins contribute to each of these optical (exciton) transitions ($N_{del} \approx 3$). However, the different realizations of the disorder for each reaction center results in the three states being delocalized over different chlorins: in other words, these three states, while exhibiting similar absorption maxima and oscillator strengths, are spatially heterogeneous.

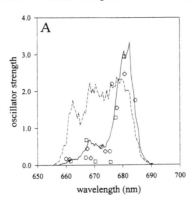

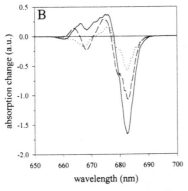

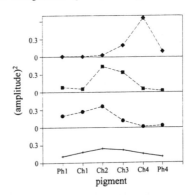

FIG. 2.   (*A*) Plot of oscillator strength (in units of $|\mu_{Ch1}|^2$) against wavelength for the exciton states calculated for three individual PS II reaction centers ($\Diamond$, $\square$, $\bigcirc$) with different values of the energetic disorder (d_n) taken from a Gaussian distribution with standard deviation $\Gamma_{inh} = 210$ cm^{-1}. Other details are as in Table 1. Also shown are the absorption spectrum (—) and density of exciton states (– – – –) generated by an ensemble average over 2000 reaction centers, with a 0.5-nm spectral resolution. (*B*) The change in reaction center absorption resulting from deletion of one chlorin: Ch2 (—), Ch1 (– – – –), or Ph1 (·····). This simulates the presence of a localized excitation (cation, anion, or triplet state) upon these pigments, neglecting excited state absorption and electrochromic shifts. a.u., arbitrary units.

FIG. 3.   Site populations $|a_{in}|^2$ of the wavefunctions $|\psi_i\rangle$ corresponding to the three lowest energy states selected from (Fig. 2*A*) ($\blacklozenge$, $\blacksquare$, $\bullet$), illustrating the delocalization of these exciton states over several reaction center chlorins. Also shown (solid line) is the result of ensemble averaging over 2000 reaction centers. Other details are as in Fig. 2.

538

Biophysics: Durrant *et al.*

Proc. Natl. Acad. Sci. USA 92 (1995)   4801

Exciton Delocalization and Energetic Disorder. The delocalization of the exciton states is driven by dipolar coupling. Our calculations indicate that this delocalization can be limited by heterogeneity in the monomer transition energies (diagonal disorder). This heterogeneity may result from site-specific shifts (due to specific pigment–protein interactions) or random disorder (inhomogeneous broadening). The width of the PS II reaction center Q_y absorption band indicates that site-specific shifts can be no more than ± 130 cm^{-1}. Our numerical simulations indicate that while such site-specific shifts will have some effect on the details of the exciton states, they are insufficient to alter the overall conclusions. Similarly, calculations for a range of values of the magnitude of the inhomogeneous disorder indicate that for all reasonable values of the disorder, significant delocalization is observed. This is illustrated in Fig. 4, which shows a plot of the mean delocalization (N_{del}) of the exciton states as a function of Γ_{inh}. It can be seen that some delocalization of the exciton states is predicted even for values of Γ_{inh} several times greater than the experimentally observed value for the inhomogeneous line width.

The exciton model used in this paper neglects electron–phonon coupling. It also assumes that the disorder is static and is therefore strictly only valid at low temperatures. However, the center of gravity of the PS II reaction center's absorption bands appears to be rather insensitive to temperature, suggesting that this model may also be applicable at room temperature. This would be consistent with recent studies of photosynthetic complexes, which have indicated that static inhomogeneous broadening is the dominant contribution to the observed spectral dynamics at room temperature (ref. 24 and references therein).

It is interesting to note that exciton theory predicts a greater delocalization of the exciton states in PS II reaction centers than in the bacterial reaction center, despite the reduced strength of the coupling in PS II. This results from the much greater spectral overlap in PS II. In the bacterial reaction center, spectral (and therefore energetic) degeneracy is broken both by the strong coupling within the special pair and by site-specific shifts of the bacteriochlorophyll and bacteriopheophytin monomer transition frequencies (2), which result in localization of the lowest energy exciton state upon the special pair.

The Multimer Model of P680: Comparison with Experiment. The calculations presented in this paper suggest that the excited singlet states of the PS II reaction center chlorins

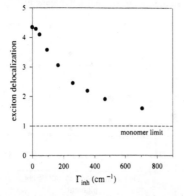

FIG. 4. Plot of the mean delocalization (N_{del}) of the reaction center exciton states as a function of the magnitude of the transition energy disorder Γ_{inh} (inhomogeneous disorder). The inhomogeneous linewidth of P680 has been determined at 4 K to be ≈ 120 cm^{-1} (17, 18) although this may be up to a 2-fold underestimate of underlying disorder due to exciton exchange narrowing (23). Other details are as in Fig. 2.

should be described by a multimer model. The term multimer is used to indicate that dipolar couplings produce exciton transitions that are delocalized over several, but not necessarily all, reaction center chlorins. In addition, the comparable magnitudes of the coupling strengths and the inhomogeneous disorder, with no two pigments being particularly strongly coupled, result in spatially heterogeneous exciton states.

Our calculations demonstrate that this multimer model is in good agreement with experimental observations, including the presence of exciton transitions near 680 nm and 665 nm, with the lower energy transitions carrying the majority of the oscillator strength. Additional calculations (not shown) indicate that the transition dipoles for the 680/665-nm exciton transitions are approximately orthogonal, which is also in agreement with experimental observations (17). Moreover our model suggests that P680 is likely to be spectrally heterogeneous, again consistent with experimental observations (13, 15, 17, 25). The low amplitude of the 670-nm shoulder in the calculated spectrum is due, at least in part, to exclusion of the peripheral weakly coupled chlorophylls from our calculations.

The inclusion of pheophytin in the P680 multimer is supported by the observation that reduction of the photoactive pheophytin results in a significant decrease in the reaction center circular dichroism spectrum (11), indicating that this pigment is excitonically coupled to other reaction center chlorins. This conclusion is also supported by the observation of a large bleaching of the pheophytin Q_x absorption band observed directly (within 300 fs) after 694 nm excitation of isolated PS II reaction centers (26). In addition our model is readily reconciled with localization of the P680 triplet state upon a chlorophyll structurally equivalent to B_L (i.e., Ch1), as has been experimentally observed (17, 18), as in this multimer model "P680" effectively includes this chlorophyll.

For "special pair" models of P680 to be meaningful, the dipole coupling between the special pair chlorins must be much greater than the dipole coupling to the other reaction center chlorins. The special pair coupling in PS II has been experimentally shown to be ≤ 140 cm^{-1} (12, 15, 16). Thus all the other dipole couplings would have to be $\ll 140$ cm^{-1} (e.g., ≤ 20 cm^{-1}). Such weak coupling strengths are clearly inconsistent with a structural model of PS II based upon the bacterial reaction center. In addition such weak coupling strengths would be inconsistent with the experimental observation that excitation energy equilibrates between the majority of reaction center pigments in ≈ 100 fs (10).

Our prediction of exciton states delocalized over several reaction center pigments is not strongly dependent upon the structure used. Indeed, any PS II reaction center structure in which several chlorins with overlapping Q_y optical transitions are placed in sufficient proximity to give rapid charge separation and 100-fs energy transfer would be likely to produce a similar conclusion. Therefore while our calculations indicate that experimental observations of PS II are consistent with a purple bacterial reaction center structure modified only in the region of the special pair, other arrangements of the pigments in PS II are possible, which would also be consistent with our model.

P680. The 870/960 nm transitions of the bacteriochlorophyll a/b primary electron donors are delocalized only over the special pair bacteriochlorophylls, as are the corresponding triplet and cation states. There is therefore no ambiguity in using the terms P870 and P960 to refer both to an optical transition and a structural component of the reaction center. The results presented in this paper suggest that term P680 should be used with considerably more caution.

The P680 triplet state, and most probably the P680 cation state, are localized upon single reaction center chlorophylls (14, 15) (it is, however, possible that these states are localized upon different chlorophylls). We suggest in this paper that the Q_y optical transitions of the reaction center are delocalized over several pigments, and therefore the spectroscopy of P680 can only be understood by considering the optical properties of the reaction center as a whole. In particular, our model suggests that PS II's primary electron donor and acceptor share common ground-state transitions, and it is therefore not surprising that a bleach at 680 nm is caused by either donor oxidation or acceptor reduction. Finally it should be pointed out that as the precursor state to primary charge separation may be delocalized over both the primary electron donor and acceptor, it is possible that the charge separation process in PS II should be described not as a conventional intermolecular electron transfer reaction but as a charge transfer reaction within a supramolecular complex.

We conclude by considering possible biological functions of the reduced exciton coupling observed for P680 relative to P870/P960. Chlorophyll monomer cations are more oxidizing than dimer cations, and the reduced dipolar coupling may be a side effect of the requirement for a localized P680 cation. The reduced coupling results in P680 being only a shallow trap for excitation energy, thereby slowing down the trapping of excitation energy from the antenna chlorophylls, would facilitate the regulation of energy transfer to the reaction center achieved by the turning on of quenching pathways in the antenna (q_E quenching) under stress conditions. Finally the high redox potential of P680$^+$ requires the trapping of a greater proportion of the incident photon energy, which might preclude a larger free energy difference between the singlet excited states of P680 and antenna chlorophylls.

J.R.D. and D.R.K. thank the Biotechnology and Biological Sciences Research Council, Research Institute of Innovative Technology for the Earth, and Royal Society of Great Britain for financial support. J.P.D., S.L.S.K., and R.v.G. thank The Netherlands Organization for Scientific Research and European Community (Grant ERB4050PL94-1071) for financial support.

1. van Grondelle, R., Dekker, J. P., Gillbro, T. & Sundstrom, V. (1994) *Biochim. Biophys. Acta* **1087**, 1–65.
2. Knapp, E. W., Fischer, S. F., Zinth, W., Sander, M., Kaiser, W., Deisenhofer, J. & Michel, H. (1985) *Proc. Natl. Acad. Sci. USA* **82**, 8463–8467.
3. Lin, V. S.-Y., Dimagno, S. G. & Therien, M. J. (1994) *Science* **264**, 1105–1111.
4. Deisenhofer, J., Epp, O., Miki, K., Huber, R. & Michel, H. (1985) *Nature (London)* **318**, 618–624.
5. Michel, H. & Deisenhofer, J. (1988) *Biochemistry* **27**, 1–7.
6. Ruffle, S. V., Donnelly, D., Blundell, T. L. & Nugent, J. H. A. (1992) *Photosynth. Res.* **34**, 287–300.
7. Svensson, B., Vass, I., Cedergren, E. & Styring, S. (1990) *EMBO J.* **9**, 2051–2059.
8. Moënne-Loccoz, P., Robert, B. & Lutz, M. (1989) *Biochemistry* **28**, 3641–3645.
9. van Gorkom, H. J. & Schelvis, J. P. M. (1993) *Photosynth. Res.* **38**, 297–301.
10. Durrant, J. R., Hastings, A., Joseph, D. M., Barber, J., Porter, G. & Klug, D. R. (1992) *Proc. Natl. Acad. Sci. USA* **89**, 11632–11636.
11. Tetenkin, V. L., Gulyaev, B. A., Seibert, M. & Rubin, A. B. (1989) *FEBS Lett.* **250**, 459–463.
12. Braun, P., Greenberg, B. M. & Scherz, A. (1990) *Biochemistry* **29**, 10376–10387.
13. Otte, S. C. M., van der Vos, R. & van Gorkom, H. J. (1992) *J. Photochem. Photobiol. B.* **15**, 5–14.
14. Durrant, J. R., Giorgi, L. B., Barber, J., Klug, D. R. & Porter, G. (1990) *Biochim. Biophys. Acta* **1017**, 167–175.
15. Kwa, S. L. S., Eijckelhoff, C., van Grondelle, R. & Dekker, J. P. (1994) *J. Phys. Chem.* **98**, 7702–7711.
16. Chang, H.-C., Jankoviak, R., Reddy, N. R. S., Yocum, C. F., Picorel, R., Seibert, M. & Small, G. J. (1994) *J. Phys. Chem.* **98**, 7725–7735.
17. van der Vos, R., van Leeuwen, P. J., Braun, P. & Hoff, A. J. (1992) *Biochim. Biophys. Acta* **1140**, 184–198.
18. van Mieghem, F. J. E., Satoh, K. & Rutherford, A. W. (1991) *Biochim. Biophys. Acta* **1058**, 379–385.
19. Noguchi, T., Inoue, Y. & Satoh, K. (1993) *Biochemistry* **32**, 7186–7195.
20. Kwa, S. L. S. (1993) Ph.D. thesis (Vrije Univerisiteit, Amsterdam).
21. Fidder, H., Knoester, J. & Wiersma, D. A. (1991) *J. Chem. Phys.* **95**, 7880–7890.
22. Vacha, F., Joseph, D. M., Durrant, J. R., Telfer, A., Klug, D. R., Porter, G. & Barber, J. (1994) *Proc. Natl. Acad. Sci. USA* **92**, 2929–2933.
23. Knapp, E. W. (1984) *Chem. Phys.* **85**, 73–82.
24. Visser, H. M., Somsen, O. J. G., van Mourik, F., Lin, S., van Stokkum, I. H. M. & van Grondelle, R. (1994) *Biophys. J.*, in press.
25. Groot, M.-L., Peterman, E. J. G., van Kan, P. J. M., van Stokkum, I. H. M., Dekker, J. P. & van Grondelle, R. (1994) *Biophys. J.* **67**, 318–330.
26. Klug, D. R., Rech, T., Joseph, D. M., Barber, J., Durrant, J. R. & Porter, G. (1994) *Chem. Phys.* **1903**, in press.

Further publications

OBSERVATION OF PHEOPHYTIN REDUCTION IN PHOTOSYSTEM TWO REACTION CENTRES USING FEMTOSECOND TRANSIENT ABSORPTION SPECTROSCOPY.
G. Hastings, J. R. Durrant, J. Barber, G. Porter and D. R. Klug,
Biochem., 1992, **31**, 7638.

FLUORESCENCE KINETICS OF D1/D2 CYTOCHROME-B559 REACTION CENTRES (XIV).
B. Crystall, P. J. Booth, J. Barber, D. R. Klug and G. Porter,
M. Baltscheffsky (ed.), Current Research in Photosynthesis, 1989, Vol. 1, 455–458.

THERMODYNAMICS OF THE PRIMARY ELECTRON TRANSFER REACTION IN D1/D2 CYTOCHROME-B559 REACTION CENTRES.
P. J. Booth, B. Crystall, J. Barber, D. R. Klug and G. Porter,
Proceedings of VIII International Congress on Photosynthesis, 1989 (XV).

CHARACTERISATION OF TRIPLET AND QUINONE INDUCED CATION RADICAL STATES IN THE ISOLATED PHOTOSYSTEM 2 REACTION CENTRE (XIII).
J. R. Durant, L. B. Giorgi, J. Barber, D. R. Klug and G. Porter,
M. Baltscheffsky (ed.), Current Research in Photosynthesis, 1989, Vol. 1, 415–418.

THERMODYNAMIC PROPERTIES OF D1/D2/CYTOCHROME B559 REACTION CENTRES INVESTIGATED BY TIME RESOLVED FLUORESCENCE MEASUREMENTS (XVII).
P. J. Booth, B. Crystal, L. B. Giorgi, J. Barber, and G. Porter,
Biochim. Biophys. Acta, 1990, **1016**, 141.

CHARACTERISATION OF TRIPLET STATES IN ISOLATED PHOTOSYSTEM II REACTION CENTRES: OXYGEN QUENCHING AS A MECHANISM FOR PHOTODAMAGE (XVIII).
J. D. Durant, L. B. Giorgi, J. Barber, D. R. Klug and G. Porter, *Biochim. Biophys. Acta*, 1990, **1017**, 167.

OBSERVATION OF MULTIPLE RADICAL PAIR STATES IN PHOTOSYSTEM TWO REACTION CENTRES.
P. J. Booth, B. Crystal, I. Ahmad, J. Barber, G. Porter and D. Klug,
Biochem., 1991, **30**, 7573.

PRIMARY RADICAL PAIR FORMATION IN PHOTOSYSTEM TWO
REACTION CENTRES.
D. R. Klug, J. R. Durrant, G. Hastings, Q. Hong, D. M. Joseph, J. Barber and G. Porter,
Proceedings 8th Ultrafast Symposium, 1992.

DETERMINATION OF P680 SINGLET STATE LIFETIMES IN
PHOTOSYSTEM TWO REACTION CENTRES.
J. R. Durrant, G. Hastings, Q. Hong, J. Barber, G. Porter and D. R. Klug,
Chem. Phys. Letters, 1992, **188**, 54.

ELECTRON AND ENERGY TRANSFER IN PHOTOSYSTEM II
REACTION CENTRES. 1. DISCRIMINATION OF THE FIVE KINETIC
COMPONENTS.
J. R. Durant, G. Hastings, D. M. Joseph, J. Barber, G. Porter and D. R. Klug,
Proceedings of the IX International Congress on Photosynthesis.
Research in Photosynthesis, Vol. II, pp. 243–246.

ELECTRON AND ENERGY TRANSFER IN PHOTOSYSTEM II
REACTION CENTRES. 1. RADICAL PAIR FORMATION.
G. Hastings, J. R. Durant, J. Barber, G. Porter and D. R. Klug.
Proceedings of the IX International Congress on Photosynthesis.
Research in Photosynthesis, Vol. II, pp. 247–250.